UNCERTAINTY AND OPTIMALITY

Probability, Statistics and Operations Research

UNCERTAINTY AND OPTIMALITY

Probability, Statistics and Operations Research

J. C. Misra

Indian Institute of Technology, Kharagpur

World Scientific

New Jersey • London • Singapore • Hong Kong

Published by

World Scientific Publishing Co. Pte. Ltd.
P O Box 128, Farrer Road, Singapore 912805
USA office: Suite 1B, 1060 Main Street, River Edge, NJ 07661
UK office: 57 Shelton Street, Covent Garden, London WC2H 9HE

British Library Cataloguing-in-Publication Data
A catalogue record for this book is available from the British Library.

UNCERTAINTY AND OPTIMALITY
Probability, Statistics and Operations Research

ISBN 981-238-082-5

Printed in Singapore by Uto-Print

Preface

This really is the golden age of Mathematics. It has been said that half the Mathematics ever created has been in the last 100 years and that half the mathematicians who have ever lived are alive today. We have seen such achievements as the resolution of the four-colour problem and Fermat's last theorem, with the latter being a special manifestation of a much more general result!

It is befitting that the golden Jubilee of the Indian Institute of Technology Kharagpur, happens to fall in the golden age of Mathematics. As a senior professor in the Department of Mathematics, I felt encouraged to bring out a series of books covering all the major areas of Mathematical Sciences during this period of historic importance.

This book is an important member of the aforesaid series and consists of chapters that deal with important topics in Biomathematics. A glance through any modern textbook or journal in the fields of ecology, genetics, physiology or biochemistry reveals that there has been an increasing use of mathematics which ranges from the solution of complicated differential equation in population studies to the use of transfer functions in the analysis of eye-tracking mechanisms. This volume deals with Applied Mathematics in Biology and Medicine and is concerned with applied mathematical models and computer simulation in the areas of Molecular and Cellular Biology, Biological Soft Tissues and Structures as well as Bio-engineering.

In this volume an attempt has been made to cover biological background and mathematical techniques whenever required. The aim has been to formulate various mathematical models on a fairly general platform, making the biological assumptions quite explicit and to perform the analysis in relatively rigorous terms. I hope, the choice and treatment of the problems will enable the readers to understand and evaluate detailed analyses of specific models and applications in the literature.

The purpose of bringing out this volume on Biomathematics dealing with interdisciplinary topics has been twofold. The objectives are to promote research in applied mathematical problems of the life sciences and to enhance cooperation and exchanges between mathematical scientists, biologists and medical researchers. This volume has both a synthetic and analytic effect. The different chapters of the volume have been mostly concerned with model building and verification in different areas of biology and the medical sciences.

I believe, people in the entire spectrum of those with an interest in ecology, from field biologists seeking a conceptual framework for their observations to mathematicians seeking fruitful areas of application, will find stimulation here. It may so happen that some readers may find some parts of this volume trivial and some of the parts incomprehensible. Keeping this in view extensive bibliographies have been given at the end of each chapter which do attempt to provide an entry to the corresponding areas of study.

For over three decades I have been engaged in teaching and research at several well-known institutions of India, Germany and North America. Publication of the series of books has been the fruit of a long period of collaboration together with relentless perseverance. It has been our endeavour to make these books useful to a large section of people interested in mathematical sciences, professional as well as amateur. The volumes have been so designed and planned that illustrative examples and exercises as well as fairly comprehensive bibliography are included in the various chapters. This will help strengthen the level of understanding of the learners. Thus the books of this series will be of interest not only to graduate students but also to instructors, research scientists and professionals. The volumes of the series might not exactly serve as textbooks, but will definitely be worthwhile supplements. Our labour will be deemed amply rewarded if at least some of those for whom the volumes are meant derive benefit from them.

I am thankful to the members of the ICRAMS committee for their kind encouragement in publishing the mathematical science series on the occasion of the Golden Jubilee of our Institute. I feel highly indebted to the contributors of all the volumes of the series who have so kindly accepted my invitation to contribute chapters. The enormous pleasure and enthusiasm with which they have accepted my invitation have touched me deeply, boosting my interest in the publication of the series.

I. I. T. Kharagpur — J. C. Misra

About the Editor

Dr. Jagadis Chandra Misra, Senior Professor at the Department of Mathematics, Indian Institute of Technology Kharagpur, received his Ph.D. degree in Applied Mathematics from Jadavpur University in 1971. Subsequently the University of Calcutta conferred on him the coveted D.Sc. degree in Applied Mathematics in year 1984. For over three decades he has been engaged in teaching and research at several distinguished institutions of India, Germany and North America. He has been the Chairman of the Department of Mathematics, IIT Kharagpur during the period 1998–2001. As a recipient of the *Humboldt Fellowship* he was at the University of Hannover during the period 1975–77 and also in 1982, where he carried out his research in the field of Biomechanics in collaboration with Professor Oskar Mahrenholtz of the University of Hannover and the Professor Christoph Hartung from the Biomedical Engineering Division of the School of Medicine, Hannover. He has held the position of Visiting Professor at the University of Calgary, Canada and also at the Indian Institute of Technology, Kanpur. He also paid brief visits to Cambridge University, Oxford University, Manchester University, Glasgow University, University of Paris, Ecole Polytechnique in France, Graz University in Austria, Delft University in the Netherlands, University of California, Los Angeles and University of California, San Diego. In 1984 he received the prestigious *Silver Jubilee Research Award* from IIT Kharagpur, his research publications having been adjudged to be *outstanding.* He has been elected a *Fellow* of the National Academy of Sciences in 1987, the Institute of Mathematics and its Applications (UK) in 1988, the Institute of Theoretical Physics in 1988, the Royal Society of Medicine, London in 1989 and the Indian National Academy of Engineering in 1999. Professor Misra has to his credit over 140 research publications in journals of international repute; he has also published several advanced level books. Most of his research investigations in the field of Biomathematics have appeared in highly prestigious journals like Journal of Mathematical Biology (USA), Bulletin of Mathematical Biology (UK), Journal of Biomechanics (UK), Journal of Biomedical Engineering (UK), Blood Vessels (Switzerland), Rheologica Acta (Germany), Biorheology (UK), International Journal of Solids and Structures (UK), International Journal of Nonlinear Mechanics (UK) etc. His publications have been well cited in the scientific literature. He has made pioneering research on mathematical modelling in each of the areas of Cardiovascular Mechanics, Mechanics of Brain Injury, Mechanics of Fracture and Remodelling of Bones and Peristaltic Flows in Physiological Systems. His theoretical findings on the compressibility of vascular tissues is a major breakthrough in the study of arterial biomechanics and were subsequently verified experimentally by Prof. Y. C. Fung of the University of California, San Diego. The Model developed by him for the study of arterial stenosis bears the potential to provide an estimate of the variation of blood viscosity as the height of the stenosis increases. The observations of the study were used by later investigators in the quantification of Doppler colour flow images from a stenosed carotid artery. Misra's theoretical study on the mechanics of cerebral concussion caused due to rotational acceleration of the head has opened up new vistas in neurological research and neurosurgery. On the basis of the study he could make some valuable predictions regarding the threshold of cerebral concussion for humans, in terms of angular accelerations. He was the first to account for the effect of material damping of osseous tissues on bone remodelling induced due to the surgical procedure of intra-medullary nailing. Misra's recent study on the effect of a magnetic field on the flow of a second-grade electrically conducting fluid serves as a very important step towards the perception of MHD flow of blood in atherosclerotic vessels. It throws sufficient light on the quantitative features of blood flow in constricted arteries.

Professor Misra has guided 20 research scholars towards their Ph.D. degree. He has been a member of the expert committee of the Council of Scientific and Industrial Research, New Delhi, of the Indira Gandhi National Open University and also of the National Science Foundation of USA as well as a member of the Technical Advisory Committee of the Indian Statistical Institute, Calcutta. In the year 1981 he delivered invited lectures at the Courant Institute of Mathematical Sciences, New York, at Cornell Medical Center, New York, at Kobe University, Japan and also at the Fourth International Congress of Biorheology held in Tokyo. Professor Misra delivered the prestigious Bhatnagar Memorial Lecture in 2001. He has been invited to deliver a lecture and to chair a session at the Fifth International Congress of Biorheology held in Germany in 1983, at the Tenth International Conference of Industrial and Applied Mathematics held in Belgium in July 2002 and also to deliver a keynote address and to chair a session at the International Conference on Applied Mathematics & Mathematical Physics held in Bangladesh in November 2002.

Contents

TOWARDS A NONSUBJECTIVE BAYESIAN PARADIGM

Jayanta K. Ghosh and Tapas Samanta

Indian Statistical Institute, Calcutta, India and Purdue university, USA

and *Indian Statistical Institute, Calcutta, India*

Abstract

We examine the historical development of the three major paradigms in Statistics and how they have influenced each other in a positive way. We then go on to argue that it is still necessary to make a choice and that the Bayesian formulation appears to be the most appropriate. Since elicitation of priors remains difficult inspite of some progress, it is suggested that nonsubjective Bayesian analysis has an important role to play. We present an overview of how nonsubjective priors are constructed and how they are used in different problems of inference involving low or high dimensional models. In particular, it is shown that many of the common perceptions or criticisms of nonsubjective Bayesian analysis are not justified.

Keywords: Hierarchical Bayes; Jeffreys prior; Parametric empirical Bayes; Probability matching; Reference prior; Uniform distributions

1. Introduction

There are three major paradigms in Statistics, namely, Data Analysis, Classical Statistics, also called Frequentist or Neyman-Pearsonian Statistics and Bayesian Analysis. We use all three names for Classical Statistics, choosing the one that best fits the context. Data Analysis may not even use any stochastic model like pre-Gaussian least squares. We assume most readers are familiar with Classical Statistics, where stochastic models are routinely used for data but the models contain unknown constants or parameters which are not treated as random variables — probabilistic calculations are applied only to repeatable uncertain events like tossing a coin but not to questions about non-repeatable uncertain events or statements like "this particular coin is fair". This restricted application of probability only in repeatable cases is called the Frequentist view of probability. In the Bayesian paradigm all uncertainties can be quantified into probability as in the case of gambling on one particular occasion. In particular, in the least squares problem both the regression coefficients β and Y in the model

$$Y = X\beta + \epsilon$$

are treated as random variables.

In addition to the three major paradigms, there are half way houses, like Conditional Frequentist Inference or Nonsubjective Bayesian Analysis, the subject of the present chapter. What is quite noticeable today is mutual tolerance and even some overlap and convergence towards a consensus. This was not the case even a couple of decades ago when there were heated controversies on foundations of Statistics. Some of these basic issues are discussed in the next section. In the subsequent sections it is shown to some extent how through some modification of the three major paradigms, some reconciliation between them is possible. However, it is not one of our premises or one of our theses, that all major differences have disappeared. That too is not the case.

We argue in this chapter that it is still important to choose a paradigm and justify its choice. Showing the chosen Bayesian paradigm in action and the fact that it does very well in applying Statistics to real life is part of the argument, not an excuse for not engaging in an argument, as suggested by the authors of two excellent books on what we regard as nonsubjective Bayesian Data Analysis, namely, Carlin and Louis ([20]) and Gelman et al. ([31]). Indeed, even Carlin and Louis ([20]), contrary to their professed view in their introduction, feel a need to reproduce some of the famous examples and arguments (e.g., [20], Ch. 1, 2) but such arguments are not explored in full, occasionally creating confusing or false expectations about nonsubjective Bayesian Analysis. To illustrate this we consider later Example 1.2 of [20] and the role of the likelihood principle in Section 6. A similar comment applies to the book by Gelman et al. ([31]), which is essentially about nonsubjective

Bayesian analysis but the excellent bibliographic note provided by on pp. 24, 25 focuses on foundational issues in the context of the subjective Bayesian paradigm. Basic references to Bayesian Analysis include [6], [16], [53] and [57].

The purpose of this chapter is to supplement books and papers in applied nonsubjective Bayesian Analysis by a critical re-examination of both the old foundational issues that dominated the sixties and seventies of the last century and specific criticisms brought against nonsubjective Bayesian methods.

We believe the Bayesian paradigm can be flexible enough to accommodate both subjective and nonsubjective Bayesians but at least for now our methods for eliciting subjective priors are so weak that most applications are nonsubjective. There can be other reasons why a future Bayesian may want to be flexible. We discuss this at the very end of the chapter.

Section 2 provides a brief history of least squares and all that came from it. It gives some idea of how the three paradigms developed with close interactions, sometimes friendly, sometimes not. In Section 3 — Why Should We All Be a Bayesian — we examine the three paradigms and argue in favour of being a Bayesian. In Sections 4 and 5 we discuss what we mean by nonsubjective priors, the motivation for using them and methods of construction. Roughly speaking, a subjective prior for a person, who may be an expert in the subject, is a quantification of uncertainty about unknown parameters in a model whereas a nonsubjective prior arises from a general algorithm applied to the particular problem in hand. A nonsubjective prior may also be validated by some introspection but introspection is not essential. Usually some Frequentist validation is sought when justifying the algorithms. Section 6 provides a critical discussion of nonsubjective priors, indicating why many of the common perceptions or criticisms are not justified. We also discuss in this section how far inferences based on these priors satisfy the Likelihood Principle and the Stopping Rule Principle. In particular, Example 1.2 of [20] is re-examined. Section 7 discusses nonsubjective Bayesian estimation and testing. High dimensional problems are briefly discussed in Section 8. The last section (Section 9) contains a discussion of some related major points.

2. How Did It All Start

Developments in geodesy and astronomy in the eighteenth century produced in each case many observations connected through a number of equations with much fewer parameters. Generally there would be n equations involving p unknowns, p being much less than n. Throughout the eighteenth century some of the best minds, including Laplace, Euler, Gauss and Legendre, considered this problem. This culminated in the discovery of the principle of least squares by Legendre in 1805. Credit is also given to Gauss who said he had discovered the principle but did not publish it. The principle determines the unknown constants in the equations so that the sum of squares of deviations between observations and values assigned by the equations with these constants is a minimum. This is a purely data analytic principle.

In 1823 Gauss provided an elegant stochastic model and proved the famous Gauss-Markov theorem that the principle of least squares leads to best linear unbiased estimates (BLUE). This may be regarded as the beginning of Classical Statistics but throughout eighteenth and nineteenth centuries the attitude to unknown parameters was ambiguous. Probability statements made about them were interpreted both as Frequentist confidence intervals and Bayesian credibility intervals given the particular data in hand.

The ambiguity has its source in the following instructive example, central to the development of much of Statistics.

Example 1. Suppose $X_1, X_2, \ldots, X_n$ are independent and identically distributed (i.i.d.) observations with normal distribution $N(\theta, 1)$. Suppose the prior for θ, namely, $\pi(\theta)$ is the (improper) uniform distribution on the parameter space $\mathbb{R}$. Let $\bar{X}$ be the sample mean and $z_{\alpha/2}$ the upper $\alpha/2$-point of $N(0,1)$, $0 < \alpha < 1$. Then the following are true.

$$\Pr\{\theta \in (\bar{X} - z_{\alpha/2}/\sqrt{n}, \bar{X} + z_{\alpha/2}/\sqrt{n})|\theta\} = 1 - \alpha \quad (1)$$

$$\Pr\{\theta \in (\bar{X} - z_{\alpha/2}/\sqrt{n}, \bar{X} + z_{\alpha/2}/\sqrt{n})|\bar{X}\} = 1 - \alpha \quad (2)$$

The first probability holds over many repetitions of samples — it is a classical

Frequentist probability. In the second equation the probability, given $\bar{X}$ is held fixed, has a Bayesian meaning. The event considered is the same but the conditioning leads to different interpretation. The prior $\pi(\theta)$ is (exactly) probability matching in the sense of [36].

The idea of a prior distribution that interacts with the likelihood or probability of observation to produce a posterior probability first appeared in Reverend Bayes's posthumous paper ([5]). The equation that shows how the prior and likelihood interact is called Bayes Theorem — an elementary result of great philosophical and methodological significance. These ideas were rediscovered and popularized by Laplace.

One of the great achievements that arose as a consequence of attention to $\bar{X}$ is Laplace's famous limit theorem ([50]) which says if X_j's are i.i.d. with mean θ and variance σ^2, then $\bar{X}$ is approximately normal with mean θ and variance σ^2/n. This implies that equation (1) can be used approximately in a great many cases. Equation (2) was used by Laplace to prove what would now be called posterior consistency as well as to find a credibility interval for θ that would be easy to interpret in a Frequentist way. It may thus be thought of as a forerunner of the Bernstein-von Mises theorem on posterior normality, just as Laplace's limit theorem is an early version of the Central Limit Theorem which occupied a central role in the theory of probability for many years. Both were rigorously proved only in the twentieth century.

Regression equations and the Least Squares principle, as we now use it, grew from the work of Francis Galton and Karl Pearson in Biometry. They took their final shape in the hands of G.U. Yule in early twentieth century.

The logical distinction between (1) and (2) was first pointed out by R.A. Fisher who began a systematic development of methods of estimation, testing and design of experiments using only classical Frequentist probability. By 1940, they had reached their present form in the hands of Neyman, Pearson and Wald. A few years later the restriction to linear estimates was removed and a new theory of minimum variance unbiased estimates was born. The major results were the Cramer-Rao inequality, the Rao-Blackwell theorem and the major tool was the notion of a complete sufficient statistic. These, along with the earlier Neyman-Pearson Lemma and Basu's Theorem on independence of complete sufficient statistics and ancillary statistics, have been the core of a first advanced undergraduate course in Classical Statistics.

With Wald had come decision theory and attention had shifted away from unbiasedness to general estimates for which minimaxity was introduced by Wald and admissibility by Lehmann. Nearly a hundred years after Gauss, we knew that under the additional assumption of Gaussian distribution, $\bar{X}$ is not only BLUE but UMVUE (uniformly minimum variance unbiased estimate), minimax and admissible too. Also Robbins introduced the Nonparametric Empirical Bayes approach and Stein proved his apparently paradoxical result that for estimating a normal mean the sample mean $\bar{X}$ ceases to be admissible for dimension greater than two. By the seventies it was clear that the Stein paradox was ubiquitous whenever one estimates many parameters having structural similarities and new insight into this was provided by the Parametric Empirical Bayes (PEB) approach of Efron and Morris. A fully Bayes approach, called Hierarchical Bayes, soon developed and calculation of posterior was made feasible by the new simulation technique of MCMC in the late eighties. This method has some advantage over PEB even in a Frequentist sense. The last decade was full of successful Hierarchical Bayes modelling of uncertainty in many, many real life high dimensional problems with calculation of posterior through some form of MCMC. These applications show that not only has the Bayesian paradigm better logical foundations than Classical Statistics but it can handle better complex practical problems as well. This must have been a major reason for the dramatic upsurge of interest in Bayesian Analysis.

This brief history indicates, among other things, how closely the development of the three different paradigms have been interlinked at different points of time.

3. Why Should We All Be a Bayesian

We have followed the growth of Least Squares and all it led to for nearly three centuries — eighteenth to twentieth — and how features of all three paradigms pervade our subject. Data Analysis provides innovative new methods and quick insight. It borrows strength from Descriptive Statistics and common sense. It does not require theoretical underpinnings or complicated mathematical justifications based on probability theory. However, a discipline based only on ad hoc data analytic techniques cannot survive for long. One of the great achievements of Classical Statistics in the twentieth century was to provide a unified logical foundation to various inferential questions of Statistics and statistical methods developed over centuries. Let us examine this aspect in some detail.

Classical Neyman-Pearsonian Statistics emerged as a paradigm in the fundamental papers of Neyman and Pearson on testing hypotheses. The starting point in this paradigm is a set of random variables $X_1, X_2, \ldots, X_n$ and a stochastic model from which one can calculate their joint density $p(x_1, x_2, \ldots, x_n)$. The model or models do not specify the density $p(x_1, \ldots, x_n)$ completely, it is allowed to depend in a well-defined way on a set of "unknown" parameters, which are treated as unknown constants rather than random variables. This is a point of departure from Bayes. Secondly, they consider a test $\phi(x_1, x_2, \ldots, x_n)$ which is a function or an algorithm that selects a particular hypothesis from the candidate hypotheses, e.g., $H_0 : \theta = 0$ vs. $H_1 : \theta > 0$ for $N(\theta, 1)$, given any data. If $\phi(x_1, \ldots, x_n) = 1$ one chooses H_1, if $\phi = 0$

one chooses H_0. If for $x_1, \ldots, x_n$, ϕ is between 0 and 1, one chooses H_1 with probability ϕ. This is a major point of departure from past practices, where statisticians, as data analysts, only considered what is to be done with the data in hand, not what would have happened with other possible data. Thirdly, a test is evaluated by its performance over all possible data. The evaluation of probability of error for a test ϕ for different values of a parameter is its risk

$$\begin{aligned} R(\phi, \theta) &= E_\theta(\phi), \text{ if } \theta = 0 \\ &= 1 - E_\theta(\phi), \text{ if } \theta > 0. \end{aligned}$$

Neyman and Pearson showed that if one looked at all ϕ with a bound on the probability of error of first kind, namely rejecting H_0 when H_0 is true, $R(\phi, 0) \leq \alpha$, then there exists ϕ_0 that minimizes $R(\phi, \theta)$ for all $\theta > 0$. Moreover, ϕ_0 is easy to describe. This is their famous UMP (uniformly most powerful) test. Very general minimax results of this type were proved later by Huber and Strassen ([46], [47]). A beautiful example of this kind is the robust version of the Neyman-Pearson lemma given in [53].

Classical Statistics introduced a new approach to decision making and inference, and evaluation and comparison of different algorithms or procedures based on performance over the whole sample space. It also provided many benchmarks and lower bounds. Thus the Gauss-Markov theorem may be treated as a precursor and the many novel developments that are described in the previous section could not have taken place if this pardigm change had not taken place in the nineteen thirties.

Some of its advantages over, say, simple minded data analysis can clarify the intellectual revolution brought about by Classical Statistics. Given any new method, Classical Statistics can test it out on many simulated stochastic models as well as compare with known benchmarks. In contrast, Data Analysis will try it out on a small list of standard available examples. Of course, if a method survives this first test, many subsequent practical applications eventually will settle if it is any good. But Classical Statistics does it faster and more systematically. Secondly, any new problem, difficult to solve in Data Analysis, may not be so difficult in a well-defined logical paradigm. We illustrate this in the next paragraph.

Often in image analysis, one would wish to merge two images. For example, they could be two pictures of the same object, say, a tumour, taken by tomography and ultrasonography, or they could be two pictures of the same subject, say, an absconding criminal, at slightly different angles. This is regarded as a challenging new problem in data analytic image analysis. Yet, in principle, it is no more difficult than the problem of combining two observations using linear models and Gauss's theorem. For an application of these ideas to image analysis see [38].

Why then does one want to move beyond Classical Statistics? There are several reasons. We list the more important ones.

1. *Flaws in foundation.* The paradigm is flawed because all its evaluations are based on averaging over the sample space, i.e., on performance for all possible data. While such measures are important, specially at the planning or design stage, they are irrelevant once the data are in hand. For example, variance of an estimate or the probability of error of a test relevant for the particular data being analyzed are the appropriate posterior risk given data — $E\{(T-\theta)^2|X_1,\ldots,X_n\}$ for an estimate T of θ or $P(H_0|X_1,\ldots,X_n)\phi + P(H_1|X_1,\ldots,X_n)(1-\phi)$ for a test ϕ — which can be calculated only in the Bayesian paradigm. It is no wonder that clients — engineers and doctors — coming to classical statisticians almost always misinterpret the quantities that are presented to them. They interpret p-values etc. as a sort of posterior risk.

2. The second reason is connected with the first reason. Quite often the variance or risk calculations are patently absurd even to classical statisticians. Here are two examples.

Example 2 (Cox, [22]). You have to take samples from $N(\theta, 1)$ and estimate θ. Suppose you toss a fair coin. If you get a head, you take a sample of size $n = 2$. If it is a tail, you take a sample of size $n = 100$. Admittedly, this is an odd sampling scheme but let us continue our analysis of this from a classical point of view.

Given $n = 2$, the classical statistician estimates θ by $T = \bar{X}_2 = (X_1 +$

$X_2)/2$ and, similarly for n =100, by $T = \bar{X}_{100}$. He says rightly, that T is an unbiased estimate for θ with variance $= \frac{1}{2} \cdot \left(\frac{1}{2} + \frac{1}{100}\right) = \frac{1}{4}$ (approximately). Suppose he actually gets a tail and has a sample of size $n = 100$. Should he quote this very large variance, namely, 1/4 or the more natural 1/100 ? Most classical statisticians confronted with this example concede they would prefer the second number.

Example 3 (Welch). You have a sample of size $n = 2$ from the uniform distribution on $(\theta - \frac{1}{2}, \theta + \frac{1}{2})$. You want a $100(1-\alpha)\%$ confidence interval for $\theta, 0 < \alpha < 1$ and usually $\alpha = 0.05$ or 0.01. It is easy to see that this being a location parameter family, you can choose $h > 0$ such that $P_\theta\{\bar{X} - h < \theta < \bar{X} + h\} = 1 - \alpha$, i.e., $\bar{X} \pm h$ is a solution to this problem. Suppose now your actual data are $X_1 = 1, X_2 = 2$. The only way such data can come as sample from the range $(\theta - 1/2, \theta + 1/2)$ is to have $X_1 = \theta - 1/2, X_2 = \theta + 1/2$ so that you know for sure $\bar{X} = \theta$. But you will say you have only $100(1-\alpha)\%$ confidence that your interval $\bar{X} \pm h$ contains θ. This is also patently absurd that classical statisticians concede this. Incidentally, the fact that $X_1 = \theta - 1/2, X_2 = \theta + 1/2$ with probability zero need not cause concern. If X_1 and X_2 are very close to $\theta \pm 1/2$ instead of being exactly equal, you would intuitively expect the true confidence to be very close to one rather than $(1-\alpha)$. You can verify by calculating conditional probability of covering θ, given $|X_2 - X_1|$. Alternatively, one can choose a discrete version of this problem as in [7].

These are examples of what are called conditionality paradoxes, first pointed out by Fisher. Fisher suggested that in each example there is a statistic, the sample size n in Example 2 and $|X_2 - X_1|$ in Example 3, whose distribution does not depend on θ but whose value seems to indicate how informative is the sample. Fisher called such a statistic ancillary and suggested one should make inference condtional on an appropriate ancillay. Such examples are treated in detail by Basu ([4]), Brown ([19]) and Kiefer ([50]). Conditional Inference has also received a lot of attention from Barndorff-Nielsen and Cox, who have shown how one can make conditional asymptotic inference based on the maximum likelihood estimate (MLE) given asymptotic ancillaries ([1] and [2]).

3. Even though most classical statisticians accept the contidionality principle (CP) (see Appendix), acceptance of CP and the sufficiency principle (SP) leads to further problems, as first pointed out by Birnbaum. The sufficiency principle (SP) says inference should be based on the (minimal) sufficient statistic, which extracts all the information in the data in a well-defined mathematical way; $P\{\text{data}|\text{minimal sufficient statistic}\}$ is free of the parameter θ and so the data cannot contain any additional information. Birnbaum showed that CP and SP together imply the likelihood principle (LP) which says that two likelihood functions $p(x|\theta)$ and $p'(x'|\theta)$ lead to the same inference about

θ if they are proportional to each other (as function of θ) and therefore, after x is observed, the inference should be based on the likelihood function for the observed x. Since classical methods are based on integration over the whole sample space rather than the data alone, most of them violate LP. It should be mentioned that Birnbaum's theorem is a mathematical theorem, in the spirit of metamathematics rather than a matter of personal philosophical belief. A proof is given in the Appendix.

We give two examples in one of which a popular classical method violates LP.

Example 4. Let $X_1, X_2, \ldots, X_n$ be i.i.d $B(1, \theta)$, i.e.,

$$P_\theta\{X_i = 1\} = \theta, P_\theta\{X_i = 0\} = 1 - \theta, 0 < \theta < 1.$$

The sample size n may be fixed (Case 1) or random, e.g., as in inverse sampling: $n =$ first i such that $X_i = 1$ (Case 2). A classical approximate 95% confidence interval for θ is

$$\hat{\theta} \pm 1.96/\hat{a}$$

where $\hat{\theta} = r/n$ is the MLE, $r = \Sigma_{i=1}^n X_i$,

$$\hat{a}^2 = -\frac{d^2 \log\ L}{d\theta^2}|_{\hat{\theta}}$$

and $\log L$ is the loglikelihood given by

$$\log L = r \log \theta + (n - r) \log(1 - \theta).$$

This is same in both Case (1) and Case (2) though they involve different sample spaces. It is easy to see that the method satisfies LP.

Example 5. Let $X_1, X_2, \ldots, X_n$ be i.i.d having a Cauchy distribution with density

$$p(x_i|\theta) = \frac{1}{\pi}.\frac{1}{1+(x_i-\theta)^2}.$$

As in Example 4, n is fixed (Case 1) or n is a random variable not depending on θ except possibly through X's.

Let $I(\theta)$ be the Fisher information defined by

$$I(\theta) = -E_\theta\left(\frac{d^2 \log p(X_i|\theta)}{d\theta^2}\right).$$

In this case $I(\theta) = I(0)$, a constant. Let $\hat{\theta}$ be the MLE of θ. A popular approximate 95% confidence interval for θ is $\hat{\theta} \pm 1.96\sqrt{1/(nI(0))}$ which violates the LP and in fact is not appropriate for Case 2. However

$$\hat{\theta} \pm 1.96\left(-\frac{d^2 \log L}{d\theta^2}|_{\hat{\theta}}\right)^{-\frac{1}{2}}$$

does not violate LP and is appropriate for both Case 1 and Case 2.

4. *Practical and Methodological Reasons for Preferring Bayesian Analysis.* So far we have been discussing somewhat abstract foundational reasons. There are also several impressive practical or methodological reasons.

The exact methods of Classical Statistics, for example, UMVUE or UMP test can only be applied in very simple cases. Slight change in the problem, for example even extra information, can cause difficulties.

Example 6. Let $X_1, X_2, \ldots, X_n$ be i.i.d. $N(\theta, 1), -\infty < \theta < \infty$. It is well-known that $\bar{X}$ is UMVUE. Suppose you have information $a \leq \theta \leq b$. The UMVUE is still $\bar{X}$ and is obviously absurd because it need not lie in the given interval $[a, b]$. The MLE, namely, $\bar{X} I(a \leq \bar{X} \leq b) + aI(\bar{X} < a) + bI(\bar{X} > b)$, is inadmissible and also somewhat absurd because it suddenly becomes flat to the left of a and right of b. In contrast Bayes estimates are in the given range, admissible and exhibit better behaviour as $\bar{X}$ crosses a and b and moves towards $\pm\infty$.

In complex real life problems all exact classical methods break down and approximate methods based on MLE have to be used. By contrast Bayesian methods are exact and can use all available information in sensible way. Also Bayesian methods generally provide better inference than MLE; see [3], [11] and [17].

5. *Axiomatic Justification of Bayesian Analysis.* Finally, there are natural rationality axioms on how one should make inference or decision under uncertainty, which force one to act like a Bayesian with a prior probability. De Finetti makes out a compelling mathematical case that unless one is coherent, i.e., has at least a finitely additive probability measure, one would be a sure loser as a gambler. This result has been extended by Heath and Sudderth ([45]) to show unless one acts as a Bayesian with a prior that is

at least finitely additive and uses the corresponding posterior, one's inference procedure would be uniformly inadmissible. Other similar rationality axioms, due to Ramsey ([61]), Savage ([63]) and others, show how a rational linearly ordered preference pattern leads logically to the existence of a prior (subjective probability measure over the θ-space) and a utility or loss function. If one is rational in any one of several possible senses of being rational, one is forced to be a Bayesian. A good exposition of these ideas can be found in [64].

6. *Decision Theoretic Reasons.* Classical statistical decision theory has two kinds of theorems which lend support to the remarks in the previous paragraph. One class of theorems show unless a decision procedure, e.g., a test or estimate, is based on a prior or a sequence of priors, it would be inadmissible. The other kind of theorems show that a class of decision procedures is complete, i.e., given any decision procedure outside it there is procedure with lower risk within the class, if it is the closure of Bayes procedures.

To sum up there are many practical, methodological and theoretical reasons why we should be a Bayesian. However, this does not mean a Bayesian has nothing to learn from the other two paradigms. Bayesians believe in Frequentist validation in the real world. One formulation of this is the prequential framework of Dawid ([27], [28]). Another, even more Frequentist formulation, is the notion of posterior consistency at a given true θ_0, due to

Laplace, von Mises, Bernstein and Freedman. This is a kind of weak validation of a Bayesian procedure against virtual, i.e., simulated reality. Diaconis and Freedman ([29]) refers to it as a sort of “What if ”, a validation through a thought experiment. The frequentist notions of bias and variance and an appropriate trade off between them in model selection remain useful in understanding Bayesian model selection. Thus a complex model reduces bias but increases variability of parameters and their estimates, often leading to inferior prediction as compared to simpler models. Other similar applications of frequentist ideas appear in later sections.

A Bayesian also finds it useful to use some of the common descriptive data analytic methods to get a quick feel for data or communicate to clients. The Bayesian answers are usually refinements of the data analytic answers — comparison of the two can lead to insight and better understanding of the former. An extremely important new book on high dimensional and often nonlinear data analysis which draws on both Classical Statistics and Bayesian ideas is [44].

4. Choice of Prior

Given a data set $X = (X_1, \ldots, X_n)$ a Bayesian has a stochastic model $p(x|\theta)$ for the joint density as in Classical Statistics. The Bayesian interprets this as likelihood or conditional density of data given θ. To set the Bayesian inference engine in motion, he needs a prior $\pi(\theta)$, namely, the prior probability density of θ. This reflects his belief or knowledge, prior to seeing data. In the light of data, his belief is quantified in the posterior density $p(\theta|X)$ given by Bayes Theorem as

$$p(\theta|X) = \frac{\pi(\theta)p(X|\theta)}{\int_{\Theta} \pi(\theta)p(X|\theta)d\theta}. \tag{3}$$

Essentially, the prior is being moved towards those values of θ which make the observed data more likely.

In relatively simple problems of inference one needs to report the posterior and some descriptive measures like posterior mean or median and posterior variance or posterior quantiles.

If one has a decision problem with an action space and a loss function, one chooses an action "a" (depending on the observed X) to minimize the average posterior loss (see Section 7).

To drive this Bayesian engine one needs the prior $\pi(\theta)$. All non-Bayesians generally agree that but for the problem of choice of the prior, the Bayesian paradigm is indeed very attractive. So let us examine critically how a prior can be chosen and what effect it has on inference.

Ideally, the prior should reflect the subjective belief or knowledge of the client or the analyst or a subject matter expert. Unfortunately, eliciting a prior from experts is not easy. Empirical studies have shown certain professions, experience and maturity help. For example, a businessman or a doctor or a lawyer will be better able to assign a probability to an uncertain event within his domain of expertize than most other people. However, usually it is not realistic to expect that one would be able to elicit more than a prior mean and variance. Occasionally, one can elicit prior covariance in an indirect way. While all Bayesians expect that this situation will improve in the future, it is hard to believe that in all but very simple situation a full subjective prior can be elicited. So it is customary to choose priors in a nonsubjective, conventional way, incorporating as much of prior information as has been elicitied.

What would be the consequences of substituting a nonsubjective prior for a subjective prior? This depends on the relative magnitude of the amount of information in data, which for i.i.d. observations may be measured by the sample size n or $nI(\theta)$, and the amount of information in the prior, which is discussed in the next section. If the former dominates, then there is hardly anything lost and in most cases of low dimensional parameter space, the situation is like that. A Bayesian would refer to it as washing away of the prior by the data. There are several mathematical theorems embodying this phenomenon. One such result is posterior normality and its refinements (see, for example, [34], [39] [52] and [64]). However, occasionally one may be very confident of certain aspects of the prior and does not wish to change it even if there is some conflict with data.

Sometimes the analyst will not have any prior information and will want to use a purely nonsubjective prior, also called a noninformative prior in the past. Such a prior may also be used to report the results of an analysis from a relatively impartial point of view. The next section describes some standard algorithms for producing purely nonsubjective priors. While such priors produced by different algorithms are not unique in general, they are very similar and even for a small data set generate nearly identical posteriors.

The older terminology of noninformative priors is no longer in favour among nonsubjective Bayesians. The older terminology leads to an expectation that such a prior reflects complete lack of information, which is impossible to define. As Poincare observed, noninformative priors do not exist. On the other hand the purely nonsubjective priors do have low information in well-defined senses of Shannon's missing entropy or non-Euclidean geometry and lead to mostly data dependent posteriors. For somewhat similar reasons the term *objective priors* used by Jeffreys can lead to somewhat wrong expectations. Another popular term is *default.* A nosubjectiver prior is, at least approximately, both noninformative and objective in the sense of not putting in much prior input but the relatively neutral term *nonsubjective* comes closest to what is meant.

5. Nonsubjective Priors

To construct nonsubjective prior on a given parameter space, one has to do one of the following things — (1) define a uniform distribution that fits the topology of the parameter space for a suitable topology induced by the Hellinger metric or a Riemannian metric arising from the Fisher information matrix, or (2) maximize a suitable measure of entropy (i.e., minimize information in this sense) or (3) choose a prior with some form of Frequentist validation since the use of a prior with little information should lead to the same sort of inference as what a Frequentist would do.

The simplest choice of a (purely) nonsubjective or so called noninformative prior is the uniform, used for different reasons by both Laplace and Bayes. It has been used in this chapter earlier in Example 1.

There are various problems with the uniform, though it still remains a reasonable choice when the other methods are not easy to apply. The three major problems with the uniform are as follows.

First, as pointed out by Fisher, it is not invariant under (continuously differentiable) one-one transformations of θ and it seems natural to require some invariance of this sort. One looks for a method that produces priors $\pi_1(\theta)$ for θ and $\pi_2(\eta)$ for any smooth one-one function $\eta(\theta)$ of θ such that one can pass from one to the other by the usual jacobian formula

$$\pi_1(\theta) = \pi_2(\eta(\theta)) \left| \frac{d\eta}{d\theta} \right| . \tag{4}$$

Secondly, it does not go naturally with the Riemannian geometry induced by the metric which is obtained through the integration of

$$\rho(d\theta) = \sum_i \sum_j I_{i,j}(\theta) d\theta_i \theta_j$$

over all curves connecting θ to θ' and minimizing over curves. This metric was introduced by C.R. Rao and is known to be "natural" in the sense that it is the only Riemannian metric that transforms as expected under continuously differentiable one-one transformations of Θ onto itself, vide [21] and [62].

Finally, the uniform seems to maximize the wrong entropy

$$H(p) = - \int_\Theta p(\theta) \log p(\theta) d\theta.$$

Shannon used $H(\cdot)$ very successfully when Θ is a finite set, where the discrete uniform is everybody's choice of a noninformative prior. But $H(\cdot)$ depends on the dominating measure (vide [15]), and is not invariant under continously differentiable one-one transformations, vide [66].

We take up the last point and define a measure of information appropriate here. The transition from the prior to the posterior distribution is an indication of the (additonal) amount of information in data $X = (X_1, \ldots, X_n)$, relative to a particular prior. The last qualification is important. The change from prior to posterior has been used by Bernardo ([15]) to measure how informative is a prior. A measure of change from the prior to the posterior is given by the expected Kullback-Leibler divergence between the posterior and the prior

$$K(p(\theta|X), \pi(\theta)) = E\left\{\log \frac{p(\theta|X)}{\pi(\theta)}\right\} \tag{5}$$

where the expectation is with respect to the joint distribution of X and θ. Bernardo suggests that if a prior is already very informative, say, degenerate at some θ_0, then the posterior is the same as prior and $K = 0$, the data cannot provide any additional information. Bernardo maximizes (5) asymptotically to get his "noninformative" prior. He calls it a reference prior, in the sense that information in other priors may be calibrated by taking Bernardo's prior as a reference point or origin.

We now present an algorithm for asymptotically maximizing (5). We assume throughout this section that X_i's are i.i.d. Fix an increasing sequence of compact sets C_i whose union is the whole parameter space. In the following initially we fix C_i and let $n \to \infty$. Then, as indicated in [36], under suitable regularity conditions,

$$\begin{aligned} K(p(\theta|X), \pi(\theta)) &= \frac{d}{2} \log \frac{n}{2\pi e} + \int_{C_i} \pi(\theta) \log\{\det I(\theta)\}^{1/2}\, d\theta \\ &\quad - \int_{C_i} \pi(\theta) \log \pi(\theta)\, d\theta + o(1) \end{aligned} \tag{6}$$

where d is the dimension of θ, $\det A$ denotes the determinant of a matrix A and $I(\theta) = [I_{i,j}(\theta)]$ is $d \times d$ Fisher information matrix given by

$$I(\theta) = E_\theta \left(\frac{\partial \log p(X_1|\theta)}{\partial \theta_i} \cdot \frac{\partial \log p(X_1|\theta)}{\partial \theta_j} \right)$$

which is assumed to be positive definte. Thus $K(p(\theta|X), \pi(\theta))$ is the sum of a constant, which does not depend on the prior, and a term which converges to the functional

$$J(\pi(\cdot), C_i) = \int_{C_i} \pi(\theta) \log \frac{\{\det I(\theta)\}^{1/2}}{\pi(\theta)} d\theta.$$

If we maximize $J(\pi(\cdot), C_i)$ with respect to all priors over C_i, we get the Jeffreys prior concentrated on C_i, i.e.,

$$\pi_i(\theta) = \begin{cases} \text{const.}\{\det I(\theta)\}^{1/2}, & \text{if } \theta \in C_i \\ 0, & \text{otherwise.} \end{cases}$$

If we now let $i \to \infty$ to make C_i tend to the whole of Θ we may regard π_i's as converging to the Jeffreys improper prior (see [34])

$$\pi_J(\theta) = \{\det I(\theta)\}^{1/2}. \tag{7}$$

Thus under suitable regularity conditions the Jeffreys prior is a reference prior.

When parameters can be arranged according to order of importance, Bernardo ([15]) and Berger and Bernardo ([9]) suggest a step by step maximization. In fact, they suggest stepwise maximization in all cases, with suitable reparametrization. This leads to a modification of the Jeffreys prior. It has worked very successfully in many examples ([9]). It is these latter priors that are now called reference priors.

In case partial information is available (vide [6], Sec. 3.4), on prior moments or quantiles, one can minimize with respect to π, the Kullback-Leibler functional

$$K(\pi, \pi_0) = \int_\Theta \pi(\theta) \log \frac{\pi(\theta)}{\pi_0(\theta)} d\theta$$

subject to

$$\int_\Theta g_i(\theta)\pi(\theta)d\theta = A_i, \; i = 1, \ldots, k.$$

Here π_0 is a nonsubjective prior, Jeffreys or reference or probability matching, that one starts with and $g_i(\theta)$ is θ^i or an indicator $I_{B_i}(\theta)$ where B_i is some interval (c_i, d_i). For example, if one wants to specify the three quartiles, one would set $B_i = (Q_{i-1}, Q_i)$, $i = 1, 2, 3$, where $Q_0 = -\infty$ and Q_i denotes the ith quartile, $i = 1, 2, 3$, and $A_1 = A_2 = A_3 = 1/4$. Let

$$\pi_1(\theta) = \text{constant.} \exp\left(\sum_1^k \lambda_i g_i(\theta)\right)$$

where λ_i's are chosen to satisfy the given constraints. Then, subject to π satisfying the constraints,

$$K(\pi, \pi_0) = \int_\Theta \pi(\theta) \log \frac{\pi_1(\theta)}{\pi_0(\theta)} d\theta + K(\pi, \pi_1) = \sum \lambda_i A_i + K(\pi, \pi_1)$$

is clearly minimized if $\pi = \pi_1$.

Sun and Berger ([71]) derive reference priors for multiparameter cases where prior information is available for some of the parameters. Given a subjective conditional prior density $\pi^s(\theta_2|\theta_1)$, where $\theta = (\theta_1, \theta_2)$, or a subjective marginal prior density $\pi^s(\theta_1)$, they derive respectively the marginal reference prior $\pi^r(\theta_1)$ or conditional reference prior $\pi^r(\theta_2|\theta_1)$. Also, a method for finding marginal reference priors is proposed when θ_1 and θ_2 are known to be independent.

We now indicate how Jeffreys prior can be obtained as a Lebesgue measure after transformation. Consider a one-one smooth transformation $\psi(\theta)$ of θ such that the information matrix I^ψ with the new parametrization ψ is identity (I) at $\psi(\theta_0)$. This means the local geometry in the ψ space is Euclidean near $\psi(\theta_0)$ and so the Lebesgue measure $d\psi$ is a suitable uniform distribution near $\psi(\theta_0)$. If we lift this back to the θ space making use of the jacobian and the elementary fact

$$\left[\frac{\partial \theta_j}{\partial \psi_i}\right] [I_{i,j}(\theta)] \left[\frac{\partial \theta_j}{\partial \psi_i}\right]' = I^\psi = I, \tag{8}$$

we get Jeffreys prior in the θ space, namely

$$d\psi \;=\; \left\{\det\left[\frac{\partial \theta_j}{\partial \psi_i}\right]\right\}^{-1} d\theta \;=\; \{\det I(\theta)\}^{1/2} d\theta.$$

An alternative way is to put a uniform distribution on a finite set of points in the parameter space that approximate well all parameter points and then put a discrete uniform prior on this. If the metric used is Hellinger then it can be shown that the discrete uniforms converge weakly as the approximation is refined more and more and the limit is the Jeffreys distribution. The same procedure can be applied to infinite dimensional cases also. For details see [32]. For a more direct derivation of the Jeffreys prior from the Hellinger metric see [43], Sec. 5.4.

One can easily see that under the usual regularity conditions the Jeffreys prior is also invariant in the sense of (4). Jeffreys prior for θ is given by (7). It is easily verified from (8) that the Jeffreys prior in the η-space given by

$$\pi_2(\eta) = \left\{ \det \left[E_\theta \left(\frac{\partial \log p(X_1|\theta)}{\partial \eta_i} \cdot \frac{\partial \log p(X_1|\theta)}{\partial \eta_j} \right) \right] \right\}^{1/2}$$

satisfies (4).

The most popular nonsubjective priors are Jeffreys, reference and uniform. Carlin and Louis ([20]) suggest some other ad hoc priors which are easier to calculate in high dimensions. However, use of nonsubjective priors, in general, needs more care than they seem to suggest. This is discussed in [23] and our Section 6.

Another popular way of generating nonsubjective priors is by matching posterior and Frequentist probabilities. This is based on the intuition that the probability statements of a Bayesian with a nonsubjective prior can be validated by a Frequentist interpretation also. These have been called probability matching by Ghosh and Mukerjee ([36]). For simplicity let us consider only a two-dimensional parameter $\theta = (\theta_1, \theta_2)$ where θ_1 is the parameter of interest. Let us first assume that the nuisance parameter θ_2 is orthogonal to θ_1 in the sense of [24]. Ghosh and Mukerjee ([36]) indicate how the Bayesian and Frequentist Bartlett corrections can be used to choose a prior π such that a likelihood ratio based confidence set has the same frequentist and posterior probability of covering the true value up to $O(n^{-2})$. We choose a confidence set $A_{1-\alpha}(X)$ for θ_1 (that depends on the prior) such that the posterior probability

$$P(\theta_1 \in A_{1-\alpha}(X)|X) = 1 - \alpha + O(n^{-2}).$$

Suppose we now wish to choose the prior π, and hence $A_{1-\alpha}$, such that the frequentist probability

$$P_\theta(\theta_1 \in A_{1-\alpha}(X)) = 1 - \alpha + O(n^{-2}) \ (\forall \alpha) \tag{9}$$

uniformly on compact sets of θ.

As indicated in [36] (see, e.g., [34] for details) a solution of (9) is given by the second-order partial differential equation for π,

$$\frac{\partial}{\partial \theta_1}\left[\frac{\pi_{10}(\theta)}{I_{11}} - \left\{\frac{K_{10.20}}{I_{11}^2} - \frac{K_{12}}{I_{11}I_{22}}\right\}\pi(\theta)\right] + \frac{\partial}{\partial \theta_2}\left\{\frac{K_{21}}{I_{11}I_{22}}\pi(\theta)\right\} = 0 \tag{10}$$

where $\pi_{10}(\theta) = \partial\pi(\theta)/\partial\theta_1$, $K_{ij} = E_\theta\left\{\partial^{i+j}\log p(X_1|\theta)/\partial\theta_1^i\partial\theta_2^j\right\}$, $I_{11} = -K_{20}$, $I_{22} = -K_{02}$ and

$$K_{ij.i'j'} = E_\theta\left[\left\{\partial^{i+j}\log p(X_1|\theta)/\partial\theta_1^i\partial\theta_2^j\right\}\left\{\partial^{i'+j'}\log p(X_1|\theta)/\partial\theta_1^{i'}\partial\theta_2^{j'}\right\}\right].$$

Example 7. Let X_i's be i.i.d. $N(\mu, \sigma)$. If μ is the parameter of interest, the probability matching equation (10) turns out to be

$$\sigma^2\frac{\partial^2\pi(\mu,\sigma)}{\partial\mu^2} + \frac{\partial}{\partial\sigma}\{\sigma\pi(\mu,\sigma)\} = 0.$$

This equation is satisfied, for example, by any prior proportional to $1/\sigma$.

In case σ is the parameter of interest, the probability matching differential equation for $\pi(\mu, \sigma)$ is

$$\frac{\partial}{\partial \sigma}\left[\frac{1}{2}\sigma^2 \frac{\partial \pi(\mu,\sigma)}{\partial \sigma} + \frac{5}{2}\sigma \pi(\mu,\sigma)\right] = 0$$

which is satisfied by a prior of the form $q(\mu)/\sigma$ where $q(\mu)$ is an arbitrary function of μ.

It is to be noted that the Jeffreys prior $\pi_J(\mu,\sigma) = 1/\sigma^2$ is not probability matching in both these cases.

If we start with one-sided confidence intervals, i.e., we choose $\theta_{1,\alpha}(X)$, depending on the prior π, such that

$$P(\theta_1 \leq \theta_{1,\alpha}(X)|X) = 1 - \alpha + O(n^{-1}),$$

and wish to choose the prior such that

(a) $P_\theta(\theta_1 \leq \theta_{1,\alpha}(X)) = 1 - \alpha + O(n^{-1}) \forall\, \alpha$

(uniformly on compact sets of θ)

or

(b) (no nuisance parameter)

$$P_{\theta_1}(\theta_1 \leq \theta_{1,\alpha}(X)) = 1 - \alpha + O(n^{-1})\ \forall\, \alpha$$

(uniformly on compact sets of θ_1).

or

(c) (integrated out nuisance parameter given conditional prior $\pi(\theta_2|\theta_1)$)

$$\int P_\theta(\theta_1 \le \theta_{1,\alpha}(X))\pi(\theta_2|\theta_1)d\theta_2 = 1 - \alpha + O(n^{-1})\ \forall\, \alpha$$

(uniformly on compact sets of θ_1).

For one-sided intervals matching is, in general, not possible beyond $O(n^{-1})$ (see [34]).

The solution for (b), due to Welch and Peers ([73]) and Stein ([69]), is the Jeffreys prior. It can be shown that no such result is available if the dimension of θ_1 is more than one. The differential equation corresponding to (a), due to Tibshirani ([72]) (see also [60]), is

$$-\partial(I^{11})^{1/2}/\partial\theta_1 = (I^{11})^{1/2}\pi_{10}(\theta)/\pi(\theta),$$

where $I^{11} = (I_{11})^{-1}$. The solution to this is

$$\pi(\theta) = \{I_{11}(\theta)\}^{1/2} q(\theta_2),$$

where $q(\theta_2)$ is an arbitrary function and, as before, $I_{11}(\theta)$ is the (per observation) Fisher information for θ_1 when θ_2 is held fixed.

Condition (c) leads to the equation

$$\int \left\{ \frac{\partial}{\partial\theta_1}\left(\log \frac{\pi(\theta)}{I_{11}^{1/2}(\theta)}\right)\right\} \frac{\pi(\theta_2|\theta_1)}{I_{11}^{1/2}(\theta)} d\theta_2 = 0 \quad \forall\, \theta_1.$$

On writing $\pi(\theta) = \pi(\theta_1)\pi(\theta_2|\theta_1)$ where $\pi(\theta_2|\theta_1)$ is given, e.g., $\pi(\theta_2|\theta_1) = I_{22}^{1/2}(\theta)$, we get the solution

$$\pi(\theta_1) = \text{constant} \times \left\{ \int \pi(\theta_2|\theta_1) I_{11}^{-1/2}(\theta) d\theta_2 \right\}^{-1}.$$

This is similar to the reference prior in this case (see, e.g., [36], Sec. 2) except that the reference prior obtained as

$$\pi(\theta_1) = \text{constant } \times \exp\left[\int \pi(\theta_2|\theta_1) \log I_{11}^{1/2}(\theta) d\theta_2\right]$$

is the geometric mean of $I_{11}^{1/2}(\theta)$ with respect to $\pi(\theta_2|\theta_1)$ whereas the (probability matching) prior obtained here is the harmonic mean of $I_{11}^{1/2}(\theta)$ with respect to $\pi(\theta_2|\theta_1)$. It is interesting to note that in the case of construction of a nonsubjective prior by taking limits of discrete uniform, as mentioned above in this section, we get square root of the arithmetic mean of $I_{11}(\theta)$ with respect to $\pi(\theta_2|\theta_1) = I_{22}^{1/2}(\theta)$ ([32]).

So far we have considered only the case with two orthogonal parameters θ_1 and θ_2 (or just a single parameter θ_1). Probability matching differential equations for the nonorthogonal cases are obtained e.g., in [25] and [30]; see also [57] and the references therein. We present below the result of Datta and Ghosh ([25]).

Let X_i's be i.i.d. with a common density involving a d-dimensional parameter θ and $g(\theta)$ be a real-valued twice continuously differentiable parametric function of interest. Consider a prior density $\pi(\theta)$ of θ with the following property of matching frequentist and posterior probability

$$P_\theta[\sqrt{n}(g(\theta)-g(\hat{\theta}))/\sqrt{v} \leq z] = P[\sqrt{n}(g(\theta)-g(\hat{\theta}))/\sqrt{v} \leq z \,|X]+O_p(n^{-1}) \quad (11)$$

for all z. In (11), $\hat{\theta}$ is the posterior mode or MLE of θ and v is the asymptotic variance of $\sqrt{n}(g(\theta) - g(\hat{\theta}))$ upto $O_p(n^{-1/2})$.

Datta and Ghosh ([25]) show that (11) holds if and only if

$$\sum_{j=1}^{d} \frac{\partial}{\partial \theta_j}\{\eta_j(\theta)\pi(\theta)\} = 0 \tag{12}$$

where $\eta(\theta) = (\eta_1(\theta), \ldots, \eta_d(\theta))'$ is defined as

$$\eta(\theta) = \frac{I^{-1}(\theta)\nabla_g}{\{\nabla_g' I^{-1}(\theta)\nabla_g\}^{1/2}} \text{ with } \nabla_g = (\frac{\partial}{\partial \theta_1} g(\theta), \cdots, \frac{\partial}{\partial \theta_d} g(\theta))'.$$

Example 7 (continued). X_i's are i.i.d. $N(\mu, \sigma)$. For the parametric function $g(\mu, \sigma) = \mu/\sigma$, the probability matching equation (12) turns out to be

$$\frac{\partial}{\partial \mu}\left[\frac{\sqrt{2}\sigma^2}{(\mu^2 + 2\sigma^2)^{1/2}}\pi(\mu, \sigma)\right] - \frac{\partial}{\partial \sigma}\left[\frac{\mu\sigma}{\sqrt{2}(\mu^2 + 2\sigma^2)^{1/2}}\pi(\mu, \sigma)\right] = 0.$$

This equation is satisfied by $\pi(\mu, \sigma) \propto 1/\sigma$.

With the cases $g(\mu, \sigma) = \mu$ and $g(\mu, \sigma) = \sigma$, the probability matching equations reduce to

$$\frac{\partial}{\partial \mu}[\sigma\pi(\mu, \sigma)] = 0 \text{ and } \frac{\partial}{\partial \sigma}[\sigma\pi(\mu, \sigma)] = 0$$

respectively. Both these equations are satisfied by $\pi(\mu, \sigma) \propto 1/\sigma$.

Two comments are in order. Very often the nonsubjective priors are improper. They can be used in an example only if the application of Bayes theorem produces a proper posterior, i.e. one needs $\int p(x_1, \ldots, x_n|\theta)\pi(\theta)d\theta < \infty$ for almost all $x_1, x_2, \ldots, x_n$. It is an odd fact that Jeffreys prior and reference priors have this property more often than the uniform or probability matching priors. An example where the Jeffreys or reference prior does not lead to proper posterior is given in [35].

It was mentioned above that a similarity exists between inference based on nonsubjective priors and that based on Frequentist ideas, i.e., Classical Statistics. Why would then one prefer nonsubjective Bayesian Analysis? Bayesian Analysis produces posteriors and data based estimates of risk, Classical Statistics lacks these.

There are several criticisms of nonsubjective Bayesian Analysis. We take them up in the next section.

The nonsubjective priors discussed so far can be used for point and interval estimation problems but can cause difficulties in model selection or testing problems, as first pointed out by Jeffreys. The source of the difficulty is that an improper prior, unlike a proper prior, is not normalized, it is defined only up to a multiplicative constant. This does not matter in estimation because the undetermined constant appears both in the numerator and denominator of the posterior and so gets cancelled. We explain below very briefly how this problem is tackled in testing.

Example 8. Let $X_1, X_2, \ldots, X_n$ be i.i.d. $N(\theta, 1)$. We wish to test $H_0 : \theta = 0$ vs. $H_1 : \theta \neq 0$. Suppose H_0 and H_1 have equal probabilities (this is a nonsubjective choice) and, given H_1, θ has a uniform distribution, i.e., $\pi(\theta|H_1) = c, -\infty < \theta < \infty$. Bayesian hypothesis testing is based on the posterior probability of H_0 which is given by

$$\frac{p(X_1, \ldots, X_n|\theta = 0)}{p(X_1, \ldots, X_n|\theta = 0) + c\int_{-\infty}^{\infty} p(X_1, \ldots, X_n|\theta)d\theta}.$$

To get rid of c, we use one of the observations, say X_1, to get a proper prior and the remaining to get the likelihood.

Start with $\pi(\theta|H_1) = c$ and calculate

$$\pi(\theta|H_1, X_1) = \frac{c\frac{1}{\sqrt{2\pi}}\exp(-\frac{1}{2}(X_1 - \theta)^2)}{\int c\frac{1}{\sqrt{2\pi}}\exp(-\frac{1}{2}(X_1 - \theta)^2)d\theta} = \frac{1}{\sqrt{2\pi}}\exp(-\frac{1}{2}(X_1 - \theta)^2).$$

Similarly, $\pi(\theta|H_0, X_1)$ remains the point mass at $\theta = 0$. Now take $\pi(\theta|H_1)$ as $\frac{1}{\sqrt{2\pi}}\exp(-\frac{1}{2}(X_1 - \theta)^2)$ and use $X_2, \ldots, X_n$ as data and recalculate the posterior probability of H_0.

It is interesting to observe that the posterior $\pi(\theta|H_1, X_1, \ldots, X_n)$ with prior $\pi(\theta|H_1) \equiv c$ is the same as the posterior $\pi(\theta|H_1, X_2, \ldots, X_n)$ obtained from the prior $\pi(\theta|H_1, X_1) = N(X_1, 1)$, so estimation under H_1 does not change.

One might ask why condition on X_1? Why not on any other X_i? Why not on $X_{i_1}, \ldots, X_{i_m}$? A new methodology which answers these questions is available in [12], [13], [14], [40], [41] [42] and [59]. This is discussed in Section 7. For reconciliation of posterior probability of H_0, when H_0 is rejected, with appropriate Frequentist evidence, see [10].

6. Nonsubjective Priors Re-examined

Having argued in favour of the nonsubjective Bayesian paradigm, we now turn to some of its criticisms. There are several of them. The most important ones are listed below as comments or questions.

1. "Noninformative priors do not exist" (Poincare, Lindley and others) (as mentioned earlier, "noninformative prior" is an older terminology for what we are calling nonsubjective prior.)

2. Nonsubjective Bayesian Analysis is ad hoc and hence no better than the ad hoc paradigms subjective Bayesian Analysis tries to replace.

3. There are too many nonsubjective priors for a problem. Which one to use?

4. Nonsubjective priors are typically improper. One should not use improper priors which do not make sense as quantification of belief (Lindley and others).

5. If the parameter θ has a uniform distribution because of lack of ignorance, then this should also be true for any smooth one-one function $\eta = g(\theta)$ (Fisher and others).

6. Why should a nonsubjective prior depend on the model of the data? (Lindley and almost all critics).

7. What are the impact of this dependence on coherence and likelihood

principle?

Our responses to these criticisms are as follows. First, The object of nonsubjective Bayesian Analysis is not to search for "nonexistent" noninformative priors but to produce posteriors which are approximations to possible posteriors that would result if one could elicit subjective priors. As it indeed leads to such posteriors it is not ad hoc. It tries to find posteriors which reflect the data "as much as possible". Definition of reference priors indicates one precise way of doing this.

Although there may be many nonsubjective priors, the posterior (and hence inference derived from it) based on nonsubjective priors usually does not change much if one switches from one prior to the other. In such cases we may hope for some sort of consensus in a "conventional prior" which is likely to be Jeffreys or reference prior.

In some cases the posterior based on an improper prior is identical with the posterior based on a proper but finitely additive (not countably additive) prior (see, e.g., [45]). In fact many characterizations of "coherence" are through finitely additive priors. Moreover, we have been stressing the posterior rather than the prior. A nonsubjective, improper prior is a convenient tool for producing a proper, mostly data driven posterior. It is the posterior which should be used to make inference, not the prior. Finally, if one feels a need to compare prior probability of two subsets, the subsets should both have finite measure. Otherwise the comparison can indeed be misleading.

As for the objection raised in (5), note that one does not look for priors which represent complete lack of information. Complete lack of information has not been defined satisfactorily. But some invariance under transformations is desirable and most nonsubjective priors possess such properties. The Jeffreys prior has this property for all smooth one-one function $g(\theta)$. Weaker invariance properties hold for reference and probability matching priors. Also, the Jeffreys prior is a uniform prior for all η's as interpreted properly in the previous section.

We turn now to the last two points, namely, (6) and (7). We have argued in the last section that the Shannon entropy is not appropriate for measuring information in a prior density. Indeed the lack of any such measure in information theory suggests that the information in a prior cannot be defined except in the context of an experiment. Further support comes from the measure of information in a prior used by Bernardo for constructing a reference prior. In view of the dependence of the notion of information on an experiment, it is but natural that nonsubjective priors should depend on the model.

As regards coherence in the sense of Heath and Sudderth there is no problem since their definition is in the context of a given model. So a (proper) prior depending on the model does not lead to incoherence.

The impact on LP is more tricky. The LP in its strict sense is violated because the prior and hence the posterior depend on the experiment as well as the likelihood function corresponding to a given data. However, for a fixed experiment, the LP is not violated, and the posterior, decision based on the posterior and posterior risk depend only on the likelihood function. The consequences are further discussed below.

Inference based on nonsubjective priors violates the stopping rule principle for different stopping rules lead to different experiments. In particular, in Example 1.2 of [20], originally suggested by Lindley and Phillips ([55]), one would get different answers according to a binomial or negative binomial model. The data consists of 9 heads and 3 tails in 12 independent tosses of a coin. The Fisher information contained in all the observations is $12/(\theta(1-\theta))$ for the binomial model and $3/(\theta(1-\theta)^2)$ for the negative binomial where θ is the probability of head in a trial. So the Jeffreys priors are similar but slightly different. Suppose we want to test the null hypothesis $H_0 : \theta = 1/2$ versus the alternative hypothesis $H_1 : \theta > 1/2$. As reported in [20, p. 4], the p-values for the binomial and negative binomial models are respectively 0.075 and 0.0325 and therefore, with the usual 5% Type I error level, the two model assumptions lead to two different decisions. A Bayes test will be based on a nonsubjective Bayes factor (BF) described in Section 7. We assume that H_0 and H_1 have equal prior probabilities (a nonsubjective choice). For the binomial model the Jeffreys prior is proportional to $\theta^{-1/2}(1-\theta)^{-1/2}$ which can be normalized to get a proper prior. For the negative binomial model the Jeffreys prior is proportional to $\theta^{-1/2}(1-\theta)^{-1}$ which is improper and so cannot be used in testing. One way out would be to treat this as data on three i.i.d. geometrically distributed random variables and find the intrinsic prior (see [12]) in this case. One can then calculate the BF under the negative binomial model also. The BF under the binomial model (with Jeffreys prior) and the BF under the negative binomial model (with the intrinsic prior) are respectively 2.073 and 2.662. They are different as were the p-values of Classical Statistics but unlike the p-values, one of which is double the other, the BF's are quite close. Incidentally, Bernardo and Smith [16, p. 249] point out that even from a subjective Bayesian point of view there is a difference between the two cases for in the case of a binomial model, θ can be one but not for a negative binomial that stops with the rth tail. See also [9, Example 4] and [74].

However, we argue the violation indicated in the previous paragraph is less serious. If the stopping time is ancillary as in Cox's example (Example 2) and the observations are i.i.d., Jeffreys, reference and probability matching priors will not depend on the stopping rule. Most deviations from fixed sample size are of this kind.

We would also suggest, not by way of a defence or justification but as a sort of apology, that the violation of the strict LP is not such a bad thing if we have to have some Frequentist validation. Surely, a paradigm that seeks such validation cannot avoid depending on the model for an experiment.

7. Nonsubjective Bayesian Estimation and Testing

We have described in Section 5 how priors can be chosen in a nonsubjective way. Having chosen the prior one uses the Bayes theorem to update it in the light of the given data and finds the posterior, on which the inference is based. We briefly discuss below Bayesian estimation and testing with nonsubjective priors.

7.1 Estimation

Posterior distribution of θ is obtained via Bayes's formula given in equation (3) of Section 4. Consider a nonsubjective prior π for which the integral in the denominator of (3) converges. This leads to a proper posterior. Even if the prior is improper, often with sufficient amount of data the posterior turns out to be proper.

Consider for simplicity the case with a real parameter θ. The usual point estimates of θ are summary measures of the posterior, such as its mean, median or mode. If estimation of θ is considered as a decision problem with a loss $L(\theta, a)$, the posterior risk in estimating θ by "a" is given by the average posterior loss

$$\psi(a|X) = \int_{\Theta} L(\theta, a)\pi(\theta|X)d\theta.$$

Given the observed data X, a Bayesian chooses "a" to minimize $\psi(a|X)$ and reports this minimizing "a" as Bayes estimate of θ and the corresponding $\psi(a|X)$ as a measure of risk for the given data. For example, for squared error loss $L(\theta, a) = (\theta - a)^2$, the Bayes estiamte is given by $E(\theta|X)$, the posterior mean and the corresponding risk is evaluated by $\mathrm{Var}(\theta|X)$, the posterior variance.

Example 9. Let $X_1, \cdots, X_n$ be i.i.d Bin $(1, \theta), 0 < \theta < 1$. Consider a Beta (α, β) prior for θ given by the density

$$\pi(\theta; \alpha, \beta) = \frac{\Gamma(\alpha+\beta)}{\Gamma(\alpha)\Gamma(\beta)} \theta^{\alpha-1}(1-\theta)^{\beta-1}, 0 < \theta < 1. \tag{13}$$

Depending on α and β this prior can take on a variety of shapes and is proper for any $\alpha, \beta > 0$.

With $\alpha = \beta = 1/2$ we have the Jeffreys prior for this problem:

$$\pi_J(\theta) = \pi(\theta; 1/2, 1/2) \propto [\theta(1-\theta)]^{-1/2}.$$

If we take $\alpha = \beta = 1$, (13) gives the uniform prior $\pi_1(\theta) = 1, 0 < \theta < 1$. Another nonsubjective prior proposed in the literature is the improper prior

$$\pi_2(\theta) = [\theta(1-\theta)]^{-1}$$

which corresponds to the case $\alpha = \beta = 0$.

The posterior obtained from the Beta (α, β) prior via Bayes's formula turns out to be a Beta distribution again with parameter $\alpha' = \alpha + S$ and

$\beta' = \beta + n - S$ where $S = \sum_{k=1}^{n} X_i$, and hence the Bayes estimate for squared error loss is the posterior mean

$$E(\theta|X) = \alpha'/(\alpha' + \beta') = (\alpha + S)/(\alpha + \beta + n).$$

It is interesting to note that the Bayes estimate may be expressed as a weighted average of the prior estimate $\alpha/(\alpha+\beta)$ and the classical estimate (MLE) S/n:

$$E(\theta|X) = (\frac{\alpha + \beta}{\alpha + \beta + n}).\frac{\alpha}{\alpha + \beta} + (\frac{n}{\alpha + \beta + n})\frac{S}{n}.$$

The Bayes estimates for the nonsubjective priors π_J, π_1, and π_2, obtained as special cases, are respectively

$$\hat{\theta}_J = \frac{S + (1/2)}{n + 1}, \ \hat{\theta}_1 = \frac{S + 1}{n + 2} \text{ and } \hat{\theta}_2 = \frac{S}{n} \text{ (MLE)} .$$

Thus the three nonsubjective priors are very similar in their answers and one of them, namely π_2, leads exactly to the classical estimate S/n. We note that the posterior obtained from the improper prior π_2 is proper if both S and $(n - S)$ are > 0, i.e., X_i's are not all zero or not all one.

7.2 Hypotheses Testing or Model Selection

A statistical hypothesis may be represented by a probability model for the given data X. Bayesian approach to hypothesis testing is based on calculation of the posterior probabilities of the models representing the hypotheses under consideration.

Consider two models M_1 and M_2 for data X with density $p_i(x|\theta_i)$ under model M_i, θ_i being an unknown parameter of dimension $d_i, i = 1, 2$. Given prior specifications $\pi_i(\theta_i)$ for parameter θ_i and prior probabilities $P(M_i)$ for model M_i, the posterior probabilities of the models can be obtained, via Bayes Theorem, as

$$P(M_i|x) = \frac{P(M_i)m_i(x)}{P(M_1)m_1(x) + P(M_2)m_2(x)}, \ i = 1, 2$$

where $m_i(x) = \int p_i(x|\theta_i)\pi_i(\theta_i)d\theta_i$ is the marginal density of x under M_i.

Bayesian hypothesis testing or model selection is achieved by comparing the posterior probabilities $P(M_i|x)$, and hence may be based on the ratio

$$\frac{P(M_2|x)}{P(M_1|x)} = \frac{P(M_2)}{P(M_1)} B_{21}(x), \tag{14}$$

where $B_{21} = B_{21}(x)$, known as the Bayes factor (BF) of M_2 to M_1 is defined as $B_{21} = m_2(x)/m_1(x)$. If the models are a priori judged equally likely, $P(M_1) = P(M_2)$ (a nonsubjective choice), the ratio in (14) is equal to the Bayes factor B_{21}.

As already mentioned and illustrated at the end of Section 5, for improper nonsubjective priors π_i which are defined only upto arbitrary multiplicative constants, the Bayes factor is indeterminate. This indeterminacy, noted by Jeffreys ([48]), has been the main motivation of new nonsubjective methods. A number of methods have been proposed in [67], [66], [12], [13], [14], [42],

[59] and others, including [48]. For a review of these methods we refer to [12], [14], [40], [41] and [42]. Below we briefly discuss only some of these methods.

The Intrinsic Bayes Factor.

A solution to the problem with improper priors is to use part of the data as a *training sample.* The idea is to use the training sample to obtain proper posterior distributions for the parameters which can then be used as priors to compute a Bayes factor with the remainder of the data. This was illustrated earlier through Example 8 of Section 5.

Let $X_1, X_2, \ldots, X_n$ constitute the whole sample. For a subsample $X_{j_1}, X_{j_2}, \ldots, X_{j_m} (1 \leq j_1 < j_2 < \ldots < j_m \leq n)$, the posterior density of θ_i given $X_{j_1}, \ldots, X_{j_m}$ under M_i is given by

$$\begin{aligned} \pi_i(\theta_i | X_{j_1}, \ldots, X_{j_m}) &= \frac{f_i(X_{j_1}, \ldots, X_{j_m} | \theta_i) \pi_i(\theta_i)}{m_i(X_{j_1}, \ldots, X_{j_m})} \\ &= \frac{f_i(X_{j_1}, \ldots, X_{j_m} | \theta_i) \pi_i(\theta_i)}{\int f_i(X_{j_1}, \ldots, X_{j_m} | \theta_i) \pi_i(\theta_i) d\theta_i}, i = 1, 2. \qquad (15) \end{aligned}$$

Berger and Pericchi ([12]) use training sample of minimal size, leaving most part of the data for model comparison. Let m be the mimimum sample size such that $\pi_i(\theta_i | X_{j_1}, \ldots, X_{j_m}), i = 1, 2$, are proper or equivalently, $m_i(X_{j_1}, \ldots, X_{j_m}), i = 1, 2$, are finite. Let $X_{j_1}, \ldots, X_{j_m}$ be such a minimal training sample. The Bayes factor with the remainder of the data using the above $\pi_i(\theta_i | X_{j_1}, \ldots, X_{j_m})$ in (15) as priors (conditional BF) is given by

$$B_{21}(j_1,\ldots,j_m) = \frac{\int \frac{f_2(X_1,\ldots,X_n|\theta_2)}{f_2(X_{j_1},\ldots,X_{j_m}|\theta_2)}\pi_2(\theta_2|X_{j_1},\ldots,X_{j_m})d\theta_2}{\int \frac{f_1(X_1,\ldots,X_n|\theta_1)}{f_1(X_{j_1},\ldots,X_{j_m}|\theta_1)}\pi_1(\theta_1|X_{j_1},\ldots,X_{j_m})d\theta_1} = B_{21}\frac{m_1(X_{j_1},\ldots,X_{j_m})}{m_2(X_{j_1},\ldots,X_{j_m})}. \quad (16)$$

It is to be noted that the arbitrary constant multiplier of B_{21} is cancelled by that of $m_1(X_{j_1},\ldots,X_{j_m})/m_2(X_{j_1},\ldots,X_{j_m})$ so that the indeterminacy of the Bayes factor is removed in (16). However, this conditional BF in (16) depends on the choice of the training sample $X_{j_1},\ldots,X_{j_m}$. Berger and Pericchi ([12]) suggest considering all possible training samples and taking average of the $\binom{n}{m}$ conditional BF's $B_{21}(j_1,\ldots,j_m)$'s to obtain what is called the *intrinsic Bayes factor* (IBF). For example, taking an arithmetic average leads to

$$AIBF_{21} = B_{21}\frac{1}{\binom{n}{m}}\sum\frac{m_1(X_{j_1},\ldots,X_{j_m})}{m_2(X_{j_1},\ldots,X_{j_m})} \quad (17)$$

while the geometric average gives

$$GIBF_{21} = B_{21}\left(\prod\frac{m_1(X_{j_1},\ldots,X_{j_m})}{m_2(X_{j_1},\ldots,X_{j_m})}\right)^{1/\binom{n}{m}}, \quad (18)$$

the sum and product in (17) and (18) being taken over the $\binom{n}{m}$ possible training samples $X_{j_1},\ldots,X_{j_m}$ with $1 \le j_1 < \ldots < j_m \le n$.

Berger and Pericchi ([12]) also suggest using trimmed averages or the median (complete trimming) of the conditional BF's when taking an average of all the conditional BFs does not seem reasonable (e.g., when the conditional

BFs vary much). AIBF and GIBF have good properties but are affected by outliers. If the sample size is very small, using a part of the sample as a training sample may be impractical and Berger and Pericchi ([12]) recommend using expected intrinsic Bayes factors that replace the averages in (17) and (18) by their expectations, evaluated at the MLE. The AIBF is justified by the possibility of its correspondence to actual Bayes factors with respect to"intrinsic" proper priors at least asymptotically. Berger and Pericchi ([12] and [14]) and Ghosh and Samanta ([41]) have argued that these intrinsic priors may be considered to be natural "default" priors for the testing problems.

The Fractional Bayes Factor

O'Hagan ([59]) proposes a solution using a fractional part of the full likelihood in place of using training samples and averaging over them. The resulting "partial" Bayes factor, called the *fractional Bayes factor* (FBF) is given by

$$FBF_{21} = \frac{m_2(X,b)}{m_1(X,b)}$$

where b is a fraction and

$$m_i(X,b) = \frac{\int f_i(X|\theta_i)\pi_i(\theta_i)d\theta_i}{\int [f_i(X|\theta_i)]^b \pi_i(\theta_i)d\theta_i}.$$

To make the FBF comparable with the IBF one may take $b = m/n$ where m is the size of a minimal training sample as defined in the case of IBF. O'Hagan also recommends other choices of b, e.g., $b = \sqrt{n}/n$ or $\log n/n$.

Example 8 (continued). Here size of a minimal training sample is one and the conditional BF, conditined on a single X_i, is

$$n^{-1/2}\exp[(1/2)(n\bar{X}^2 - X_i^2)].$$

The IBF's are obtained by averaging these n conditional BF's. The AIBF in this case approximately equals the BF with a $N(0,2)$ prior for θ.

The FBF, with fraction b is given by

$$\sqrt{b}\exp[n(1-b)\bar{X}^2/2]$$

and is exactly equal to the BF with a $N(0,(b^{-1}-1)/n)$ prior.

8. High Dimensional Problems, PEB and HB

The nonsubjective Bayesian methods discussed earlier do not work well when the dimension of θ is large. However, two satisfactory nonsubjective Bayesian methods have been developed, namely, Parametric Empirical Bayes (PEB) and Hierarchical Bayes (HB). Basically these are methods for handling high dimensional random effects.

Suppose we have p similar but not identical populations with densities $f(x,\theta_1),\ldots,f(x,\theta_p)$ and from the jth we have samples $X_{j1},\ldots,X_{jr}, j = 1,2,\ldots,p$. The total number of observations is $n = pr$. These p populations may correspond to p clinical studies at p hospitals or p villages or countries etc.

Often it seems natural to model $\theta_1, \ldots, \theta_p$ as exchangeable and hence given a hyperparameter vector $\boldsymbol{\eta}$, i.i.d. For fixed $\boldsymbol{\eta}$, one may choose one of the partially specified nonsubjective priors for θ_i — e.g., conjugate priors or their mixtures.

Thus for fixed $\boldsymbol{\eta}$, θ_j's are i.i.d. $\pi(\theta_i|\boldsymbol{\eta})$ and given $\boldsymbol{\eta}$ and θ_i's, X_{ji}'s are independent, $X_{ji} \sim f(x_{ji}|\theta_j)$.

We illustrate with a simple but illuminating example. The p populations are $N(\theta_j, \sigma^2)$, σ^2 assumed known for simplicity, $r = 1$ so that the observations are simply $X_1 \equiv X_{11}, X_2 \equiv X_{21}, \ldots, X_p \equiv X_{p1}$, and θ_j's are i.i.d. $N(\eta, \tau^2)$ where again τ^2 is assumed known for simplicity. Let η have uniform distribution on $\mathbb{R}$. The following facts are easy to verify.

(1) Given η, (integrating out θ's) X_j's are i.i.d. $N(\eta, \sigma^2 + \tau^2)$

(2) $\pi(\theta_j|\eta, X_1, \ldots, X_p) = \pi(\theta_j|\eta, X_j) = N\left(\frac{X_j\tau^2+\eta\sigma^2}{\tau^2+\sigma^2}, \frac{\sigma^2\tau^2}{\tau^2+\sigma^2}\right)$

(3) $\pi(\eta|X_1, \ldots, X_p) = N(\bar{X}, \sigma^2 + \tau^2)$

(4) $\pi(\theta_j|X_1, \ldots, X_p) = \int \pi(\theta_j|\eta, X_j)N(\bar{X}, \sigma^2 + \tau^2)(d\eta)$

The last relation in this Hierarchical Bayesian Analysis describes how the Bayesian inference engine provides inference about θ_j given the full data set. It may seem puzzling that inference about θ_j should depend not only on X_j but the full data set. This happens because the full data set is used for η via (3) and then used in the integral apprearing in (4). The hyperparameter η

captures some aspect common in the p similar populations. Use of it through (3) makes (4) superior to the posterior $\pi(\theta_j|X_j)$ depending only on X_j. If we had put a noninformative prior for θ_j's our estimates would have been X_j for θ_j. The HB method produces instead an estimate that shrinks the estiamte of θ_j towards $\bar{X}$. Note

$$E(\theta_j|X_1, \ldots, X_p) = \int E(\theta_j|\eta, X_j)N(\bar{X}, \sigma^2 + \tau^2)d\eta = (\bar{X}\sigma^2 + X_j\tau^2)/(\sigma^2 + \tau^2).$$

Parametric Empricial Bayes does not put a prior on η but replaces η by an estimate $\hat{\eta}$, e.g., MLE or UMVUE of η based on (1), i.e., in this example η is replaced by $\bar{X}$. Then (2) can be used instead of (4) with η replaced by $\bar{X}$.

The point estimates for θ_j are almost indistinguishable in the two methods described above but the variances of the estimate of θ_j can differ substantially. The HB uses

(5) $E\{(\theta_j - \tilde{\theta}_j)^2|X_1, \ldots, X_p\}$ where $\tilde{\theta}_j = E(\theta_j|X_1, \ldots, X_p)$

whereas (naive) PEB uses

(6) $E\{(\theta_j - \hat{\theta}_j)^2|\hat{\eta}, X_j\}$ where $\hat{\theta}_j = E(\theta_j|\hat{\eta}, X_j)$.

The second expression is somewhat inappropriate because it does not provide for the variation of η around $\hat{\eta}$ either in a Bayesian or a Frequentist sense.

In this particular example (6) tends to be an underestimate. Hence (naive) confidence (credibility) intervals for θ_j of the form $\hat{\theta}_j \pm z_{\alpha/2} \times \sqrt{(6)}$ cover θ_j with probability less then $1 - \alpha$. Morris ([56]) has provided an ad hoc approximation to (5) and suggested the use of this approximation instead of (6) in the PEB confidence interval. He has conjectured that the PEB coverage probability will then be $\geq 1 - \alpha$. Subsequent developments for this example as well as the general case are discussed with great clarity and detail in [20, Ch. 3]. Morris's conjecture is re-examined in [37]. See also [26]. Both these papers make use of the techniques of higher order asymptotics and probability matching discussed in Section 5.

In complicated problems the posteriors cannot be written down as easily as in this example. One has to use MCMC. Good sources for methods and discussion are [20], [31] and [64]. In such problems the estimate $\hat{\eta}$ required by PEB is not available in explicit form. The last two books contain a good discussion of how $\hat{\eta}$ can be found numerically by a judicious application of the EM algorithm.

A few general remarks are in order. There is a lot of information on η for moderately large p, as is evident from (3) and posterior normality. It is less clear but true that there is a lot of information in the empirical distribution of X_i's which can be used to guess the approximate form of $\pi(\theta_j|\eta, X_j)$, provided mixtures are identifiable. In particular, one should be able to assess whether the assumed normality in the likelihood and prior is valid for X_j's or only after a suitable transformation. The methods just discussed, namely, PEB and HB do so well compared to classical Frequentist intervals for θ_j based on X_j because of these two facts. That the improvement can be very dramatic is evident from Table 3.4 of [20, p. 101]. Average length of 95% confidence intervals goes down from 39 to about 5 for naive PEB and about 8 for adjusted PEB. An adjustment is needed to get the confidence coefficient right.

9. Concluding Remarks

A nonsubjective Bayesian accepts subjective input but faces the fact that it is often not available at all and even when available specifies only parts of the prior, so that nonsubjective priors need to be constructed and used for posterior analysis. He also believes in a certain amount of Frequentist validation. We believe as the Bayesian paradigm becomes the central paradigm of our subject and is applied to all kinds of old and new data, there will be no alternative to being more flexible without losing a hard core of subjectivity, namely, that inference takes place through interaction of data and the analyst's knowledge and belief. Using nonsubjective priors is part of such flexibility.

We have tried to justify a move towards a nonsubjective Bayesian paradigm, away from both Data Analysis and Classical Statistics. The new paradigm has the strengths of the last two but avoids their weaknesses. We have sketched out briefly how the approach works in low and high dimensional problems and pointed out how one can ensure Frequentist validation as well as data based posterior (rather than integrated) risk estimates. Cox ([23]) provides brief critical overview of some aspects of these methods from the point of view of Classical Statistics.

For lack of space we have not discussed any of these issues in the infinite dimensional case of Bayesian nonparametrics. Frequentist validation is considerably weaker in this context and consists in checking posterior consistency and optimum rates of its convergence (in a Frequentist sense), see, e.g., [33] and the references therein. Hopefully, future work will also lead to Bernstein-von Mises theorems on posterior normality for many interesting functionals. Ghosal et al. ([32]) present a general procedure for getting a uniform distribution in infinite dimensional cases that leads to the Jeffreys prior and analogues of reference priors in the finite dimensional parametric cases. For more details on all these points we refer the reader to [39].

No discussion of nonsubjective Bayesian Analysis can be complete without some observations on Bayesian robustness, or more precisely, robustness with respect to prior. Robustness is taken care of in different ways for different purposes.

The minimum that needs to be done is to do some analysis of sensitivity of posterior with respect to prior. Most MCMC programmes can easily accomodate calculation of posterior quantities for different priors. Both Carlin and Louis ([20]) and Gelman et al. ([31]) discuss this aspect in detail with quite specific advice about how to handle outliers. They also discuss in detail model assessment through residuals and cross validations. One leaves out a part of the data and uses the rest to produce a predictive distribution. The predictive distribution is then tested on the first set. One can also approach robustness from a theoretical point of view with general nonparametric contamination classes of priors

$$\pi_c = \pi_0(1-\epsilon) + \epsilon\pi_G,$$

π_G belongs to some nonparametric class $\mathcal{G}$. A good, comprehensive discussion can be found in [6] and [8].

At a third level one may think of robust Bayesian Analysis as an alternative, nonsubjective Bayesian paradigm which rests on relaxed rationality axioms. The preference ordering is assumed to be a partial rather than linear order. One then gets a (subjective) family of priors rather than a single subjective prior. This class leads to quantification of uncertainty via upper and lower probability. It has striking similarities with a theory of such probabilities due to A.P. Dempster and Glenn Shafer. For details see [65] and [49]. The method of lower and upper probabilities, once quite popular among engineers, seems to be less used now because it is not easy to implement and can lead to counter intuitive inference.

Appendix: Birnbaum's Theorem on Likelihood Principle

The object of this appendix is to rewrite the usual proof (e.g., as given in [4]) using only mathematical statements and carefully defining all symbols and the domain of discourse.

Let $\theta \in \Theta$ be the parameter of interest. A statistical experiment $\mathcal{E}$ is performed to generate a sample x. An experiment $\mathcal{E}$ is given by the triplet $(\mathcal{X}, \mathcal{A}, p)$ where $\mathcal{X}$ is the sample space, $\mathcal{A}$ is a class of (measurable) subsets of $\mathcal{X}$ and $p = \{p(\cdot|\theta), \theta \in \Theta\}$ is a family of probability functions on $(\mathcal{X}, \mathcal{A})$, indexed by the parameter space Θ. For simplicity, we assume both $\mathcal{X}$ and Θ are finite sets; $\mathcal{A}$ is taken to be the class of all subsets. Below we consider experiments with a fixed parameter space Θ.

A (finite) mixture of experiments $\mathcal{E}_1, \ldots, \mathcal{E}_k$ with mixture probabilities $\pi_1, \ldots, \pi_k$ (nonnegative numbers free of θ, summing to unity), which may be written as $\sum_{i=1}^{k} \pi_i \mathcal{E}_i$, is defined as a two stage experiment where one first selects $\mathcal{E}_i$ with probability π_i and then observe $x_i \in \mathcal{X}_i$ by performing the experiment $\mathcal{E}_i$.

Consider now a class of experiments closed under the formation of (finite) mixtures. We use equivalence relations to represent different principles. Let $\mathcal{E} = (\mathcal{X}, \mathcal{A}, p)$ and $\mathcal{E}' = (\mathcal{X}', \mathcal{A}', p')$ be two experiments and $x \in \mathcal{X}, x' \in \mathcal{X}'$. By equivalence of the two points $(\mathcal{E}, x)$ and $(\mathcal{E}', x')$, we mean one makes the same inference on θ if one performs $\mathcal{E}$ and observes x or performs $\mathcal{E}'$ and observes x', and we denote this as

$$(\mathcal{E}, x) \sim (\mathcal{E}', x').$$

We now consider the following principles.

The likelihood principle (LP): We say that the equivalence relation "$\sim$" obeys the likelihood principle if $(\mathcal{E}, x) \sim (\mathcal{E}', x')$ whenever

$$p(x|\theta) = c\, p'(x'|\theta) \text{ for all } \theta \in \Theta$$

for some constant $c > 0$.

The weak conditionality principle (WCP): An equivalence relation "$\sim$" satisfies WCP if for a mixture of experiments $\mathcal{E} = \sum_{i=1}^{k} \pi_i \mathcal{E}_i$,

$$(\mathcal{E}, (i, x_i)) \sim (\mathcal{E}_i, x_i)$$

for any $i \in \{1, \ldots, k\}$ and $x_i \in \mathcal{X}_i$.

The sufficiency principle (SP): An equivalence relation "$\sim$" satisfies SP if $(\mathcal{E}, x) \sim (\mathcal{E}, x')$ whenever $S(x) = S(x')$ for some sufficient statistic S (for θ).

The weak sufficiency principle (WSP): An equivalence relation "$\sim$" satisfies WSP if $(\mathcal{E}, x) \sim (\mathcal{E}, x')$ whenever $p(x|\theta) = p(x'|\theta)$ for all θ.

If follows that SP implies WSP which can be seen by noting that

$$S(x) = \left\{ \frac{p(x|\theta)}{\sum_{\theta' \in \Theta} p(x|\theta')}, \ \theta \in \Theta \right\}$$

is a (minimal) sufficient statistic. We assume without loss of generality that

$$\sum_{\theta \in \Theta} p(x|\theta) > 0 \text{ for all } x \in \mathcal{X}.$$

We now state and prove Birnbaum's theorem on Likelihood principle ([18]).

Theorem. WCP and WSP together imply LP, i.e., if an equivalence relation satisfies WCP and WSP then it also satisfies LP.

Proof. Suppose an equivalence relation "$\sim$" satisfies WCP and WSP. Consider two experiments $\mathcal{E}_1 = (\mathcal{X}_1 \mathcal{A}_1, p_1)$ and $\mathcal{E}_2 = (\mathcal{X}_2, \mathcal{A}_2, p_2)$ with same Θ and samples $x_i \in \mathcal{X}_i$, $i = 1, 2$, such that

$$p_1(x_1|\theta) = c p_2(x_2|\theta) \text{ for all } \theta \in \Theta \tag{A1}$$

for some $c > 0$.

We are to show that $(\mathcal{E}_1, x_1) \sim (\mathcal{E}_2, x_2)$. Consider the mixture experiment $\mathcal{E}$ of $\mathcal{E}_1$ and $\mathcal{E}_2$ with mixture probabilities $1/(1+c)$ and $c/(1+c)$ respectively, i.e.,

$$\mathcal{E} = \frac{1}{1+c}\mathcal{E}_1 + \frac{c}{1+c}\mathcal{E}_2.$$

The points $(1, x_1)$ and $(2, x_2)$ in the sample space of $\mathcal{E}$ have probabilities $p_1(x_1|\theta)/(1+c)$ and $p_2(x_2|\theta)c/(1+c)$ respectively, which are the same by (A1). WSP then implies that

$$(\mathcal{E}, (1, x_1)) \sim (\mathcal{E}, (2, x_2)). \qquad (A2)$$

Also, by WCP

$$(\mathcal{E}, (1, x_1)) \sim (\mathcal{E}_1, x_1) \text{ and } (\mathcal{E}, (2, x_2)) \sim (\mathcal{E}_2, x_2). \qquad (A3)$$

From (A2) and (A3) we have $(\mathcal{E}_1, x_1) \sim (\mathcal{E}_2, x_2)$.

Acknowledgement. Parts of this review, specifically Sections 3 and 6, have been presented before at Bayesian conferences at Teheran and Amravati (India).

References

1. Barndorff-Nielsen, O. *Parametric Statistical Models and Likelihood*, Lecture Notes in Statistics, v. 50, Springer-Verlag, New York (1988).

2. Barndorff-Nielsen, O. and Cox, D.R. *Asymptotic Techniques for Use in Statistics*, Chapman and Hall, London (1989).

3. Basu, D. Statistical information and likelihood (with discussion). *Sankhya, Ser. A* **37**, 1-71 (1975).

4. Basu, D. *Statistical Information and Likelihood.* A collection of critical essays by Dr. D. Basu (Edited by J.K. Ghosh), Lecture Notes in Statistics, Springer-Verlag, New york (1988).

5. Bayes, Thomas. An essay towards solving a problem in the doctrine of chances. *Phil. Trans. Roy. Soc.* **53**, 370-418 (1763).

6. Berger, J.O. *Statistical Decision Theory and Bayesian Analysis* (2nd edn.), Springer-Verlag, New York (1985).

7. Berger, J.O. The frequentist viewpoint and conditioning. In *Proceedings of the Berkeley Conference in Honor of Jerzy Neyman and Jack Kiefer, Vol. I* (Edited by L.M. Le Cam and R.A. Olshen), 15-44, Wadsworth, Inc., Monterey, California (1985).

8. Berger, J.O. An overview of robust Bayesian analysis (with discussion). *Test* **3**, 5-124 (1994).

9. Berger J.O. and Bernardo, J.M. On the development of the reference prior method. In *Bayesian Statistics* **4**, (Edited by J.M. Bernardo *et al.*), 35-60, Oxford University press, London (1992).

10. Berger, J.O., Brown, L.D. and Wolpert, R. A unified conditional frequentist and Bayesian test for fixed and sequential hypothesis

testing. *Ann. Statist.* **22**, 1787-1807 (1994).

11. Berger, J.O., Liseo, B. and Wolpert, R. Integrated likelihood methods for eliminating nuisance parameters (with discussion). *Statistical Science* **14**, 1-28 (1999).

12. Berger, J.O., and Pericchi, L.R. The intrinsic Bayes factor for model selection and prediction. *J. Amer. Statist. Assoc.* **91**, 109-122 (1996).

13. Berger, J.O. and L.R. Pericchi. The intrinsic Bayes factor for linear models (with discussion). In *Bayesian Statistics* **5**, (Edited by J.M. Bernardo *et al.*), 25-44, Oxford University press, London (1996).

14. Berger, J.O. and Pericchi, L.R. Objective Bayesian methods for model selection: introduction and comparison. IMS Lecture Note Series (Edited by P. Lahiri) (2001).

15. Bernardo, J.M. Reference posterior distributions for Bayesian inference. *J. Roy. Statist. Soc. B* **41**, 113-147 (1979).

16. Bernardo, J.M. and Smith, A.F.M. *Bayesian Theory*, Wiley, Chichester (1994).

17. Bhanja, J. Estimation of the common probability of success for several binomials — a simulation study. *Calcutta Statist. Assoc. Bulletin* **48**, 61-72 (1998).

18. Birnbaum, A. On the foundations of statistical inference (with discussion). *J. Amer. Statist. Assoc.* **57**, 269-326 (1962).

19. Brown, L.D. A contribution to Kiefer's theory of conditional confidence procedures. *Ann. Statist.* **6**, 59-71 (1978).

20. Carlin, B.P. and Louis, T.A. *Bayes and Empirical Bayes Methods for Data Analysis*, Chapman and Hall, London (1996).

21. Cencov, N.N. *Statistical Decision Rules and Optimal Inference*, American Mathematical Society, Providence, R.I., Translation from Russian edited by Lev L. Leifman (1982).

22. Cox, D.R. Some problems connected with statistical inference. *Ann. Math. Statist.* **29**, 357-372 (1958).

23. Cox, D.R. The five faces of Bayesian statistics. *Calcutta Statist. Assoc. Bulletin* **50**, 127-136 (2000).

24. Cox, D.R. and Reid, N. Parameter orthogonality and approximate conditional inference (with discussion). *J. Roy. Statist. Soc. Ser. B* **49**, 1-39 (1987).

25. Datta, G.S. and Ghosh, J.K. On priors providing frequentist validity for Bayesian inference. *Biometrika*, **82**, 37-45 (1995).

26. Datta, G.S., Ghosh, M. and Mukerjee, R. Some new results on probability matching priors. *Calcutta Statist. Assoc. Bulletin* **50**,

179-192 (2000).

27. Dawid, A.P. Statistical theory. The prequential approach *J. Roy. Statist. Soc. Ser. A*, 278-292 (1984).

28. Dawid, A.P. Fisherian inference in likelihood and prequential frames of reference (with discussion). *J. Roy. Statist. Soc. Ser. B* **53**, 79-109 (1991).

29. Diaconis, P. and Freedman, D. On the consistency of Bayes estimates (with discussion). *Ann. Statist.* **14**, 1-67 (1986).

30. DiCiccio, T.J. and Stern, S.E. Frequentist and Bayesian Bartlett correction of test statistics based on adjusted profile likelihoods. *J. Roy. Statist. Soc. B* **56**, 397-408 (1994).

31. Gelman, A., Carlin, J.B., Stern, H.S. and Rubin, D.B. *Bayesian Data Analysis*, Chapman and Hall, London (1995).

32. Ghosal, S., Ghosh, J.K. and Ramamoorthi, R.V. Noninformative priors via sieves and packing numbers. In *Advances in Statistical Decision Theory and Applications* (Edited by S. Panchapakesan and N. Balakrishnan), 119-132, Birkhauser, Boston (1997).

33. Ghosal, S., Ghosh, J.K. and van der Vaart, A.W. Convergence rates of posterior distributions. *Ann. Statist.* **28**, 500-531 (2000).

34. Ghosh, J.K. *Higher Order Asymptotics*, NSF-CBMS Regional Con-

ference Series in Probability and Statistics, Vol. 4 (1994).

35. Ghosh, J.K. Discussion of "Noninformative priors do not exist: a discussion" by Bernardo, J.M. *J. Statist. Plann. Inference* **65**, 159-189 (1997).

36. Ghosh, J.K. and Mukerjee, R. Non-informative priors (with discussion). In *Bayesian Statistics* **4**, (Edited by J.M. Bernardo *et al.*), 195-210, Oxford University press, London (1992).

37. Ghosh, J.K. and Mukerjee, R. Empirical Bayes confidence intervals (under preparation) (2001).

38. Ghosh, J.K. and Murthy, C.A. Discussion of "A penalized likelihood approach to image warping" by Glasbey, C.A. and Mardia, K.V. To appear in *J. Roy. Statist. Soc. Ser. B* (2001).

39. Ghosh, J.K. and Ramamoorthi, R.V. *Bayesian Nonparametrics* (under preparation) (2001).

40. Ghosh, J.K. and Samanta, T. Model selection – an overview. *Current Science* **80**, 1135-1144 (2001).

41. Ghosh, J.K. and Samanta, T. Discussion of "Objective Bayesian methods for model selection: introduction and comparison" by Berger, J.O. and Pericchi, L.R. IMS Lecture Note Series (Edited by P. Lahiri) (2001).

42. Ghosh, J.K. and Samanta, T. Nonsubjective Bayes testing – an overview. To appear in *J. Statist. Plann. Inf.* (2001).

43. Hartigan, J.A. *Bayes Theory*, Springer-Verlag, New York (1983).

44. Hastie, T., Tibshirani, R. and Friedman, J. *The Elements of Statistical Learning (Data Mining, Inference and Prediction)*, Springer, New York (2001).

45. Heath, D. and Sudderth, W. On finitely additive priors, coherence, and extended admissibility. *Ann. Statist.* **6**, 333-345 (1978).

46. Huber, P.J. and Strassen, V. Minimax tests and the Neyman-Pearson lemma for capacities. *Ann. Statist.* **1**, 251-263 (1973).

47. Huber, P.J. and Strassen, V. Corrections: "Minimax tests and the Neyman-Pearson lemma for capacities"(*Ann. Statist.* **1**, 251-263 (1973)). *Ann. Statist.* **2**, 223-224 (1974).

48. Jeffreys, H. *Theory of Probability.* Oxford University press, London (1961).

49. Kadane, J.B. and Wasserman, L. Symmetric, coherent, Choquet capacities. Ann. Statist. **24**, 1250-1264 (1996).

50. Kiefer, J. Conditional confidence statements and confidence estimators (with discussion). *J. Amer. Statist. Assoc.* **72**, 789-827 (1977).

51. Laplace, P.S. Memoire sur les approximations des formules qui sont fonctions de tres grands nombres et sur leur application aux probabilites. *Memoires de l'Academie des sciences de Paris*, 353-415, 559-565 (1810).

52. Le Cam, L. and Yang, G.L. *Asymptotics in Statistics.* (2nd edn.), Springer, New york (2000).

53. Lehmann, E.L. *Testing Statistical Hypotheses* (2nd edn.), John Wiley and Sons, New york (1986).

54. Lindley, D.V. *Bayesian Statistics, A Review*, SIAM, Philadelphia (1971).

55. Lindley, D.V. and Phillips, L.D. Inference for a Bernoulli process (a Bayesian view). *Amer. Statist.* **30**, 112-119 (1976).

56. Morris, C. Parametric empirical Bayes inference: Theory and applications. *J. Amer. Statist. Assoc.* **78**, 47-65 (1983).

57. Mukerjee, R. and Reid, N. Second-order probability matching priors for a parametric function with application to Bayesian tolerance limits. *Biometrika* **88**, 587-592 (2001).

58. O'Hagan, A. *Kendall's Advanced Theory of Statistics, Volume 2b: Bayesian Inference*, Edward Arnold, London (1994).

59. O'Hagan, A. Fractional Bayes factors for model comparisons (with

discussion). *J. Roy. Statist. Soc. Ser. B* **57**, 99-138 (1995).

60. Peers, H.W. On confidence sets and Bayesian probability points in the case of several parameters. *J. Roy. Statist. Soc. Ser. B* **27**, 9-16 (1965).

61. Ramsey, F.P. Truth and probability. Reprinted in *Studies in Subjective Probability* (Edited by H.E. Kyburg and H.E. Smokler), Wiley, New York (1926).

62. Rao, C.R. Differential metrics in probability spaces. In *Differential Geometry in Statistical Inference.* Lecture Notes-Monograph Series, Vol. 10, Institute of Math. Stat., New York (1987).

63. Savage, L.J. *The Foundations of Statistics*, Wiley, New York (1954).

64. Schervish, M.J. *Theory of Statistics.* Springer-Verlag, New York (1995).

65. Seidenfeld, T., Schervish, M.J. and Kadane J.B. A representation of partially ordered preferences. *Ann. Statist.* **23**, 2168-2217 (1995).

66. Shannon, C. A mathematical theory of communication. *Bell System Tech. J.* **27**, 379-423 and 623-656 (1948).

67. Smith, A.F.M. and Spiegelhalter, D.J. Bayes factors and choice criteria for linear models. *J. Roy. Statist. Soc. Ser. B* **42**, 213-220 (1980).

68. Spiegelhalter, D.J. and Smith, A.F.M. (1982). Bayes factors for linear and log-linear models with vague prior information. *J. Roy. Statist. Soc. Ser. B* **44**, 377-387 (1982).

69. Stein, C. On the coverage probability of confidence sets based on a prior distribution. *Sequential Meth. Statist.* **16**, 485-514 (1985).

70. Stigler, S.M. *The History of Statistics: The Measurement of Uncertainty before 1900*, The Belknap Press of Harvard University Press, Cambridge (1986).

71. Sun, D. and Berger, J.O. Reference priors with partial information. *Biometrika* **85**, 55-71 (1998).

72. Tibshirani, R. Noninformative priors for one parameter of many. *Biometrika* **76**, 604-608 (1989).

73. Welch, B.L. and Peers, H.W. On formulae for confidence points based on integrals of weighted likelihoods. *J. Roy. Statist. Soc. B* **25**, 318-329 (1963).

74. Ye, K.Y. *Noninformative Priors in Bayesian Analysis.* Ph.D. Thesis, Department of Statistics, Purdue University, USA (1990).

ON SOME PROBLEMS OF ESTIMATION FOR SOME STOCHASTIC PARTIAL DIFFERENTIAL EQUATIONS

B.L.S. Prakasa Rao

Indian Statistical Institute, New Delhi

Abstract

Stochastic partial differential equations (SPDE) are used for stochastic modelling , for instance, in the study of neuronal behaviour in neurophysiology, in modelling sea surface temparature and sea surface height in physical oceanography , in building stochastic models for turbulence and in modelling environmental pollution. Probabilistic theory underlying the subject of SPDE is discussed in Ito [2] and more recently in Kallianpur and Xiong [11] among others. The study of statistical inference for the parameters involved in SPDE is more recent. Asymptotic theory of maximum likelihood estimators for a class of SPDE is discussed in Huebner, Khasminskii and Rozovskii [7] and Huebner and Rozovskii [8] following the methods in Ibragimov and Khasminskii [9]. Bayes estimation problems for such a class of SPDE are investigated in Prakasa Rao [21,25] following the techniques developed in Borwanker et al. [2]. An analogue of the Bernstein-von Mises theorem for parabolic stochastic partial differential equations is proved in Prakasa Rao [21]. As a consequence, the asymptotic properties of the Bayes estimators of the parameters are investigated. Asymptotic properties of estimators obtained by the method of minimum distance estimation are discussed in Prakasa Rao [30]. Nonparametric estimation of a linear multiplier for some classes of SPDE is studied in Prakasa Rao [26,27] by the kernel method of density estimation following the techniques in Kutoyants [12]. In all the papers cited above , it was assumed that a continuous observation of the random field satisfying the SPDE is available. It is obvious that this assumption is not tenable in practice for various reasons. The question is how to study problem of estimation when there is only a discrete sampling on the random field. A simplified version of this problem is investigated in Prakasa Rao [28,29,30,31]. A review of these and related results is given.

Key words: Bernstein-von Mises theorem, Stochastic partial differential equation, Maximum likelihood estimation, Bayes estimation, Minimum distance estimation, Parametric inference, Nonparametric inference, Linear multiplier, Continuous sampling, Discrete sampling.

AMS Subject classification (2000): Primary 62M40; Secondary 60H15, 35 R 60.

1 Introduction

Stochastic partial differential equations(SPDE) are used for stochastic modelling, for instance, in the study of neuronal behviour in neurophysiology , in modelling sea surface temperature and sea surface height in physical oceanography and in building stochastic models for the behaviour of turbulence and in modelling environmental pollution(cf. Kallianpur and Xiong [11]). The probabilistic theory of SPDE is investigated in Ito [2], Rozovskii [33], Kallianpur and Xiong [11] and De Prato and Zabczyk [3] among others. Huebner et al. [7] started the investigation of maximum likelihood estimation of parameters for a class of SPDE and extended their results to parabolic SPDE in Huebner and Rozovskii [8] following the approach of Ibragimov and Khasminskii [9]. Bernstein -von Mises type theorems were developed for such SPDE in Prakasa Rao [21, 25] following the techniques in Borwanker et al. [2] and Prakasa Rao [18]. Asymptotic properties of the Bayes estimators of parameters for SPDE were discussed in Prakasa Rao [21,25]. Statistical inference for diffusion type processes and semimartingales in general is studied in Prakasa Rao [22,23]. As a consequence, the asymptotic properties of the Bayes estimators of the parameters are investigated using the asymptotic properties of maximum likelihood estimators proved in Huebner and Rozovskii [8]. Asymptotic properties obtained by the method of minimum distance estimation are discussed in Prakasa Rao [30]. Nonparametric estimation of a linear multiplier for some classes of SPDE are studied in Prakasa Rao [26,27] by the kernel method of density estimation following the techniques in Kutoyants [12]. In all the papers cited above , it was assumed that a continuous observation of the random field satisfying the SPDE is available. It is obvious that this assumption is not tenable in practice for various reasons. The question is how to study the problem of estimation when there is only a discrete sampling on the underlying random field. A simplified version of this problem is discussed in Prakasa Rao [28,29] and in Prakasa Rao [30,31] .

Our aim in this paper is to review some of our earlier work and to present some new results.

2 Stochastic modelling

Any problem of statistical inference based on data can be termed as data assimilation or summarization. The problem is to develop suitable models to study the underlying phenomenon, estimate the unknown coefficients in the model, predict the future observations based on the model, validate the model by comparing the predicted values with actual observations and revise the model based on the experience so obtained and continue this cycle of operations.

As Kallianpur and Xiong [11] indicate , stochastic partial differential equations arise from attempts to introduce randomness in a meaningful way into phenomena regarded as deterministic. Examples of such modelling occur in chemical-reactor diffusions, neurophysiology, physical oceanography, study of turbulence and more recently in modelling of environmental pollution. Hodgkin and Huxley [6] studied the electrical behaviour of neuronal membranes and the role of ionic currents. They modeled the flow of current through the surface membrane of the giant axon from a Loligo Squid through partial differential equations. Kallianpur and Xiong [11] point out that, in a realistic description of neuronal activity, one needs to take into account synaptic inputs occuring randomly in time and at different sites on the neurons' surface leading to a SPDE. Another area of stochastic modeling by SPDE occurs in physical oceanography, for instance, in the study of modeling sea surface temperature and sea surface height (Piterbarg and Rozovskii [15]). In both these problems and in any other problem involved in modelling by an SPDE, the problem of estimation of coefficients involved in the SPDE from the observed data is of paramount importance.

We will now study the problems of estimation for some classes of parabolic SPDE which are amenable for statistical inference.

3 Parametric Estimation for Stochastic PDE with linear drift (Absolutely continuous case) (Continuous sampling)

Let $(\Omega, \mathcal{F}, P)$ be a probability space and consider the process $u_\varepsilon(t,x), 0 \leq x \leq 1, 0 \leq t \leq T$ governed by the stochastic partial differential equation

$$du_\varepsilon(t,x) = (\Delta u_\varepsilon(t,x) + \theta u_\varepsilon(t,x))dt + \varepsilon dW_Q(t,x) \tag{3. 1}$$

where $\Delta = \frac{\partial^2}{\partial x^2}$. Suppose that $\varepsilon \to 0$ and $\theta \in \Theta \subset R$. Suppose the initial and the boundary conditions are given by

$$\left\{\begin{array}{l} u_\varepsilon(0,x) = f(x), f \in L_2[0,1] \\ u_\varepsilon(t,0) = u_\varepsilon(t,1) = 0, 0 \le t \le T \end{array}\right. \tag{3. 2}$$

and Q is the nuclear covariance operator for the Wiener process $W_Q(t,x)$ taking values in $L_2[0,1]$ so that

$$W_Q(t,x) = Q^{1/2}W(t,x)$$

and $W(t,x)$ is a cylindrical Brownian motion in $L_2[0,1]$. Then, it is known that (cf. Rozovskii [33])

$$W_Q(t,x) = \sum_{i=1}^{\infty} q_i^{1/2} e_i(x) W_i(t) \text{ a.s.} \tag{3. 3}$$

where $\{W_i(t), 0 \le t \le T\}, i \ge 1$ are independent one - dimensional standard Wiener processes and $\{e_i\}$ is a complete orthonormal system in $L_2[0,1]$ consisting of eigen vectors of Q and $\{q_i\}$ eigen values of Q.

Let us consider a special covariance operator Q with $e_k = \sin k\pi x, k \ge 1$ and $\lambda_k = (\pi k)^2, k \ge 1$. Then $\{e_k\}$ is a complete orthonormal system with eigen values $q_i = (1 + \lambda_i)^{-1}, i \ge 1$ for the operator Q and $Q = (I - \Delta)^{-1}$. Further more

$$dW_Q = Q^{1/2}dW.$$

We define a solution $u_\varepsilon(t,x)$ of (3.1) as a formal sum

$$u_\varepsilon(t,x) = \sum_{i=1}^{\infty} u_{i\varepsilon}(t) e_i(x) \tag{3. 4}$$

(cf. Rozovskii [33]). It is known that the Fourier coefficient $u_{i\varepsilon}(t)$ satisfies the stochastic differential equation

$$du_{i\varepsilon}(t) = (\theta - \lambda_i)u_{i\varepsilon}(t)dt + \frac{\varepsilon}{\sqrt{\lambda_i + 1}} dW_i(t),\ 0 \le t \le T \tag{3. 5}$$

with the initial condition

$$u_{i\varepsilon}(0) = v_i,\ v_i = \int_0^1 f(x)e_i(x)dx. \tag{3. 6}$$

It is further known that $u_\varepsilon(t,x)$ as defined above belongs to $L_2([0,T] \times \Omega; L_2[0,1])$ together with its derivative in t. Further more $u_\varepsilon(t,x)$ is the only solution to (3.1) under the boundary condition (3.2). Let $P_\theta^{(\varepsilon)}$ be the measure generated by u_ε when θ is the true parameter.

Suppose θ_0 is the true parameter .It has been shown by Huebner et al. [7] that the family of measures $\{P_\theta^{(\varepsilon)}, \theta \in \Theta\}$ are mutually absolutely continuous and

$$(3.\ 7)\quad \log \frac{dP_\theta^{(\varepsilon)}}{dP_{\theta_0}^{(\varepsilon)}}(u_\varepsilon) = \sum_{i=1}^{\infty} \frac{\lambda_i+1}{\varepsilon^2}[(\theta-\theta_0)\int_0^T u_{i\varepsilon}(t)du_{i\varepsilon}(t) - \frac{1}{2}\{(\theta-\lambda_i)^2-(\theta_0-\lambda_i)^2\}\int_0^T u_{i\varepsilon}^2(t)dt].$$

Maximum Likelihood Estimation

It can be checked that the MLE $\hat{\theta}_\varepsilon$ of θ based on u_ε satisfies the likelihood equation

$$(3.\ 8)\qquad \alpha_\varepsilon = \varepsilon^{-1}(\hat{\theta}_\varepsilon - \theta_0)\beta_\varepsilon$$

when θ_0 is the true parameter where

$$(3.\ 9)\qquad \alpha_\varepsilon = \sum_{i=1}^{\infty} \sqrt{\lambda_i+1}\int_0^T u_{i\varepsilon}(t)dW_i(t)$$

and

$$(3.\ 10)\qquad \beta_\varepsilon = \sum_{i=1}^{\infty}(\lambda_i+1)\int_0^T u_{i\varepsilon}^2(t)dt.$$

Huebner et al. [7] proved that the estimator $\hat{\theta}_\varepsilon$ is consistent and asymptotically $N(0, I(\theta)^{-1})$ and asymptotically efficient in the Hajek - Le Cam sense. They proved that

$$(3.\ 11)\qquad \lim_{\varepsilon\to 0}\sup_{|\theta-\theta_0|<\delta} E_{\theta,\varepsilon} w(\varepsilon^{-1}(\theta_\varepsilon^* - \theta)) \geq Ew(\xi)$$

where ξ is $N(0, I(\theta)^{-1})$ for any estimator θ_ε^* based on $u_\varepsilon(t,x)$ for a class of loss functions $w(x)$ which are bounded, symmetric with $w(0) = 0$ and $w(x)$ monotone for $x \geq 0$. Here

$$(3.\ 12)\qquad I(\theta) = \frac{1}{2}\sum_{i=1}^{\infty}\frac{\lambda_i+1}{\lambda_i-\theta}v_i^2(1-e^{-2(\theta-\lambda_i)T}).$$

Bernstein-Von Mises Theorem

Suppose that Λ is a prior probability measure on $(\Theta, \mathcal{B})$ where $\mathcal{B}$ is the σ-algebra of Borel subsets of an open set $\Theta \subset R$. Further suppose that Λ has the density $\lambda(\cdot)$ with respect to the Lebesgue measure and the density $\lambda(\cdot)$ is continuous and positive in an open neighborhood of θ_0, the true parameter. The posterior density of θ given $u_\varepsilon(t,x), 0 < x < 1, 0 \leq t \leq T$ is

$$(3.\ 13)\qquad p(\theta|u_\varepsilon) = \frac{\frac{dP_\theta^{(\varepsilon)}}{dP_{\theta_0}^{(\varepsilon)}}(u_\varepsilon)\lambda(\theta)}{\int_\Theta \frac{dP_\theta^{(\varepsilon)}}{dP_{\theta_0}^{(\varepsilon)}}(u_\varepsilon)\lambda(\theta)d\theta}.$$

Let $\tau = \varepsilon^{-1}(\theta - \hat{\theta}_\varepsilon)$ and

(3. 14)
$$p^*(\tau|u_\varepsilon) = \varepsilon\, p(\hat{\theta}_\varepsilon + \varepsilon\tau|u_\varepsilon).$$

Then $p^*(\tau|u_\varepsilon)$ is the posterior density of $\varepsilon^{-1}(\theta - \hat{\theta}_\varepsilon)$. Let

(3. 15)
$$\nu_\varepsilon(\tau) \equiv \frac{dP^{(\varepsilon)}_{\hat{\theta}_\varepsilon+\varepsilon\tau}}{dP^{(\varepsilon)}_{\theta_0}} / \frac{dP^{(\varepsilon)}_{\hat{\theta}_\varepsilon}}{dP^{(\varepsilon)}_{\theta_0}} = \frac{dP^{(\varepsilon)}_{\hat{\theta}_\varepsilon+\varepsilon\tau}}{dP^{(\varepsilon)}_{\hat{\theta}_\varepsilon}} \text{ a.s.}$$

In view of (3.7), it follows that

(3. 16)
$$\begin{aligned} \log \nu_\varepsilon(\tau) &= \tau\alpha_\varepsilon - \tau\frac{\alpha_\varepsilon}{\beta_\varepsilon}\beta_\varepsilon - \frac{\tau^2}{2}\beta_\varepsilon \text{ a.s.} \\ &= -\frac{\tau^2}{2}\beta_\varepsilon \text{ a.s.} \end{aligned}$$

from the equations (3.5) and (3.8) . Let

(3. 17)
$$C_\varepsilon = \int_{-\infty}^{\infty} \nu_\varepsilon(\tau)\lambda(\hat{\theta}_\varepsilon + \varepsilon\tau)d\tau.$$

It can be seen that

(3. 18)
$$p^*(\tau|u_\varepsilon) = C_\varepsilon^{-1}\nu_\varepsilon(\tau)\lambda(\hat{\theta}_\varepsilon + \varepsilon\tau).$$

Suppose the following conditions hold.

(C1) There exists a constant $\beta > 0$ such that $\beta_\varepsilon = \sum_{i=1}^{\infty}(\lambda_i + 1)\int_0^T u_{i\varepsilon}^2(t)dt \to \beta > 0$ *a.s.* $[P_{\theta_0}]$ as $\varepsilon \to 0$.

(C2) The maximum likelihood estimator $\hat{\theta}_\varepsilon$ is strongly consistent, that is ,

$$\hat{\theta}_\varepsilon \to \theta_0 \text{ a.s. } [P_{\theta_0}] \text{ as } \varepsilon \to 0;$$

and

(C3) $K(\cdot)$ is a nonnegative function such that, for some $0 < \gamma < \beta$,

$$\int_{-\infty}^{\infty} K(\tau)e^{-\frac{1}{2}\tau^2(\beta-\gamma)}d\tau < \infty.$$

We have now the following main theorem which is an analogue of the Bernstein -von Mises theorem in Borwanker et al. [2]. For proof, see Prakasa Rao [25].

Theorem 3.1 : Suppose the conditions (C1) to (C4) hold where $\lambda(\cdot)$ is a prior density which is continuous and positive in an open neighborhood of θ_0, the true parameter. Then

$$\lim_{\varepsilon\to 0}\int_{-\infty}^{\infty} K(\tau)|p^*(\tau|u_\varepsilon) - (\frac{\beta}{2\pi})^{1/2}e^{-\frac{1}{2}\beta\tau^2}|d\tau = 0 \text{ a.s. } [P_{\theta_0}] \tag{3. 19}$$

As a consequence of Theorem 3.1, it is easy to see that the following result holds (cf. Borwanker et al. [2]).

Theorem 3.2 : Suppose the following conditions hold :

(D1) $\hat{\theta}_\varepsilon \to \theta_0$ a.s. $[P_{\theta_0}]$ as $\varepsilon \to 0$;

(D2) $\beta_\varepsilon \to \beta > 0$ a.s. $[P_{\theta_0}]$ as $\varepsilon \to 0$;

(D3) $\lambda(\cdot)$ is a prior density which is continuous and positive in an open neighborhood of θ_0, the true parameter ; and

(D4) $\int_{-\infty}^{\infty}|\theta|^m\lambda(\theta)d\theta < \infty$ for some integer $m \geq 0$.

Then

$$\lim_{\varepsilon\to 0}\int_{-\infty}^{\infty} |\tau|^m|p^*(\tau|u_\varepsilon) - (\frac{\beta}{2\pi})^{1/2}e^{-\frac{1}{2}\beta\tau^2}|d\tau = 0 \text{ a.s. } [P_{\theta_0}]. \tag{3. 20}$$

Remark: It is clear that the condition (D4) holds for $m = 0$. Suppose the conditions (D1) to (D3) hold. Then it follows that

$$\lim_{\varepsilon\to 0}\int_{-\infty}^{\infty} |p^*(\tau|u_\varepsilon) - (\frac{\beta}{2\pi})^{1/2}e^{-\frac{1}{2}\beta\tau^2}|d\tau = 0 \text{ a.s.}[P_{\theta_0}]. \tag{3. 21}$$

This is the analogue of the Bernstein - von Mises Theorem in the classical statistical inference. As a special case of Theorem 3.2, we obtain that

$$E_{\theta_0}[\varepsilon^{-1}(\hat{\theta}_\varepsilon - \theta_0)]^m \to E[Z^m] \text{ as } \varepsilon \to 0 \tag{3. 22}$$

where Z is $N(0, \beta^{-1})$.

Bayes Estimation

We define a *Bayes estimator* $\tilde{\theta}_\varepsilon$ of θ, based on the path u_ε and the prior density $\lambda(\theta)$, to be a minimizer of the function

$$B_\varepsilon(\phi) = \int_\Theta \tilde{L}(\theta,\phi)p(\theta|u_\varepsilon)d\theta, \phi \in \Theta \tag{3. 23}$$

where $\tilde{L}(\theta,\phi)$ is a given loss function defined on $\Theta \times \Theta$. Suppose there exists a Bayes estimator $\tilde{\theta}_\varepsilon$. Further suppose that the loss function satisfies the following conditions :

(E1) $\tilde{L}(\theta, \phi) = L(|\theta - \phi|) \geq 0$;

(E2) $L(t)$ is non decreasing for $t \geq 0$;

(E3) there exist nonnegative functions $R_\varepsilon, K(\tau)$ and $G(\tau)$ such that
(a) $R_\varepsilon L(\tau\varepsilon) \leq G(\tau)$ for all $\varepsilon \geq 0$,
(b) $R_\varepsilon L(\tau\varepsilon) \to K(\tau)$ as $\varepsilon \to 0$ uniformly on bounded intervals of τ,
(c) the function $\int_{-\infty}^{\infty} K(\tau + m)e^{-\frac{1}{2}\beta\tau^2} d\tau$ achieves its minimum at $m = 0$, and
(d) $G(\tau)$ satisfies the conditions akin to (C3) and (C4) .

The following result can be proved by arguments similar to those given in Borwanker et al. [2]).

Theorem 3.3 : Suppose the conditions, (D1) - (D3) of Theorem 3.2 hold. In addition suppose that the loss function $\tilde{L}(\theta, \phi)$ satisfies the conditions (E1) - (E3) stated above. Then

(3. 24)
$$\varepsilon^{-1}(\hat{\theta}_\varepsilon - \tilde{\theta}_\varepsilon) \to 0 \text{ a.s. } [P_{\theta_0}] \text{ as } \varepsilon \to 0$$

and

(3. 25)
$$\begin{aligned} \lim_{\varepsilon \to 0} R_\varepsilon B_\varepsilon(\tilde{\theta}_\varepsilon) &= \lim_{\varepsilon \to 0} R_\varepsilon B_\varepsilon(\hat{\theta}_\varepsilon) \\ &= \left(\frac{\beta}{2\pi}\right)^{1/2} \int_{-\infty}^{\infty} K(\tau) e^{-\frac{1}{2}\beta\tau^2} d\tau \text{ a.s. } [P_{\theta_0}]. \end{aligned}$$

Relations (3.8) to (3.10) and the central limit theorem for stochastic integrals prove that

(3. 26)
$$\varepsilon^{-1}(\hat{\theta}_\varepsilon - \theta_0) \xrightarrow{\mathcal{L}} N(0, \beta^{-1}) \text{ as } \varepsilon \to 0$$

under the probability measure P_{θ_0}. As a consequence of Theorem 3.3 and the condition (D1), it follows that

(3. 27)
$$\tilde{\theta}_\varepsilon \to \theta_0 \text{ a.s. } [P_{\theta_0}] \text{ as } \varepsilon \to 0$$

and

(3. 28)
$$\varepsilon^{-1}(\tilde{\theta}_\varepsilon - \theta_0) \xrightarrow{\mathcal{L}} N(0, \beta^{-1}) \text{ as } \varepsilon \to 0.$$

In other words, the Bayes estimator of the parameter θ in the SPDE given by (3.1) is asymptotically normal and asymptotically efficient under the conditions stated in Theorem 3.3.

Minimum Distance Estimation

We have discussed asymptotic properties of the maximum likelihood estimators (MLE) and the Bayes estimators and it is known that these estimators are consistent, asymptotically normal and asymptotically efficient. In spite of having such nice properties, the MLE have some short comings. Their calculation is cumbersome and difficult as the expressions for MLE involve stochastic integrals which need good approximants for computation. Further more the MLE are not robust in the sense that a slight perturbation in the noise component, say, from a Wiener process to a Gaussian process with finite variation will change the properties of the MLE . In order to circumvent this problem, an alternate approach to estimate the parameter θ can be adapted and that is estimation by the method of minimum distance. The theory of minimum distance estimation in a general frame work is given in Millar [14]. Observe that the parameter θ in the SPDE (3.1) can be estimated from the equation (3.5). We now apply the minimum distance approach adapted by Kutoyants and Pilibossian [13] to estimate the parameter θ satisfying the equation (3.5). We define the minimum L_1-norm estimate $\tilde{\theta}_{i\varepsilon T}$ by the relation

$$\tilde{\theta}_{i\varepsilon T} = \lambda_i + \arg\inf_{\theta\in\Theta} \int_0^T |u_{i\varepsilon}(t) - u_i(t,\theta)| \; dt$$

where $u_i(t,\theta)$ is the solution of the ordinary differential equation

$$\frac{du_i(t)}{dt} = (\theta - \lambda_i)u_i(t), u_i(0,\theta) = v_i.$$

It is easy to see that

$$u_i(t,\theta) = v_i e^{(\theta-\lambda_i)t}.$$

Let

$$g_i(\delta) = \inf_{|\theta-\theta_0|>\delta} \int_0^T |u_i(t,\theta) - u_i(t,\theta_0)|dt.$$

The following theorem is a consequence of Theorem 1 of Kutoyants and Plibossian [13].

Theorem 3.4 : For any $\delta > 0$,

$$P_{\theta_0}^{(\varepsilon)}(|\tilde{\theta}_{i\varepsilon T} - \theta_0| \geq \delta) \leq 2\exp\{-k_i(\lambda_i + 1)g_i^2(\delta)\varepsilon^{-2}\}$$

where

$$k_i = \exp\{-2|\theta_0 - \lambda_i|T\}/(2T)^3.$$

Let

$$Y_i(t) = e^{(\theta_0-\lambda_i)t} \int_0^t e^{-(\theta_0-\lambda_i)s} \; dW_i(s).$$

Note that the process $Y_i(t)$ is a gaussian process. Define

$$\zeta_{iT} = \arg\inf_u \int_0^T |Y_i(t) - utv_i e^{(\theta_0-\lambda_i)t}| \ dt.$$

The following theorem is again a consequence of Theorems 2 and 3 of Kutoyants and Plibossian [13].

Theorem 3.5 : For any fixed $T > 0$,

$$(\frac{\varepsilon}{\sqrt{\lambda_i + 1}})^{-1}(\tilde{\theta}_{i\varepsilon T} - \theta_0) \xrightarrow{p} \zeta_{iT} \ \text{as} \ \varepsilon \to 0$$

where θ_0 is the true parameter. Further more if $\theta_0 > \lambda_i$, then

$$\zeta_{iT} T v_i \sqrt{2(\theta_0 - \lambda_i)} \xrightarrow{\mathcal{L}} N(0,1) \ \text{as} \ T \to \infty.$$

We now state and prove a lemma.

Lemma 3.6 : Suppose that for every $T > 0$,

$$X_{\varepsilon T} \xrightarrow{p} Y_T \ \text{as} \ \varepsilon \to 0$$

and further suppose that

$$Y_T \xrightarrow{\mathcal{L}} Y \ \text{as} \ T \to \infty.$$

Then

$$X_{\varepsilon T} \xrightarrow{\mathcal{L}} Y \ \text{as} \ \varepsilon \to 0 \ \text{and} \ T \to \infty.$$

Proof: Let F be a closed set and $F_\delta = \{x : \rho(x, F) \leq \delta\}$ where $\rho(x, F)$ denotes the distance between the point x and the closed set F. Note that F_δ decreases to the set F as δ decreases to zero. Then

(3. 29) $$P(X_{\varepsilon T} \in F) \leq P(Y_T \in F_\delta) + P(|X_{\varepsilon T} - Y_T| \geq \delta).$$

Hence

$$\begin{aligned}\limsup_{\varepsilon \to 0} P(X_{\varepsilon T} \in F) &\leq P(Y_T \in F_\delta) + \limsup_{\varepsilon \to 0} P(|X_{\varepsilon T} - Y_T| \geq \delta) \\ &= P(Y_T \in F_\delta)\end{aligned}$$

since $X_{\varepsilon T} \xrightarrow{p} Y_T$ as $\varepsilon \to 0$. Taking limit as $T \to \infty$ in the above inequalities, we get that

$$\begin{aligned}\limsup_{T \to \infty} \limsup_{\varepsilon \to 0} P(X_{\varepsilon T} \in F) &\leq \limsup_{T \to \infty} P(Y_T \in F_\delta) \\ &\leq P(Y \in F_\delta)\end{aligned}$$

since the set F_δ is closed and $Y_T \xrightarrow{\mathcal{L}} Y$ as $T \to \infty$. Let $\delta \to 0$. Then we have

$$\limsup_{T\to\infty}\limsup_{\varepsilon\to 0} P(X_{\varepsilon T} \in F) \leq P(Y \in F)$$

for every closed set F. Hence, by the standard results from the theory of weak convergence, it follows that

$$X_{\varepsilon T} \xrightarrow{\mathcal{L}} Y \text{ as } \varepsilon \to 0 \text{ and } T \to \infty. \tag{3. 30}$$

Let

$$X_{\varepsilon T} = \left(\frac{\varepsilon}{\sqrt{\lambda_i + 1}}\right)^{-1}(\tilde{\theta}_{i\varepsilon T} - \theta_0)T v_i \sqrt{2(\theta_0 - \lambda_i)}, \tag{3. 31}$$

$$Y_T = \zeta_{iT} T v_i \sqrt{\theta_0 - \lambda_i} \tag{3. 32}$$

and Y be a standard normal random variable. Applying the Lemma 3.6, we get the following result.

Theorem 3.7 : If $\theta_0 > \lambda_i$,then

$$\left(\frac{\varepsilon}{\sqrt{\lambda_i + 1}}\right)^{-1}(\tilde{\theta}_{i\varepsilon T} - \theta_0)T v_i \sqrt{2(\theta_0 - \lambda_i)} \xrightarrow{\mathcal{L}} N(0,1) \text{ as } \varepsilon \to 0 \text{ and } T \to \infty. \tag{3. 33}$$

In view of Theorem 3.7, the variance of the limiting normal distribution of estimator $\tilde{\theta}_{i\varepsilon T}$ is proportional to $[v_i^2(\theta_0 - \lambda_i)(\lambda_i + 1)]^{-1}$. Note that the estimators $\tilde{\theta}_{i\varepsilon T}, i \geq 1$ are independent estimators of the parameter θ since the processes $\{W_i(t), t \geq 0\}, i \geq 1$ are independent standard Wiener processes. We will now construct an optimum estimator out of the estimators $\tilde{\theta}_{i\varepsilon T}, 1 \leq i \leq N$ for any $N \geq 1$.

Let $\tilde{\theta}_{\varepsilon T} = \sum_{i=1}^N \alpha_i \tilde{\theta}_{i\varepsilon T}$ where $\alpha_i, 1 \leq i \leq N$ is a nonrandom sequence of coefficients to be chosen. Note that

$$\tilde{\theta}_{\varepsilon T} \xrightarrow{p} [\sum_{i=1}^N \alpha_i]\theta_0 \text{ as } \varepsilon \to 0 \text{ and } T \to \infty$$

by Theorem 3.7 and hence $\tilde{\theta}_{\varepsilon T}$ is a consistent estimator for θ_0 as $\varepsilon \to 0$ and as $T \to \infty$ provided $\sum_{i=1}^N \alpha_i = 1$. Further more

$$\varepsilon^{-1}T(\tilde{\theta}_{\varepsilon T} - \theta_0) \xrightarrow{\mathcal{L}} N(0, \sum_{i=1}^N \alpha_i^2[2v_i^2(\theta_0 - \lambda_i)(\lambda_i + 1)]^{-1}) \text{ as } \varepsilon \to 0 \text{ and } T \to \infty.$$

This follows again by Theorem 3.7 and the independence of the estimators $\{\tilde{\theta}_{i\varepsilon T}, 1 \leq i \leq N\}$. We now obtain the optimum combination of the coefficients $\{\alpha_i, 1 \leq i \leq N\}$ by minimizing the asymptotic variance

$$\sum_{i=1}^{N} \alpha_i^2 [2v_i^2(\theta_0 - \lambda_i)(\lambda_i + 1)]^{-1}$$

subject to the condition $\sum_{i=1}^{N} \alpha_i = 1$. It is easy to see that α_i is proportional to $[(\theta_0 - \lambda_i)(\lambda_i + 1)]$ and the optimal choice of $\{\alpha_i, 1 \leq i \leq N\}$ leads to the "estimator"

$$\theta^*_{\varepsilon T} = \frac{\sum_{i=1}^{N} v_i^2(\theta_0 - \lambda_i)(\lambda_i + 1)\tilde{\theta}_{i\varepsilon T}}{\sum_{i=1}^{N} v_i^2(\theta_0 - \lambda_i)(\lambda_i + 1)}. \tag{3. 34}$$

It is easy to see that

$$\theta^*_{\varepsilon T} \xrightarrow{p} \theta_0 \text{ as } \varepsilon \to 0 \text{ and } T \to \infty$$

and

$$\varepsilon^{-1}T(\hat{\theta}^*_{\varepsilon T} - \theta_0) \xrightarrow{\mathcal{L}} N(0, [\sum_{i=1}^{N} 2v_i^2(\theta_0 - \lambda_i)(\lambda_i + 1)]^{-1}) \text{ as } \varepsilon \to 0 \text{ and } T \to \infty$$

again due to the independence of the estimators $\tilde{\theta}_{i\varepsilon T}, 1 \leq i \leq N$. However the random variable $\theta^*_{\varepsilon T}$ cannot be considered as an estimator of the parameter θ_0 since it depends on the unknown parameter θ_0. In order to avoid this problem, we consider a modified estimator

$$\hat{\theta}_{\varepsilon T} = \frac{\sum_{i=1}^{N} v_i^2(\tilde{\theta}_{i\varepsilon T} - \lambda_i)(\lambda_i + 1)\tilde{\theta}_{i\varepsilon T}}{\sum_{i=1}^{N} v_i^2(\tilde{\theta}_{i\varepsilon T} - \lambda_i)(\lambda_i + 1)}. \tag{3. 35}$$

which is obtained from $\theta^*_{\varepsilon T}$ by substituting the estimator $\tilde{\theta}_{i\varepsilon T}$ for the unknown parameter θ_0 in the i-th term in the numerator and the denominator in (3.34). In view of the independence, consistency and asymptotic normality of the estimators $\tilde{\theta}_{i\varepsilon T}, 1 \leq i \leq N$, it follows that the estimator $\hat{\theta}_{\varepsilon T}$ is consistent and asymptotically normal for the parameter θ_0 and we have the following result.

Theorem 3.8: Under the probability measure P_{θ_0},

$$\hat{\theta}_{\varepsilon T} \xrightarrow{p} \theta_0 \text{ as } \varepsilon \to 0 \text{ and } T \to \infty$$

and if $\theta_0 > N^2\pi^2$, then

$$\varepsilon^{-1}T(\hat{\theta}_{\varepsilon T} - \theta_0) \xrightarrow{\mathcal{L}} N(0, [\sum_{i=1}^{N} 2v_i^2(\theta_0 - \lambda_i)(\lambda_i + 1)]^{-1}) \text{ as } \varepsilon \to 0 \text{ and } T \to \infty.$$

4 Parametric Estimation for Stochastic PDE with linear drift (Singular case) (Continuous sampling)

Let $(\Omega, \mathcal{F}, P)$ be a probability space and consider the process $u_\varepsilon(t,x), 0 \le x \le 1, 0 \le t \le T$ governed by the stochastic partial differential equation

$$du_\varepsilon(t,x) = \theta \, \Delta u_\varepsilon(t,x)dt + \varepsilon(I-\Delta)^{-1/2}dW(t,x) \tag{4. 1}$$

where $\theta > 0$ satisfying the initial and the boundary conditions

$$\begin{aligned} u_\varepsilon(0,x) &= f(x),\ 0<x<1,\ f \in L_2[0,1], \\ u_\varepsilon(t,0) &= u_\varepsilon(t,1) = 0,\ 0 \le t \le T. \end{aligned} \tag{4. 2}$$

Here I is the identity operator, $\Delta = \frac{\partial^2}{\partial x^2}$ as defined in Section 3 and the process $W(t,x)$ is the cylindrical Brownian motion in $L_2[0,1]$.. In analogy with (3.5), the Fourier coefficients $u_{i\varepsilon}(t)$ satisfy the stochastic differential equations

$$du_{i\varepsilon}(t) = -\theta\lambda_i u_{i\varepsilon}(t)dt + \frac{\varepsilon}{\sqrt{\lambda_i+1}}dW_i(t),\ 0 \le t \le T, \tag{4. 3}$$

with

$$u_{i\varepsilon}(0) = v_i, v_i = \int_0^1 f(x)e_i(x)dx. \tag{4. 4}$$

Let $P_\theta^{(\varepsilon)}$ be the measure generated by u_ε when θ is the true parameter. It can be shown that the family of measures $\{P_\theta^{(\varepsilon)}, \theta \in \Theta\}$ *do not* form a family of equivalent probability measures. In fact, $P_\theta^{(\varepsilon)}$ is singular with respect to $P_{\theta'}^{(\varepsilon)}$ whenever $\theta \ne \theta'$ in Θ (cf. Huebner et al. [7]).

Let $u_\varepsilon^{(N)}(t,x)$ be the projection of $u_\varepsilon(t,x)$ onto the subspace spanned by $\{e_1, \cdots, e_N\}$ in $L_2[0,1]$. In other words

$$u_\varepsilon^{(N)}(t,x) = \sum_{i=1}^N u_{i\varepsilon}(t)e_i(x). \tag{4. 5}$$

Let $P_\theta^{(\varepsilon,N)}$ be the probability measure generated by $u_\varepsilon^{(N)}$ on the subspace spanned by $\{e_1, \cdots, e_N\}$ in $L_2[0,1]$. It can be shown that the measures $\{P_\theta^{(\varepsilon,N)}, \theta \in \Theta\}$ from an equivalent family and

(4. 6)

$$\begin{aligned} &\log \frac{dP_\theta^{(\varepsilon,N)}}{dP_{\theta_0}^{(\varepsilon,N)}}(u_\varepsilon^{(N)}) \\ &= -\frac{1}{\varepsilon^2}\sum_{i=1}^N \lambda_i(\lambda_i+1)[(\theta-\theta_0)\int_0^T u_{i\varepsilon}(t)(du_{i\varepsilon}(t)+\theta_0\lambda_i u_{i,\varepsilon}(t)dt) + \frac{1}{2}(\theta-\theta_0)^2\lambda_i\int_0^T u_{i\varepsilon}^2(t)dt]. \end{aligned}$$

Maximum Likelihood Estimation

It can be checked that the MLE $\hat{\theta}_{\varepsilon,N}$ of θ based on $u_\varepsilon^{(N)}$ satisfies the likelihood equation

$$\alpha_{\varepsilon,N} = -\varepsilon^{-1}(\hat{\theta}_{\varepsilon,N} - \theta_0)\beta_{\varepsilon,N} \tag{4. 7}$$

when θ_0 is the true parameter where

$$\alpha_{\varepsilon,N} = \sum_{i=1}^{N} \lambda_i \sqrt{\lambda_i + 1} \int_0^T u_{i\varepsilon}(t) dW_i(t) \tag{4. 8}$$

and

$$\beta_{\varepsilon,N} = \sum_{i=1}^{N} (\lambda_i + 1)\lambda_i^2 \int_0^T u_{i,\varepsilon}^2(t) dt. \tag{4. 9}$$

Huebner et al. [7] prove that, for any fixed $N \geq 1$, the estimator $\hat{\theta}_{\varepsilon,N}$ is consistent and asymptotically $N(0, I_N(\theta_0)^{-1})$ under $P_{\theta_0}^{(\varepsilon,N)}$ as $\varepsilon \to 0$ where

$$I_N(\theta) = \frac{1}{2\theta} \sum_{i=1}^{N} \lambda_i(\lambda_i + 1) v_i^2 (1 - e^{-2\theta\lambda_i T}). \tag{4. 10}$$

They further prove that, for any fixed $\varepsilon > 0$,

$$\hat{\theta}_{\varepsilon,N} \xrightarrow{p} \theta_0 \text{ under } P_{\theta_0}^{(\varepsilon)} \text{ as } N \to \infty \tag{4. 11}$$

and

$$Q_{N,\varepsilon}^{-1}(\theta_0)(\hat{\theta}_{\varepsilon,N} - \theta_0) \xrightarrow{\mathcal{L}} N(0,1) \text{ under } P_{\theta_0}^{(\varepsilon)} \text{ as } N \to \infty \tag{4. 12}$$

where

$$Q_{N,\varepsilon}(\theta) = \left(\sum_{i=1}^{N} \lambda_i^2 (\lambda_i + 1) E \int_0^T u_{i\varepsilon}^2(t) dt \right)^{-1/2}. \tag{4. 13}$$

In addition, they show that, for any fixed ε and for any estimator $\theta_{\varepsilon,N}^*$ based on $u_\varepsilon^{(N)}(t,x)$,

$$\lim_{N \to \infty} \sup_{|\theta - \theta_0| < \delta} E_{\theta,\varepsilon}^N \{ w(Q_{N,\varepsilon}^{-1}(\theta)(\theta_{\varepsilon,N}^* - \theta)) \} \geq E w(\zeta) \tag{4. 14}$$

where ζ is $N(0,1)$ for a class of loss functions $w(x)$ which are bounded, symmetric with $w(0) = 0$ and $w(x)$ monotone for $x \geq 0$. Here $E_{\theta,\varepsilon}^N$ denotes the expectation under the probability measure $P_\theta^{(\varepsilon,N)}$.

We will now investigate the asymptotic behaviour of the Bayes estimators of θ as $\varepsilon \to 0$ for fixed N and as $N \to \infty$ for fixed $\varepsilon > 0$. The former case is similar to the results discussed in Section 2.

Bernstein - von Mises Theorem (when N is fixed as $\varepsilon \to 0$)

Suppose that Λ is a prior probability measure on $(\Theta, \mathcal{B})$ where $\mathcal{B}$ is the σ-algebra of Borel subsets of an open set $\Theta \subset R$. Further suppose that Λ has a density $\lambda(\cdot)$ with respect to the Lebesgue measure and the density $\lambda(\cdot)$ is continuous and positive in an open neighbourhood of θ_0, the true parameter.

The posterior density of θ given $u_\varepsilon^{(N)}$ is

$$p(\theta|u_\varepsilon^{(N)}) = \frac{\frac{dP_\theta^{(\varepsilon,N)}}{dP_{\theta_0}^{(\varepsilon,N)}}(u_\varepsilon^{(N)})\ \lambda(\theta)}{\int\limits_\Theta \frac{dP_\theta^{(\varepsilon,N)}}{dP_{\theta_0}^{(\varepsilon,N)}}(u_\varepsilon^{(N)})\ \lambda(\theta)d\theta}. \tag{4. 15}$$

Let

$$\tau = \varepsilon^{-1}(\theta - \hat{\theta}_{\varepsilon,N}) \tag{4. 16}$$

and

$$p^*(\tau|u_\varepsilon^{(N)}) = \varepsilon\ p(\hat{\theta}_{\varepsilon,N} + \varepsilon\tau|u_\varepsilon^{(N)}). \tag{4. 17}$$

Then $p^*(\tau|u_\varepsilon^{(N)})$ is the posterior density of $\varepsilon^{-1}(\theta - \hat{\theta}_{\varepsilon,N})$. Let

$$\nu_{\varepsilon,N}(\tau) = \frac{dP_{\hat{\theta}_{\varepsilon,N}+\varepsilon\tau}^{(\varepsilon,N)}}{dP_{\theta_0}^{(\varepsilon,N)}} \Bigg/ \frac{dP_{\hat{\theta}_{\varepsilon,N}}^{(\varepsilon,N)}}{dP_{\theta_0}^{(\varepsilon,N)}} \quad \text{a.s. } [P_{\theta_0}^{(\varepsilon,N)}]. \tag{4. 18}$$

It is easy to see that

$$\log \nu_{\varepsilon,N}(\tau) = -\frac{\tau^2}{2}\beta_{\varepsilon,N} \text{ a.s. } [P_{\theta_0}^{(\varepsilon,N)}] \tag{4. 19}$$

in view of (4.7). Suppose the following conditions hold:

(C1)′ $\beta_{\varepsilon,N} = \sum_{i=1}^N (\lambda_i + 1)\lambda_i^2 \int_0^T u_{i\varepsilon}^2(t)dt \to \beta_N > 0$ a.s. under $\{P_{\theta_0}^{(\varepsilon,N)}\}$ as $\varepsilon \to 0$;

(C2)′ the maximum likelihood estimator $\hat{\theta}_{\varepsilon,N}$ is strongly consistent as $\varepsilon \to 0$, that is,

$$\hat{\theta}_{\varepsilon,N} \to \theta_0 \text{ a.s. under } \{P_{\theta_0}^{(\varepsilon,N)}\} \text{ as } \varepsilon \to 0;$$

(C3)′ $K(\cdot)$ is a nonnegative function such that, for some $0 < \gamma < \beta_N$,

$$\int_{-\infty}^{\infty} K(\tau)e^{-\frac{1}{2}\tau^2(\beta_N - \gamma)}d\tau < \infty;$$

and

(C4)′ for every $\eta > 0$ and $\delta > 0$

$$e^{-\eta\varepsilon^{-2}} \int_{|\tau|>\delta} K(\tau\varepsilon^{-1})\lambda(\hat{\theta}_{\varepsilon,N}+\tau)d\tau \to 0 \text{ a.s.}$$

under $\{P_{\theta_0}^{(\varepsilon,N)}\}$ as $\varepsilon \to 0$.

Under the conditions $(C1)' - (C4)'$, the following theorems can be proved by arguments analogous to those given in the proofs of Theorems 3.1 and Theorem 3.2.

Theorem 4.1 : Suppose the conditions $(C1)' - (C4)'$ hold where $\lambda(\cdot)$ is a prior density which is continuous and positive in an open neighbourhood of θ_0, the true parameter. Then

(4. 20)
$$\lim_{\varepsilon\to 0} \int_{-\infty}^{\infty} K(\tau)\left|p^*(\tau|u_\varepsilon^{(N)}) - (\frac{\beta_N}{2\pi})^{1/2}e^{-\frac{1}{2}\beta_N\tau^2}\right| d\tau = 0 \text{ a.s.}$$

under $\{P_{\theta_0}^{(\varepsilon,N)}\}$.

Theorem 4.2 : Suppose the following conditions hold:

(D1)′ $\hat{\theta}_{\varepsilon,N} \to \theta_0$ a.s. under $P_{\theta_0}^{(\varepsilon,N)}$ as $\varepsilon \to 0$;

(D2)′ $\beta_{\varepsilon,N} \to \beta_N > 0$ a.s. under $\{P_{\theta_0}^{(\varepsilon,N)}\}$ as $\varepsilon \to 0$;

(D3)′ $\lambda(\cdot)$ is a prior density which is continuous and positive in an open neighbourhood of θ_0, the true parameter; and

(D4)′ $\int_{-\infty}^{\infty} |\theta|^m \lambda(\theta)d\theta < \infty$ for some integer $m \geq 0$.

Then

(4. 21)
$$\lim_{\varepsilon\to 0} \int_{-\infty}^{\infty} |\tau|^m \left|p^*(\tau|u_\varepsilon^{(N)}) - (\frac{\beta_N}{2\pi})^{1/2}e^{-\frac{1}{2}\beta_N\tau^2}\right| d\tau = 0 \text{ a.s..}$$

under $\{P_{\theta_0}^{(\varepsilon,N)}\}$.

Bayes Estimation (when N is fixed and $\varepsilon \to 0$)

We define a Bayes estimator $\tilde{\theta}_{\varepsilon,N}$ of θ based on the path $u_\varepsilon^{(N)}$ and the prior density $\lambda(\theta)$ as an estimator which minimizes

(4. 22)
$$B_{\varepsilon,N}(\phi) = \int_{\Theta} \tilde{L}(\theta,\phi)p(\theta|u_\varepsilon^{(N)})d\theta$$

where $\tilde{L}(\theta,\phi)$ is a loss function satisfying the properties (E1)-(E3) stated in Section 3. One can prove the following theorem as an application of Theorem 4.2.

Theorem 4.3 : Suppose the conditions $(D1)' - (D3)'$ of Theorem 4.2 hold. In addition suppose the loss function $\tilde{L}(\theta,\phi)$ satisfies the conditions (E1)-(E3) stated in Section 3. Then

$$\varepsilon^{-1}(\hat{\theta}_{\varepsilon,N} - \tilde{\theta}_{\varepsilon,N}) \to 0 \text{ a.s. under } \{P_{\theta_0}^{(\varepsilon,N)}\} \text{ as } \varepsilon \to 0 \tag{4. 23}$$

and

$$\begin{aligned}\lim_{\varepsilon\to 0} R_\varepsilon B_{\varepsilon,N}(\tilde{\theta}_{\varepsilon,N}) &= \lim_{\varepsilon\to 0} R_\varepsilon B_{\varepsilon,N}(\hat{\theta}_{\varepsilon,N}) \\ &= \left(\frac{\beta_N}{2\pi}\right)^{1/2} \int_{-\infty}^{\infty} K(\tau)e^{-\frac{1}{2}\beta_N\tau^2}d\tau \text{ a.s.}\end{aligned} \tag{4. 24}$$

under $\{P_{\theta_0}^{(\varepsilon,N)}\}$.

In particular, it follows that

$$\tilde{\theta}_{\varepsilon,N} \to \theta_0 \text{ a.s. under } \{P_{\theta_0}^{(\varepsilon,N)}\} \text{ as } \varepsilon \to 0 \tag{4. 25}$$

and

$$\varepsilon^{-1}(\tilde{\theta}_{\varepsilon,N} - \theta_0) \xrightarrow{\mathcal{L}} N(0,\beta_N^{-1}) \text{ as } \varepsilon \to 0 \tag{4. 26}$$

giving the asymptotic properties of the Bayes estimator $\tilde{\theta}_{\varepsilon,N}$.

Let us now consider the problem of Bayes estimation for the stochastic PDE given by (4.1) as $N \to \infty$ for any fixed $\varepsilon > 0$.

Bernstein - von Mises theorem and Bayes estimation (when ε is fixed and $N \to \infty$)

Let

$$Q_N^{(\varepsilon)}(\theta) = \left(\sum_{i=1}^{N} \lambda_i^2(\lambda_i+1)E_{\varepsilon,N}\int_0^T u_{i\varepsilon}^2(t)dt\right)^{-1/2} \tag{4. 27}$$

and suppose that

$(D0) Q_N^{(\varepsilon)}(\theta) \to 0$ as $N \to \infty$ for any fixed $\varepsilon > 0$. Let

$$\tau = Q_N^{(\varepsilon)}(\theta)^{-1}(\theta - \hat{\theta}_{\varepsilon,N}), \tag{4. 28}$$

(4. 29) $$\tilde{p}(\tau|u_\varepsilon^{(N)}) = Q_N^{(\varepsilon)}(\theta)p(\hat{\theta}_{\varepsilon,N} + Q_N^{(\varepsilon)}(\theta)\tau|u_\varepsilon^{(N)}),$$

and

(4. 30) $$\tilde{\nu}_{\varepsilon,N}(\tau) = \frac{dP^{(\varepsilon,N)}_{\hat{\theta}_{\varepsilon,N}+Q_N^{(\varepsilon)}(\theta)\tau}}{dP^{(\varepsilon,N)}_{\theta_0}} \Big/ \frac{dP^{(\varepsilon,N)}_{\hat{\theta}_{\varepsilon,N}}}{dP^{(\varepsilon,N)}_{\theta_0}} \text{ a.s. } [P^{(\varepsilon,N)}_{\theta_0}].$$

It can be checked that

(4. 31) $$\log \tilde{\nu}_{\varepsilon,N}(\tau) = -\frac{Q_N^{(\varepsilon)}(\theta_0)^2}{2\varepsilon^2}\tau^2\beta_{\varepsilon,N} \text{ a.s. } [P^{(\varepsilon,N)}_{\theta_0}]$$

in view of (4.7). Note that ε is a fixed positive constant in the present discussison. Suppose the following conditions hold:

(C1)″ $\frac{Q_N^{(\varepsilon)}(\theta_0)^2\beta_{\varepsilon,N}}{\varepsilon^2} \to 1$ a.s. under $\{P^{(\varepsilon,N)}_{\theta_0}\}$ as $N \to \infty$;

(C2)″ the maximum likelihood estimator $\hat{\theta}_{\varepsilon,N}$ is strongly consistent as $N \to \infty$, that is

$$\hat{\theta}_{\varepsilon,N} \to \theta_0 \text{ a.s. under } \{P^{(\varepsilon,N)}_{\theta_0}\} \text{ as } N \to \infty;$$

(C3)″ the function $K(\cdot)$ is a nonnegative function such that for some $0 < \gamma < 1$,

$$\int_{-\infty}^{\infty} K(\tau)e^{-\frac{1}{2}\tau^2(1-\gamma)}d\tau < \infty;$$

and

(C4)″ for every $\eta > 0$ and $\delta > 0$

$$e^{-\eta Q_N^{(\varepsilon)}(\theta_0)^{-2}} \int_{-\infty}^{\infty} K(\tau Q_N^{(\varepsilon)^{-1}}(\theta_0))\lambda(\hat{\theta}_{\varepsilon,N} + \tau)d\tau \to 0 \text{ a.s. under } \{P^{(\varepsilon,N)}_{\theta_0}\} \text{ as } N \to \infty.$$

The following analogues of Theorem 3.1 to 3.3 hold under the conditions (C1)″-(C4)″. We omit the details.

Theorem 4.4 : Suppose the conditions (C1)″ - (C4)″ hold where $\lambda(\cdot)$ is a prior density which is continuous and positive in an open neighbourhood of θ_0, the true parameter. Then, for any fixed $\varepsilon > 0$,

(4. 32) $$\lim_{N\to\infty} \int_{-\infty}^{\infty} K(\tau)|\tilde{p}(\tau|u_\varepsilon^{(N)}) - (\frac{1}{2\pi})^{1/2}e^{-\frac{1}{2}\tau^2}|d\tau = 0 \text{ a.s. } [P^{(\varepsilon)}_{\theta_0}].$$

Theorem 4.5 : Suppose the following conditions hold, for a fixed $\varepsilon > 0$, in addition to the condition (D0) stated above:

(D1)″ $\hat{\theta}_{\varepsilon,N} \to \theta_0$ a.s. $[P_{\theta_0}^{(\varepsilon)}]$ as $N \to \infty$;

(D2)″ $\frac{\beta_{\varepsilon,N} Q_N^{(\varepsilon)}(\theta_0)^2}{\varepsilon^2} \to 1$ a.s. $[P_{\theta_0}^{(\varepsilon)}]$ as $N \to \infty$;

(D3)″ $\lambda(\cdot)$ is a prior density which is continuous and positive in an open neighbourhood of θ_0, the true parameter ; and

(D4)″ $\int_{-\infty}^{\infty} |\theta|^m \lambda(\theta) d\theta < \infty$ for some integer $m \geq 0$.

Then

(4. 33) $$\lim_{n\to\infty} \int_{-\infty}^{\infty} |\tau|^m |\tilde{p}(\tau|u_\varepsilon^{(N)}) - (\frac{1}{2\pi})^{1/2} e^{-\frac{1}{2}\tau^2}| d\tau = 0 \text{ a.s. under } \{P_{\theta_0}^{(\varepsilon,N)}\}.$$

Theorem 4.6 : Suppose the conditions (D1)″ - (D3)″ of Theorem 4.5 hold. In addition suppose the loss function $\tilde{L}(\theta,\phi)$ satisfies the conditions (E1)-(E3) stated in Section 3. Then, for any fixed $\varepsilon > 0$,

(4. 34) $$Q_N^{(\varepsilon)^{-1}}(\theta_0)(\hat{\theta}_{\varepsilon,N} - \tilde{\theta}_{\varepsilon,N}) \to 0 \text{ a.s. } [P_{\theta_0}^{(\varepsilon)}] \text{ as } N \to \infty,$$

and

(4. 35) $$\begin{aligned} \lim_{N\to\infty} R_{Q_N^{(\varepsilon)}(\theta_0)} B_{\varepsilon,N}(\tilde{\theta}_{\varepsilon,N}) &= \lim_{N\to\infty} R_{Q_N^{(\varepsilon)}(\theta_0)} B_{\varepsilon,N}(\hat{\theta}_{\varepsilon,N}) \\ &= (\frac{1}{2\pi})^{1/2} \int_{-\infty}^{\infty} K(\tau) e^{-\frac{1}{2}\tau^2} d\tau \text{ a.s. } [P_{\theta_0}^{(\varepsilon)}]. \end{aligned}$$

As a consequence of Theorem 4.6 and the relations (4.11) and (4.12) it follows that, for any fixed $\varepsilon > 0$,

(4. 36) $$\tilde{\theta}_{\varepsilon,N} \to \theta_0 \text{ a.s. under } \{P_{\theta_0}^{(\varepsilon,N)}\} \text{ as } N \to \infty$$

and

(4. 37) $$Q_N^{(\varepsilon)^{-1}}(\theta_0)(\tilde{\theta}_{\varepsilon,N} - \theta_0) \xrightarrow{\mathcal{L}} N(0,1) \text{ as } N \to \infty.$$

Minimum Distance Estimation

An alternate approach for the estimation of the parameter θ is by the minimum distance method. Observe that the parameter θ can be estimated from the equation (4.3). We now again apply the minimum distance approach adapted by Kutoyants and Pilibossian [13] as before to estimate the parameter θ satisfying the equation (3.3). We define the minimum L_1-norm estimate $\tilde{\theta}_{i\varepsilon T}$ by the relation

$$\tilde{\theta}_{i\varepsilon T} = -\lambda_i^{-1} \arg\inf_{\theta\in\Theta} \int_0^T |u_{i\varepsilon}(t) - u_i(t,\theta)| \ dt$$

where $u_i(t,\theta)$ is the solution of the ordinary differential equation

$$\frac{du_i(t)}{dt} = -\theta\lambda_i u_i(t), u_i(0,\theta) = v_i.$$

It is easy to see that

$$u_i(t,\theta) = v_i e^{-\theta\lambda_i t}.$$

Let

$$g_i(\delta) = \inf_{|\theta-\theta_0|>\delta\lambda_i^{-1}} \int_0^T |u_i(t,\theta) - u_i(t,\theta_0)|dt.$$

The following theorem is a consequence of Theorem 1 of Kutoyants and Pilibossian [13].

Theorem 4.7 : For any $\delta > 0$,

$$P_{\theta_0}^{(\varepsilon)}(|\tilde{\theta}_{i\varepsilon T} - \theta_0| \geq \delta\lambda_i^{-1}) \leq 2\exp\{-k_i(\lambda_i+1)g_i^2(\delta)\varepsilon^{-2}\}$$

where

$$k_i = \exp\{-2|\theta_0|T\lambda_i\}/(2T)^3.$$

Let

$$Y_i(t) = e^{-\theta_0\lambda_i t}\int_0^t e^{\theta_0\lambda_i s}\ dW_i(s).$$

Note that the process $Y_i(t)$ is a gaussian process. Define

$$\eta_{iT} = \arg\inf_u \int_0^T |Y_i(t) - utv_i e^{-\theta_0\lambda_i t}|\ dt.$$

The following theorem is again a consequence of Theorems 2 and 3 of Kutoyants and Pilibossian [13].

Theorem 4.8 : For any fixed $T > 0$,

$$(\frac{\varepsilon}{\sqrt{\lambda_i+1}})^{-1}(\tilde{\theta}_{i\varepsilon T} - \theta_0)\lambda_i \xrightarrow{p} -\eta_{iT} \text{ as } \varepsilon \to 0$$

where θ_0 is the true parameter. Further more if $\theta_0 < 0$, then

$$\eta_{iT} T v_i \sqrt{-2\theta_0\lambda_i} \xrightarrow{\mathcal{L}} N(0,1) \text{ as } T \to \infty.$$

Applying the Lemma 3.6, we get the following result.

Theorem 4.9 : Under the probability measure P_{θ_0}, if $\theta_0 < 0$, then

$$(4.38)\quad \left(\frac{\varepsilon}{\sqrt{\lambda_i+1}}\right)^{-1}\lambda_i(\tilde{\theta}_{i\varepsilon T}-\theta_0)Tv_i\sqrt{-2\theta_0\lambda_i} \xrightarrow{\mathcal{L}} N(0,1) \text{ as } \varepsilon\to 0 \text{ and } T\to\infty.$$

In view of Theorem 4.9, the variance of the limiting normal distribution of estimator $\tilde{\theta}_{i\varepsilon T}$ is proportional to $[-\theta_0 v_i^2\lambda_i^3(\lambda_i+1)]^{-1}$. Note that the estimators $\tilde{\theta}_{i\varepsilon T}, i\geq 1$ are independent estimators of the parameter θ since the processes $\{W_i(t), t\geq 0\}, i\geq 1$ are independent Wiener processes. We will now construct an optimum estimator out of the estimators $\tilde{\theta}_{i\varepsilon T}, 1\leq i\leq N$ for any $N\geq 1$.

Let $\tilde{\theta}_{\varepsilon T}=\sum_{i=1}^N \alpha_i\tilde{\theta}_{i\varepsilon T}$ where $\alpha_i, 1\leq i\leq N$ is a nonrandom sequence of coefficients to be chosen. Note that

$$\tilde{\theta}_{\varepsilon T}\xrightarrow{p}[\sum_{i=1}^N\alpha_i]\theta_0 \text{ as } \varepsilon\to 0 \text{ and } T\to\infty$$

by Theorem 4.9 and hence $\tilde{\theta}_{\varepsilon T}$ is a consistent estimator for θ_0 as $\varepsilon\to 0$ and as $T\to\infty$ provided $\sum_{i=1}^N\alpha_i=1$. Further more

$$\varepsilon^{-1}T(\tilde{\theta}_{\varepsilon T}-\theta_0)\xrightarrow{\mathcal{L}} N(0,\sum_{i=1}^N\alpha_i^2[-2\theta_0 v_i^2\lambda_i^3(\lambda_i+1)]^{-1}) \text{ as } \varepsilon\to 0 \text{ and } T\to\infty.$$

This follows again by Theorem 4.9 and the independence of the estimators $\{\tilde{\theta}_{i\varepsilon T}, 1\leq i\leq N\}$. We now obtain the optimum combination of the coefficients $\{\alpha_i, 1\leq i\leq N\}$ by minimizing the asymptotic variance

$$\sum_{i=1}^N\alpha_i^2[-2\theta_0 v_i^2\lambda_i^3(\lambda_i+1)]^{-1}$$

subject to the condition $\sum_{i=1}^N\alpha_i=1$. It is easy to see that α_i is proportional to $[-\theta_0\lambda_i^3(\lambda_i+1)]$ and the optimal choice of $\{\alpha_i, 1\leq i\leq N\}$ leads to the estimator

$$(4.39)\quad \theta^*_{\varepsilon T}=\frac{\sum_{i=1}^N v_i^2\lambda_i^3(\lambda_i+1)\tilde{\theta}_{i\varepsilon T}}{\sum_{i=1}^N v_i^2\lambda_i^3(\lambda_i+1)}.$$

It is easy to see that

$$\theta^*_{\varepsilon T}\xrightarrow{p}\theta_0 \text{ as } \varepsilon\to 0 \text{ and } T\to\infty$$

and

$$\varepsilon^{-1}T(\theta^*_{\varepsilon T}-\theta_0)\xrightarrow{\mathcal{L}} N(0,[-2\sum_{i=1}^N\theta_0 v_i^2\lambda_i^3(\lambda_i+1)]^{-1}) \text{ as } \varepsilon\to 0 \text{ and } T\to\infty$$

again due to the independence of the estimators $\tilde{\theta}_{i\varepsilon T}, 1\leq i\leq N$ and we have the following result.

Theorem 4.10: Under the probability measure P_{θ_0},

$$\theta^*_{\varepsilon T} \xrightarrow{p} \theta_0 \text{ as } \varepsilon \to 0 \text{ and } T \to \infty$$

and, if $\theta_0 < 0$, then

$$\varepsilon^{-1}T(\theta^*_{\varepsilon T} - \theta_0) \xrightarrow{\mathcal{L}} N(0, [-2\sum_{i=1}^{N} \theta_0 v_i^2 \lambda_i^3(\lambda_i + 1)]^{-1}) \text{ as } \varepsilon \to 0 \text{ and } T \to \infty$$

for any fixed $N \geq 1$.

5 Parametric Estimation for Parabolic SPDE (Continuous sampling)

Let $(\Omega, \mathcal{F}, P)$ be a probability space and consider a stochastic partial differential equation (SPDE) of the form

$$du^\theta(t,x) = A^\theta u^\theta(t,x)dt + dW(t,x), 0 \leq t \leq T, x \in G \tag{5. 1}$$

where $A^\theta = \theta A_1 + A_0$, A_1 and A_0 being partial differential operators, $\theta \in \Theta \subset R$ and $W(t,x)$ is a cylindrical Brownian motion in $L_2(G), G$ being a bounded domain in R^d with the boundary ∂G as a C^∞-manifold of dimension $(d-1)$ and locally G is totally on one side of ∂G. For the definition of cylindrical Brownian motion, see, Kallianpur and Xiong [11], p.93.

The order $Ord(A)$ of a partial differential operator A is defined to be the order of the highest partial derivative in A. Let m_0 and m_1 be the orders of the operators A_0 and A_1 respectively. We assume that the operators A_0 and A_1 commute, m_1 is even and

$$\text{(C0)} \qquad m_1 \geq \frac{1}{2}(Ord(A^\theta) - d)$$

in the following discussion.

Suppose the solution $u^\theta(t,x)$ of (5.1) has to satisfy the boundary conditions

$$u^\theta(0,x) = u_0(x) \tag{5. 2}$$

and

$$D^\gamma u^\theta(t,x)|_{\partial G} = 0 \tag{5. 3}$$

for all multiindices γ such that $|\gamma| = m - 1$ where $2m = \max(m_1, m_0)$. Here

$$D^{\gamma} f(\boldsymbol{x}) = \frac{\partial^{|\gamma|}}{\partial x_1^{\gamma_1} \cdots \partial x_d^{\gamma_d}} f(\boldsymbol{x}) \tag{5. 4}$$

with $|\gamma| = \gamma_1 + \cdots + \gamma_d$. Suppose that

$$A_i(\boldsymbol{x})u = - \sum_{|\alpha|,|\beta| \leq m_i} (-1)^{|\alpha|} D^{\alpha}(a_i^{\alpha\beta}(\boldsymbol{x}) D^{\beta} u) \tag{5. 5}$$

where

$$a_i^{\alpha\beta}(\boldsymbol{x}) \in C^{\infty}(\bar{G}). \tag{5. 6}$$

Let

$$a^{\alpha\beta}(\theta, x) = \theta a_1^{\alpha\beta}(x) + a_0^{\alpha\beta}(x). \tag{5. 7}$$

Suppose θ_0 is the true parameter.

We follow the notation introduced in Huebner and Rozovskii [8]. Assume that the following conditions hold.

(H1) The operators A_0 and A_1 satisfy the condition

$$\int_G A_i u v dx = \int_G u A_i v dx, u, v \in C_0^{\infty}(G), i = 0, 1.$$

(H2) There is a compact neighbourghood Θ of θ_0 so that $\{A_\theta, \theta \in \Theta\}$ is a family of uniformly strongly elliptic operators of order $2m = max(m_1, m_0)$.

For $s > 0$, denote the closure of $C_0^{\infty}(G)$ in the Sobolev space $W^{s,2}(G)$ by $W_0^{s,2}$.

The operator A^{θ} with boundary conditions defined by (5.2) and (5.3) can be extended to a closed self-adjoint operator $\mathcal{L}_\theta$ on $L_2(G)$ (Shimakura [34]). In view of the condition (H2) , the operator $\mathcal{L}_\theta$ is lower semibounded, that is there exists a constant $k(\theta)$ such that $-\mathcal{L}_\theta + k(\theta)I > 0$ and the resolvent $(k(\theta)I - \mathcal{L}_\theta)^{-1}$ is compact. Let $\Lambda_\theta = (k(\theta)I - \mathcal{L}_\theta)^{\frac{1}{2m}}$. Let $h_i(\theta)$ be an orthonormal system of eigen functions of Λ_θ. We assume that the following condition holds.

(H3) There exists a complete orthonormal system $\{h_i, i \geq 1\}$ *independent* of θ such that

$$\Lambda_\theta h_i = \lambda_i(\theta) h_i, \theta \in \Theta.$$

The elements of the basis $\{h_i, i \geq 1\}$ are also eigen functions for the operator $\mathcal{L}_\theta$, that is

$$\mathcal{L}_\theta h_i = \mu_i^{\theta} h_i$$

where

$$\mu_i^\theta = -\lambda_i^{2m}(\theta) + k(\theta).$$

For $s \geq 0$, define H_θ^s to be the set of all $u \in L_2(G)$ such that

$$||u||_{s,\theta} = (\sum_{j=1}^{\infty} \lambda_j^{2s}(\theta)|(u, h_j)_{L_2(G)}|^2)^{1/2} < \infty.$$

For $s < 0$, H_θ^s is defined to be the closure of $L_2(G)$ in the norm $||u||_{s,\theta}$ given above. Then H_θ^s is a Hilbert space with respect to the inner product $(.,.)_{s,\theta}$ associated with the norm $||.||_{s,\theta}$ and the functions $h_{i,\theta}^s = \lambda_i^{-s}(\theta)h_i, i \geq 1$ form an orthonormal basis in H_θ^s. Condition (H2) imples that for every s ,the spaces H_θ^s are equivalent for all θ. We identify the spaces H_θ^s (denoted by H^s) and the norms $||.||_{s,\theta}$ for different $\theta \in \Theta$.

In addition to the conditions (H1)-(H3), we assume that
(H4)$u_0 \in H^{-\alpha}$ where $\alpha > \frac{d}{2}$. Note that $u_0 \in L_2(G)$,
(H5) the operator A_1 is uniformly strongly elliptic of even order m_1 and has the same system of eigen functions $\{h_i, i \geq 1\}$ as $\mathcal{L}_\theta$.

The conditons (H1)-(H5) described above are the same as those in Huebner and Rozovskii (1995).

Note that $u_0 \in H^{-\alpha}$. For $\theta \in \Theta$, define

(5. 8) $$u_{0i}^\theta = (u_o, h_{i\theta}^{-\alpha})_{-\alpha}.$$

Then the random field

(5. 9) $$u^\theta(t,x) = \sum_{i=1}^{\infty} u_i^\theta(t) h_{i\theta}^{-\alpha}(x)$$

is the solution of (5.1) subject to the boundary conditions (5.2) and (5.3) where $u_i^\theta(t)$ is the unique solution of the stochastic differential equation

(5. 10) $$du_i^\theta(t) = \mu_i^\theta u_i^\theta(t)dt + \lambda_i^{-\alpha}(\theta)dW_i(t), 0 \leq t \leq T,$$

(5. 11) $$u_i^{(\theta)}(0) = u_{0i}^\theta.$$

Let π^N be the orthogonal projection operator of $H^{-\alpha}$ onto the subspace spanned by $\{h_{i\theta}^{-\alpha}, 1 \leq i \leq N\}$. Let

(5. 12) $$\begin{aligned} u^{N,\theta}(t,x) &= \pi^N u^\theta(t,x) \\ &= \sum_{i=1}^{N} u_i^\theta(t) h_{i\theta}^{-\alpha}(x) \end{aligned}$$

where $u_i^\theta(t)$ is the solution of (5.10) subject to (5.11). Note that

$$du^{N,\theta}(t,x) = A^\theta u^{N,\theta}(t,x)dt + dW^N(t,x), 0 \le t \le T, x \in G \tag{5. 13}$$

with

$$u^{N,\theta}(0,x) = \pi^N u_0(x) \tag{5. 14}$$

and

$$W^N(t,x) = \sum_{i=1}^{N} \lambda_i^{-\alpha} W_i(t) h_{i\theta}^{-\alpha}(x). \tag{5. 15}$$

Here $\{W_i(t), t \ge 0\}, i \ge 1$ are independent standard Wiener processes.

Let P_θ^N be the probability measure generated by $u^{N,\theta}$ on $C([0,T]; R^N)$. Let $h_i^{-\alpha}$ denote $h_{i,\theta_0}^{-\alpha}$, u^N denote u^{N,θ_0} and u denote u^{θ_0} when θ_0 is the true parameter. It is known that, for any $\theta \in \Theta$, the measures P_θ^N and $P_{\theta_0}^N$ are absolutely continuous with respect to each other and

$$\begin{aligned} \log \frac{dP_\theta^N}{dP_{\theta_0}^N}(u^N) &= (\theta - \theta_0) \int_0^T (A_1 u^N(s), du^N(s))_0 - \frac{(\theta^2 - \theta_0^2)}{2} \int_0^T \|A_1 u^N(s)\|_0^2 ds \\ &\quad -(\theta - \theta_0) \int_0^T (A_1 u^N(s), A_0 u^N(s))_0 ds. \end{aligned} \tag{5. 16}$$

Maximum Likelihood Estimation

It is easy check that (cf. Huebner and Rozovskii [8])

$$\hat{\theta}_N - \theta_0 = \frac{\int_0^T (A_1 u^N(s), dW^N(s))_0}{\int_0^T \|A_1 u^N(s)\|_0^2 ds} \tag{5. 17}$$

where $\hat{\theta}_N$ is the maximum likelihood estimator of θ_0. Huebner and Rozovskii [8] studied the asymptotic properties of this estimator under the conditions (H1)-(H5). Further more the Fisher information is given by

$$I_N = E \int_0^T \|A_1 u^{N,\theta_0}(s)\|_0^2 ds. \tag{5. 18}$$

Note that $I_N \to \infty$ as $N \to \infty$ from the Lemma 2.1 of Huebner and Rozovskii [8].

Bernstein-Von Mises Theorem

Suppose that Λ is a prior probability measure on $(\Theta, \mathcal{B})$ where $\mathcal{B}$ is the σ-algebra of Borel subsets of set $\Theta \subset R$. We assume that the true parameter $\theta_0 \in \Theta^0$, the interior of Θ. Further suppose that Λ has the density $\lambda(\cdot)$ with respect to the Lebesgue measure and the density $\lambda(\cdot)$ is continuous and positive in an open neighbourhood of θ_0, the true parameter.

Let

(5. 19) $$\tau = I_N^{1/2}(\theta - \hat{\theta}_N)$$

and

(5. 20) $$p^*(\tau|u^N) = I_N^{-1/2} p(\hat{\theta}_N + \tau I_N^{-1/2}|u^N)$$

where $p(\theta|u^N)$ is the posterior density of θ given u^N. Note that

(5. 21) $$p(\theta|u^N) = \frac{\frac{dP_\theta^N}{dP_{\theta_0}^N}(u^N)\lambda(\theta)}{\int\limits_\Theta \frac{dP_{\theta}^N}{dP_{\theta_0}^N}(u^N)\lambda(\theta)d\theta}$$

and let $p^*(\tau|u^N)$ denote the posterior density of $I_N^{1/2}(\theta - \hat{\theta}_N)$. Let

(5. 22) $$\begin{aligned} \nu_N(\tau) &= \frac{dP^N_{\hat{\theta}_N + \tau I_N^{-1/2}}}{dP^N_{\theta_0}} \Big/ \frac{dP^N_{\hat{\theta}_N}}{dP^N_{\theta_0}} \\ &= \frac{dP^N_{\hat{\theta}_N + \tau I_N^{-1/2}}}{dP^N_{\hat{\theta}_N}} \quad \text{a.s.} \end{aligned}$$

In view of (5.16), it follows that

(5. 23) $$\log \nu_N(\tau) = -\frac{1}{2}\tau^2 I_N^{-1} \int_0^T ||A_1 u^N(s)||_0^2 ds$$

since

(5. 24) $$\hat{\theta}_N = \frac{\int_0^T (A_1 u^N(s), du^N(s) - A_0(s)u^N(s)ds)_0}{\int_0^T ||A_1 u^N(s)||_0^2 ds}.$$

Let

(5. 25) $$C_N = \int_{-\infty}^{\infty} \nu_N(\tau)\lambda(\hat{\theta}_N + \tau I_N^{-1/2})d\tau.$$

It can be checked that

(5. 26) $$p^*(\tau|u^N) = C_N^{-1}\nu_N(\tau)\lambda(\hat{\theta}_N + \tau I_N^{-1/2}).$$

Note that

(C1) $\beta_N = I_N^{-1} \int_0^T \|A_1 u^N(s)\|_0^2 ds \to 1$ a.s. $[P_{\theta_0}]$ as $N \to \infty$

from the Lemma 2.2 of Huebner and Rozovskii [8]. Then the following relations hold:

(i) $\lim_{N\to\infty} \nu_N(\tau) = \exp(-\frac{1}{2}\tau^2)$ a.s. $[P_{\theta_0}]$,

(ii) for any $0 < \gamma < 1$,

$$\log \nu_N(\tau) \leq -\frac{1}{2}\tau^2(1-\gamma)$$

for every τ for sufficiently large N, and

(iii) for every $\delta > 0$, there exists $\gamma' > 0$ such that

$$\sup_{|\tau|>\delta I_N^{1/2}} \nu_N(\tau) \leq \exp\{-\frac{1}{4}\gamma' I_N^{-1}\}$$

as $N \to \infty$.

Further more

(C2) the maximum likelihood estimator $\hat{\theta}_N$ is strongly consistent, that is

$$\hat{\theta}_N \to \theta_0 \text{ a.s. } [P_{\theta_0}] \text{ as } N \to \infty$$

from the Lemmas 2.1 and 2.2 in Huebner and Rozovskii [8]. Suppose that

(C3) $K(\cdot)$ is a nonnegative function such that, for some $0 < \gamma < 1$,

$$\int_{-\infty}^{\infty} K(\tau) e^{-\frac{1}{2}\tau^2(1-\gamma)} d\tau < \infty.$$

(C4) For every $\eta > 0$ and $\delta > 0$,

$$e^{-\eta I_N^{-1}} \int_{|\tau|>\delta} K(\tau I_N^{-1/2}) \lambda(\hat{\theta}_N + \tau) d\tau \to 0 \text{ a.s } [P_{\theta_0}]$$

as $N \to \infty$.

We now have the following main theorem which is an analogue of the Bernstein-von Mises theorem (cf.Prakasa Rao [18,19]) for diffusion processes and diffusion fields. A special case of this result for some classes of SPDE's was proved in Prakasa Rao [25].

Theorem 5.1: Suppose the conditions (C3) and (C4) hold in addition to the conditions (H1)-(H5) stated earlier where $\lambda(\cdot)$ is a prior density which is continuous and positive in an open neighbourhood of θ_0, the true parameter. Then

(5. 27)
$$\lim_{N\to\infty} \int_{-\infty}^{\infty} K(\tau)|p^*(\tau|u^N) - \left(\frac{1}{2\pi}\right)^{1/2} e^{-\frac{1}{2}\tau^2}|d\tau = 0 \text{ a.s. } [P_{\theta_0}].$$

As a consequence of Theorem 5.1, it is easy to get the following result.

Theorem 5.2: Suppose the conditions (H1)-(H5) hold. In addition suppose that:

(D1) $\lambda(\cdot)$ is a prior density which is continuous and positive in an open neighbourhood of θ_0, the true parameter; and

(D2) $\int_{-\infty}^{\infty} |\theta|^m \lambda(\theta) d\theta < \infty$ for some integer $m \geq 0$.

Then

(5. 28)
$$\lim_{N\to\infty} \int_{-\infty}^{\infty} |\tau|^m |p^*(\tau|u^N) - \left(\frac{1}{2\pi}\right)^{\frac{1}{2}} e^{-\frac{1}{2}\tau^2}|d\tau = 0 \text{ a.s. } [P_{\theta_0}].$$

Remarks: It is obvious that the condition (D2) holds for $m = 0$. Suppose the condition (D1) holds. Then it follows that

(5. 29)
$$\lim_{N\to\infty} \int_{-\infty}^{\infty} |p^*(\tau|u^N) - \left(\frac{1}{2\pi}\right)^{1/2} e^{-\frac{1}{2}\tau^2}|d\tau = 0 \text{ a.s. } [P_{\theta_0}].$$

This is the analogue of the Bernstein-von Mises theorem in the classical statistical inference. As a particular case of Theorem 5.2, we obtain that

(5. 30)
$$E_{\theta_0}[I_N^{1/2}(\hat{\theta}_N - \theta_0)]^m \to E[Z]^m \text{ as } N \to \infty$$

where Z is $N(0,1)$.

For proofs of Theorems 5.1 and 5.2, see Prakasa Rao [21].

Bayes estimation

We define an estimator $\tilde{\theta}_N$ for θ to be a Bayes estimator based on the path u^N corresponding to the loss function $\tilde{L}(\theta,\varphi)$ and the prior density $\lambda(\theta)$ if it is an estimator which minimizes the function

$$B_N(\varphi) = \int \tilde{L}(\theta,\varphi)p(\theta|u^N)d\theta, \varphi \in \Theta$$

where $\tilde{L}(\theta, \varphi)$ is defined on $\Theta \times \Theta$. Suppose there exist a Bayes estimator $\tilde{\theta}_N$. Further suppose that the loss function $\tilde{L}(\theta, \varphi)$ satisfies the following conditions:

(E1) $\tilde{L}(\theta, \varphi) = L(|\theta - \varphi|) \geq 0$;

(E2) $L(t)$ is nondecreasing for $t \geq 0$;

(E3) there exists nonnegative functions $R_N, K(\tau)$ and $G(\tau)$ such that

(a) $R_N L(\tau I_N^{-1/2}) \leq G(\tau)$ for all $N \geq 1$;

(b) $R_N L(\tau I_N^{-1/2}) \to K(\tau)$ as $N \to \infty$ uniformly on bounded intervals of τ;

(c) the function

$$\int_{-\infty}^{\infty} K(\tau + m) e^{-\frac{1}{2}\tau^2} d\tau$$

achieves its minimum at $m = 0$, and

(d) $G(\tau)$ satisfies the conditions similar to (C3) and (C4).

The following result can be proved by arguments similar to those given in Borwanker et al. [2]. We omit the proof.

Theorem 5.3: Suppose the conditions (D1)-(D2) of Theorem 5.2 hold in addition to (H1)-(H5) stated earlier. In addition , suppose that the loss function $\tilde{L}(\theta, \varphi)$ satisfies the conditions (E1) - (E3) stated above. Then

(5. 31)
$$I_N^{1/2}(\hat{\theta}_N - \tilde{\theta}_N) \to 0 \text{ a.s. } [P_{\theta_0}] \text{ as } N \to \infty$$

and

(5. 32)
$$\begin{aligned} \lim_{N \to \infty} R_N B_N(\tilde{\theta}_N) &= \lim_{N \to \infty} R_N B_N(\hat{\theta}_N) \\ &= \left(\frac{1}{2\pi}\right)^{1/2} \int_{-\infty}^{\infty} K(\tau) e^{-\frac{1}{2}\tau^2} d\tau. \end{aligned}$$

Huebner and Rozovskii [8] proved that

(5. 33)
$$\hat{\theta}_N \to \theta_0 \text{ a.s. } [P_{\theta_0}] \text{ as } N \to \infty$$

and

(5. 34)
$$I_N^{1/2}(\hat{\theta}_N - \theta_0) \xrightarrow{\mathcal{L}} N(0,1) \text{ as } N \to \infty$$

under the conditions (H1)-(H5). As a consequence of Theorem 5.3, it follows that

(5. 35) $$\tilde{\theta}_N \to \theta_0 \text{ a.s } [P_{\theta_0}] \text{ as } N \to \infty$$

and

(5. 36) $$I_N^{1/2}(\tilde{\theta}_N - \theta_0) \xrightarrow{\mathcal{L}} N(0,1) \text{ as } N \to \infty.$$

In other words the Bayes estimator $\tilde{\theta}_N$ of the paramaeter θ in the parabolic SPDE given by (5.1) is strongly consistent, asymptotically normal and asymptotically efficient as $N \to \infty$ under the conditions (H1)-(H5) of Huebner and Rozovskii [8] and the conditions stated in Theorem 5.3.

Remarks: A general approach for the study of asymptotic properties of maximum likelihood estimators and Bayes estimators is by proving the local asymptotic normality of the loglikelihood ratio process as was done in Prakasa Rao [17], Ibragimov and Khasminskii [9] in the classical i.i.d. cases and by Huebner and Rozovskii [8] for some classes of SPDE. Our approach for Bayes estimation, via the comparison of the rates of convergence of the difference between the maximum likelihood estimator and the Bayes estimator, is a consequence of the the Bernstein - Von Mises type theorem.

Minimum Distance Estimation

We now apply the minimum distance approach for the estimation of the parameter θ in the SPDE (5.1). Observe that the parameter θ can be estimated from the equation (5.10). We now again apply the minimum distance approach adapted by Kutoyants and Pilibossian [13] as before to estimate the parameter θ satisfying the equation (5.10). We define the minimum L_1-norm estimate $\tilde{\theta}_{i\varepsilon T}$ by the relation

$$\mu_i(\tilde{\theta}_{i\varepsilon T}) = \arg\inf_{\theta\in\Theta} \int_0^T |u_{i\varepsilon}^{\theta}(t) - u_i^*(t,\theta)| \; dt$$

where $u_i^*(t,\theta)$ is the solution of the ordinary differential equation

$$\frac{du_i^*(t)}{dt} = \mu_i(\theta)u_i^*(t), u_i^*(0,\theta) = v_i.$$

It is easy to see that

$$u_i^*(t,\theta) = v_i e^{\mu_i(\theta)t}.$$

Let

$$h_i(\delta) = \inf_{|\mu_i(\theta)-\mu_i(\theta_0)|>\delta} \int_0^T |u_i^*(t,\theta) - u_i^*(t,\theta_0)|dt.$$

The following theorem is a consequence of Theorem 1 of Kutoyants and Pilibossian [13].

Theorem 5.4 : For any $\delta > 0$,

$$P_{\theta_0}(|\mu_i(\tilde{\theta}_{i\varepsilon T} - \mu_i(\theta_0)| \geq \delta) \leq 2\exp\{-q_i\lambda_i^{2\alpha}(\theta_0)h_i^2(\delta)\varepsilon^{-2}\}$$

where

$$q_i = \exp\{-2|\mu_i(\theta_0)|T\}/(2T)^3.$$

Let

$$J_i(t) = e^{\mu_i(\theta_0)t}\int_0^t e^{-\mu_i(\theta_0)s}\, dW_i(s).$$

Note that the process $J_i(t)$ is a gaussian process. Define

$$\gamma_{iT} = \arg\inf_u \int_0^T |J_i(t) - utv_ie^{\mu_i(\theta_0)t}|\, dt.$$

The following theorem is again a consequence of Theorems 2 and 3 of Kutoyants and Pilibossian [13].

Theorem 5.5 : For any fixed $T > 0$,

$$(\varepsilon\lambda_i^{-\alpha})^{-1}(\mu_i(\tilde{\theta}_{i\varepsilon T}) - \mu_i(\theta_0)) \xrightarrow{p} J_{iT} \text{ as } \varepsilon \to 0$$

when θ_0 is the true parameter. Further more if $\mu_i(\theta_0) > 0$, then

$$J_{iT}Tv_i\sqrt{2\mu_i(\theta_0)} \xrightarrow{\mathcal{L}} N(0,1) \text{ as } T \to \infty.$$

Applying the Lemma 3.6, we get the following result.

Theorem 5.6 : Under the probability measure P_{θ_0}, if $\mu_i(\theta_0) > 0$, then

(5. 37) $(\varepsilon\lambda_i^{-\alpha})^{-1}v_iT(\mu_i(\tilde{\theta}_{i\varepsilon T}) - \mu_i(\theta_0))\sqrt{2\mu_i(\theta_0)} \xrightarrow{\mathcal{L}} N(0,1)$ as $\varepsilon \to 0$ and $T \to \infty$.

In addition to the conditions (H1)-(H5), suppose that

(H6) the functions $\mu_i(\theta)$ are differentiable with respect to θ with nonzero derivatives.

Let $\mu_i'(\theta)$ denote the derivative of the function $\mu_i(\theta)$ with respect to θ. Applying the delta method, we obtain the following result.

Theorem 5.7 : Under the probability measure P_{θ_0}, if $\mu_i(\theta_0) > 0$, then

(5. 38) $(\varepsilon\lambda_i^{-\alpha})^{-1}v_iT(\tilde{\theta}_{i\varepsilon T} - \theta_0))\sqrt{2\mu_i(\theta_0)} \xrightarrow{\mathcal{L}} N(0,[\mu_i'(\theta_0)]^{-2})$ as $\varepsilon \to 0$ and $T \to \infty$.

In view of Theorem 5.7, the variance of the limiting normal distribution of estimator $\tilde{\theta}_{i\varepsilon T}$ is proportional to

$$\{v_i^2\mu_i(\theta_0)\lambda_i^{2\alpha}(\theta_0)[\mu_i'(\theta_0)]^2\}^{-1}.$$

Note that the estimators $\tilde{\theta}_{i\varepsilon T}, i \geq 1$ are independent estimators of the parameter θ_0 since the processes $\{W_i(t), t \geq 0\}, i \geq 1$ are independent Wiener processes. We will now construct an optimum estimator out of the estimators $\tilde{\theta}_{i\varepsilon T}, 1 \leq i \leq N$ for any $N \geq 1$.

Let $\tilde{\theta}_{\varepsilon T} = \sum_{i=1}^N \alpha_i \tilde{\theta}_{i\varepsilon T}$ where $\alpha_i, 1 \leq i \leq N$ is a nonrandom sequence of coefficients to be chosen. Note that

$$\tilde{\theta}_{\varepsilon T} \xrightarrow{p} [\sum_{i=1}^N \alpha_i]\theta_0 \text{ as } \varepsilon \to 0 \text{ and } T \to \infty$$

by Theorem 5.7 and hence $\tilde{\theta}_{\varepsilon T}$ is a consistent estimator for θ_0 as $\varepsilon \to 0$ and as $T \to \infty$ provided $\sum_{i=1}^N \alpha_i = 1$. Further more

$$\varepsilon^{-1}T(\tilde{\theta}_{\varepsilon T} - \theta_0) \xrightarrow{\mathcal{L}} N(0, \sum_{i=1}^N \alpha_i^2\{2v_i^2\mu_i(\theta_0)\lambda_i^{2\alpha}(\theta_0)[\mu_i'(\theta_0)]^2\}^{-1}) \text{ as } \varepsilon \to 0 \text{ and } T \to \infty.$$

This follows again by Theorem 5.7 and the independence of the estimators $\{\tilde{\theta}_{i\varepsilon T}, 1 \leq i \leq N\}$. We now obtain the optimum combination of the coefficients $\{\alpha_i, 1 \leq i \leq N\}$ by minimizing the asymptotic variance

$$\sum_{i=1}^N \alpha_i^2\{2v_i^2\mu_i(\theta_0)\lambda_i^{2\alpha}(\theta_0)[\mu_i'(\theta_0)]^2\}^{-1}$$

subject to the condition $\sum_{i=1}^N \alpha_i = 1$. It is easy to see that α_i is proportional to

$$\{v_i^2\mu_i(\theta_0)\lambda_i^{2\alpha}(\theta_0)[\mu_i'(\theta_0)]^2\}$$

and the optimal choice of $\{\alpha_i, 1 \leq i \leq N\}$ leads to the estimator

$$\theta^*_{\varepsilon T} = \frac{\sum_{i=1}^N v_i^2\mu_i(\theta_0)\lambda_i^{2\alpha}(\theta_0)[\mu_i'(\theta_0)]^2\tilde{\theta}_{i\varepsilon T}}{\sum_{i=1}^N v_i^2\mu_i(\theta_0)\lambda_i^{2\alpha}(\theta_0)[\mu_i'(\theta_0)]^2]}. \tag{5. 39}$$

It is easy to see that

$$\theta^*_{\varepsilon T} \xrightarrow{p} \theta_0 \text{ as } \varepsilon \to 0 \text{ and } T \to \infty$$

and

$$\varepsilon^{-1}T(\theta^*_{\varepsilon T} - \theta_0) \xrightarrow{\mathcal{L}} N(0, [\sum_{i=1}^N \{2v_i^2\mu_i(\theta_0)\lambda_i^{2\alpha}(\theta_0)[\mu_i'(\theta_0)]^2]^{-1}) \text{ as } \varepsilon \to 0 \text{ and } T \to \infty$$

again due to the independence of the estimators $\tilde{\theta}_{i\varepsilon T}, 1 \leq i \leq N$. However the random variable $\theta^*_{\varepsilon T}$ cannot be considered as an estimator of the parameter θ_0 since it depends on the unknown parameter θ_0. In order to avoid this problem, we consider a modified estimator

$$\hat{\theta}_{\varepsilon T} = \frac{\sum_{i=1}^N v_i^2 \mu_i(\tilde{\theta}_{i\varepsilon T})\lambda_i^{2\alpha}(\tilde{\theta}_{i\varepsilon T})[\mu_i'(\tilde{\theta}_{i\varepsilon T})]^2 \tilde{\theta}_{i\varepsilon T}}{\sum_{i=1}^N v_i^2 \mu_i(\tilde{\theta}_{i\varepsilon T})\lambda_i^{2\alpha}(\tilde{\theta}_{i\varepsilon T})[\mu_i'(\tilde{\theta}_{i\varepsilon T})]^2}. \tag{5. 40}$$

which is obtained from $\theta^*_{\varepsilon T}$ by substituting the estimator $\tilde{\theta}_{i\varepsilon T}$ for the unknown parameter θ_0 in the i-th term in the numerator and the denominator in (5.39). In view of the independence, consistency and asymptotic normality of the estimators $\tilde{\theta}_{i\varepsilon T}, 1 \leq i \leq N$, it follows that the estimator $\hat{\theta}_{\varepsilon T}$ is consistent and asymptotically normal for the parameter θ_0 and we have the following result.

Theorem 5.8 : Under the probability measure P_{θ_0},

$$\hat{\theta}_{\varepsilon T} \xrightarrow{p} \theta_0 \text{ as } \varepsilon \to 0 \text{ and } T \to \infty$$

and, if $\mu_i(\theta_0) < 0, 1 \leq i \leq N$, then

$$\varepsilon^{-1} T(\hat{\theta}_{\varepsilon T} - \theta_0) \xrightarrow{\mathcal{L}} N(0, \{2\sum_{i=1}^N v_i^2 \mu_i(\theta_0)\lambda_i^{2\alpha}(\theta_0)[\mu_i'(\theta_0)]^2\}^{-1}) \text{ as } \varepsilon \to 0 \text{ and } T \to \infty$$

for any fixed $N \geq 1$.

6 Nonparametric Estimation for Stochastic PDE with linear multiplier (Continuous sampling)

Example I

Let $(\Omega, \mathcal{F}, P)$ be a probability space and consider the process $u_\varepsilon(t,x), 0 \leq x \leq 1, 0 \leq t \leq T$ governed by the stochastic partial differential equation

$$du_\varepsilon(t,x) = (\Delta u_\varepsilon(t,x) + \theta(t)u_\varepsilon(t,x))dt + \varepsilon dW_Q(t,x) \tag{6. 1}$$

where $\Delta = \frac{\partial^2}{\partial x^2}$. Suppose that $\varepsilon \to 0$ and $\theta \in \Theta$ where Θ is a class of real valued functions $\theta(t), 0 \leq t \leq T$ uniformly bounded , k times continuously differentiable and suppose that the k-th derivative $\theta^{(k)}(.)$ satisfies the Lipschitz condition of order $\alpha \in (0,1]$, that is,

$$|\theta^{(k)}(t) - \theta^{(k)}(s)| \leq |t-s|^\alpha, \beta = k + \alpha. \tag{6. 2}$$

Further suppose the initial and the boundary conditions are given by

$$(6.\ 3)\qquad \begin{cases} u_\varepsilon(0,x) = f(x), f \in L_2[0,1] \\ u_\varepsilon(t,0) = u_\varepsilon(t,1) = 0, 0 \le t \le T \end{cases}$$

and Q is the nuclear covariance operator for the Wiener process $W_Q(t,x)$ taking values in $L_2[0,1]$ so that

$$W_Q(t,x) = Q^{1/2}W(t,x)$$

and $W(t,x)$ is a cylindrical Brownian motion on $L_2[0,1]$. Then, it is known that (cf. Rozovskii [33], Kallianpur and Xiong [11])

$$(6.\ 4)\qquad W_Q(t,x) = \sum_{i=1}^{\infty} q_i^{1/2} e_i(x) W_i(t) \text{ a.s.}$$

where $\{W_i(t), 0 \le t \le T\}, i \ge 1$ are independent one - dimensional standard Wiener processes and $\{e_i\}$ is a complete orthonormal system in $L_2[0,1]$ consisting of eigen vectors of Q and $\{q_i\}$ eigen values of Q.

We assume that the operator Q is a special covariance operator Q with $e_k = sin(k\pi x), k \ge 1$ and $\lambda_k = (\pi k)^2, k \ge 1$. Then $\{e_k\}$ is a complete orthonormal system with eigen values $q_i = (1+\lambda_i)^{-1}, i \ge 1$ for the operator Q and $Q = (I - \Delta)^{-1}$. Note that

$$(6.\ 5)\qquad dW_Q = Q^{1/2} dW.$$

We define a solution $u_\varepsilon(t,x)$ of (6.1) as a formal sum

$$(6.\ 6)\qquad u_\varepsilon(t,x) = \sum_{i=1}^{\infty} u_{i\varepsilon}(t) e_i(x)$$

(cf. Rozovskii [33]). It can be checked that the Fourier coefficient $u_{i\varepsilon}(t)$ satisfies the stochastic differential equation

$$(6.\ 7)\qquad du_{i\varepsilon}(t) = (\theta(t) - \lambda_i) u_{i\varepsilon}(t)dt + \frac{\varepsilon}{\sqrt{\lambda_i + 1}} dW_i(t),\ 0 \le t \le T$$

with the initial condition

$$(6.\ 8)\qquad u_{i\varepsilon}(0) = v_i,\ v_i = \int_0^1 f(x) e_i(x) dx.$$

We assume that the initial function f in (6.3) is such that

$$v_i = \int_0^1 f(x) e_i(x) dx > 0,\ i \ge 1.$$

Estimation of linear multiplier

We now consider the problem of estimation of the function $\theta(t), 0 \leq t \leq T$ based on the observation of the Fourier coefficients $u_{i\varepsilon}(t), 1 \leq i \leq N$ over $[0,T]$ or equivalently the projection $u_\varepsilon^{(N)}(t,x)$ of the process $u_\varepsilon(t,x)$ onto the subspace spanned by $\{e_1,\ldots,e_N\}$ in $L_2[0,1]$.

We will at first construct an estimator of $\theta(.)$ based on the path $\{u_{i\varepsilon}(t), 0 \leq t \leq T\}$. Our technique follows the methods in Kutoyants [12], p.155.

Let us suppose that

(6. 9) $$\sup_{\theta\in\Theta}\sup_{0\leq t\leq T}|\theta(t)| \leq L_0.$$

Consider the differential equation

(6. 10) $$\frac{du_i(t)}{dt} = (\theta(t)-\lambda_i))u_i(t), u_i(0) = v_i, 0 \leq t \leq T.$$

It is easy to see that

$$u_i(t) = v_i e^{\int_0^t(\theta(s)-\lambda_i)ds}, 0 \leq t \leq T$$

and hence

(6. 11) $$u_i(t) \geq v_i e^{-M_i t}, 0 \leq t \leq T$$

where

(6. 12) $$M_i = L_0 + \lambda_i.$$

From the Lemma 1.13 of Kutoyants [12], it follows that

(6. 13) $$\sup_{0\leq s\leq T}|u_{i\varepsilon}(s) - u_i(s)| \leq \frac{\varepsilon}{\sqrt{\lambda_i+1}} e^{M_i t} \sup_{0\leq s\leq T}|W_i(s)|$$

almost surely. Let

(6. 14) $$A_t^{(i)} = \{\omega : \inf_{0\leq s\leq t} u_{i\varepsilon}(s) \geq \frac{1}{2} v_i e^{-M_i t}\}$$

and $A_i = A_T^{(i)}$. Note that $A_t^{(i)}$ contains the set A_i for $0 \leq t \leq T$.

Define the process $\{Y_{i\varepsilon}(t), 0 \leq t \leq T\}$ by the stochastic differential equation

$$dY_{i\varepsilon}(t) = -\frac{\varepsilon^2}{2(\lambda_i+1)} u_{i\varepsilon}^{-2}(t)\chi(A_t^{(i)})dt$$

(6. 15) $$+u_{i\varepsilon}^{-1}(t)\chi(A_t^{(i)})du_{i\varepsilon}(t), 0 \le t \le T$$

where $\chi(E)$ denotes the indicator function of a set E. Let $\phi_\varepsilon \to 0$ as $\varepsilon \to 0$ and define

(6. 16) $$\hat{\theta}_{i\varepsilon}(t) = \lambda_i + \chi(A_i)\phi_\varepsilon^{-1}\int_0^T G(\frac{t-s}{\phi_\varepsilon})dY_{i\varepsilon}(s)$$

where $G(.)$ is a bounded kernel with finite support , that is , there exists constants a and b such that

(6. 17) $$\int_a^b G(u)du = 1, G(u) = 0 \text{ for } u < a \text{ and } u > b.$$

We suppose that $a < 0$ and $b > 0$. Further suppose that the kernel $G(.)$ satisfies the additional condition

(6. 18) $$\int_{-\infty}^{\infty} G(u)u^j du = 0, j = 1, \dots, k.$$

Note that $\hat{\theta}_{i\varepsilon}(t), 1 \le i \le N$ are independent estimators of $\theta(t)$ since the processes $W_i, 1 \le i \le N$ are independent Wiener processes.

Let $\gamma_\varepsilon = \varepsilon^{-\frac{2\beta}{2\beta+1}}$. Suppose that Under some additional conditions , it can be shown (cf. Prakasa Rao [24]) that the estimators $\hat{\theta}_{i\varepsilon}(t), 1 \le i \le N$ are independent estimators of $\theta(t)$ such that

(6. 19) $$\sup_{1\le i\le N} |E[\hat{\theta}_{i\varepsilon}(t) - \theta(t)]| \le C_1 \varepsilon^{\frac{2\beta}{2\beta+1}}$$

and

(6. 20) $$\sup_{1\le i\le N} E[\hat{\theta}_{i\varepsilon}(t) - \theta(t)]^2 \le C_2 \varepsilon^{\frac{4\beta}{2\beta+1}}$$

where C_1 and C_2 are constants depending on the kernel $G(.)$, the Lipschitz constant L_0 and N. Note that the estimators $\hat{\theta}_{i\varepsilon}(t), 1 \le i \le N$ are the best estimators of $\theta(t)$ as far as the rate of mean square error are concerned by Theorem 4.6 in Kutoyants [12]. We now combine these estimators in an optimum fashion to get an estimator using all the information available. Define

(6. 21) $$\tilde{\theta}_{N\varepsilon}(t) = \frac{\sum_{i=1}^N \hat{\theta}_{i\varepsilon}(t)(\lambda_i+1)u_i^2(t)}{\sum_{i=1}^N (\lambda_i+1)u_i^2(t)}.$$

Note that the random variable $\tilde{\theta}_{N\varepsilon}(t)$ is not an estimator of $\theta(t)$ as the functions $u_i(t)$ depend on the function $\theta(t)$. However the random variable $\tilde{\theta}_{N\varepsilon}(t)$ is a linear function of independent random variables $\hat{\theta}_{i\varepsilon}(t), 1 \leq i \leq N$. From the earlier calculations , it can be checked that

$$\begin{aligned} E(\tilde{\theta}_{N\varepsilon}(t)-\theta(t))^2 &= Var(\tilde{\theta}_{N\varepsilon}(t))+(E(\tilde{\theta}_{N\varepsilon}(t)-\theta(t))^2 \\ &\leq C_3\varepsilon^{\frac{4\beta}{2\beta+1}}+C_4\varepsilon^{\frac{4\beta}{2\beta+1}} \\ &\leq C_5\varepsilon^{\frac{4\beta}{2\beta+1}}. \end{aligned} \tag{6. 22}$$

Let

$$\theta^*_{N\varepsilon}(t)=\frac{\sum_{i=1}^N \hat{\theta}_{i\varepsilon}(t)(\lambda_i+1)\hat{u}^2_{i\varepsilon}(t)}{\sum_{i=1}^N(\lambda_i+1)\hat{u}^2_{i\varepsilon}(t)} \tag{6. 23}$$

where

$$\hat{u}_{i\varepsilon}(t)=v_i e^{\int_0^t(\hat{\theta}_{i\varepsilon}(s)-\lambda_i)ds}. \tag{6. 24}$$

The following results can be proved. For details, see Prakasa Rao [24].

Theorem 6.1: For $0<t<T$,

$$\theta^*_{N\varepsilon}(t)\xrightarrow{p}\theta(t) \text{ as } \varepsilon\to 0. \tag{6. 25}$$

Since the estimators $\hat{\theta}_{i\varepsilon}(t), 1\leq i\leq N$ are independent random variables, it follows that the estimator $\theta^*_{N\varepsilon}(t)$ is asymptotically normal and we have the following theorem. For details, see Prakasa Rao [24].

Theorem 6.2: For $0<t<T$,

$$\gamma_\varepsilon(\theta^*_{N\varepsilon}(t)-\theta(t))\xrightarrow{\mathcal{L}}N(0,\sigma^2(t)) \text{ as } \varepsilon\to 0 \tag{6. 26}$$

where

$$\gamma_\varepsilon=\varepsilon^{-\frac{2\beta}{2\beta+1}} \tag{6. 27}$$

and

$$\sigma^2(t)=\frac{1}{\sum_{i=1}^N u_i^2(t)(\lambda_i+1)}\int_{-\infty}^{\infty}G^2(u)du. \tag{6. 28}$$

Remarks 1: If $k = 0$ and $\beta = 1$, that is, the function $\theta(.) \in \Theta$ where Θ is the class of uniformly bounded functions which are Lipschitzian of order one, then it follows that

$$\varepsilon^{-\frac{2}{3}}(\theta^{*}_{N\varepsilon}(t) - \theta(t)) \xrightarrow{\mathcal{L}} N(0, \sigma^2(t)) \text{ as } \varepsilon \to 0. \tag{6. 29}$$

Remarks 2: It is known that the probability measures generated by stochastic processes satisfying the SPDE given by (6.1) are absolutely continuous with respect to each other when $\theta(.)$ is a constant (cf. Huebner et al. [7]). There are classes of SPDE which generate probability measures which are singular with respect to each other when $\theta(.)$ is a constant. We now study the problem of nonparametric inference for a linear multiplier for such a class of SPDE by the above methods(cf. Prakasa Rao [27]).

Example II

Let $(\Omega, \mathcal{F}, P)$ be a probability space and consider the process $u_\varepsilon(t,x), 0 \le x \le 1, 0 \le t \le T$ governed by the stochastic partial differential equation

$$du_\varepsilon(t,x) = \theta(t)\Delta u_\varepsilon(t,x)dt + \varepsilon dW_Q(t,x) \tag{6. 30}$$

where $\Delta = \frac{\partial^2}{\partial x^2}$. Suppose that $\varepsilon \to 0$ and $\theta \in \Theta$ where Θ is a class of positive valued functions $\theta(t), 0 \le t \le T$ uniformly bounded , k times continuously differentiable and that the k-th derivative $\theta^{(k)}(.)$ satisfies the Lipschitz condition of order $\alpha \in (0,1]$, that is,

$$|\theta^{(k)}(t) - \theta^{(k)}(s)| \le |t-s|^\alpha, \beta = k + \alpha. \tag{6. 31}$$

Further suppose the initial and the boundary conditions are given by

$$\begin{cases} u_\varepsilon(0,x) = f(x), f \in L_2[0,1] \\ u_\varepsilon(t,0) = u_\varepsilon(t,1) = 0, 0 \le t \le T \end{cases} \tag{6. 32}$$

and Q is the nuclear covariance operator for the Wiener process $W_Q(t,x)$ taking values in $L_2[0,1]$ so that

$$W_Q(t,x) = Q^{1/2}W(t,x)$$

and $W(t,x)$ is a cylindrical Brownian motion in $L_2[0,1]$. Then, it is known that (cf. Rozovskii [33], Kallianpur and Xiong [11])

$$W_Q(t,x) = \sum_{i=1}^{\infty} q_i^{1/2} e_i(x) W_i(t) \text{ a.s.} \tag{6. 33}$$

where $\{W_i(t), 0 \leq t \leq T\}, i \geq 1$ are independent one - dimensional standard Wiener processes and $\{e_i\}$ is a complete orthonormal system in $L_2[0,1]$ consisting of eigen vectors of Q and $\{q_i\}$ eigen values of Q.

We assume that the operator Q is a special covariance operator Q with $e_k = sin(k\pi x), k \geq 1$ and $\lambda_k = (\pi k)^2, k \geq 1$. Then $\{e_k\}$ is a complete orthonormal system with eigen values $q_i = (1+\lambda_i)^{-1}, i \geq 1$ for the operator Q and $Q = (I - \Delta)^{-1}$. Note that

$$dW_Q = Q^{1/2} dW. \tag{6. 34}$$

We define a solution $u_\varepsilon(t,x)$ of (6.29) as a formal sum

$$u_\varepsilon(t,x) = \sum_{i=1}^{\infty} u_{i\varepsilon}(t) e_i(x) \tag{6. 35}$$

(cf. Rozovskii [33]). It can be checked that the Fourier coefficient $u_{i\varepsilon}(t)$ satisfies the stochastic differential equation

$$du_{i\varepsilon}(t) = -\theta(t)\lambda_i u_{i\varepsilon}(t)dt + \frac{\varepsilon}{\sqrt{\lambda_i + 1}} dW_i(t), \; 0 \leq t \leq T \tag{6. 36}$$

with the initial condition

$$u_{i\varepsilon}(0) = v_i, \; v_i = \int_0^1 f(x) e_i(x) dx. \tag{6. 37}$$

We assume that the initial function f in (6.32) is such that

$$v_i = \int_0^1 f(x) e_i(x) dx > 0, \; i \geq 1.$$

Estimation

We now consider the problem of estimation of the function $\theta(t), 0 \leq t \leq T$ based on the observation of the Fourier coefficients $u_{i\varepsilon}(t), 1 \leq i \leq N$ over $[0,T]$ or equivalently the projection $u_\varepsilon^{(N)}(t,x)$ of the process $u_\varepsilon(t,x)$ onto the subspace spanned by $\{e_1, \ldots, e_N\}$ in $L_2[0,1]$.

We will at first construct an estimator of $\theta(.)$ based on the path $\{u_{i\varepsilon}(t), 0 \leq t \leq T\}$. Our technique follows the methods in Kutoyants [12], p.155 as before.

Let us suppose that

$$\sup_{\theta \in \Theta} \sup_{0 \leq t \leq T} \theta(t) \leq L_0. \tag{6. 38}$$

Consider the differential equation

$$\frac{du_i(t)}{dt} = -\theta(t)\lambda_i u_i(t), u_i(0) = v_i, 0 \leq t \leq T. \tag{6. 39}$$

It is easy to see that

$$u_i(t) = v_i e^{-\lambda_i \int_0^t \theta(s)ds}, 0 \leq t \leq T$$

and hence

$$u_i(t) \geq v_i e^{-M_i t}, 0 \leq t \leq T \tag{6. 40}$$

where

$$M_i = L_0 \lambda_i. \tag{6. 41}$$

From Lemma 1.13 of Kutoyants [12], it follows that

$$\sup_{0 \leq s \leq t} |u_{i\varepsilon}(s) - u_i(s)| \leq \frac{\varepsilon}{\sqrt{\lambda_i + 1}} e^{M_i t} \sup_{0 \leq s \leq t} |W_i(s)| \tag{6. 42}$$

almost surely. Let

$$A_t^{(i)} = \{\omega : \inf_{0 \leq s \leq t} u_{i\varepsilon}(s) \geq \frac{1}{2} v_i e^{-M_i t}\} \tag{6. 43}$$

and $A_i = A_T^{(i)}$. Note that $A_t^{(i)}$ contains the set A_i for $0 \leq t \leq T$.
Define the process $\{Y_{i\varepsilon}(t), 0 \leq t \leq T\}$ by the stochastic differential equation

$$\begin{aligned} dY_{i\varepsilon}(t) &= -\frac{\varepsilon^2}{2(\lambda_i + 1)} u_{i\varepsilon}^{-2}(t)\chi(A_t^{(i)})dt \\ &\quad + u_{i\varepsilon}^{-1}(t)\chi(A_t^{(i)})du_{i\varepsilon}(t), 0 \leq t \leq T \end{aligned} \tag{6. 44}$$

where $\chi(E)$ denotes the indicator function of a set E. Let $\phi_\varepsilon \to 0$ as $\varepsilon \to 0$ and define

$$\hat{\theta}_{i\varepsilon}(t)\lambda_i = -\{\chi(A_i)\phi_\varepsilon^{-1} \int_0^T G(\frac{t-s}{\phi_\varepsilon})dY_{i\varepsilon}(s)\} \tag{6. 45}$$

where $G(.)$ is a bounded kernel with finite support , that is , there exists constants a and b such that

$$\int_a^b G(u)du = 1, G(u) = 0 \text{ for } u < a \text{ and } u > b. \tag{6. 46}$$

We suppose that $a < 0$ and $b > 0$. Further suppose that the kernel $G(.)$ satisfies the additional condition

$$\int_{-\infty}^{\infty} G(u)u^j du = 0, j = 1, \ldots, k. \tag{6. 47}$$

Note that $\hat{\theta}_{i\varepsilon}(t), 1 \leq i \leq N$ are independent estimators of $\theta(t)$ since the processes $W_i, 1 \leq i \leq N$ are independent Wiener processes.

Let $\gamma_\varepsilon = \varepsilon^{\frac{-2\beta}{2\beta+1}}$. Under some conditions, it can be shown that the estimators $\hat{\theta}_{i\varepsilon}(t), 1 \leq i \leq N$ are independent estimators of $\theta(t)$ such that

$$\sup_{1 \leq i \leq N} |E[\hat{\theta}_{i\varepsilon}(t) - \theta(t)]| \leq C_6 \varepsilon^{\frac{2\beta}{2\beta+1}} \tag{6. 48}$$

and

$$\sup_{1 \leq i \leq N} E[\hat{\theta}_{i\varepsilon}(t) - \theta(t)]^2 \leq C_7 \varepsilon^{\frac{4\beta}{2\beta+1}} \tag{6. 49}$$

where C_6 and C_7 are constants depending on the kernel $G(.)$, the Lipschitz constant L_0 and N. Note that the estimators $\hat{\theta}_{i\varepsilon}(t), 1 \leq i \leq N$ are the best estimators of $\theta(t)$ as far as the rate of mean square error are concerned by Theorem 4.6 in Kutoyants [12]. We now combine these estimators in an optimum fashion to get an estimator using all the information available.

Define

$$\tilde{\theta}_{N\varepsilon}(t) = \frac{\sum_{i=1}^{N} \hat{\theta}_{i\varepsilon}(t) \lambda_i^2 (\lambda_i + 1) u_i^2(t)}{\sum_{i=1}^{N} \lambda_i^2 (\lambda_i + 1) u_i^2(t)}. \tag{6. 50}$$

Note that the random variable $\tilde{\theta}_{N\varepsilon}(t)$ is not an estimator of $\theta(t)$ as the functions $u_i(t)$ depend on the function $\theta(t)$. However the random variable $\tilde{\theta}_{N\varepsilon}(t)$ is a linear function of independent random variables $\hat{\theta}_{i\varepsilon}(t), 1 \leq i \leq N$. It can be checked that

$$\begin{aligned} E(\tilde{\theta}_{N\varepsilon}(t) - \theta(t))^2 &= Var(\tilde{\theta}_{N\varepsilon}(t)) + (E(\tilde{\theta}_{N\varepsilon}(t) - \theta(t))^2 \\ &\leq C_8 \varepsilon^{\frac{4\beta}{2\beta+1}} + C_9 \varepsilon^{\frac{4\beta}{2\beta+1}} \\ &\leq C_{10} \varepsilon^{\frac{4\beta}{2\beta+1}}. \end{aligned} \tag{6. 51}$$

As a consequence, we have the following result.

Theorem 6.3: For $0 < t < T$,

(i) $\tilde{\theta}_{N\varepsilon}(t) \xrightarrow{p} \theta(t)$ as $\varepsilon \to 0$;

(ii) $E(\tilde{\theta}_{N\varepsilon}(t)) \to \theta(t)$ as $\varepsilon \to 0$;

(iii) $\lim_{\varepsilon \to 0} E(\tilde{\theta}_{N\varepsilon}(t) - \theta(t))^2 \to$ as $\varepsilon \to 0$;

(iv) $\limsup_{\varepsilon \to 0} E(\tilde{\theta}_{N\varepsilon}(t) - \theta(t))^2 \varepsilon^{\frac{-4\beta}{2\beta+1}} < \infty$;

(v) $\varepsilon^{\frac{-2\beta}{2\beta+1}}(\tilde{\theta}_{N\varepsilon}(t) - \theta(t)) \xrightarrow{\mathcal{L}} N(0, \sigma^2(t))$ as $\varepsilon \to 0$

where $N(0, \sigma^2(t))$ denotes the normal distribution with mean zero and variance $\sigma^2(t))$ given by

$$\sigma^2(t) = \frac{1}{\sum_{i=1}^N u_i^2(t)\lambda_i^2(\lambda_i + 1)} \int_{-\infty}^{\infty} G^2(u)du. \tag{6. 52}$$

Let

$$\theta^*_{N\varepsilon}(t) = \frac{\sum_{i=1}^N \hat{\theta}_{i\varepsilon}(t)\lambda_i^2(\lambda_i + 1)\hat{u}^2_{i\varepsilon}(t)}{\sum_{i=1}^N \lambda_i^2(\lambda_i + 1)\hat{u}^2_{i\varepsilon}(t)} \tag{6. 53}$$

where

$$\hat{u}_{i\varepsilon}(t) = v_i e^{-\lambda_i \int_0^t \hat{\theta}_{i\varepsilon}(s)ds}. \tag{6. 54}$$

Theorem 6.4: For $0 < t < T$,

$$\theta^*_{N\varepsilon}(t) \xrightarrow{p} \theta(t) \text{ as } \varepsilon \to 0. \tag{6. 55}$$

Note that

$$\begin{aligned} \gamma_\varepsilon[\theta^*_{N\varepsilon}(t) - \theta(t)] &= \gamma_\varepsilon\left[\frac{\sum_{i=1}^N \hat{\theta}_{i\varepsilon}(t)\lambda_i^2(\lambda_i + 1)\hat{u}^2_{i\varepsilon}(t)}{\sum_{i=1}^N \lambda_i^2(\lambda_i + 1)\hat{u}^2_{i\varepsilon}(t)} - \theta(t)\right] \\ &= \frac{\sum_{i=1}^N \gamma_\varepsilon(\hat{\theta}_{i\varepsilon}(t) - \theta(t))\lambda_i^2(\lambda_i + 1)\hat{u}^2_{i\varepsilon}(t)}{\sum_{i=1}^N \lambda_i^2(\lambda_i + 1)\hat{u}^2_{i\varepsilon}(t)}. \end{aligned}$$

Since

(i) $\gamma_\varepsilon(\hat{\theta}_{i\varepsilon}(t) - \theta(t)) \xrightarrow{\mathcal{L}} N(0, \sigma^2(t))$ as $\varepsilon \to 0$ for $1 \le i \le N$,

(ii) $\hat{u}_{i\varepsilon}(t) \xrightarrow{p} u_i(t)$ as $\varepsilon \to 0$ for $1 \le i \le N$,

for $0 < t < T$, and since the estimators $\hat{\theta}_{i\varepsilon}(t), 1 \le i \le N$ are independent random variables, it follows that the estimator $\theta^*_{N\varepsilon}(t)$ is asymptotically normal and we have the following theorem.

Theorem 6.5: For $0 < t < T$,

$$\gamma_\varepsilon(\theta^*_{N\varepsilon}(t) - \theta(t)) \xrightarrow{\mathcal{L}} N(0, \sigma^2(t)) \text{ as } \varepsilon \to 0 \tag{6. 56}$$

where

(6. 57) $$\gamma_\varepsilon = \varepsilon^{-\frac{2\beta}{2\beta+1}}$$

and

(6. 58) $$\sigma^2(t) = \frac{1}{\sum_{i=1}^N u_i^2(t)\lambda_i^2(\lambda_i+1)} \int_{-\infty}^{\infty} G^2(u)du.$$

For details, see Prakasa Rao [27].

Remarks :(1)If $k = 0$ and $\beta = 1$,that is, the function $\theta(.) \in \Theta$ where Θ is the class of uniformly bounded nonegative functions which are Lipschitzian of order one, then it follows that, for $0 < t < T$,

(6. 59) $$\varepsilon^{-\frac{2}{3}}(\theta^*_{N\varepsilon}(t) - \theta(t)) \xrightarrow{\mathcal{L}} N(0, \sigma^2(t)) \text{ as } \varepsilon \to 0.$$

(2) It is well known that if $\alpha_i, 1 \leq i \leq N$ are unbiased estimators of a parameter θ with variances $\sigma_i^2, 1 \leq i \leq N$ respectively, then a better estimator, in the sense of smaller variance, can be obtained by taking a linear combination of $\alpha_i, 1 \leq i \leq N$ with the coefficient of α_i inversely proportional to the variance σ_i^2 and adjusting the proprtionality constant so that the new estimator is also unbiased. Here the estimators $\hat{\theta}_{i\varepsilon}(t), 1 \leq i \leq N$ are asymptotically unbiased estimators of $\theta(t)$ and the estimator $\theta^*_{N\varepsilon}(t)$ is obtained following the above procedure so that this estimator has smaller asymptotic variance compared to the asymptotic variances of $\hat{\theta}_{i\varepsilon}(t), 1 \leq i \leq N$.

(3)It should be possible to study nonparametric estimation of the function $\theta(t)$ in the examples I and II and by other methods of estimation such as the Method of Sieves or the Method of Wavelets (cf. Prakasa Rao [22]) and recover the function $\theta(t)$ either by keeping ε fixed and letting $N \to \infty$ or by linking ε and N such that $N = N(\varepsilon) \to \infty$.

7 Parameter Estimation for Stochastic PDE with linear drift (Absolutely continuous case) (Discrete sampling)

In all the earlier sections, it was assumed that a continuous observation of a random field $u_\varepsilon(x,t)$ satisfying a SPDE over the region $[0,1] \times [0,T]$ is available. It is obvious that this assumption is not tenable in practice and the problem of interest is to develop methods of estimation of the parameters from a random field $u_\varepsilon(x,t)$ observed at discrete times t and at discrete positions x or from the Fourier coefficients $u_{i\varepsilon}(t)$ observed at discrete time instants. We will discuss the latter problem in this paper for two types of SPDE's. Prakasa Rao [20] discusses statistical inference from sampled data for stochastic processes in general and the methods of statistical inference for the special class of diffusion type processes is investigated extensively in Prakasa Rao [22].

Let us consider the SPDE (3.1) discussed in Section 3.

Suppose the collection of observations consists of $\{u_{i\varepsilon}(j\Delta), 0 \leq j \leq n, 1 \leq i \leq N\}$ where $\Delta > 0$. The problem of estimation of the parameter θ can be considered from three different angles (i) discretise the likelihood equation obtained from the continuous observation by approximating the terms α_ε and β_ε in the likelihood equation by suitable approximating sums and then solve for an approximate maximum likelihood estimator, (ii) discretise the maximum likelihood estimator obtained from the continuous observation and (iii) compute the likelihood function based on the discrete set of observations and the maximize the corresponding likelihood. All these approaches have been studied for the estimation of parameters for diffusion type processes (cf.Prakasa Rao [22]).

We now approach the problem following the techniques in Bibby and Sorensen [1]. The Fourier coefficients $u_{i\varepsilon}(t)$ of the random field $u_\varepsilon(t,x)$ satisfy the stochastic differential equation

(7. 1) $$du_{i\varepsilon}(t) = (\theta - \lambda_i)u_{i\varepsilon}(t)dt + \frac{\varepsilon}{\sqrt{\lambda_i + 1}}dW_i(t),\ 0 \leq t \leq T$$

with the initial condition

(7. 2) $$u_{i\varepsilon}(0) = v_i,\ v_i = \int_0^1 f(x)e_i(x)dx.$$

Note that the process $\{u_{i\varepsilon}(t), 0 \leq t \leq T\}$ is the Ornstein-Uhlenbeck process and it is well known that the conditional distribution of $u_{i\varepsilon}(\Delta)$ given $u_{i\varepsilon}(0)$ is normal with mean $v_i e^{(\theta-\lambda_i)\Delta}$ and variance $\frac{\varepsilon^2(e^{2(\theta-\lambda_i)\Delta}-1)}{2(\theta-\lambda_i)(\lambda_i+1)}$. It can be shown that

(7. 3) $$G_n(\theta) = \frac{\lambda_i + 1}{\varepsilon^2}\sum_{j=1}^{n} u_{i\varepsilon}((j-1)\Delta)(u_{i\varepsilon}(j\Delta) - u_{i\varepsilon}((j-1)\Delta)e^{\theta\Delta})$$

is proportional to the optimal estimating function for the estimation of the parameter $\theta - \lambda_i$ (cf. Bibby and Sorensen [1]) and an estimator for θ is of the form

(7. 4) $$\hat{\theta}_{i\varepsilon} = \lambda_i + \Delta^{-1}\log\frac{\sum_{j=1}^{n} u_{i\varepsilon}(j\Delta)u_{i\varepsilon}((j-1)\Delta)}{\sum_{j=1}^{n} u_{i\varepsilon}^2((j-1)\Delta)}.$$

This estimator is also the maximum likelihood estimator of θ based on the discrete data $u_{i\varepsilon}(j\Delta), 0 \leq j \leq n$.

Since $\theta < 0$, it is well known that the solution of (7.1) is ergodic with the stationary measure with density μ_θ given by the normal distribution with mean zero and variance $\beta_i^2(\theta) = \varepsilon^2\{2(\lambda_i - \theta)(\lambda_i + 1)\}^{-1}$. Further more we have already noted that the transition probability density π_Δ^θ of $u_{i\varepsilon}(\Delta)$ given that $u_{i\varepsilon}(0) = x$ is the normal probability density with mean $xe^{(\theta-\lambda_i)\Delta}$ and variance $\frac{\varepsilon^2(e^{2(\theta-\lambda_i)\Delta}-1)}{2(\theta-\lambda_i)(\lambda_i+1)}$.

The following result can be proved. For details, see Prakasa Rao [28].

Theorem 7.1: The estimator $\hat{\theta}_{i\varepsilon}$ converges in probability to θ as $n \to \infty$ and

$$n^{1/2}\Delta(\hat{\theta}_{i\varepsilon} - \theta) \xrightarrow{\mathcal{L}} N(0, e^{-2\Delta(\theta-\lambda_i)} - 1) \text{ as } n \to \infty.$$

Remarks: Note that the estimators $\{\hat{\theta}_{i\varepsilon}, 1 \leq i \leq N\}$ are *independent*, consistent and asymptotically normal for the parameter θ in the stochastic partial differential equation (3.1). We will now discuss a method for combining these estimators to get an improved estimator.

Let $\tilde{\theta}_\varepsilon = \sum_{i=1}^N \alpha_i \hat{\theta}_{i\varepsilon}$ where $\alpha_i, 1 \leq i \leq N$ is a nonrandom sequence of coefficients to be chosen. Note that

$$\tilde{\theta}_\varepsilon \xrightarrow{p} [\sum_{i=1}^N \alpha_i]\theta \text{ as } n \to \infty$$

by the Theorem 7.1 and hence $\tilde{\theta}_\varepsilon$ is consistent for θ provided $\sum_{i=1}^N \alpha_i = 1$. Further more

$$\sqrt{n}\Delta(\tilde{\theta}_\varepsilon - \theta) \xrightarrow{\mathcal{L}} N(0, \sum_{i=1}^N \alpha_i^2(e^{-2\Delta(\theta-\lambda_i)} - 1)) \text{ as } n \to \infty.$$

This follows from the Theorem 7.1 and the independence of the estimators $\{\hat{\theta}_{i\varepsilon}, 1 \leq i \leq N\}$. We now obtain the optimum combination of the coefficients $\{\alpha_i, 1 \leq i \leq N\}$ by minimising the asymptotic variance

$$\sum_{i=1}^N \alpha_i^2(e^{-2\Delta(\theta-\lambda_i)} - 1)$$

subject to the condition $\sum_{i=1}^N \alpha_i = 1$. It is easy to show that α_i is proportional to $(e^{2\Delta(\lambda_i-\theta)} - 1)^{-1}$ and the optimum choice of $\alpha_i, 1 \leq i \leq N$ leads to the "estimator"

$$\theta_\varepsilon^* = \frac{\sum_{i=1}^N (e^{2\Delta(\lambda_i-\theta)} - 1)^{-1}\hat{\theta}_{i\varepsilon}}{\sum_{i=1}^N (e^{2\Delta(\lambda_i-\theta)} - 1)^{-1}}. \tag{7. 5}$$

It is easy to see that

$$\theta_\varepsilon^* \xrightarrow{p} \theta \text{ as } n \to \infty$$

and

$$\sqrt{n}\Delta(\theta_\varepsilon^* - \theta) \xrightarrow{\mathcal{L}} N(0, (\sum_{i=1}^{N}(e^{-2\Delta(\theta-\lambda_i)} - 1)^{-1})^{-1}) \text{ as } n \to \infty$$

again due to the independence of the estimators $\{\hat{\theta}_{i\varepsilon}, 1 \leq i \leq N\}$. However the random variable θ_ε^* cannot be considered as an estimator of θ in the true sense since it depends on the unknown parameter θ. In order to avoid this problem, we can consider a modified estimator

$$\hat{\theta}_\varepsilon = \frac{\sum_{i=1}^{N}(e^{2\Delta(\lambda_i - \hat{\theta}_{i\varepsilon})} - 1)^{-1}\hat{\theta}_{i\varepsilon}}{\sum_{i=1}^{N}(e^{2\Delta(\lambda_i - \hat{\theta}_{i\varepsilon})} - 1)^{-1}} \tag{7. 6}$$

which is obtained from θ_ε^* by substituting the estimator $\hat{\theta}_{i\varepsilon}$ for the unknown parameter θ in the i-th term in the numerator and denominator in (7.5). In view of the independence, consistency and asymptotic normality of the estimators $\hat{\theta}_{i\varepsilon}, 1 \leq i \leq N$, it follows that the estimator $\hat{\theta}_\varepsilon$ is consistent and asymptotically normal for the parameter θ and we have the following result.

Theorem 7.2: Under the conditions stated above,

$$\hat{\theta}_\varepsilon \xrightarrow{p} \theta \text{ as } n \to \infty$$

and

$$n^{1/2}\Delta(\hat{\theta}_\varepsilon - \theta)) \xrightarrow{\mathcal{L}} N(0, (\sum_{i=1}^{N}(e^{-2\Delta(\theta-\lambda_i)} - 1)^{-1})^{-1}) \text{ as } n \to \infty$$

for any fixed $N \geq 1$.

8 Parametric Estimation for Stochastic PDE with linear drift (Singular case) (Discrete sampling)

Let us consider the SPDE (4.1) in the Section 4.

Suppose the collection of observations consists of $\{u_{i\varepsilon}(j\Delta), 0 \leq j \leq n, 1 \leq i \leq N\}$ where $\Delta > 0$. As discussed in the previous section, consider the stochastic differential equation

$$du_{i\varepsilon}(t) = -\theta\lambda_i u_{i\varepsilon}(t)dt + \frac{\varepsilon}{\sqrt{\lambda_i + 1}}dW_i(t),\ 0 \leq t \leq T \tag{8. 1}$$

with the initial condition

$$(8.\ 2)\qquad u_{i\varepsilon}(0) = v_i,\ v_i = \int_0^1 f(x)e_i(x)dx.$$

Note that the process $\{u_{i\varepsilon}(t), 0 \le t \le T\}$ is the Ornstein-Uhlenbeck process and it is well known that the conditional distribution of $u_{i\varepsilon}(\Delta)$ given $u_{i\varepsilon}(0)$ is normal with mean $v_i e^{-\theta\lambda_i\Delta}$ and variance $\frac{\varepsilon^2(e^{-2\theta\lambda_i\Delta}-1)}{(-2\theta\lambda_i)(\lambda_i+1)}$. It can be shown that

$$(8.\ 3)\qquad H_n(\theta) = \frac{\lambda_i+1}{\varepsilon^2}\sum_{j=1}^{n} u_{i\varepsilon}((j-1)\Delta)(u_{i\varepsilon}(j\Delta) - u_{i\varepsilon}((j-1)\Delta)e^{-\theta\lambda_i\Delta})$$

is proportional to the optimal estimating function for the estimation of the parameter θ (cf. Bibby and Sorensen [1]) and an estimator for θ is of the form

$$(8.\ 4)\qquad \hat{\theta}_{i\varepsilon} = -\lambda_i^{-1}\Delta^{-1}\log\frac{\sum_{j=1}^{n} u_{i\varepsilon}(j\Delta)u_{i\varepsilon}((j-1)\Delta)}{\sum_{j=1}^{n} u_{i\varepsilon}^2((j-1)\Delta)}.$$

This estimator is also the maximum likelihood estimator of θ based on the discrete data $u_{i\varepsilon}(j\Delta), 0 \le j \le n$.

Since $\theta > 0$, it is well known that the solution of (8.1) is ergodic with the stationary measure with density ν_θ given by the normal distribution with mean zero and variance $\zeta_i^2(\theta) = \varepsilon^2\{2\lambda_i\theta(\lambda_i+1)\}^{-1}$. Further more we have already noted that the transition probability density π_Δ^θ of $u_{i\varepsilon}(\Delta)$ given that $u_{i\varepsilon}(0) = x$ is the normal probability density with mean $xe^{(-\theta\lambda_i)\Delta}$ and variance $\frac{\varepsilon^2(e^{2(-\theta\lambda_i)\Delta}-1)}{2(-\theta\lambda_i)(\lambda_i+1)}$. Let X be a random variable with stationary measure ν_θ and Y be a random variable such that the conditional density of Y given $X = x$ is given by π_Δ^θ. Note that

$$(8.\ 5)\qquad \begin{aligned} E[XY] &= E[X\,E(Y|X)] = E[X\,Xe^{(-\theta\lambda_i)\Delta}] \\ &= \zeta_i^2(\theta)e^{-\theta\lambda_i\Delta} \end{aligned}$$

and

$$(8.\ 6)\qquad E[X^2] = \zeta_i^2(\theta).$$

It is easy to check that the Condition 3.1 in Bibby and Sorensen [1] holds in this case and applying the Lemma 3.1 in Bibby and Sorensen [1] (cf. Florens-Zmirou [4]), we obtain that

$$\frac{1}{n}\sum_{j=1}^{n} u_{i\varepsilon}(j\Delta)u_{i\varepsilon}((j-1)\Delta) \to E[XY] \text{ in probability as } n \to \infty$$

and

$$\frac{1}{n}\sum_{j=1}^{n} u_{i\varepsilon}^2((j-1)\Delta) \to E[X^2] \text{ in probability as } n \to \infty.$$

The above relations imply that

$$\frac{\sum_{j=1}^{n} u_{i\varepsilon}(j\Delta)u_{i\varepsilon}((j-1)\Delta)}{\sum_{j=1}^{n} u_{i\varepsilon}^2((j-1)\Delta)} \to \frac{E[XY]}{E[X^2]} \text{ in probability as } n \to \infty.$$

The following result follows as a consequence of the above observation and the relations (8.5) and (8.6).

Theorem 8.1: The estimator $\hat{\theta}_{i\varepsilon}$ converges in probability to θ as $n \to \infty$.

Let $\psi_i(\theta) = e^{-\theta\lambda_i\Delta}$ Note that

$$\psi_i(\hat{\theta}_{i\varepsilon}) = \frac{\sum_{j=1}^{n} u_{i\varepsilon}(j\Delta)u_{i\varepsilon}((j-1)\Delta)}{\sum_{j=1}^{n} u_{i\varepsilon}^2((j-1)\Delta)}.$$

Hence

$$\sqrt{n}\{\psi_i(\hat{\theta}_{i\varepsilon}) - \psi_i(\theta)\} = \frac{n^{-1/2}\sum_{j=1}^{n}[u_{i\varepsilon}(j\Delta)u_{i\varepsilon}((j-1)\Delta) - \psi_i(\theta)u_{i\varepsilon}^2((j-1)\Delta)]}{n^{-1}\sum_{j=1}^{n} u_{i\varepsilon}^2((j-1)\Delta)}.$$

Since

$$E[u_{i\varepsilon}(j\Delta)u_{i\varepsilon}((j-1)\Delta)|u_{i\varepsilon}((j-1)\Delta)] = \psi_i(\theta)u_{i\varepsilon}^2((j-1)\Delta),$$

it follows by the Lemma 3.1 of Bibby and Sorensen [1] that

$$n^{-1/2}\sum_{j=1}^{n}[u_{i\varepsilon}(j\Delta)u_{i\varepsilon}((j-1)\Delta) - \psi_i(\theta)u_{i\varepsilon}^2((j-1)\Delta)]$$

converges in distribution to the normal distribution with mean zero and variance equal to $\tau_i(\theta) = E[XY - E(XY|X)]^2$ where the random variables X and Y are as defined above. It can be checked that

$$\tau_i(\theta) = \frac{\varepsilon^2(e^{-2\theta\lambda_i\Delta} - 1)}{2(-\theta\lambda_i)(\lambda_i + 1)}\zeta_i^2(\theta).$$

Applying the δ - method , we obtain that

$$n^{1/2}\Delta(\hat{\theta}_{i\varepsilon} - \theta)$$

converges in distribution to the normal distribution with mean zero and variance $\lambda_i^{-2}e^{2\Delta\theta\lambda_i}\tau_i(\theta)\zeta_i^{-4}(\theta)$ and we have the following theorem.

Theorem 8.2: Under the conditions stated above

$$n^{1/2}\Delta(\hat{\theta}_{i\varepsilon}-\theta) \xrightarrow{\mathcal{L}} N(0, \lambda_i^{-2}(e^{2\Delta\theta\lambda_i}-1)) \text{ as } n\to\infty.$$

Remarks: Note that the estimators $\{\hat{\theta}_{i\varepsilon}, 1 \le i \le N\}$ are *independent*, consistent and asymptotically normal for the parameter θ in the stochastic partial differential equation (4.1). We will now discuss a method for combining these estimators to get an improved estimator.

Let $\tilde{\theta}_\varepsilon = \sum_{i=1}^N \alpha_i\hat{\theta}_{i\varepsilon}$ where $\alpha_i, 1 \le i \le N$ is a nonrandom sequence of coefficients to be chosen. Note that

$$\tilde{\theta}_\varepsilon \xrightarrow{p} [\sum_{i=1}^N \alpha_i]\theta \text{ as } n\to\infty$$

by the Theorem 8.1 and hence $\tilde{\theta}_\varepsilon$ is consistent for θ provided $\sum_{i=1}^N \alpha_i = 1$. Further more

$$\sqrt{n}\Delta(\tilde{\theta}_\varepsilon-\theta) \xrightarrow{\mathcal{L}} N(0, \sum_{i=1}^N \alpha_i^2\lambda_i^{-2}(e^{2\Delta\theta\lambda_i}-1)) \text{ as } n\to\infty.$$

This follows from the Theorem 8.2 and the independence of the estimators $\{\hat{\theta}_{i\varepsilon}, 1\le i\le N\}$. We now obtain the optimum combination of the coefficients $\{\alpha_i, 1\le i\le N\}$ by minimising the asymptotic variance

$$\sum_{i=1}^N \alpha_i^2\lambda_i^{-2}(e^{2\Delta\theta\lambda_i}-1)$$

subject to the condition $\sum_{i=1}^N \alpha_i = 1$. It is easy to show that α_i is proportional to $\lambda_i^2(e^{2\Delta\lambda_i\theta}-1)^{-1}$ and the optimum choice of $\alpha_i, 1\le i\le N$ leads to the "estimator"

$$\theta_\varepsilon^* = \frac{\sum_{i=1}^N \lambda_i^2(e^{2\Delta\lambda_i\theta}-1)^{-1}\hat{\theta}_{i\varepsilon}}{\sum_{i=1}^N \lambda_i^2(e^{2\Delta\lambda_i\theta}-1)^{-1}}. \tag{8. 7}$$

It is easy to see that

$$\theta_\varepsilon^* \xrightarrow{p} \theta \text{ as } n\to\infty$$

and

$$\sqrt{n}\Delta(\theta_\varepsilon^*-\theta) \xrightarrow{\mathcal{L}} N(0, (\sum_{i=1}^N \lambda_i^2(e^{2\Delta\theta\lambda_i}-1)^{-1})^{-1}) \text{ as } n\to\infty$$

again due to the independence of the estimators $\{\hat{\theta}_{i\varepsilon}, 1\le i\le N\}$. However the random variable θ_ε^* cannot be considered as an estimator of θ since it depends on the unknown parameter θ. In order to avoid this problem, we can consider a modified estimator

$$\hat{\theta}_\varepsilon = \frac{\sum_{i=1}^N \lambda_i^2(e^{2\Delta\lambda_i\hat{\theta}_{i\varepsilon}}-1)^{-1}\hat{\theta}_{i\varepsilon}}{\sum_{i=1}^N \lambda_i^2(e^{2\Delta\lambda_i\hat{\theta}_{i\varepsilon}}-1)^{-1}} \tag{8. 8}$$

which is obtained from θ_ε^* by substituting the estimator $\hat{\theta}_{i\varepsilon}$ for the unknown parameter θ in the i-th term in the numerator and denominator in (8.7). In view of the independence, consistency and asymptotic normality of the estimators $\hat{\theta}_{i\varepsilon}, 1 \leq i \leq N$, it follows that the estimator $\hat{\theta}_\varepsilon$ is consistent and asymptotically normal for the parameter θ and we have the following result.

Theorem 8.3: Under the conditions stated above,

$$\hat{\theta}_\varepsilon \xrightarrow{p} \theta \text{ as } n \to \infty$$

and

$$n^{1/2}\Delta(\hat{\theta}_\varepsilon - \theta)) \xrightarrow{\mathcal{L}} N(0, (\sum_{i=1}^{N} \lambda_i^2(e^{2\Delta\theta\lambda_i} - 1)^{-1})^{-1}) \text{ as } n \to \infty$$

for any fixed $N \geq 1$.

9 Parametric Estimation for Some SPDE (Discrete sampling)

We have discussed the problem of estimation of a parameter θ when it is present only in the "trend" part of an SPDE. We will now discuss the problem of estimation when the parameter involved occurs in the "trend" part as well as in the "forcing" part of the SPDE. Prakasa Rao [20] discusses statistical inference from sampled data for stochastic processes in general and the methods of statistical inference for the special class of diffusion type processes is investigated extensively in Prakasa Rao [22]. Piterbarg and Rozovskii [16] studied the properties of maximum likelihood estimators based on discrete time sampling for parameters invoved in parabolic stochastic partial differential equations when the " trend" part of the SPDE involves the parameter but not the " forcing part of the SPDE.

Example I

Let $(\Omega, \mathcal{F}, P)$ be a probability space and consider the process $u_\varepsilon(t,x), 0 \leq x \leq 1, 0 \leq t \leq T$ governed by the stochastic partial differential equation

(9. 1) $$du_\varepsilon(t,x) = (\Delta u_\varepsilon(t,x) + b(\theta)u_\varepsilon(t,x))dt + \varepsilon\sigma(\theta)dW_Q(t,x)$$

where $\Delta = \frac{\partial^2}{\partial x^2}$. Suppose that $\theta \in \Theta \subset R$. Assume that the function $b(\theta) < 0$ for all $\theta \in \Theta$. Further suppose that the functional form of the function $b(\theta)$ is known and it is differentiable with respect to θ with non zero derivative. In addition assume that the function $\sigma(\theta) > 0$ is known but the parameter $\theta \in \Theta$ is unknown. Suppose the initial and the boundary conditions are given by

$$\begin{cases} u_\varepsilon(0,x) = f(x), f \in L_2[0,1] \\ u_\varepsilon(t,0) = u_\varepsilon(t,1) = 0, 0 \leq t \leq T \end{cases} \tag{9. 2}$$

and Q is the nuclear covariance operator for the Wiener process $W_Q(t,x)$ taking values in $L_2[0,1]$ so that

$$W_Q(t,x) = Q^{1/2}W(t,x)$$

and $W(t,x)$ is a cylindrical Brownian motion in $L_2[0,1]$. Then, it is known that (cf. Rozovskii [33])

$$W_Q(t,x) = \sum_{i=1}^{\infty} q_i^{1/2} e_i(x) W_i(t) \text{ a.s.} \tag{9. 3}$$

where $\{W_i(t), 0 \leq t \leq T\}, i \geq 1$ are independent one - dimensional standard Wiener processes and $\{e_i\}$ is a complete orthonormal system in $L_2[0,1]$ consisting of the eigen vectors of Q and $\{q_i\}$ the eigen values of Q.

Let us consider a special covariance operator Q with $e_k = \sin k\pi x, k \geq 1$ and $\lambda_k = (\pi k)^2, k \geq 1$. Then $\{e_k\}$ is a complete orthonormal system with the eigen values $q_i = (1+\lambda_i)^{-1}, i \geq 1$ for the operator Q and $Q = (I - \Delta)^{-1}$. Further more

$$dW_Q = Q^{1/2}dW.$$

We define a solution $u_\varepsilon(t,x)$ of (9.1) as a formal sum

$$u_\varepsilon(t,x) = \sum_{i=1}^{\infty} u_{i\varepsilon}(t) e_i(x) \tag{9. 4}$$

(cf. Rozovskii [33]). It can be checked that the Fourier coefficient $u_{i\varepsilon}(t)$ satisfies the stochastic differential equation

$$du_{i\varepsilon}(t) = (b(\theta) - \lambda_i)u_{i\varepsilon}(t)dt + \frac{\varepsilon}{\sqrt{\lambda_i + 1}}\sigma(\theta)dW_i(t), \; 0 \leq t \leq T \tag{9. 5}$$

with the initial condition

$$u_{i\varepsilon}(0) = v_i, \; v_i = \int_0^1 f(x)e_i(x)dx. \tag{9. 6}$$

Suppose the collection of observations consists of $\{u_{i\varepsilon}(j\Delta), 0 \leq j \leq n, 1 \leq i \leq N\}$ where $\Delta > 0$. We now approach the problem following the techniques in Bibby and Sorensen [1] using the method of estimating functions.

Note that the process $\{u_{i\varepsilon}(t), 0 \leq t \leq T\}$ is the Ornstein-Uhlenbeck process and it is well known that the conditional distribution of $u_{i\varepsilon}(\Delta)$ given $u_{i\varepsilon}(0)$ is normal with mean $v_i e^{(b(\theta)-\lambda_i)\Delta}$ and variance $\frac{\varepsilon^2\sigma^2(\theta)(e^{2(b(\theta)-\lambda_i)\Delta}-1)}{2(b(\theta)-\lambda_i)(\lambda_i+1)}$. It can be shown that

$$G_n(\theta) = \frac{\lambda_i + 1}{\sigma^2(\theta)\varepsilon^2}\sum_{j=1}^{n} b'(\theta)u_{i\varepsilon}((j-1)\Delta)(u_{i\varepsilon}(j\Delta) - u_{i\varepsilon}((j-1)\Delta)e^{(b(\theta)-\lambda_i)\Delta}) \tag{9. 7}$$

is proportional to the optimal estimating function for the estimation of the parameter θ (cf. Bibby and Sorensen [1]) and an estimator for $b(\theta)$ is of the form

$$\hat{b}_{i\varepsilon} = \lambda_i + \Delta^{-1}\log\frac{\sum_{j=1}^{n} u_{i\varepsilon}(j\Delta)u_{i\varepsilon}((j-1)\Delta)}{\sum_{j=1}^{n} u_{i\varepsilon}^2((j-1)\Delta)}. \tag{9. 8}$$

Since $b(\theta) < 0$, it is well known that the solution of (9.5) is ergodic with the stationary measure with the density μ_θ given by the normal distribution with mean zero and variance $\beta_i^2(\theta) = \varepsilon^2\sigma^2(\theta)\{2(\lambda_i - b(\theta))(\lambda_i+1)\}^{-1}$. Further more we have already noted that the transition probability density π_Δ^θ of $u_{i\varepsilon}(\Delta)$ given that $u_{i\varepsilon}(0) = x$ is the normal probability density with mean $xe^{(b(\theta)-\lambda_i)\Delta}$ and variance $\frac{\varepsilon^2\sigma^2(\theta)(e^{2(b(\theta)-\lambda_i)\Delta}-1)}{2(b(\theta)-\lambda_i)(\lambda_i+1)}$. Let X be a random variable with the stationary measure μ_θ and Y be a random variable such that the conditional density of Y given $X = x$ is given by π_Δ^θ. Note that

$$\begin{aligned} E[XY] &= E[XE(Y|X)] = E[XXe^{(b(\theta)-\lambda_i)\Delta}] \\ &= \beta_i^2(\theta)e^{(b(\theta)-\lambda_i)\Delta} \end{aligned} \tag{9. 9}$$

and

$$E[X^2] = \beta_i^2(\theta). \tag{9. 10}$$

It is easy to check that the Condition 3.1 in Bibby and Sorensen [1] holds in this case and applying the Lemma 3.1 in Bibby and Sorensen [1] (cf. Florens-Zmirou [4]), we obtain that

$$\frac{1}{n}\sum_{j=1}^{n} u_{i\varepsilon}(j\Delta)u_{i\varepsilon}((j-1)\Delta) \to E[XY] \text{ in probability as } n \to \infty$$

and

$$\frac{1}{n}\sum_{j=1}^{n} u_{i\varepsilon}^2((j-1)\Delta) \to E[X^2] \text{ in probability as } n \to \infty.$$

The above relations imply that

$$\frac{\sum_{j=1}^{n} u_{i\varepsilon}(j\Delta)u_{i\varepsilon}((j-1)\Delta)}{\sum_{j=1}^{n} u_{i\varepsilon}^2((j-1)\Delta)} \rightarrow \frac{E[XY]}{E[X^2]} \text{ in probability as } n \rightarrow \infty.$$

The following result follows as a consequence of the above observation and the relations (9.9) and (9.10).

Theorem 9.1: The estimator $\hat{b}_{i\varepsilon}$ converges in probability to $b(\theta)$ as $n \rightarrow \infty$.

Since the function $b(\theta)$ has a continuous inverse function, the following result is a consequence of Theorem 9.1.

Theorem 9.2: The estimator $\hat{\theta}_{i\varepsilon} = b^{-1}(\hat{b}_{i\varepsilon})$ converges in probability to θ as $n \rightarrow \infty$.

Let $\psi_i(\theta) = e^{\Delta(b(\theta)-\lambda_i)}$. Note that

$$\psi_i(\hat{\theta}_{i\varepsilon}) = \frac{\sum_{j=1}^{n} u_{i\varepsilon}(j\Delta)u_{i\varepsilon}((j-1)\Delta)}{\sum_{j=1}^{n} u_{i\varepsilon}^2((j-1)\Delta)}.$$

Hence

$$\sqrt{n}\{\psi_i(\hat{\theta}_{i\varepsilon}) - \psi_i(\theta)\} = \frac{n^{-1/2}\sum_{j=1}^{n}[u_{i\varepsilon}(j\Delta)u_{i\varepsilon}((j-1)\Delta) - \psi_i(\theta)u_{i\varepsilon}^2((j-1)\Delta)]}{n^{-1}\sum_{j=1}^{n} u_{i\varepsilon}^2((j-1)\Delta)}.$$

Since

$$E[u_{i\varepsilon}(j\Delta)u_{i\varepsilon}((j-1)\Delta)|u_{i\varepsilon}((j-1)\Delta)] = \psi_i(\theta)u_{i\varepsilon}^2((j-1)\Delta)$$

it follows by Lemma 3.1 of Bibby and Sorensen (1995) that

$$n^{-1/2}\sum_{j=1}^{n}[u_{i\varepsilon}(j\Delta)u_{i\varepsilon}((j-1)\Delta) - \psi_i(\theta)u_{i\varepsilon}^2((j-1)\Delta)]$$

converges in distribution to the normal distribution with mean zero and variance equal to $\gamma_i(\theta) = E[XY - E(XY|X)]^2$ where the random variables X and Y are as defined above. It can be checked that

$$\begin{aligned}\gamma_i(\theta) &= \frac{\varepsilon^2\sigma^2(\theta)(e^{2(b(\theta)-\lambda_i)\Delta} - 1)}{2(b(\theta)-\lambda_i)(\lambda_i+1)}\beta_i^2(\theta) \\ &= (1 - e^{2(b(\theta)-\lambda_i)\Delta})\beta_i^4(\theta).\end{aligned}$$

Applying the δ - method , we obtain that

$$n^{1/2}\Delta(b(\hat{\theta}_{i\varepsilon}) - b(\theta))$$

converges in distribution to the normal distribution with mean zero and variance $e^{-2\Delta(b(\theta)-\lambda_i)}\gamma_i(\theta)\beta_i^{-4}(\theta)$ and we have the following theorem.

Theorem 9.3: Under the conditions stated above

$$n^{1/2}\Delta(b(\hat{\theta}_{i\varepsilon}) - b(\theta)) \xrightarrow{\mathcal{L}} N(0, e^{-2\Delta(b(\theta)-\lambda_i)} - 1) \text{ as } n \to \infty.$$

Applying the δ-method once again, we obtain that

$$n^{1/2}\Delta(\hat{\theta}_{i\varepsilon} - \theta)$$

converges in distribution to the normal distribution with mean zero and variance $b'(\theta)^{-2}e^{-2\Delta(b(\theta)-\lambda_i)}\gamma_i(\theta)\beta_i^{-4}(\theta)$ and we have the following theorem.

Theorem 9.4: Under the conditions stated above

$$n^{1/2}\Delta(\hat{\theta}_{i\varepsilon} - \theta) \xrightarrow{\mathcal{L}} N(0, (b'(\theta))^{-2}(e^{-2\Delta(b(\theta)-\lambda_i)} - 1)) \text{ as } n \to \infty.$$

Remarks: Note that the estimators $\{\hat{\theta}_{i\varepsilon}, 1 \leq i \leq N\}$ are *independent*, consistent and asymptotically normal for the parameter θ in the stochastic partial differential equation (9.1). We will now discuss a method for combining these estimators to get an improved estimator.

Let $\tilde{\theta}_\varepsilon = \sum_{i=1}^N \alpha_i\hat{\theta}_{i\varepsilon}$ where $\alpha_i, 1 \leq i \leq N$ is a nonrandom sequence of coefficients to be chosen. Note that

$$\tilde{\theta}_\varepsilon \xrightarrow{p} [\sum_{i=1}^N \alpha_i]\theta \text{ as } n \to \infty$$

by the Theorem 9.2 and hence $\tilde{\theta}_\varepsilon$ is consistent for θ provided $\sum_{i=1}^N \alpha_i = 1$. Further more

$$\sqrt{n}\Delta(\tilde{\theta}_\varepsilon - \theta) \xrightarrow{\mathcal{L}} N(0, (b'(\theta))^{-2}\sum_{i=1}^N \alpha_i^2(e^{-2\Delta(b(\theta)-\lambda_i)} - 1)) \text{ as } n \to \infty.$$

This follows from the Theorem 9.4 and the independence of the estimators $\{\hat{\theta}_{i\varepsilon}, 1 \leq i \leq N\}$. We now obtain the optimum combination of the coefficients $\{\alpha_i, 1 \leq i \leq N\}$ by minimising the asymptotic variance

$$\sum_{i=1}^N \alpha_i^2(e^{-2\Delta(b(\theta)-\lambda_i)} - 1)$$

subject to the condition $\sum_{i=1}^{N}\alpha_i = 1$. It is easy to show that α_i is proportional to $(e^{2\Delta(\lambda_i - b(\theta))} - 1)^{-1}$ and the optimum choice of $\alpha_i, 1 \le i \le N$ leads to the "Estimator"

$$\theta_\varepsilon^* = \frac{\sum_{i=1}^{N}(e^{2\Delta(\lambda_i - b(\theta))} - 1)^{-1}\hat{\theta}_{i\varepsilon}}{\sum_{i=1}^{N}(e^{2\Delta(\lambda_i - b(\theta))} - 1)^{-1}}. \tag{9. 11}$$

It is easy to see that

$$\theta_\varepsilon^* \xrightarrow{p} \theta \text{ as } n \to \infty$$

and

$$\sqrt{n}\Delta(\theta_\varepsilon^* - \theta) \xrightarrow{\mathcal{L}} N(0, (b'(\theta))^{-2}(\sum_{i=1}^{N}(e^{-2\Delta(b(\theta)-\lambda_i)} - 1)^{-1})^{-1}) \text{ as } n \to \infty$$

again due to the independence of the estimators $\{\hat{\theta}_{i\varepsilon}, 1 \le i \le N\}$. However the random variable θ_ε^* cannot be considered as an estimator of θ in the true sense since it depends on the unknown parameter θ. In order to avoid this problem, we can consider a modified estimator

$$\hat{\theta}_\varepsilon = \frac{\sum_{i=1}^{N}(e^{2\Delta(\lambda_i - \hat{b}_{i\varepsilon})} - 1)^{-1}\hat{\theta}_{i\varepsilon}}{\sum_{i=1}^{N}(e^{2\Delta(\lambda_i - \hat{b}_{i\varepsilon})} - 1)^{-1}} \tag{9. 12}$$

which is obtained from θ_ε^* by substituting the estimator $\hat{\theta}_{i\varepsilon}$ for the unknown parameter θ in the i-th term in the numerator and denominator in (9.11). In view of the independence, consistency and asymptotic normality of the estimators $\hat{\theta}_{i\varepsilon}, 1 \le i \le N$, it follows that the estimator. $\hat{\theta}_\varepsilon$ is consistent and asymptotically normal for the parameter θ and we have the following result.

Theorem 9.5: Under the conditions stated above,

$$\hat{\theta}_\varepsilon \xrightarrow{p} \theta \text{ as } n \to \infty$$

and

$$n^{1/2}\Delta(\hat{\theta}_\varepsilon - \theta)) \xrightarrow{\mathcal{L}} N(0, (b'(\theta))^{-2}(\sum_{i=1}^{N}(e^{-2\Delta(b(\theta)-\lambda_i)} - 1)^{-1})^{-1}) \text{ as } n \to \infty$$

for any fixed $N \ge 1$.

Example II

Let $(\Omega, \mathcal{F}, P)$ be a probability space and consider the process $u_\varepsilon(t,x), 0 \le x \le 1, 0 \le t \le T$ governed by the stochastic partial differential equation

$$du_\varepsilon(t,x) = b(\theta)\,\Delta u_\varepsilon(t,x)dt + \varepsilon\sigma(\theta)(I - \Delta)^{-1/2}dW(t,x) \tag{9. 13}$$

Suppose that $\theta \in \Theta \subset R$. Assume that the function $b(\theta) > 0$ for all $\theta \in \Theta$. Further suppose that the functional form of the function $b(\theta)$ is known and it is differentiable with respect to θ with non zero derivative. In addition assume that the function $\sigma(\theta) > 0$ is known but the parameter $\theta \in \Theta$ is unknown. Suppose further the following the initial and the boundary conditions

$$
\begin{aligned}
u_\varepsilon(0,x) &= f(x),\ 0 < x < 1,\ f \in L_2[0,1], \\
u_\varepsilon(t,0) &= u_\varepsilon(t,1) = 0,\ 0 \le t \le T.
\end{aligned} \tag{9. 14}
$$

hold. Here I is the identity operator, $\Delta = \frac{\partial^2}{\partial x^2}$ and the process $W(t,x)$ is the cylindrical Brownian motion in $L_2[0,1]$. The Fourier coefficients $u_{i\varepsilon}(t)$ satisfy the stochastic differential equations

$$du_{i\varepsilon}(t) = -b(\theta)\lambda_i u_{i\varepsilon}(t)dt + \sigma(\theta)\frac{\varepsilon}{\sqrt{\lambda_i + 1}}dW_i(t),\ 0 \le t \le T, \tag{9. 15}$$

with

$$u_{i\varepsilon}(0) = v_i, v_i = \int_0^1 f(x)e_i(x)dx. \tag{9. 16}$$

Suppose the collection of observations consists of $\{u_{i\varepsilon}(j\Delta), 0 \le j \le n, 1 \le i \le N\}$ where $\Delta > 0$.

Note that the process $\{u_{i\varepsilon}(t), 0 \le t \le T\}$ is the Ornstein-Uhlenbeck process and it is well known that the conditional distribution of $u_{i\varepsilon}(\Delta)$ given $u_{i\varepsilon}(0)$ is normal with mean $v_i e^{-b(\theta)\lambda_i\Delta}$ and variance $\sigma^2(\theta)\frac{\varepsilon^2(e^{-2b(\theta)\lambda_i\Delta}-1)}{(-2b(\theta)\lambda_i)(\lambda_i+1)}$. It can be shown that

$$H_n(\theta) = -\frac{\lambda_i + 1}{\varepsilon^2\sigma^2(\theta)}\sum_{j=1}^n b'(\theta)\lambda_i u_{i\varepsilon}((j-1)\Delta)(u_{i\varepsilon}(j\Delta) - u_{i\varepsilon}((j-1)\Delta)e^{-b(\theta)\lambda_i\Delta}) \tag{9. 17}$$

is proportional to the optimal estimating function for the estimation of the parameter θ (cf. Bibby and Sorensen [1]) and an estimator for $b(\theta)$ is of the form

$$\hat{b}_{i\varepsilon} = -\lambda_i^{-1}\Delta^{-1}\log\frac{\sum_{j=1}^n u_{i\varepsilon}(j\Delta)u_{i\varepsilon}((j-1)\Delta)}{\sum_{j=1}^n u_{i\varepsilon}^2((j-1)\Delta)}. \tag{9. 18}$$

Since $b(\theta) > 0$, it is well known that the solution of (9.15) is ergodic with the stationary measure with density ν_θ given by the normal distribution with mean zero and variance $\zeta_i^2(\theta) = \varepsilon^2\sigma^2(\theta)\{2\lambda_i b(\theta)(\lambda_i + 1)\}^{-1}$. Further more we have already noted that the transition probability density π_Δ^θ of $u_{i\varepsilon}(\Delta)$ given that $u_{i\varepsilon}(0) = x$ is the normal probability density with mean $xe^{-b(\theta)\lambda_i\Delta}$ and variance $\sigma^2(\theta)\frac{\varepsilon^2(e^{2(-b(\theta)\lambda_i)\Delta}-1)}{2(-b(\theta)\lambda_i)(\lambda_i+1)}$. Let X be a random variable with stationary measure ν_θ and Y be a random variable such that the conditional density of Y given $X = x$ is given by π_Δ^θ. Note that

$$\text{(9. 19)} \qquad \begin{aligned} E[XY] &= E[XE(Y|X)] = E[XXe^{-b(\theta)\lambda_i\Delta}] \\ &= \zeta_i^2(\theta)e^{-b(\theta)\lambda_i\Delta} \end{aligned}$$

and

$$\text{(9. 20)} \qquad E[X^2] = \zeta_i^2(\theta).$$

It is easy to check that the Condition 3.1 in Bibby and Sorensen [1] holds in this case and applying the Lemma 3.1 in Bibby and Sorensen [1] (cf. Florens-Zmirou [4]), we obtain that

$$\frac{1}{n}\sum_{j=1}^{n} u_{i\varepsilon}(j\Delta)u_{i\varepsilon}((j-1)\Delta) \to E[XY] \text{ in probability as } n \to \infty$$

and

$$\frac{1}{n}\sum_{j=1}^{n} u_{i\varepsilon}^2((j-1)\Delta) \to E[X^2] \text{ in probability as } n \to \infty.$$

The above relations imply that

$$\frac{\sum_{j=1}^n u_{i\varepsilon}(j\Delta)u_{i\varepsilon}((j-1)\Delta)}{\sum_{j=1}^n u_{i\varepsilon}^2((j-1)\Delta)} \to \frac{E[XY]}{E[X^2]} \text{ in probability as } n \to \infty.$$

The following result follows as a consequence of the above observation and the relations (9.19) and (9.20).

Theorem 9.6: The estimator $\hat{b}_{i\varepsilon}$ converges in probability to $b(\theta)$ as $n \to \infty$.

Since the function $b(\theta)$ has a continuous inverse function, the following result is a consequence of Theorem 9.6.

Theorem 9.7: The estimator $\hat{\theta}_{i\varepsilon} = b^{-1}(\hat{b}_{i\varepsilon})$ converges in probability to θ as $n \to \infty$.

Let $\psi_i(\theta) = e^{-b(\theta)\lambda_i\Delta}$. Note that

$$\psi_i(\hat{\theta}_{i\varepsilon}) = \frac{\sum_{j=1}^n u_{i\varepsilon}(j\Delta)u_{i\varepsilon}((j-1)\Delta)}{\sum_{j=1}^n u_{i\varepsilon}^2((j-1)\Delta)}.$$

Hence

$$\sqrt{n}\{\psi_i(\hat{\theta}_{i\varepsilon}) - \psi_i(\theta)\} = \frac{n^{-1/2}\sum_{j=1}^n [u_{i\varepsilon}(j\Delta)u_{i\varepsilon}((j-1)\Delta) - \psi_i(\theta)u_{i\varepsilon}^2((j-1)\Delta)]}{n^{-1}\sum_{j=1}^n u_{i\varepsilon}^2((j-1)\Delta)}.$$

Since

$$E[u_{i\varepsilon}(j\Delta)u_{i\varepsilon}((j-1)\Delta)|u_{i\varepsilon}((j-1)\Delta)] = \psi_i(\theta)u_{i\varepsilon}^2((j-1)\Delta),$$

it follows by the Lemma 3.1 of Bibby and Sorensen [1] that

$$n^{-1/2}\sum_{j=1}^{n}[u_{i\varepsilon}(j\Delta)u_{i\varepsilon}((j-1)\Delta) - \psi_i(\theta)u_{i\varepsilon}^2((j-1)\Delta)]$$

converges in distribution to the normal distribution with mean zero and variance equal to $\tau_i(\theta) = E[XY - E(XY|X)]^2$ where the random variables X and Y are as defined above. It can be checked that

$$\begin{aligned}\tau_i(\theta) &= \sigma^2(\theta)\frac{\varepsilon^2(e^{-2b(\theta)\lambda_i\Delta}-1)}{2(-b(\theta)\lambda_i)(\lambda_i+1)}\zeta_i^2(\theta) \\ &= (1-e^{-2b(\theta)\lambda_i\Delta})\zeta_i^4(\theta).\end{aligned}$$

Applying the δ - method , we obtain that

$$n^{1/2}\Delta(b(\hat{\theta}_{i\varepsilon}) - b(\theta))$$

converges in distribution to the normal distribution with mean zero and variance $\lambda_i^{-2}e^{2\Delta b(\theta)\lambda_i}\tau_i(\theta)\zeta_i^{-4}(\theta)$ and we have the following theorem.

Theorem 9.8: Under the conditions stated above

$$n^{1/2}\Delta(b(\hat{\theta}_{i\varepsilon}) - b(\theta)) \xrightarrow{\mathcal{L}} N(0, \lambda_i^{-2}(e^{2\Delta b(\theta)\lambda_i} - 1)) \text{ as } n \to \infty.$$

Applying the δ-method once again, we obtain that

$$n^{1/2}\Delta(\hat{\theta}_{i\varepsilon} - \theta)$$

converges in distribution to the normal distribution with mean zero and variance $(b'(\theta))^{-2}\lambda_i^{-2}(e^{2\Delta b(\theta)\lambda_i})\tau_i(\theta)\beta_i^{-4}(\theta)$ and we have the following theorem.

Theorem 9.9: Under the conditions stated above

$$n^{1/2}\Delta(\hat{\theta}_{i\varepsilon} - \theta) \xrightarrow{\mathcal{L}} N(0, (b'(\theta))^{-2}\lambda_i^{-2}(e^{2\Delta b(\theta)\lambda_i} - 1)) \text{ as } n \to \infty.$$

Remarks: Note that the estimators $\{\hat{\theta}_{i\varepsilon}, 1 \le i \le N\}$ are *independent*, consistent and asymptotically normal for the parameter θ in the stochastic partial differential equation (9.13). We will now discuss a method for combining these estimators to get an improved estimator.

Let $\tilde{\theta}_\varepsilon = \sum_{i=1}^N \alpha_i \hat{\theta}_{i\varepsilon}$ where $\alpha_i, 1 \leq i \leq N$ is a nonrandom sequence of coefficients to be chosen. Note that

$$\tilde{\theta}_\varepsilon \xrightarrow{p} [\sum_{i=1}^N \alpha_i]\theta \text{ as } n \to \infty$$

by the Theorem 9.7 and hence $\tilde{\theta}_\varepsilon$ is consistent for θ provided $\sum_{i=1}^N \alpha_i = 1$. Further more

$$\sqrt{n}\Delta(\tilde{\theta}_\varepsilon - \theta) \xrightarrow{\mathcal{L}} N(0, \sum_{i=1}^N \alpha_i^2 (b'(\theta))^{-2} \lambda_i^{-2} (e^{2\Delta b(\theta)\lambda_i} - 1)) \text{ as } n \to \infty.$$

This follows from the Theorem 9.9 and the independence of the estimators $\{\hat{\theta}_{i\varepsilon}, 1 \leq i \leq N\}$. We now obtain the optimum combination of the coefficients $\{\alpha_i, 1 \leq i \leq N\}$ by minimising the asymptotic variance

$$\sum_{i=1}^N \alpha_i^2 (b'(\theta))^{-2} \lambda_i^{-2} (e^{2\Delta b(\theta)\lambda_i} - 1)$$

subject to the condition $\sum_{i=1}^N \alpha_i = 1$. It is easy to show that α_i is proportional to $\lambda_i^2 (e^{2\Delta\lambda_i b(\theta)} - 1)^{-1}$ and the optimum choice of $\alpha_i, 1 \leq i \leq N$ leads to the "Estimator"

$$\theta_\varepsilon^* = \frac{\sum_{i=1}^N \lambda_i^2 (e^{2\Delta\lambda_i b(\theta)} - 1)^{-1} \hat{\theta}_{i\varepsilon}}{\sum_{i=1}^N \lambda_i^2 (e^{2\Delta\lambda_i b(\theta)} - 1)^{-1}}. \tag{9. 21}$$

It is easy to see that

$$\theta_\varepsilon^* \xrightarrow{p} \theta \text{ as } n \to \infty$$

and

$$\sqrt{n}\Delta(\theta_\varepsilon^* - \theta) \xrightarrow{\mathcal{L}} N(0, ((b'(\theta))^{-2} \sum_{i=1}^N \lambda_i^2 (e^{2\Delta b(\theta)\lambda_i} - 1)^{-1})^{-1}) \text{ as } n \to \infty$$

again due to the independence of the estimators $\{\hat{\theta}_{i\varepsilon}, 1 \leq i \leq N\}$. However the random variable θ_ε^* cannot be considered as an estimator of θ since it depends on the unknown parameter θ. In order to avoid this problem, we can consider a modified estimator

$$\hat{\theta}_\varepsilon = \frac{\sum_{i=1}^N \lambda_i^2 (e^{2\Delta\lambda_i \hat{b}_{i\varepsilon}} - 1)^{-1} \hat{\theta}_{i\varepsilon}}{\sum_{i=1}^N \lambda_i^2 (e^{2\Delta\lambda_i \hat{b}_{i\varepsilon}} - 1)^{-1}} \tag{9. 22}$$

which is obtained from θ_ε^* by substituting the estimator $\hat{\theta}_{i\varepsilon}$ for the unknown parameter θ in the i-th term in the numerator and denominator in (9.21). In view of the independence, consistency and asymptotic normality of the estimators $\hat{\theta}_{i\varepsilon}, 1 \leq i \leq N$, it follows that the estimator $\hat{\theta}_\varepsilon$ is consistent and asymptotically normal for the parameter θ and we have the following result.

Theorem 9.10: Under the conditions stated above,

$$\hat{\theta}_\varepsilon \xrightarrow{p} \theta \text{ as } n \to \infty$$

and

$$n^{1/2}\Delta(\hat{\theta}_\varepsilon - \theta)) \xrightarrow{\mathcal{L}} N(0, ((b'(\theta))^{-2} \sum_{i=1}^{N} \lambda_i^2 (e^{2\Delta b(\theta)\lambda_i} - 1)^{-1})^{-1}) \text{ as } n \to \infty$$

for any fixed $N \geq 1$.

Remarks : In the two examples discussed above, we have assumed that the drift coefficient and the diffusion coefficient are known except for the parameter θ which is to be estimated from the data. Since the estimating functions considered above are linear martingale estimating functions , the function $b(\theta)$ in the diffusion coefficient is only involved and not the function $\sigma(\theta)$ which makes the procedure inefficient. However we get explicit solution for the estimator if we make use of linear martingale estimating functions. If one uses the quadratic martingale estimating functions as described by Sorensen [35] (cf . Prakasa Rao [22]), the estimating function involves both the functions $b(\theta)$ and $\sigma(\theta)$ but the resulting equations are too complex to solve for a user. Since the discretely observed Ornstein -Uhlenbeck processes encountered in both the examples are autoregressive processes, the likelihood function can be computed and the maximum likelihood estimator can be obtained which is asymptotically efficient. The estimators $\hat{b}_{i\varepsilon}$ described in the equations (9.8) and (9.18) *are the the maximum likelihood estimators* in the natural parametrization, that is when the reparametrization by means of the functions $b(\theta)$ and $\sigma(\theta)$ is done away with, and the scalar parameter θ in the drift and the scalar parameter σ in the diffusion term are allowed to vary *independently* keeping the σ fixed eventually, as the interest centers around the parameter θ. In such a case, the combined estimator $\hat{\theta}_\varepsilon$, described in (9.22), of the parameter θ in the drift term is asymptotically efficient.

10 Nonparametric Estimation for some Special SPDE (Discrete sampling)

We now discuss nonparametric estimation of a function $\theta(t)$ involved in the "forcing" term for a class of SPDE's. The problem of estimation of the diffusion coefficient in a SDE from discrete observations has attracted lot of attention recently in view of the applications in mathematical finance especially for modelling interest rates. Our work here deals with a similar probem for a SPDE. A short review of recent results on parametric and nonparametric inference for SPDE's is given in Prakasa Rao [24].

Example I

Let $(\Omega, \mathcal{F}, P)$ be a probability space and consider the process $u_\varepsilon(t,x), 0 \le x \le 1, 0 \le t \le T$ governed by the stochastic partial differential equation

$$du_\varepsilon(t,x) = (\Delta u_\varepsilon(t,x) + u_\varepsilon(t,x))dt + \varepsilon\ \theta(t)\ dW_Q(t,x) \tag{10. 1}$$

where $\Delta = \frac{\partial^2}{\partial x^2}$. Suppose that $\theta(.)$ is a positive valued function with $\theta(t) \in C^m([0,\infty))$ for some $m \ge 1$. Further suppose that $\theta^2(.) \in L^2(R)$ and that the function $\theta(.)$ has a compact support contained in the interval $[-\varepsilon, T+\varepsilon]$ for some $\varepsilon > 0$.

Further suppose the initial and the boundary conditions are given by

$$\begin{cases} u_\varepsilon(0,x) = f(x), f \in L_2[0,1] \\ u_\varepsilon(t,0) = u_\varepsilon(t,1) = 0, 0 \le t \le T \end{cases} \tag{10. 2}$$

and Q is the nuclear covariance operator for the Wiener process $W_Q(t,x)$ taking values in $L_2[0,1]$ so that

$$W_Q(t,x) = Q^{1/2}W(t,x)$$

and $W(t,x)$ is a cylindrical Brownian motion in $L_2[0,1]$. Then, it is known that (cf. Rozovskii [33], Kallianpur and Xiong [11])

$$W_Q(t,x) = \sum_{i=1}^{\infty} q_i^{1/2} e_i(x) W_i(t) \text{ a.s.} \tag{10. 3}$$

where $\{W_i(t), 0 \le t \le T\}, i \ge 1$ are independent one - dimensional standard Wiener processes and $\{e_i\}$ is a complete orthonormal system in $L_2[0,1]$ consisting of eigen vectors of Q and $\{q_i\}$ eigen values of Q.

We assume that the operator Q is a special covariance operator Q with $e_k = sin(k\pi x), k \ge 1$ and $\lambda_k = (\pi k)^2, k \ge 1$. Then $\{e_k\}$ is a complete orthonormal system with the eigen values $q_i = (1+\lambda_i)^{-1}, i \ge 1$ for the operator Q and $Q = (I - \Delta)^{-1}$. Note that

$$dW_Q = Q^{1/2} dW. \tag{10. 4}$$

We define a solution $u_\varepsilon(t,x)$ of (10.1) as a formal sum

$$u_\varepsilon(t,x) = \sum_{i=1}^{\infty} u_{i\varepsilon}(t) e_i(x) \tag{10. 5}$$

(cf. Rozovskii [33]). It can be checked that the Fourier coefficient $u_{i\varepsilon}(t)$ satisfies the stochastic differential equation

$$du_{i\varepsilon}(t) = (1-\lambda_i)u_{i\varepsilon}(t)dt + \frac{\varepsilon}{\sqrt{\lambda_i+1}}\theta(t)dW_i(t),\ 0 \le t \le T \tag{10. 6}$$

with the initial condition

$$u_{i\varepsilon}(0) = v_i,\ v_i = \int_0^1 f(x)e_i(x)dx. \tag{10. 7}$$

Estimation

We now consider the problem of estimation of the function $\theta(t), 0 \le t \le T$ based on the observation of the Fourier coefficients $u_{i\varepsilon}(t_j), t_j = j2^{-n}, j = 0,1,\ldots,[2^nT], 1 \le i \le N$, or equivalently based on the observations $u_\varepsilon^{(N)}(t_j,x), t_j = j2^{-n}, j = 0,1,\ldots,[2^nT]$ of the projection of the process $u_\varepsilon(t,x)$ onto the subspace spanned by $\{e_1,\ldots,e_N\}$ in $L_2[0,1]$. Here $[x]$ denotes the largest integer less than or equal to x.

We will at first construct an estimator of $\theta(.)$ based on the observations $\{u_{i\varepsilon}(t_j), t_j = j2^{-n}, j = 0,1,\ldots,[2^nT]\}$. Our technique follows the methods in Genon-Catalot et al. [5].

Let $\{V_j, -\infty < j < \infty\}$ be an increasing sequence of closed subspaces of $L^2(R)$. Suppose the family $\{V_j, -\infty < j < \infty\}$ is an r-regular multiresolution analysis of $L^2(R)$ such that the associated scale function ϕ and wavelet function ψ are compactly supported and belong to $C^r(R)$. For a short introduction to the properties of wavelets and multiresolution analysis, see Prakasa Rao [22].

Let W_j be the subspace defined by

$$V_{j+1} = V_j \oplus W_j \tag{10. 8}$$

and define

$$\phi_{j,k}(x) = 2^{j/2}\phi(2^jx-k), -\infty < j,k < \infty \tag{10. 9}$$

$$\psi_{j,k}(x) = 2^{j/2}\psi(2^jx-k), -\infty < j,k < \infty. \tag{10. 10}$$

Then (i) for all $-\infty < j < \infty$, the collection of functions $\{\phi_{j,k}, -\infty < k < \infty\}$ is an orthonormal basis of V_j ; (ii) for all $-\infty < j < \infty$, the collection of functions $\{\psi_{j,k}, -\infty < k < \infty\}$ is an orthonormal basis of W_j ; and (iii) the collection of functions $\{\psi_{j,k}, -\infty < j,k < \infty\}$ is an orthonormal basis of $L^2(R)$.

In view of the earlier assumptions made on the function $\theta(t)$, it follows that the function $\theta(t)$ belongs to the Sobolev space $H^m(R)$. Let $j(n)$ be an increasing sequence of positive integers tending to infinity as $n \to \infty$. The space $L^2(R)$ has the following decomposition:

(10. 11) $$L^2(R) = V_{j(n)} \oplus (\oplus_{j \geq j(n)} W_j).$$

The function $\theta^2(t)$ can be represented in the form

(10. 12) $$\theta^2(t) = \sum_{k=-\infty}^{\infty} \mu_{j(n),k}\phi_{j(n),k}(t) + \sum_{j \geq j(n), -\infty<k<\infty} \nu_{j,k}\psi_{j,k}(t)$$

where

(10. 13) $$\mu_{j,k} = \int_R \theta^2(t)\phi_{j,k}(t)dt$$

and

(10. 14) $$\nu_{j,k} = \int_R \theta^2(t)\psi_{j,k}(t)dt.$$

We will now define estimators of the coefficients $\mu_{j,k}$ based on the observations $\{u_{i\varepsilon}(t_r), t_r = r2^{-n}, j = 0, 1, \ldots, [2^n T]\}$. Define

(10. 15) $$\hat{\mu}_{j,k}^{(i)} = \frac{\lambda_i + 1}{\varepsilon^2} \sum_{r=0}^{M-1} \phi_{j,k}(t_r)(u_{i\varepsilon}(t_{r+1}) - u_{i\varepsilon}(t_r))^2$$

where $M = [2^n T]$.

The subspace V_j is not finite dimensional. However, the functions θ^2 and the functions ϕ are compactly supported. Hence, for each resolution j, the set of all k such that $\mu_{j,k} \neq 0$ and the set of all k such that $\hat{\mu}_{j,k} \neq 0$ is a finite set L_j depending only on the constant T and the support of ϕ and the cardinality of the set is $O(2^j)$.

Define the estimator of $\theta^2(t)$ by

(10. 16) $$\hat{\theta}_i^2(t) = \sum_{k \in L_{j(n)}} \hat{\mu}_{j(n),k}^{(i)}\phi_{j(n),k}(t)$$

(10. 17) $$= \sum_{-\infty<k<\infty} \hat{\mu}_{j(n),k}^{(i)}\phi_{j(n),k}(t).$$

Note that for any function f such that

$$\int_0^T f(t)\theta^2(t)dt < \infty,$$

it can be shown that

$$\sum_{r=0}^{M-1} f(t_r)(u_{i\varepsilon}(t_{r+1}) - u_{i\varepsilon}(t_r))^2 \xrightarrow{p} \frac{\varepsilon^2}{\lambda_i + 1} \int_0^T f(t)\theta^2(t)dt \text{ as } n \to \infty.$$

Hence

(10. 18)
$$\hat{\mu}^{(i)}_{j,k} \xrightarrow{p} \mu_{j,k} \text{ as } n \to \infty.$$

Let $h(.)$ be a continuous function on $[0,T]$ with compact support contained in $(0,T)$ and belonging to the Sobolev space $H^{m'}(R)$ with $m' > \frac{1}{2}$. Let h_j be the projection of h on the space V_j. Further more suppose that

(10. 19)
$$r \wedge m + r \wedge m' > 2, j(n) = [\alpha n]$$

with

(10. 20)
$$(2(r \wedge m + r \wedge m'))^{-1} \leq \alpha < \frac{1}{4}.$$

Note that r is the regularity of the multiresolution analysis, m is the exponent of the Soblev space to which θ^2 belongs to and m' is the exponent of the Soblev space to which h belongs to. Applying the Proposition 3.1 of Genon-Catalot et al. [5], we obtain that the following representation holds:

$$\begin{aligned} J_{in} &\equiv 2^{n/2} \int_0^T h(t)(\hat{\theta}_i^2(t) - \theta^2(t))dt \\ &= 2^{n/2} \sum_{r=0}^{M-1} h_{j(n)}(t_r)[(\int_{t_r}^{t_{r+1}} \theta(s)\ dW_i(s))^2 - \int_{t_r}^{t_{r+1}} \theta^2(s)\ ds] + R_{in} \end{aligned}$$

where $R_{in} = o_p(1)$ as $n \to \infty$. Further more

(10. 21)
$$J_{in} \xrightarrow{\mathcal{L}} \mathcal{N}(0, 2\int_0^T h^2(t)\theta^4(t)\ dt) \text{ as } n \to \infty$$

by Theorem 3.1 of Genon-Catalot et al. [5]. Note the estimators $\{\hat{\theta}_i(t), i \geq 1\}$ are independent estimators of $\theta(t)$ for any fixed t since the processes $\{W_i, i \geq 1\}$ are independent Wiener processes.

Let $\gamma(t)$ be a nonegative continuous function with support contained in the interval $[0,T]$. Define

(10. 22)
$$Q_{in} = E\{\int_0^T \gamma(t)(\hat{\theta}_i^2(t) - \theta^2(t))^2 dt\}.$$

Note that Q_{in} is the integrated mean square error of the estimator $\hat{\theta}_i^2(t)$ of the function $\theta^2(t)$ corresponding to the weight function $\gamma(t)$. It can be written in the form

(10. 23)
$$Q_{in} = B_{in}^2 + V_{in}$$

where

$$B_{in}^2 = \int_0^T \gamma(t)(E\hat{\theta}_i^2(t) - \theta^2(t))^2 dt \tag{10. 24}$$

is the integrated square of the bias term with the weight function $\gamma(t)$ and

$$V_{in} = E\{\int_0^T \gamma(t)(\hat{\theta}_i^2(t) - E\hat{\theta}_i^2(t))^2 dt\} \tag{10. 25}$$

is the integrated square of the variance term with the weight function $\gamma(t)$. Let

$$D_{in} = E\{\int_0^T (\hat{\theta}_i^2(t) - E\hat{\theta}_i^2(t))^2 dt\} \tag{10. 26}$$

and suppose that $\sup\{\gamma(t) : t \in [0,T]\} \leq K$. Further suppose that $j(n) - \frac{n}{2} \to -\infty$. Then it follows, by Theorem 4.1 of Genon-Catalot et al. [5], that there exists a constant C_i depending on ε, λ_i and the functions ϕ, γ and θ^2 such that

$$B_{in}^2 \leq C_i(2^{4j(n)-2n} + 2^{-2j(n)(m\wedge r)} + 2^{-n}) \tag{10. 27}$$

and

$$D_{in} = 2^{j(n)-n}2\int_0^T \theta^4(t)dt + o(2^{j(n)-n}). \tag{10. 28}$$

Further more

$$V_{in} \leq KD_{in}. \tag{10. 29}$$

Let

$$\tilde{\theta}_N^2(t) = \frac{1}{N}\sum_{i=1}^N \hat{\theta}_i^2(t). \tag{10. 30}$$

It is obvious that, for any function h satisfying the conditions stated above, and for any fixed integer $N \geq 1$,

$$\begin{aligned}
&2^{n/2}\int_0^T h(t)(\tilde{\theta}_N^2(t) - \theta^2(t))dt \\
&= N^{-1}\sum_{i=1}^N J_{in} \\
&= N^{-1}\sum_{i=1}^N\{2^{n/2}\sum_{r=0}^{M-1} h_{j(n)}(t_r)[(\int_{t_r}^{t_{r+1}} \theta(s)\ dW_i(s))^2 - \int_{t_r}^{t_{r+1}} \theta^2(s)\ ds]\} + N^{-1}\sum_{i=1}^N R_{in} \\
&= 2^{n/2}\sum_{r=0}^{M-1} h_{j(n)}(t_r)\{N^{-1}\sum_{i=1}^N[(\int_{t_r}^{t_{r+1}} \theta(s)\ dW_i(s))^2 - \int_{t_r}^{t_{r+1}} \theta^2(s)\ ds]\} + N^{-1}\sum_{i=1}^N R_{in}.
\end{aligned}$$

From the independence of the estimators $\hat{\theta}_i(t), 1 \leq i \leq N$, it follows from the Theorem 3.1 of Genon-Catalot et al. [5] that

$$(10.\ 31)\quad 2^{n/2} \int_0^T h(t)(\tilde{\theta}_N^2(t) - \theta^2(t))dt \xrightarrow{\mathcal{L}} \mathcal{N}(0, 2N^{-1} \int_0^T h^2(t)\theta^4(t)\ dt) \text{ as } n \to \infty.$$

We have the following theorem.

Theorem 10.1 : Under the conditions stated above , the estimator $\tilde{\theta}_N^2(t)$ of $\theta^2(t)$ satisfies the following property for any function $h(t)$ as defined earlier:

$$(10.\ 32)\quad 2^{n/2} \int_0^T h(t)(\tilde{\theta}_N^2(t) - \theta^2(t))dt \xrightarrow{\mathcal{L}} \mathcal{N}(0, 2N^{-1} \int_0^T h^2(t)\theta^4(t)\ dt) \text{ as } n \to \infty.$$

Let $\gamma(t)$ be a nonegative continuous function with support contained in the interval $[0,T]$. Define

$$(10.\ 33)\qquad Q_n = E\{\int_0^T \gamma(t)(\tilde{\theta}_N^2(t) - \theta^2(t))^2 dt\}.$$

Note that Q_n is the integrated mean square error of the estimator $\tilde{\theta}_N^2(t)$ of the function $\theta^2(t)$ corresponding to the weight function $\gamma(t)$. It can be written in the form

$$(10.\ 34)\qquad Q_n = B_n^2 + V_n$$

where

$$(10.\ 35)\qquad B_n^2 = \int_0^T \gamma(t)(E\tilde{\theta}_N^2(t) - \theta^2(t))^2 dt$$

is the integrated square of the bias term with the weight function $\gamma(t)$ and

$$(10.\ 36)\qquad V_n = E\{\int_0^T \gamma(t)(\tilde{\theta}_N^2(t) - E\tilde{\theta}^2(t))^2 dt\}$$

is the integrated square of the variance term with the weight function $\gamma(t)$. Let

$$(10.\ 37)\qquad D_n = E\{\int_0^T (\tilde{\theta}_N^2(t) - E\tilde{\theta}_i^2(t))^2 dt\}.$$

We have the following theorem from the estimates on $\{B_{in}, 1 \leq i \leq N\}$ and on $\{D_{in}, 1 \leq i \leq N\}$ given above.

Theorem 10.2: Suppose that $j(n) - \frac{n}{2} \to -\infty$. Then there exists a constant C_N depending on $N, \phi, \gamma, \theta^2$ such that

$$(10.\ 38)\qquad B_n^2 \leq C_N\ (2^{4j(n)-2n} + 2^{-2j(n)(m \wedge r)} + 2^{-n})$$

and

(10. 39)
$$D_n = N^{-1} 2^{j(n)-n} 2 \int_0^T \theta^4(t)dt + o(N^{-1}\, 2^{j(n)-n}).$$

Further more

(10. 40)
$$V_n \leq K D_n$$

where $K = \sup\{\gamma(t) : 0 \leq t \leq T\}$.

Example II

Let $(\Omega, \mathcal{F}, P)$ be a probability space and consider the process $u_\varepsilon(t,x), 0 \leq x \leq 1, 0 \leq t \leq T$ governed by the stochastic partial differential equation

(10. 41)
$$du_\varepsilon(t,x) = \Delta u_\varepsilon(t,x)dt + \varepsilon\ \theta(t)\ (I - \Delta)^{-1/2}\ dW(t,x)$$

where $\Delta = \frac{\partial^2}{\partial x^2}$. Suppose that $\theta(.)$ is a positive valued function with $\theta(t) \in C^m([0,\infty])$ for some $m \geq 1$. Further suppose that $\theta^2(.) \in L^2(R)$ and that the function $\theta(.)$ has a compact support contained in the interval $[-\varepsilon, T+\varepsilon]$ for some $\varepsilon > 0$.

Further suppose the initial and the boundary conditions are given by

(10. 42)
$$\begin{cases} u_\varepsilon(0,x) = f(x), f \in L_2[0,1] \\ u_\varepsilon(t,0) = u_\varepsilon(t,1) = 0, 0 \leq t \leq T \end{cases}$$

We define a solution $u_\varepsilon(t,x)$ of (10.41) as a formal sum

(10. 43)
$$u_\varepsilon(t,x) = \sum_{i=1}^{\infty} u_{i\varepsilon}(t)e_i(x)$$

(cf. Rozovskii [33]). It can be checked that the Fourier coefficient $u_{i\varepsilon}(t)$ satisfies the stochastic differential equation

(10. 44)
$$du_{i\varepsilon}(t) = -\lambda_i u_{i\varepsilon}(t)dt + \frac{\varepsilon}{\sqrt{\lambda_i + 1}}\theta(t)dW_i(t),\ 0 \leq t \leq T$$

with the initial condition

(10. 45)
$$u_{i\varepsilon}(0) = v_i,\ v_i = \int_0^1 f(x)e_i(x)dx.$$

Estimation

We now consider the problem of estimation of the function $\theta(t), 0 \leq t \leq T$ based on the observation of the Fourier coefficients $u_{i\varepsilon}(t_j), t_j = j2^{-n}, j = 0, 1, \ldots, [2^n T], 1 \leq i \leq N$, or equivalently based on discrete observations $u_\varepsilon^{(N)}(t_j,x), t_j = j2^{-n}, j = 0,1,\ldots,[2^nT]$ of the projection of the process $u_\varepsilon(t,x)$ onto the subspace spanned by $\{e_1,\ldots,e_N\}$ in $L_2[0,1]$.

We will at first construct an estimator of $\theta(.)$ based on the observations $\{u_{i\varepsilon}(t_j), t_j = j2^{-n}, j = 0,1,\ldots,[2^nT]\}$. Our technique again follows the methods in Genon-Catalot et al. [5] using the method of wavelets.

In view of the earlier assumptions made on the function $\theta(t)$, it follows that the function $\theta(t)$ belongs to the Sobolev space $H^m(R)$. Let $j(n)$ be an increasing sequence of positive integers tending to infinity as $n \to \infty$. The space $L^2(R)$ has the following decomposition:

(10. 46)
$$L^2(R) = V_{j(n)} \oplus (\oplus_{j\geq j(n)} W_j).$$

The function $\theta^2(t)$ can be represented in the form

(10. 47)
$$\theta^2(t) = \sum_{k=-\infty}^{\infty} \mu_{j(n),k}\phi_{j(n),k}(t) + \sum_{j\geq j(n), -\infty<k<\infty} \nu_{j,k}\psi_{j,k}(t)$$

where

(10. 48)
$$\mu_{j,k} = \int_R \theta^2(t)\phi_{j,k}(t)dt$$

and

(10. 49)
$$\nu_{j,k} = \int_R \theta^2(t)\psi_{j,k}(t)dt.$$

We will now define estimators of the coefficients $\mu_{j,k}$ based on the observations $\{u_{i\varepsilon}(t_r), t_r = r2^{-n}, j = 0,1,\ldots,[2^nT]\}$. Define

(10. 50)
$$\hat{\mu}_{j,k}^{(i)} = \frac{\lambda_i + 1}{\varepsilon^2}\sum_{r=0}^{M-1} \phi_{j,k}(t_r)(u_{i\varepsilon}(t_{r+1}) - u_{i\varepsilon}(t_r))^2$$

where $M = [2^nT]$.

The subspace V_j is not finite dimensional. However, the functions θ^2 and the functions ϕ are compactly supported. Hence, for each resolution j, the set of all k such that $\mu_{j,k} \neq 0$ and the set of all k such that $\hat{\mu}_{j,k} \neq 0$ is a finite set L_j depending only on the constant T and the support of ϕ and the cardinality of the set is $O(2^j)$.

Define the estimator of $\theta^2(t)$ by

(10. 51)
$$\hat{\theta}_i^2(t) = \sum_{k\in L_{j(n)}} \hat{\mu}_{j(n),k}^{(i)}\phi_{j(n),k}(t)$$

(10. 52)
$$= \sum_{-\infty<k<\infty} \hat{\mu}_{j(n),k}^{(i)}\phi_{j(n),k}(t).$$

Note that for any function f such that

$$\int_0^T f(t)\theta^2(t)dt < \infty,$$

it can be shown that

$$\sum_{r=0}^{M-1} f(t_r)(u_{i\varepsilon}(t_{r+1}) - u_{i\varepsilon}(t_r))^2 \xrightarrow{p} \frac{\varepsilon^2}{\lambda_i + 1}\int_0^T f(t)\theta^2(t)dt \text{ as } n \to \infty.$$

Hence

(10. 53) $$\hat{\mu}_{j,k}^{(i)} \xrightarrow{p} \mu_{j,k} \text{ as } n \to \infty.$$

Let $h(.)$ be a continuous function on $[0,T]$ with compact support contained in $(0,T)$ and belonging to the Sobolev space $H^{m'}(R)$ with $m' > \frac{1}{2}$. Let h_j be the projection of h on the space V_j. Further more suppose that

(10. 54) $$r \wedge m + r \wedge m' > 2, j(n) = [\alpha n]$$

with

(10. 55) $$(2(r \wedge m + r \wedge m'))^{-1} \leq \alpha < \frac{1}{4}.$$

Note that r is the regularity of the multiresolution analysis, m is the exponent of the Soblev space to which θ^2 belongs to and m' is the exponent of the Soblev space to which h belongs to. Applying the Proposition 3.1 of Genon-Catalot et al. [5], we obtain that the following representation holds:

$$\begin{aligned}\tilde{J}_{in} &\equiv 2^{n/2}\int_0^T h(t)(\hat{\theta}_i^2(t) - \theta^2(t))dt \\ &= 2^{n/2}\sum_{r=0}^{M-1} h_{j(n)}(t_r)[(\int_{t_r}^{t_{r+1}} \theta(s)\ dW_i(s))^2 - \int_{t_r}^{t_{r+1}} \theta^2(s)\ ds] + \tilde{R}_{in}\end{aligned}$$

where $\tilde{R}_{in} = o_p(1)$ as $n \to \infty$. Further more

(10. 56) $$\tilde{J}_{in} \xrightarrow{\mathcal{L}} \mathcal{N}(0, 2\int_0^T h^2(t)\theta^4(t)\ dt) \text{ as } n \to \infty$$

by Theorem 3.1 of Genon-Catalot et al. [5]. Note the estimators $\{\hat{\theta}_i(t), i \geq 1\}$ are independent estimators of $\theta(t)$ for any fixed t since the processes $\{W_i, i \geq 1\}$ are independent Wiener processes.

Let $\gamma(t)$ be a nonegative continuous function with support contained in the interval $[0,T]$. Define

$$\tilde{Q}_{in} = E\{\int_0^T \gamma(t)(\hat{\theta}_i^2(t) - \theta^2(t))^2 dt\}. \tag{10. 57}$$

Note that $\tilde{Q}_{in}$ is the integrated mean square error of the estimator $\hat{\theta}_i^2(t)$ of the function $\theta^2(t)$ corresponding to the weight function $\gamma(t)$. It can be written in the form

$$\tilde{Q}_{in} = \tilde{B}_{in}^2 + \tilde{V}_{in} \tag{10. 58}$$

where

$$\tilde{B}_{in}^2 = \int_0^T \gamma(t)(E\hat{\theta}_i^2(t) - \theta^2(t))^2 dt \tag{10. 59}$$

is the integrated square of the bias term with the weight function $\gamma(t)$ and

$$\tilde{V}_{in} = E\{\int_0^T \gamma(t)(\hat{\theta}_i^2(t) - E\hat{\theta}_i^2(t))^2 dt\} \tag{10. 60}$$

is the integrated square of the variance term with the weight function $\gamma(t)$. Let

$$\tilde{D}_{in} = E\{\int_0^T (\hat{\theta}_i^2(t) - E\hat{\theta}_i^2(t))^2 dt\} \tag{10. 61}$$

and suppose that $\sup\{\gamma(t) : t \in [0,T]\} \leq K$. Further suppose that $j(n) - \frac{n}{2} \to -\infty$. Then it follows, by Theorem 4.1 of Genon-Catalot et al. [5], that there exists a constant $\tilde{C}_i$ depending on ε, λ_i and the functions ϕ, γ and θ^2 such that

$$\tilde{B}_{in}^2 \leq \tilde{C}_i(2^{4j(n)-2n} + 2^{-2j(n)(m\wedge r)} + 2^{-n}) \tag{10. 62}$$

and

$$\tilde{D}_{in} = 2^{j(n)-n} 2\int_0^T \theta^4(t)dt + o(2^{j(n)-n}). \tag{10. 63}$$

Further more

$$\tilde{V}_{in} \leq K\tilde{D}_{in}. \tag{10. 64}$$

Let

$$\tilde{\theta}_N^2(t) = \frac{1}{N}\sum_{i=1}^N \hat{\theta}_i^2(t). \tag{10. 65}$$

It is obvious that, for any function h satisfying the conditions stated above, and for any fixed integer $N \geq 1$,

$$\begin{aligned} & 2^{n/2}\int_0^T h(t)(\tilde{\theta}_N^2(t) - \theta^2(t))dt \\ = \;& N^{-1}\sum_{i=1}^N \tilde{J}_{in} \end{aligned}$$

$$= N^{-1}\sum_{i=1}^{N}\{2^{n/2}\sum_{r=0}^{M-1} h_{j(n)}(t_r)[(\int_{t_r}^{t_{r+1}} \theta(s)\ dW_i(s))^2 - \int_{t_r}^{t_{r+1}} \theta^2(s)\ ds]\} + N^{-1}\sum_{i=1}^{N}\tilde{R}_{in}$$

$$= 2^{n/2}\sum_{r=0}^{M-1} h_{j(n)}(t_r)\{N^{-1}\sum_{i=1}^{N}[(\int_{t_r}^{t_{r+1}} \theta(s)\ dW_i(s))^2 - \int_{t_r}^{t_{r+1}} \theta^2(s)\ ds]\} + N^{-1}\sum_{i=1}^{N}\tilde{R}_{in}.$$

From the independence of the estimators $\hat{\theta}_i(t), 1 \leq i \leq N$, it follows from the Theorem 3.1 of Genon-Catalot et al. [5] that

(10. 66) $2^{n/2}\int_0^T h(t)(\tilde{\theta}_N^2(t) - \theta^2(t))dt \xrightarrow{\mathcal{L}} \mathcal{N}(0, 2N^{-1}\int_0^T h^2(t)\theta^4(t)\ dt)$ as $n \to \infty$.

We have the following theorem.

Theorem 10.3: Under the conditions stated above , the estimator $\tilde{\theta}_N^2(t)$ of $\theta^2(t)$ satisfies the following property for any function $h(t)$ as defined earlier:

(10. 67) $2^{n/2}\int_0^T h(t)(\tilde{\theta}_N^2(t) - \theta^2(t))dt \xrightarrow{\mathcal{L}} \mathcal{N}(0, 2N^{-1}\int_0^T h^2(t)\theta^4(t)\ dt)$ as $n \to \infty$.

Let $\gamma(t)$ be a nonegative continuous function with support contained in the interval $[0, T]$. Define

(10. 68)
$$\tilde{Q}_n = E\{\int_0^T \gamma(t)(\tilde{\theta}_N^2(t) - \theta^2(t))^2 dt\}.$$

Note that $\tilde{Q}_n$ is the integrated mean square error of the estimator $\tilde{\theta}_N^2(t)$ of the function $\theta^2(t)$ corresponding to the weight function $\gamma(t)$. It can be written in the form

(10. 69)
$$\tilde{Q}_n = \tilde{B}_n^2 + \tilde{V}_n$$

where

(10. 70)
$$\tilde{B}_n^2 = \int_0^T \gamma(t)(E\tilde{\theta}_N^2(t) - \theta^2(t))^2 dt$$

is the integrated square of the bias term with the weight function $\gamma(t)$ and

(10. 71)
$$\tilde{V}_n = E\{\int_0^T \gamma(t)(\tilde{\theta}_N^2(t) - E\tilde{\theta}^2(t))^2 dt\}$$

is the integrated square of the variance term with the weight function $\gamma(t)$. Let

(10. 72)
$$\tilde{D}_n = E\{\int_0^T (\tilde{\theta}_N^2(t) - E\tilde{\theta}_i^2(t))^2 dt\}.$$

We have the following theorem from the estimates on $\{\tilde{B}_{in}, 1 \leq i \leq N\}$ and on $\{\tilde{D}_{in}, 1 \leq i \leq N\}$ given above.

Theorem 10.4: Suppose that $j(n) - \frac{n}{2} \to -\infty$. Then there exists a constant $\tilde{C}_N$ depending on $N, \phi, \gamma, \theta^2$ such that

(10. 73)
$$\tilde{B}_n^2 \le \tilde{C}_N \left(2^{4j(n)-2n} + 2^{-2j(n)(m\wedge r)} + 2^{-n}\right)$$

and

(10. 74)
$$\tilde{D}_n = N^{-1}2^{j(n)-n}2\int_0^T \theta^4(t)dt + o(N^{-1}\, 2^{j(n)-n}).$$

Further more

(10. 75)
$$\tilde{V}_n \le K\tilde{D}_n$$

where $K = \sup\{\gamma(t) : 0 \le t \le T\}$.

Remarks : It can be seen, from the Theorems 10.1 and 10.2 and from the Theorems 10.3 and 10.4, that the limiting behaviour of the estimator $\tilde{\theta}_N^2(t)$ of $\theta^2(t)$ does not depend on the "trend" terms in the SPDE's discussed in both the examples as long as the "trend" terms in the SDE's satisfied by the Fourier coefficients do not depend on the function $\theta(t)$ or any other unknown functions. This has also been pointed out by Genon-Catalot et al. [5] in their work on the estimation of the diffusion coefficient for SDE's.

11 Nonparametric Estimation for Parabolic SPDE (Discrete sampling)

Let $(\Omega, \mathcal{F}, P)$ be a probability space and consider a stochastic partial differential equation (SPDE) of the form

(11. 1)
$$du(t,x) = Au(t,x)dt + \theta(t)dW(t,x), 0 \le t \le T, x \in G$$

where A is a partial differential operator, $\theta(t)$ is a positive valued function with $\theta(t) \in C^m([0,\infty)]$ for some $m \ge 1$ and $W(t,x)$ is a cylindrical Brownian motion in $L_2(G), G$ being a bounded domain in R^d with the boundary ∂G as a C^∞-manifold of dimension $(d-1)$ and locally G is totally on one side of ∂G. For the definition of cylindrical Brownian motion, see, Kallianpur and Xiong [11], p.93.

Suppose the solution $u(t,x)$ of (11.1) has to satisfy the boundary conditions

(11. 2)
$$u(0,x) = u_0(x)$$

and

(11. 3)
$$D^\gamma u(t,x)|_{\partial G} = 0$$

for all multiindices γ such that $|\gamma| = \frac{1}{2}\ell - 1$ where ℓ is positive even integer. Here

(11. 4) $$D^{\gamma} f(\boldsymbol{x}) = \frac{\partial^{|\gamma|}}{\partial x_1^{\gamma_1} \cdots \partial x_d^{\gamma_d}} f(\boldsymbol{x})$$

with $|\gamma| = \gamma_1 + \cdots + \gamma_d$. Suppose that

(11. 5) $$A(\boldsymbol{x})u = - \sum_{|\alpha|,|\beta| \leq \ell} (-1)^{|\alpha|} D^{\alpha}(a^{\alpha\beta}(\boldsymbol{x}) D^{\beta} u)$$

where

(11. 6) $$a^{\alpha\beta}(\boldsymbol{x}) \in C^{\infty}(\bar{G}).$$

We follow the notation introduced in Huebner and Rozovskii [8]. Assume that the following conditions hold.

(H1) The operator A satisfies the condition

$$\int_G Auv dx = \int_G uAv dx, u, v \in C_0^{\infty}(G).$$

(H2) A is a uniformly strongly elliptic operator of order ℓ.

For $s > 0$, denote the closure of $C_0^{\infty}(G)$ in the Sobolev space $W^{s,2}(G)$ by $W_0^{s,2}$.

The operator A with boundary conditions defined above can be extended to a closed self-adjoint operator $\mathcal{L}$ on $L_2(G)$ (Shimakura [34]). In view of the condition (H2) , the operator $\mathcal{L}$ is lower semibounded, that is there exists a constant k such that $-\mathcal{L} + kI > 0$ and the resolvent $(kI - \mathcal{L})^{-1}$ is compact. Let $\Lambda = (kI - \mathcal{L})^{\frac{1}{\ell}}$ where $m = Ord(A)$. Let h_i be an orthonormal system of eigen functions of Λ. We assume that the following condition holds.

(H3) There exists a complete orthonormal system $\{h_i, i \geq 1\}$ such that

$$\Lambda h_i = \lambda_i h_i.$$

The elements of the basis $\{h_i, i \geq 1\}$ are also eigen functions for the operator $\mathcal{L}$, that is

$$\mathcal{L} h_i = \mu_i h_i$$

where

$$\mu_i = -\lambda_i^{\ell} + k.$$

For $s \geq 0$, define H^s to be the set of all $u \in L_2(G)$ such that

$$||u||_s = (\sum_{j=1}^{\infty} \lambda_j^{2s} |(u, h_j)_{L_2(G)}|^2)^{1/2} < \infty.$$

For $s < 0$, H^s is defined to be the closure of $L_2(G)$ in the norm $||u||_s$ given above. Then H^s is a Hilbert space with respect to the inner product $(.,)_s$ associated with the norm $||.||_s$ and the functions $h_i^s = \lambda_i^{-s} h_i, i \geq 1$ form an orthonormal basis in H^s.

In addition to the conditions (H1)-(H3), we assume that

(H4)$u_0 \in H^{-\alpha}$ where $\alpha > \frac{d}{2}$. Note that $u_0 \in L_2(G)$.

(H5) The operator A is a uniformly strongly elliptic of even order ℓ and has the same system of eigen functions $\{h_i, i \geq 1\}$ as $\mathcal{L}$.

The conditons (H1)-(H5) described above are similar as those in Huebner and Rozovskii [8].

Note that $u_0 \in H^{-\alpha}$. Define

(11. 7) $$u_{0i} = (u_o, h_i^{-\alpha})_{-\alpha}.$$

Then the random field

(11. 8) $$u(t,x) = \sum_{i=1}^{\infty} u_i(t) h_i^{-\alpha}(x)$$

is the solution of (11.1) subject to the boundary conditions (11.2) and (11.3) where $u_i(t)$ is the unique solution of the stochastic differential equation

(11. 9) $$du_i(t) = \mu_i u_i(t)dt + \lambda_i^{-\alpha}\theta(t)dW_i(t), 0 \leq t \leq T,$$

(11. 10) $$u_i(0) = u_{0i}.$$

Let π^N be the orthogonal projection operator of $H^{-\alpha}$ onto the subspace spanned by $\{h_i^{-\alpha}, 1 \leq i \leq N\}$. Let

(11. 11) $$\begin{aligned} u^N(t,x) &= \pi^N u(t,x) \\ &= \sum_{i=1}^{N} u_i(t) h_i^{-\alpha}(x) \end{aligned}$$

where $u_i(t)$ is the solution of (11.9) subject to (11.10). Note that

(11. 12) $$du^N(t,x) = Au^N(t,x)dt + \theta(t)dW^N(t,x), 0 \leq t \leq T, x \in G$$

with

(11. 13) $$u^N(0,x) = \pi^N u_0(x)$$

and

$$W^N(t,x) = \sum_{i=1}^{N} \lambda_i^{-\alpha} W_i(t) h_i^{-\alpha}(x). \tag{11. 14}$$

Here $\{W_i(t), t \geq 0\}, i \geq 1$ are independent standard Wiener processes.

We now consider the problem of estimation of the function $\theta(t), 0 \leq t \leq T$ based on the observation of the Fourier coefficients $u_i(t_j), t_j = j2^{-n}, j = 0, 1, \ldots, [2^n T], 1 \leq i \leq N$, or equivalently based on the observations $u^N(t_j, x), t_j = j2^{-n}, j = 0, 1, \ldots, [2^n T]$. Here $[x]$ denotes the largest integer less than or equal to x.

We will at first construct an estimator of $\theta(.)$ based on the observations $\{u_i(t_j), t_j = j2^{-n}, j = 0, 1, \ldots, [2^n T]\}$. Our technique follows the methods in Genon-Catalot et al. [5] as before.

Let $\{V_j, -\infty < j < \infty\}$ be an increasing sequence of closed subspaces of $L^2(R)$. Suppose the family $\{V_j, -\infty < j < \infty\}$ is an r-regular multiresolution analysis of $L^2(R)$ such that the associated scale function ϕ and wavelet function ψ are compactly supported and belong to $C^r(R)$. For a short introduction to the properties of wavelets and multiresolution analysis, see Prakasa Rao [22].

Let W_j be the subspace defined by

$$V_{j+1} = V_j \oplus W_j \tag{11. 15}$$

and define

$$\phi_{j,k}(x) = 2^{j/2}\phi(2^j x - k), -\infty < j, k < \infty \tag{11. 16}$$

$$\psi_{j,k}(x) = 2^{j/2}\psi(2^j x - k), -\infty < j, k < \infty. \tag{11. 17}$$

Then (i) for all $-\infty < j < \infty$, the collection of functions $\{\phi_{j,k}, -\infty < k < \infty\}$ is an orthonormal basis of V_j ; (ii) for all $-\infty < j < \infty$, the collection of functions $\{\psi_{j,k}, -\infty < k < \infty\}$ is an orthonormal basis of W_j ; and (iii) the collection of functions $\{\psi_{j,k}, -\infty < j, k < \infty\}$ is an orthonormal basis of $L^2(R)$.

In view of the earlier assumptions made on the function $\theta(t)$, it follows that the function $\theta(t)$ belongs to the Sobolev space $H^m(R)$. Let $j(n)$ be an increasing sequence of positive integers tending to infinity as $n \to \infty$. The space $L^2(R)$ has the following decomposition:

$$L^2(R) = V_{j(n)} \oplus (\oplus_{j \geq j(n)} W_j). \tag{11. 18}$$

The function $\theta^2(t)$ can be represented in the form

$$\theta^2(t) = \sum_{k=-\infty}^{\infty} \mu_{j(n),k}\phi_{j(n),k}(t) + \sum_{j \geq j(n), -\infty < k < \infty} \nu_{j,k}\psi_{j,k}(t) \tag{11. 19}$$

where

$$\mu_{j,k} = \int_R \theta^2(t)\phi_{j,k}(t)dt \tag{11. 20}$$

and

$$\nu_{j,k} = \int_R \theta^2(t)\psi_{j,k}(t)dt. \tag{11. 21}$$

We will now define estimators of the coefficients $\mu_{j,k}$ based on the observations $\{u_i(t_r), t_r = r2^{-n}, j = 0, 1, \ldots, [2^n T]\}$. Define

$$\hat{\mu}_{j,k}^{(i)} = \lambda_i^{2\alpha} \sum_{r=0}^{M-1} \phi_{j,k}(t_r)(u_i(t_{r+1}) - u_i(t_r))^2 \tag{11. 22}$$

where $M = [2^n T]$.

The subspace V_j is not finite dimensional. However, the functions θ^2 and the functions ϕ are compactly supported. Hence, for each resolution j, the set of all k such that $\mu_{j,k} \neq 0$ and the set of all k such that $\hat{\mu}_{j,k} \neq 0$ is a finite set L_j depending only on the constant T and the support of ϕ and the cardinality of the set is $O(2^j)$.

Define the estimator of $\theta^2(t)$ by

$$\hat{\theta}_i^2(t) = \sum_{k \in L_{j(n)}} \hat{\mu}_{j(n),k}^{(i)} \phi_{j(n),k}(t) \tag{11. 23}$$

$$= \sum_{-\infty < k < \infty} \hat{\mu}_{j(n),k}^{(i)} \phi_{j(n),k}(t). \tag{11. 24}$$

Note that for any function f such that

$$\int_0^T f(t)\theta^2(t)dt < \infty,$$

it can be shown that

$$\sum_{r=0}^{M-1} f(t_r)(u_i(t_{r+1}) - u_i(t_r))^2 \xrightarrow{p} \frac{\varepsilon^2}{\lambda_i + 1} \int_0^T f(t)\theta^2(t)dt \quad \text{as } n \to \infty.$$

Hence

$$\hat{\mu}_{j,k}^{(i)} \xrightarrow{p} \mu_{j,k} \quad \text{as } n \to \infty. \tag{11. 25}$$

Let $h(.)$ be a continuous function on $[0, T]$ with compact support contained in $(0, T)$ and belonging to the Sobolev space $H^{m'}(R)$ with $m' > \frac{1}{2}$. Let h_j be the projection of h on the space V_j. Further more suppose that

$$r \wedge m + r \wedge m' > 2, j(n) = [\alpha n] \tag{11. 26}$$

with

(11. 27) $$(2(r \wedge m + r \wedge m'))^{-1} \leq \alpha < \frac{1}{4}.$$

Note that r is the regularity of the multiresolution analysis, m is the exponent of the Soblev space to which θ^2 belongs to and m' is the exponent of the Soblev space to which h belongs to. Applying the Proposition 3.1 of Genon-Catalot et al. [5], we obtain that the following representation holds:

$$\begin{aligned} J_{in} &\equiv 2^{n/2} \int_0^T h(t)(\hat{\theta}_i^2(t) - \theta^2(t))dt \\ &= 2^{n/2} \sum_{r=0}^{M-1} h_{j(n)}(t_r)[(\int_{t_r}^{t_{r+1}} \theta(s)\ dW_i(s))^2 - \int_{t_r}^{t_{r+1}} \theta^2(s)\ ds] + R_{in} \end{aligned}$$

where $R_{in} = o_p(1)$ as $n \to \infty$. Further more

(11. 28) $$J_{in} \xrightarrow{\mathcal{L}} \mathcal{N}(0, 2\int_0^T h^2(t)\theta^4(t)\ dt) \text{ as } n \to \infty$$

by Theorem 3.1 of Genon-Catalot et al. [5]. Note the estimators $\{\hat{\theta}_i(t), i \geq 1\}$ are independent estimators of $\theta(t)$ for any fixed t since the processes $\{W_i, i \geq 1\}$ are independent Wiener processes.

Let $\gamma(t)$ be a nonegative continuous function with support contained in the interval $[0, T]$. Define

(11. 29) $$Q_{in} = E\{\int_0^T \gamma(t)(\hat{\theta}_i^2(t) - \theta^2(t))^2 dt\}.$$

Note that Q_{in} is the integrated mean square error of the estimator $\hat{\theta}_i^2(t)$ of the function $\theta^2(t)$ corresponding to the weight function $\gamma(t)$. It can be written in the form

(11. 30) $$Q_{in} = B_{in}^2 + V_{in}$$

where

(11. 31) $$B_{in}^2 = \int_0^T \gamma(t)(E\hat{\theta}_i^2(t) - \theta^2(t))^2 dt$$

is the integrated square of the bias term with the weight function $\gamma(t)$ and

(11. 32) $$V_{in} = E\{\int_0^T \gamma(t)(\hat{\theta}_i^2(t) - E\hat{\theta}_i^2(t))^2 dt\}$$

is the integrated square of the variance term with the weight function $\gamma(t)$. Let

(11. 33) $$D_{in} = E\{\int_0^T (\hat{\theta}_i^2(t) - E\hat{\theta}_i^2(t))^2 dt\}$$

and suppose that $\sup\{\gamma(t) : t \in [0,T]\} \leq K$. Further suppose that $j(n) - \frac{n}{2} \to -\infty$. Then it follows, by Theorem 4.1 of Genon-Catalot et al. [5], that there exists a constant C_i depending on ε, λ_i and the functions ϕ, γ and θ^2 such that

$$B_{in}^2 \leq C_i(2^{4j(n)-2n} + 2^{-2j(n)(m\wedge r)} + 2^{-n}) \tag{11. 34}$$

and

$$D_{in} = 2^{j(n)-n}2\int_0^T \theta^4(t)dt + o(2^{j(n)-n}). \tag{11. 35}$$

Further more

$$V_{in} \leq K D_{in}. \tag{11. 36}$$

Let

$$\tilde{\theta}_N^2(t) = \frac{1}{N}\sum_{i=1}^{N}\hat{\theta}_i^2(t). \tag{11. 37}$$

It is obvious that, for any function h satisfying the conditions stated above, and for any fixed integer $N \geq 1$,

$$\begin{aligned}
&2^{n/2}\int_0^T h(t)(\tilde{\theta}_N^2(t) - \theta^2(t))dt \\
&= N^{-1}\sum_{i=1}^{N} J_{in} \\
&= N^{-1}\sum_{i=1}^{N}\{2^{n/2}\sum_{r=0}^{M-1} h_{j(n)}(t_r)[(\int_{t_r}^{t_{r+1}} \theta(s)\ dW_i(s))^2 - \int_{t_r}^{t_{r+1}} \theta^2(s)\ ds]\} + N^{-1}\sum_{i=1}^{N} R_{in} \\
&= 2^{n/2}\sum_{r=0}^{M-1} h_{j(n)}(t_r)\{N^{-1}\sum_{i=1}^{N}[(\int_{t_r}^{t_{r+1}} \theta(s)\ dW_i(s))^2 - \int_{t_r}^{t_{r+1}} \theta^2(s)\ ds]\} + N^{-1}\sum_{i=1}^{N} R_{in}.
\end{aligned}$$

From the independence of the estimators $\hat{\theta}_i(t), 1 \leq i \leq N$, it follows from the Theorem 3.1 of Genon-Catalot et al. [5] that

$$2^{n/2}\int_0^T h(t)(\tilde{\theta}_N^2(t) - \theta^2(t))dt \xrightarrow{\mathcal{L}} \mathcal{N}(0, 2N^{-1}\int_0^T h^2(t)\theta^4(t)\ dt) \text{ as } n \to \infty. \tag{11. 38}$$

We have the following theorem.

Theorem 11.1 : Under the conditions stated above , the estimator $\tilde{\theta}_N^2(t)$ of $\theta^2(t)$ satisfies the following property for any function $h(t)$ as defined earlier:

$$2^{n/2}\int_0^T h(t)(\tilde{\theta}_N^2(t) - \theta^2(t))dt \xrightarrow{\mathcal{L}} \mathcal{N}(0, 2N^{-1}\int_0^T h^2(t)\theta^4(t)\ dt) \text{ as } n \to \infty. \tag{11. 39}$$

Let $\gamma(t)$ be a nonegative continuous function with support contained in the interval $[0,T]$. Define

(11. 40)
$$Q_n = E\{\int_0^T \gamma(t)(\tilde{\theta}_N^2(t) - \theta^2(t))^2 dt\}.$$

Note that Q_n is the integrated mean square error of the estimator $\tilde{\theta}_N^2(t)$ of the function $\theta^2(t)$ corresponding to the weight function $\gamma(t)$. It can be written in the form

(11. 41)
$$Q_n = B_n^2 + V_n$$

where

(11. 42)
$$B_n^2 = \int_0^T \gamma(t)(E\tilde{\theta}_N^2(t) - \theta^2(t))^2 dt$$

is the integrated square of the bias term with the weight function $\gamma(t)$ and

(11. 43)
$$V_n = E\{\int_0^T \gamma(t)(\tilde{\theta}_N^2(t) - E\tilde{\theta}^2(t))^2 dt\}$$

is the integrated square of the variance term with the weight function $\gamma(t)$. Let

(11. 44)
$$D_n = E\{\int_0^T (\tilde{\theta}_N^2(t) - E\tilde{\theta}_i^2(t))^2 dt\}.$$

We have the following theorem from the estimates on $\{B_{in}, 1 \leq i \leq N\}$ and on $\{D_{in}, 1 \leq i \leq N\}$ given above.

Theorem 11.2: Suppose that $j(n) - \frac{n}{2} \to -\infty$. Then there exists a constant C_N depending on $N, \phi, \gamma, \theta^2$ such that

(11. 45)
$$B_n^2 \leq C_N \ (2^{4j(n)-2n} + 2^{-2j(n)(m \wedge r)} + 2^{-n})$$

and

(11. 46)
$$D_n = N^{-1} 2^{j(n)-n} 2 \int_0^T \theta^4(t) dt + o(N^{-1}\, 2^{j(n)-n}).$$

Further more

(11. 47)
$$V_n \leq K D_n$$

where $K = \sup\{\gamma(t) : 0 \leq t \leq T\}$.

12 Remarks

We have considered the problems of nonparametric and parametric estimation of coefficients involved in a special class of parabolic SPDE's which can be reduced to infinite systems of stochastic differential equations. It is possible such a decoupling might not be possible for several classes of SPDE's which occur in stochastic modelling. The problem is to develop statistical methods of inference for such SPDE's in the case of continuous observation of the random field $u(t,x)$ over time and space variables. In our discussion on estimation problems for SPDE's based on discrete sampling, we have assumed that discrete data on the Fourier coefficients of the random field $u(t,x)$ are available. This assumption is also more of mathematical convenience and it amounts to continuous observation of the random field over the space variable x. A more realistic problem is to estimate the coefficients of the SPDE satisfied from observations on the random field observed at discrete times $t_i, 1 \leq i \leq N$ and at discrete points $x_j, 1 \leq j \leq M$. As far as the author is aware, this problem has not been studied. The problems of finding methods for simulation as well as methods for numerical approximation of general SPDE's has not been studied much in the literature

References

1 Bibby, B.M. and Sorensen, M., Martingale estimating functions for discretely observed diffusion processes, *Bernoulli*, **1**, 17-39 (1995).

2 Borwanker, J.D., Kallianpur, G. and Prakasa Rao, B.L.S., The Bernstein-von Mises theorem for Markov processes, *Ann. Math Statist.* **42**, 1241-1253 (1971).

3 Da Prato, G. and Zabczyk, J., *Stochastic Equations in Infinite Dimensions*, Cambridge University Press, Cambridge (1992).

4 Florens-Zmirou, D., Approximate discrete-time schemes for statistics of diffusion processes, *Statistics*, **20**, 547-557 (1989).

5 Genon-Catalot. V., Laredo, C., and Picard, D., Noparametric estimation of the diffusion coefficient by wavelets method, *Scand. J. Statist.*, **19**, 317-335 (1992).

6 Hodgkin, A.L., and Huxley, A.F., A quantitive description of membrance current and its application to conduction and excitation in nerve, *Journal of Physiology*, **117**, 500-544 (1952).

7 Huebner, M., Khasminskii, R. and Rozovskii. B.L., Two examples of parameter estimation for stochastic partial differential equations. In *Stochastic Processes : A Festschrift in Honour of Gopinath Kallianpur*, (Edited by S.Cambanis, J.K.Ghosh, R.L.Karandikar, P.K.Sen), pp. 149-160, Springer, New York (1993).

8 Huebner, M., and Rozovskii, B.L., On asymptotic properties of maximum likelihood estimators for parabolic stochastic SPDE's. *Prob. Theory and Relat. Fields*, **103**, 143-163 (1995).

9 Ibragimov, I.A., and Khasminskii, R. *Statistical Estimation: Asymptotic Theory*, Springer-Verlag, Berlin (1981).

10 Ito, K. *Foundations of Stochastic Differential Equations in Infinite Dimensional Spaces*, Vol. 47 of CBMS Notes, SIAM, Baton Rouge (1984).

11 Kallianpur, G., and Xiong, J. *Stochastic Differential Equations in Infinite Dimensions* , Vol. **26**, IMS Lecture Notes, Hayward, California (1995).

12 Kutoyants, Yu. *Identification of Dynamical Systems with Small Noise* , Kluwer Academic Publishers, Dordrecht (1994).

13 Kutoyants, Yu. and Pilibossian, P., On minimum L_1-norm estimate of the parameter of the Ornstein-Uhlenbeck process. *Statist. Probab. Lett.*, **20**, 117-123 (1994).

14 Millar, P.W., A general approach to the optimality of the minimum distance estimators, *Trans. Amer. Math. Soc.*, **286**, 249-272 (1984).

15 Piterbarg, L., and Rozovskii, B., Maximum likelihood estimators in the equations of physical oceanography. In *Stochastic Modelling in Physical Oceanography*, (Edited by R.J. Adler, P. Muller, and B. Rozovskii), pp.397-421, Birkhauser, Boston (1996).

16 Piterbarg, L., and Rozovskii, B., On asymptotic problems of parameter estimation in stochastic PDE's: Discrete time sampling. *Mathematical Methods of Statistics*, **6**, 200-223 (1997).

17 Prakasa Rao, B.L.S., Estimation of the location of the cusp of a continuous density, *Ann. Math. Statist.*, **39**, 76-87 (1968).

18 Prakasa Rao, B.L.S., The Bernstein - von Mises theorem for a class of diffusion processes,*Teor. Sluch. Proc.*,**9**, 95-101 (1981) (In Russian).

19 Prakasa Rao, B.L.S., On Bayes estimation for diffusion fields. In *Statistics : Applications and New Directions*, (Edited by J.K. Ghosh and J. Roy), pp. 504-511, Statistical Publishing Society, Calcutta (1984).

20 Prakasa Rao, B.L.S., Statistical inference from sampled data for stochastic processes.In *Contemporary Mathematics*,**80** (Edited by N.U.Prabhu), pp. 249-284, American Math. Soc., Providence, (1988).

21 Prakasa Rao, B.L.S., Bernstein - von Mises theorem for parabolic stochastic partial differential equations, Preprint, Indian Statistical Institute, New Delhi (1998).

22 Prakasa Rao, B.L.S. *Statistical Inference for Diffusion type Processes*, Arnold, London and Oxford university Press, New York (1999).

23 Prakasa Rao, B.L.S. *Semimartingales and their Statistical Inference*, CRC Press, Boca Raton , Florida and Chapman and Hall, London (1999).

24 Prakasa Rao, B.L.S., Statistical inference for stochastic partial differential equations. In *Selected Papers: Proc. Symp. Infer. for Stoch. Proc.*, IMS Lecture Notes, Hayward, California (2001) (To appear).

25 Prakasa Rao, B.L.S., Bayes estimation for some stochastic partial differential equations, *J. Statist. Plan. Infer.* , **91**, 511-524 (2000).

26 Prakasa Rao, B.L.S., Nonparametric inference for a class of stochastic partial differential equations, Tech. Report No. 293, Dept. of Statistics and Actuarial Science, University of Iowa (2000).

27 Prakasa Rao, B.L.S., Nonparametric inference for a class of stochastic partial differential equations II, *Statist. Infer. for Stoch. Proc.*, **4**, 41-52 (2001).

28 Prakasa Rao, B.L.S., Estimation for some stochastic partial differential equations based on discrete observations, *Calcutta Stat. Assoc. Bull.*, **50**, 193-200 (2000).

29 Prakasa Rao, B.L.S., Estimation for some stochastic partial differential equations based on discrete observations II, Preprint, Indian Statistical Institute, New Delhi (2000).

30 Prakasa Rao, B.L.S., Minimum distance estimation for some stochastic partial differential equations, Preprint, Indian Statistical Institute, New Delhi (2001).

31 Prakasa Rao, B.L.S., Nonparametric inference for a class of stochastic partial differential equations based on discrete observations, *Sankhya Ser.A* (To appear) .

32 Prakasa Rao, B.L.S., Nonparametric inference for parabolic stochastic partial differential equations, *Random Operators and Stochastic Equations*, **9**, (2001) (To appear).

33 Rozovskii, B.L. *Stochastic Evolution Systems*, Kluwer, Dordrecht (1990).

34 Shimakura, N. *Partial Differential Operators of Elliptic Type*, AMS Transl. Vol. **99**, American Mathematical Society, Providence (1992).

35 Sorensen, M., Estimating functions for discretely observed diffusions: A Review. In *Selected Proceedings of the Symposium on Estimating Functions* , (Edited by I.V.Basawa, V.P. Godambe and R.L. Taylor) ,pp. 305-325, Lecture Notes-Monograph Series, Vol.32, Institute of Mathematical Statistics. Hayward (1997).

SOME UNIFYING TECHNIQUES IN THE THEORY OF ORDER STATISTICS

H. A. David
Department of Statistics, Iowa State University
Ames, IA 50011.
e-mail: hadavid@iastate.edu

Abstract

Let $X_1, \ldots, X_n$ be any n random variables (rv's) and let $X_{1:n} \leq \ldots \leq X_{n:n}$ denote the same variables arranged in nondecreasing order. Then $X_{r:n}$ is called the rth order statistic, $r = 1, \ldots, n$. When one of the X's is dropped at random, there results a simple relation between the order statistics in the original and the reduced samples. This "dropping" argument will be shown to provide a unified approach to establishing recurrence relations between moments of order statistics, whatever the dependence structure of the observations. Also useful in studying recurrence relations is the classical theorem on the probability of occurrence of r events out of n.

It will also be shown that a simple general method of obtaining universal bounds for linear functions of order statistics in terms of the sample standard deviation can be based on Cauchy's inequality coupled with convexity arguments.

Keywords: Recurrence relations; "dropping" argument; universal bounds; Cauchy's inequality; convexity.

1 INTRODUCTION

I greatly appreciate being asked to contribute a paper to this volume. An article on order statistics is particularly appropriate, since Indian statisticians have made major contributions to the subject. Two massive multi-authored volumes, [1] and [2], recently published, summarize much of the theory and applications of order statistics. Their indexes list 70 distinct areas of application including life testing and reliability, the treatment of outlying observations, median and order-statistics filters in signal or image processing, the estimation of parameters and hypothesis testing, etc. The theory of order statistics draws on a rich variety of mathematical techniques. Beyond the familiar mathematical branches, authors have used the theories of convexity and majorization, stochastic orderings and inequalities, functional and integral equations, the calculus of variations, linear programming, combinatorial analysis, and no doubt other techniques.

In this paper we deal with two quite different aspects of order statistics, but show in each case how a few techniques illuminate, unify, and provide easy proofs of key results.

2 RECURRENCE RELATIONS

We need a few preliminaries. Let

$$F_{r:n}(x) = Pr(X_{r:n} \leq x)$$

denote the cumulative distribution function (cdf) of $X_{r:n}$, and

$$\mu_{r:n} = E(X_{r:n})$$

the expectation or mean or first moment of $X_{r:n}$. Continuing to the joint cdf of $X_{r:n}$ and $X_{s:n} (r < s)$, we define

$$F_{r,s:n}(x, y) = Pr(X_{r:n} \leq x, X_{s:n} \leq y)$$

and the product moments

$$\mu_{r,s:n} = E(X_{r:n}X_{s:n}).$$

Finally, we say the random variables (rv's) $X_1, \ldots, X_n$ are *exchangeable* if the joint distribution of any subset depends only on the size of the subset and not on which X's are in the subset. Exchangeable variates are a first generalization of independent, identically distributed (iid) rv's, much used in statistics.

Many authors have studied recurrence relations between the moments of order statistics, both for their intrinsic interest and for their usefulness in reducing the number of independent calculations required for the evaluation of the moments.

Relation 1. Let $X_1, \ldots, X_n$ be exchangeable variates with $Pr\{X_{r:n} \leq x\} = F_{r:n}(x)$ and $E(X_{r:n}) = \mu_{r:n}, r = 1, \ldots, n$. Then subject to the existence of all terms involved,

$$(n-r)\mu_{r:n} + r\mu_{r+1:n} = n\mu_{r:n-1}.$$

Proof. We use the following "dropping" argument, [10]: Drop one of $X_1, \ldots, X_n$ at random and suppose this is $X_{i:n}(i = 1, \ldots, n)$. The resulting ordered variate $X_{r:n-1}$ in the sample of $n-1$ exchangeable rv's is then given by

$$\begin{aligned} X_{r:n-1} &= X_{r+1:n} \text{ for } i = 1, \ldots, r \qquad (A) \\ &= X_{r:n} \text{ for } i = r+1, \ldots, n \qquad (B) \end{aligned}$$

since for (A) the rv with rank $r+1$ in the sample of n has rank r in the sample of $n-1$, etc. But the events A and B have respective probabilities r/n and $(n-r)/n$, so that

$$\begin{aligned} Pr\{X_{r:n-1} \leq x\} &= Pr\{A\}Pr\{X_{r:n-1} \leq x|A\} \\ &+ Pr\{B\}Pr\{X_{r:n-1} \leq x|B\} \\ &= \frac{r}{n}Pr\{X_{r+1:n} \leq x\} + \frac{n-r}{n}Pr\{X_{r:n} \leq x\} \end{aligned}$$

or

$$nF_{r:n-1}(x) = rF_{r+1:n}(x) + (n-r)F_{r:n}(x). \quad (1)$$

Relation 1 follows.

Comments.

1. Relation 1 was first obtained in [6] for iid continuous variables.

2. The "dropping" argument generalizes easily to give the well-known relation for $1 \le r < s \le n$

$$r\mu_{r+1,s+1:n} + (s-r)\mu_{r,s+1:n} + (n-s)\mu_{r,s:n} = n\mu_{r,s:n-1},$$

first proved in the iid continuous case in [12].

Generalization of Relation 1 to any variates $X_1, \ldots, X_n$. In [18] it is shown that even when the X_i have an arbitrary joint distribution, a generalization of (1) is possible, namely

$$rF_{r+1:n}(x) + (n-r)F_{r:n}(x) = \sum_{i=1}^{n} F_{r:n-1}^{(i)}(x), \quad (2)$$

where $F_{r:n-1}^{(i)}(x)$ denotes the cdf of $X_{r:n-1}$ when X_i has been dropped from $X_1, \ldots, X_n (i = 1, \ldots, n)$.

The "dropping" argument provides a simple proof of (2), [8]. Consider dropping at random one of $X_{1:n}, \ldots, X_{n:n}$. The resulting variates may be denoted by $X_{1:n-1}^{(J)}, \ldots, X_{n-1:n-1}^{(J)}$, where J has a discrete uniform distribution over $j = 1, \ldots, n$. Then (A) and (B) will still apply if $X_{r:n-1}$ is replaced by $X_{r:n-1}^{(J)}$. Eq. (2) now follows, since by conditioning on $J = i$, we see that the cdf of $X_{r:n-1}^{(J)}$ is

$$\frac{1}{n}\sum_{i=1}^{n} F_{r:n-1}^{(i)}(x).$$

Relation 2. In the situation of Relation 1

$$\mu_{r:n}^{(k)} = \sum_{i=r}^{n} \binom{i-1}{r-1}\binom{n}{i}(-1)^{i-r}\mu_{i:i}^{(k)}.$$

Thus the moments of $X_{r:n}$ are expressible in terms of the simpler moments of the maxima in samples of $r, r+1, \ldots, n$.

This relation can be established by repeated application of Relation 1, or algebraically. We again use a probabilistic argument, [9], that lends itself to generalization. By a classical result in probability theory, the probability $p_{r,n}$ of the realization of at least r out of the n events $A_1, \ldots, A_n$ is given by (e.g., [11, p. 99])

$$p_{r,n} = \sum_{j=r}^{n} (-1)^{j-r} \binom{j-1}{r-1} S_j. \tag{3}$$

where

$$S_j = \sum_{1 \le i_1 < \ldots < i_j \le n} Pr\{A_{i_1} \ldots A_{i_j}\}.$$

If A_i is the event $X_i \le x, i = 1, \ldots, n$, then $p_{r:n} = Pr\{X_{r:n} \le x\} = F_{r:n}(x)$. Eq. (3) clearly becomes

$$F_{r:n}(x) = \sum_{j=r}^{n} (-1)^{j-r} \binom{j-1}{r-1} \binom{n}{j} F_{j:j}(x), \tag{4}$$

which gives Relation 2 as before.

Duality Principle. If A_i in (3) is taken to be the event $X_i > x, i = 1, \ldots, n$, then $p_{r,n} = Pr\{X_{n-r+1:n} > x\}$. With $\bar{F}(x) = 1 - F(x)$, eq. (3) now gives

$$\bar{F}_{n-r+1:n}(x) = \sum_{j=r}^{n} (-1)^{j-r} \binom{j-1}{r-1} \binom{n}{j} \bar{F}_{1:j}(x).$$

Setting $x = -\infty$, we have

$$1 = \sum_{j=r}^{n} (-1)^{j-r} \binom{j-1}{r-1} \binom{n}{j}$$

and hence obtain the "dual" of (4)

$$F_{n-r+1:n}(x) = \sum_{j=r}^{n} (-1)^{j-r} \binom{j-1}{r-1} \binom{n}{j} F_{1:j}(x) \tag{5}$$

Clearly, to reach (5) from (4) we simply need to change $F_{a:b}(x)$ to $F_{b-a+1:b}(x), 1 \le a \le b \le n$.

This argument may be extended to the general case of dependent rv's $X_1, \ldots, X_n$. The result is made explicit in [5], where the "duality principle" is introduced through a different, slightly less general approach.

Interdependence of Linear Relations. It is interesting to note that eq. (1) can be deduced by applying (4) to each term of (1). Since each of (1) and (4) can be obtained from the other, they must be equivalent. More generally, any recurrence relation linear in the $F_{i:j}$ must be of the form

$$\sum_{j=1}^{n}\sum_{i=1}^{j} a_{ij}F_{i:j}(x) = \sum_{j=1}^{n}\sum_{i=1}^{j} b_{ij}F_{i:j}(x), \tag{6}$$

where the a_{ij} and b_{ij} are constants. By (4) each side of (6) must equal the same linear function $\sum_{j=1}^{n} c_j F_{j:j}(x)$, say, since for arbitrary $F(x)$ there can be no linear identity linking $F_{1:1}(x), \ldots, F_{n:n}(x)$ for all x (except in the trivial case $Pr\{X_1 = \ldots = X_n\} = 1$). In other words, any linear recurrence relation for arbitrary $F(x)$ must be deducible from (4) and therefore also from (1). If proved in the simple case when $X_1, \ldots, X_n$ are iid and continuous, it must automatically hold also when the X's are exchangeable, whether continuous or not.

Generalizations to any variates $X_1, \ldots, X_n$. By repeated application of (2), eq. (4) is generalized in [3] to

$$F_{r:n}(x) = \sum_{j=r}^{n}(-1)^{j-r}\binom{j-1}{r-1}\binom{n}{j}F_{j:j}^{[n-j]}(x), \tag{7}$$

where, with an extension of the "dropping" notation,

$$\binom{n}{j}F_{j:j}^{[n-j]}(x) = \sum_{1\le i_1<\ldots<i_{n-j}\le n} F_{j:j}^{(i_1,\ldots,i_{n-j})}(x). \tag{8}$$

It is seen that (7) also follows immediately from (3), since S_j equals the RHS of (8). Hence the dual of (7) is, in generalization of (5),

$$F_{n-r+1:n}(x) = \sum_{j=r}^{n}(-1)^{j-r}\binom{j-1}{r-1}\binom{n}{j}F_{1:j}^{[n-j]}(x).$$

Further extensions to the joint cdf of two order statistics are given in [9].

3 BOUNDS FOR LINEAR FUNCTIONS OF ORDER STATISTICS

Let $x_1, \ldots, x_n$ be any n observations and $c_1, \ldots, c_n$ any n constants. With Σ denoting summation from 1 to n, we show how Cauchy's inequality may be used to unify the construction of bounds for $\Sigma c_i x_{i:n}$ in terms of the sample mean $\bar{x}$ and sample standard deviation $s = [\sum(x_i - \bar{x})^2/(n-1)]^{1/2}$. Hölder's inequality could be used similarly to give more general results, but we confine ourselves to this most important special case. Several examples illustrate the usefulness of the bounds.

Since $\Sigma c_i(x_{i:n} - \bar{x}) = \Sigma(c_i - \bar{c})(x_{i:n} - \bar{x})$, we have from Cauchy's inequality that

$$|\Sigma c_i(x_{i:n} - \bar{x})| \leq [\Sigma(c_i - \bar{c})^2 \Sigma(x_{i:n} - \bar{x})^2]^{1/2}. \tag{9}$$

Focusing for definiteness on finding upper bounds, we take $\Sigma c_i(x_{i:n} - \bar{x}) \geq 0$, so that from (9)

$$\Sigma c_i x_{i:n} \leq \bar{x}\Sigma c_i + [(n-1)\Sigma(c_i - \bar{c})^2]^{1/2} s. \tag{10}$$

Equality holds iff for some constant k (which must be positive)

$$x_{i:n} - \bar{x} = k(c_i - \bar{c}) = x_i' \text{ (say)} \quad i = 1, \ldots, n. \tag{11}$$

It follows that the c_i must be nondecreasing in i for (10) to give sharp bounds.

Example 1. For the internally studentized extreme deviate from the sample mean

$$d_n = (x_{n:n} - \bar{x})/s$$

we have $c_i = \ldots = c_{n-1} = -\frac{1}{n}, c_n = 1 - \frac{1}{n}$, so that $\Sigma c_i = 0$ and from (10)

$$d_n \leq (n-1)/\sqrt{n} = d_n'.$$

From (11) we see that the maximizing configuration is given by $x_1' = \ldots = x_{n-1}' = -\frac{k}{n}, x_n' = \frac{n-1}{n}k$.

⋮ 1 n-1 · n=5

x'_1 0 x'_n

This is the oldest result of this type, already obtained by a different method in [14]. In fact, also obtained there is a bound d'_{n-1} for $d_{n-1} = (x_{n-1:n} - \bar{x})/s$. The authors noted that for $x \geq d'_{n-1}$

$$Pr(D_n > x) = nPr((X_1 - \bar{X})/S > x),$$

enabling them to find exact upper percentage points of D_n in the normal case for $n \leq 14$ (5%) and $n \leq 19$ (1%) by using the relation of $(X_1 - \bar{X})/S$ to the t-distribution. They noted also that if there are two equal outliers in a sample of $n \leq 14$, neither can be detected by this test at the 5% level, a phenomenon later termed the "masking effect".

See [7] for other examples where the c_i are nondecreasing. In such cases $\ell = \Sigma c_i x_{i:n}$ is a convex function. A function ϕ of n variables is *convex* in a region R_n if for any two points $\mathbf{x}$ and $\mathbf{y}$ in R_n and $0 \leq \alpha \leq 1$.

$$\phi(\alpha\mathbf{x} + (1-\alpha)\mathbf{y}) \leq \alpha\phi(\mathbf{x}) + (1-\alpha)\phi(\mathbf{y}).$$

Clearly, $x_{n:n}$ is convex, since

$$(\alpha\mathbf{x} + (1-\alpha)\mathbf{y})_{n:n} \leq \alpha x_{n:n} + (1-\alpha)y_{n:n},$$

and so is any ϕ expressible as a maximum. In particular, $S_i = x_{n:n} + \ldots + x_{n-i+1:n}, i = 1, \ldots n$, is convex, since $S_i = max_{1 \leq j_1 < \ldots < j_i \leq n}(x_{j1} + \ldots + x_{ji})$. But

$$\ell = c_1 S_n + (c_2 - c_1)S_{n-1} + \ldots + (c_n - c_{n-1})S_1$$

and such a combination of convex functions with nonnegative coefficients is itself convex (see [13, p. 451]).

Correspondingly, ℓ is *concave* if the c_i are nonincreasing. Finding the maximum c_i in this case requires a special approach, as does finding the minimum when ℓ is convex (see [19] or [7] for a method due to C.L. Mallows). All other situations can be handled as in the following examples. We first need a simple result.

Suppose $c_{i+1} < c_i$. Then, writing $c_i' = c_{i+1}' = \frac{1}{2}(c_i + c_{i+1})$, we have

$$c_i x_{i:n} + c_{i+1} x_{i+1:n} - c_i' x_{i:n} - c_{i+1}' x_{i+1:n} = \frac{1}{2}(x_{i+1:n} - x_{i:n})(c_{i+1} - c_i) \leq 0.$$

Thus,

$$c_i x_{i:n} + c_{i+1} x_{i+1:n} \leq c_i' x_{(i:n)} + c_{i+1}' x_{i+1:n}, \tag{12}$$

with equality holding iff $x_{i:n} = x_{i+1:n}$. Continuing this process until convexity of $\sum c_i' x_{i:n}$ is achieved requires averaging two or more c_i in nonconvex subgroups. The upper bound of $\Sigma c_i x_{i:n}$ is then the attainable upper bound of $\Sigma c_i' x_{i:n}$.

Example 2. $d_r = (x_{r:n} - \bar{x})/s \quad r = 2, \ldots, n-1$. For $x_1, \ldots, x_n$ arbitrary, d_r is neither convex nor concave. To make d_r convex, we simply set $x_{r:n} = x_{r+1:n} = \ldots = x_{n:n}$, thereby changing the c_i to c_i' as follows:

$$\begin{aligned} c_i' &= c_i = -\frac{1}{n} & i &= 1, \ldots, r-1 \\ c_i' &= \frac{r-1}{n(n-r+1)} & i &= r, \ldots, n \end{aligned}$$

Hence, by (10),

$$d_r \leq \left[\frac{(n-1)(r-1)}{n(n-r+1)}\right]^{1/2} = d_r' \text{ (say)} \quad r = 2, \ldots, n-1$$

which holds for $r = n$ also. The maximizing configuration is given by $x_1' = \ldots = x_{r-1}' = -k/n, x_r' = \ldots = x_n' = k(r-1)/[n(n-r+1)]$

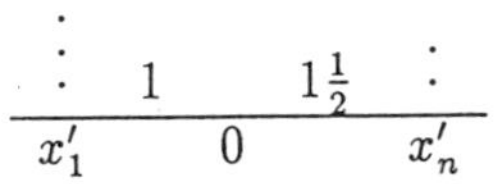

n=5, r=3

Example 3. $q_r = (x_{m:n} + x_{m+1:n} - x_{r:n} - x_{n+1-r:n})/s \quad r = 1, \ldots, m-1; m = n/2 = 2, 3, \ldots$. The numerator is 2 (median-midquasirange). To make the c_i'

nondecreasing we must take $c_1' = \ldots = c_r' = -1/r, c_{r+1}' = \ldots = c_{m-1}' = 0, c_m' = \ldots = c_n' = 1/(m+1)$. This gives

$$\Sigma c_i'^2 = \frac{1}{r} + \frac{1}{m+1} = \frac{n+2+2r}{r(n+2)}$$

Hence, from (10) and symmetry considerations, we have

$$|q_r| \leq \left[\frac{(n-1)(n+2+2r)}{r(n+2)}\right]^{1/2}$$

For $n = 8, r = 2$ the maximizing configuration is proportional to

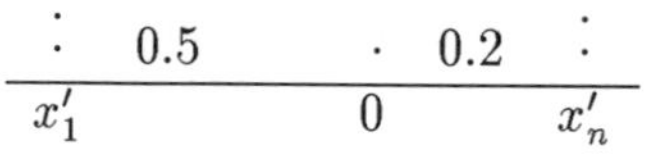

Example 2 and other examples of this convexity creating approach were given in [7], but inequality (12), which provides a formal justification of the procedure used, is new here. In the meantime, Rychlik in [15], using quite different arguments, had arrived at essentially the same results and had formalized obtaining the c_i'. See also the related review paper [16] and the monograph [17] for further unifying results.

Acknowledgement This paper is a modified version of a presentation to the International Conference on Order Statistics and Extreme Values, Mysore, India, December 2000. I am indebted to T. Rychlik for drawing my attention to his 1992 paper [15].

References

[1] Balakrishnan, N. and Rao, C.R. (Eds.) (1998a). Order Statistics: Theory and Methods. *Handbook of Statistics* **16**. North-Holland, Amsterdam.

[2] Balakrishnan, N. and Rao, C.R. (Eds.) (1998b). Order Statistics: Applications. *Handbook of Statistics* **17**. North-Holland, Amsterdam.

[3] Balakrishnan, N., Bendre, S.M., and Malik, H.J. (1992). General relations and identities for order statistics from non-independent non-identical variables. *Ann. Inst. Statist. Math.* **44**, 177-183.

[4] Balasubramanian, K. and Balakrishnan, N. (1992). Indicator method for a recurrence relation for order statistics. *Statist. Probab. Lett.* **14**, 67-69.

[5] Balasubramanian, K. and Balakrishnan, N. (1993). Duality principle in order statistics. *J. R. Statist. Soc. B* **55**, 687-691.

[6] Cole, R.H. (1951). Relations between moments of order statistics. *Ann. Math. Statist.* **22**, 308-310.

[7] David, H.A. (1988). General bounds and inequalities in order statistics. *Commun. Statist.- Theory Meth.* **17**, 2119-2134.

[8] David, H.A. (1993). A note on order statistics for dependent variates. *Amer. Statist.* **47**, 198-199.

[9] David, H.A. (1995). On recurrence relations for order statistics. *Statist. Probab. Lett.* **24**, 133-138.

[10] David, H.A. and Joshi, P.C. (1968). Recurrence relations between moments of order statistics for exchangeable variates. *Ann. Math. Statist.* **39**, 272-274.

[11] Feller, W. (1957). *An Introduction to Probability Theory and Its Applications*, Vol. 1, 2nd edn. Wiley, New York.

[12] Govindarajulu, Z. (1963). On moments of order statistics and quasi-ranges from normal populations. *Ann. Math. Statist.* **34**, 633-651.

[13] Marshall, A.W. and Olkin, I. (1979). *Inequalities: Theory ofMajorization and Its Applications.* Academic Press, New York.

[14] Pearson, E.S. and Chandra Sekar, C. (1936). The efficiency of statistical tools and a criterion for the rejection of outlying observations. *Biometrika* **28**, 308-320.

[15] Rychlik, T. (1992). Sharp inequalities for linear combinations of elements of monotone sequences. *Bull. Polish Acad. Sci. Math.* **40**, 247.254.

[16] Rychlik, T. (1998). Bounds for expectations of *L*-estimates. In [1, pp. 105-145].

[17] Rychlik, T. (2001). *Projecting Statistical Functionals.* Lecture Notes in Statistics, **160**. Springer.

[18] Sathe, Y.S. and Dixit, V.J. (1990). On a recurrence relation for order statistics. *Statist. Probab. Lett.* **9**, 1-4.

[19] Shapiro, S.S. and Wilk, M.B. (1965). An analysis of variance test for normality (complete samples). *Biometrika* **52**, 591-611.

Stochastic Orderings among Order Statistics and Sample Spacings

Baha-Eldin Khaledi
Dept. of Statistics, College of Science,
Razi University, Kermanshah, Iran
e-mail: bkhaledi@hotmail.com

Subhash Kochar
Indian Statistical Institute, 7, SJS Sansanwal Marg
New Delhi-110016, India
e-mail : kochar@isid.ac.in

Abstract

In this paper we review some of the results obtained recently in the area of stochastic comparisons of order statistics and sample spacings. We consider the cases when the parent observations are identically as well as non-identically distributed. But most of the time we shall be assuming that the observations are independent. The case of independent exponentials with unequal scale parameters is discussed in detail.

1 Introduction

The simplest and the most common way of comparing two random variables is through their means and variances. It may happen that in some cases the median of X is larger than that of Y, while the mean of X is smaller than the mean of Y. However, this confusion will not arise if the random variables are stochastically ordered. Similarly, the same may happen if one would like to compare the variability of X with that of Y based only on numerical measures like standard deviation etc. Besides, these characteristics of distributions might not exist in some cases. In most cases one can express various forms of knowledge about the underlying distributions in terms of their survival functions, hazard rate functions, mean residual functions, quantile functions and other suitable functions of probability distributions. These methods are much more informative than those based only on few numerical characteristics of distributions. Comparisons of random variables based on such functions usually establish partial orders among them. We call them as stochastic orders.

Stochastic models are usually sufficiently complex in various fields of statistics, particularly in reliability theory. Obtaining bounds and approximations for their characteristics is of practical importance. That is, the approximation of a stochastic model either by a simpler model or by a model with simple constituent components might lead to convenient bounds and approximations for some particular and desired characteristics of the model. The study of changes in the properties of a model, as the constituent components vary, is also of great interest. Accordingly, since the stochastic components of models involve random variables, the topic of stochastic orders among random variables plays an important role in these areas.

Order statistics and spacings are of great interest in many areas of statistics and they have received a lot of attention from many researchers. Let $X_1, \ldots, X_n$ be n random variables. The ith order statistic, the ith smallest of X_i's, is denoted by

$X_{i:n}$. A k-out-of-n system of n components functions if at least k of n components function. The time of a k-out-of-n system of n components with life times $X_1, \ldots, X_n$ corresponds to the $(n-k+1)$th order statistic. Thus, the study of lifetimes of k-out-of-n systems is equivalent to the study of the stochastic properties of order statistics. Spacings, the differences between successive order statistics, and their functions are also important in statistics, in general, and in particular in the context of life testing and reliability models. Lot of work has been done in the literature on different aspects of order statistics and spacings. For a glimpse of this, see the books by David (1981), and Arnold, Balakrishnan and Nagaraja (1992); and two volumes of papers on this topic by Balakrishnan and Rao (1998 a and b). But most of this work has been confined to the case when the observations are i.i.d. In many practical situations, like in reliability theory, the observations are not necessarily i.i.d. Because of the complicated nature of the problem, not much work has been done for the non i.i.d. case. Some references for this case are Sen (1970), David (1981, p.22), Shaked and Tong (1984), Bapat and Beg (1989), Boland et al. (1996), Kochar (1996), and Nappo and Spizzichino (1998), among others.

Some interesting partial ordering results on order statistics and spacings from independent but non-identically random variables have been obtained by Pledger and Proschan (1971), Proschan and Sethuraman (1976), Bapat and Kochar (1994), Boland, El-Neweihi, and Proschan (1994), Kochar and Kirmani (1995), Kochar and Korwar (1996), Kochar and Rojo (1996), Dykstra, Kochar, and Rojo (1997), Kochar and Ma (1999), Bon and Paltanea (1999), Kochar (1999), Khaledi and Kochar (1999), Khaledi and Kochar (2000 a,b,c), and Khaledi and Kochar (2001).

In this chapter, we discuss some newly obtained results on stochastic comparisons of order statistics and spacings. Kochar (1998) and Boland, Shaked and Shanthikumar (1998) have given comprehensive reviews on this topic upto 1998. In Section 2, we introduce the required notation and definitions. Section 3 and 4 are devoted to stochastic comparisons of order statistics in one-sample and two-sample problems, respectively. In Sections 5, we discuss the stochastic ordering among spacings in one-sample problem and two sample problem. Section 6 is devoted to stochastic properties of sample range Throughout this chapter *increasing* means *nondecreasing* and *decreasing* means *nonincreasing*; and we shall be assuming that all distributions under study are absolutely continuous.

2 Definitions

Let X and Y be univariate random variables with distribution functions F and G, survival functions $\overline{F}$ and $\overline{G}$, density functions f and g; and hazard rates r_F $(= f/\overline{F})$ and r_G $(= g/\overline{G})$, respectively. Let l_X (l_Y) and u_X (u_Y) be the left and the right endpoints of the support of X (Y).

Stochastic orderings

Definition 2.1 *X is said to be stochastically smaller than Y (denoted by $X \leq_{st} Y$) if $\overline{F}(x) \leq \overline{G}(x)$ for all x.*

This is equivalent to saying that $Eg(X) \leq Eg(Y)$ for any increasing function g for which expectations exist.

Definition 2.2 *X is said to be smaller than Y in hazard rate ordering (denoted by $X \leq_{hr} Y$) if $\overline{G}(x)/\overline{F}(x)$ is increasing in $x \in (-\infty, max(u_X, u_Y))$.*

It is worth noting that $X \leq_{hr} Y$ is equivalent to the inequalities

$$P[X - t > x | X > t] \leq P[Y - t > x | Y > t], \quad \text{for all } x \geq 0 \text{ and } t.$$

In other words, the conditional distributions, given that the random variables are at least of a certain size, are all stochastically ordered (in the standard sense) in the same direction. Thus, if X and Y represent the survival times of different models of an appliance that satisfy this ordering, one model is better (in the sense of stochastic ordering) when the appliances are new, the same appliance is better when both are one month old, and in fact is better no matter how much time has elapsed. It is clearly useful to know when this strong type of stochastic ordering holds since quantities judgements are then easy to make. In case the hazard rates exist, it is easy to see that $X \leq_{hr} Y$, if and only if, $r_G(x) \leq r_F(x)$ for every x. The hazard rate ordering is also known as uniform stochastic ordering in the literature.

Definition 2.3 *X is said to be smaller than Y in likelihood ratio ordering (denoted by $X \leq_{lr} Y$) if $g(x)/f(x)$ is increasing in $x \in (l_X, u_X) \cup (l_Y, u_Y)$.*

When the supports of X and Y have a common left end-point, we have the following chain of implications among the above stochastic orders :

$$X \leq_{lr} Y \Rightarrow X \leq_{hr} Y \Rightarrow X \leq_{st} Y. \tag{2.1}$$

Definition 2.4 *The random vector $\mathbf{X} = (X_1, \ldots, X_n)$ is smaller than the random vector $\mathbf{Y} = (Y_1, \ldots, Y_n)$ in the multivariate stochastic order (denoted by $\mathbf{X} \overset{st}{\preceq} \mathbf{Y}$) if $h(\mathbf{X}) \leq_{st} h(\mathbf{Y})$ for all increasing functions h.*

It is easy to see that multivariate stochastic ordering implies component-wise usual stochastic ordering. For more details on stochastic orderings, see Chapters 1 and 4 of Shaked and Shanthikumar (1994).

One of the basic criteria for comparing variability in probability distributions is that of dispersive ordering. Let F^{-1} and G^{-1} be the right continuous inverses (quantile functions) of F and G, respectively. We say that X is less *dispersed* than Y (denoted by $X \leq_{disp} Y$) if $F^{-1}(\beta) - F^{-1}(\alpha) \leq G^{-1}(\beta) - G^{-1}(\alpha)$, for all $0 \leq \alpha \leq \beta \leq 1$. From this one can easily obtain that

$$X \leq_{disp} Y \Longleftrightarrow g(x) \leq f\left(F^{-1}G(x)\right) \quad \forall\ x, \tag{2.2}$$

when the random variables X and Y admit densities. A consequence of $X \leq_{disp} Y$ is that $|X_1 - X_2| \leq_{st} |Y_1 - Y_2|$ and which in turn implies $var(X) \leq var(Y)$ as well as $E[|X_1 - X_2|] \leq E[|Y_1 - Y_2|]$, where $X_1, X_2\,(Y_1, Y_2)$ are two independent copies of X (Y). For details, see Saunders and Moran (1978), Lewis and Thompson (1981), Deshpande and Kochar (1983), Bagai and Kochar (1986), Bartoszewicz (1986, 1987); and Section 2.B of Shaked and Shanthikumar (1994).

Notions of Majorization and related orderings

One of the basic tools in establishing various inequalities in statistics and probability is the notion of majorization.

Let $\{x_{(1)} \leq x_{(2)} \leq \ldots \leq x_{(n)}\}$ denote the increasing arrangement of the components of the vector $\mathbf{x} = (x_1, x_2, \ldots, x_n)$.

Definition 2.5 *The vector* $\mathbf{x}$ *is said to majorize the vector* $\mathbf{y}$ *(written* $\mathbf{x} \overset{m}{\succeq} \mathbf{y}$*) if* $\sum_{i=1}^{j} x_{(i)} \leq \sum_{i=1}^{j} y_{(i)}$ *for* $j = 1, \ldots, n-1$ *and* $\sum_{i=1}^{n} x_{(i)} = \sum_{i=1}^{n} y_{(i)}$.

Functions that preserve the majorization ordering are called Schur-convex functions. The vector $\mathbf{x}$ is said to majorize the vector $\mathbf{y}$ weakly (written $\mathbf{x} \overset{w}{\succeq} \mathbf{y}$) if $\sum_{i=1}^{j} x_{(i)} \leq \sum_{i=1}^{j} y_{(i)}$ for $j = 1, \ldots, n$. Marshall and Olkin (1979) provides extensive and comprehensive details on the theory of majorization and its applications in statistics.

Recently Bon and Paltanea (1999) have considered a pre-order on $\mathbb{R}^{+^n}$, which they call as a *p-larger order.*

Definition 2.6 *A vector* $\mathbf{x}$ *in* $\mathbb{R}^{+^n}$ *is said to be p-larger than another vector* $\mathbf{y}$ *also in* $\mathbb{R}^{+^n}$ *(written* $\mathbf{x} \overset{p}{\succeq} \mathbf{y}$*) if* $\prod_{i=1}^{j} x_{(i)} \leq \prod_{i=1}^{j} y_{(i)}, j = 1, \ldots, n$.

Let $\log(\mathbf{x})$ denote the vector of logarithms of the coordinates of $\mathbf{x}$. It is easy to verify that

$$\mathbf{x} \overset{p}{\succeq} \mathbf{y} \Leftrightarrow \log(\mathbf{x}) \overset{w}{\succeq} \log(\mathbf{y}). \tag{2.3}$$

It is known that $\mathbf{x} \overset{m}{\succeq} \mathbf{y} \Longrightarrow (g(x_1), \ldots, g(x_n)) \overset{w}{\succeq} (g(y_1), \ldots, g(y_n))$ for all concave functions g (cf. Marshal and Olkin, 1979, p. 115). From this and (2.3), it follows that when $\mathbf{x}, \mathbf{y} \in I\!R^{+^n}$

$$\mathbf{x} \overset{m}{\succeq} \mathbf{y} \Longrightarrow \mathbf{x} \overset{p}{\succeq} \mathbf{y}.$$

The converse is, however, not true. For example, the vectors $(0.2, 1, 5) \overset{p}{\succeq} (1, 2, 3)$ but majorization does not hold between these two vectors.

Notions of Aging

Let X be a random variable with distribution function F and let X_t denote a random variable with the same distribution as that of $X - t|X > t$. We will use the following notions of aging in this article.

(*a*) X is said to have an increasing failure rate (denoted by IFR) distribution if $X_t \leq_{st} X_{t'}$, for $t > t'$. This is equivalent to saying that $\overline{F}(x+t)/\overline{F}(t)$ decreasing in t for $x > 0$. It is easy to see that in case the random variable X admits density, F is IFR if and only if, the hazard rate $r_F(t) = f(t)/\overline{F}(t)$ is increasing in t.

(*b*) X is said to have a decreasing failure rate (denoted by DFR) distribution if $X_t \geq_{st} X_{t'}$, for $t > t'$. This is equivalent to $\overline{F}(x+t)/\overline{F}(t)$ increasing in t for $x > 0$.

Next theorem due to Bagai and Kochar (1986) and Bartoszewicz (1987) establishes a connection between dispersive ordering and hazard rate ordering.

THEOREM **2.1** *Let X and Y be random variables with distribution function F and G, respectively. Then,*

(a) $X \leq_{hr} Y$ and F or G being DFR implies $X \leq_{disp} Y$;

(b) $X \leq_{disp} Y$ and F or G being IFR implies $X \leq_{hr} Y$.

3 Stochastic Comparisons of Order Statistics in one-sample problem

Let $X_1, \ldots, X_n$ be a set of independent random variables. It is easy to see that $X_{i:n} \leq_{st} X_{j:n}$, for all $i < j$. Boland, El-Neweihi and Proschan (1994) extended this result from usual stochastic order to hazard rate order. Using the definition of likelihood ratio ordering, it is not hard to prove that $X_{i:n} \leq_{lr} X_{j:n}$ for $i < j$. Shaked and Shanthikumar (1994) considered the problem of comparing order statistics from samples with possibly unequal sample sizes. They showed that if random variables X_i's are iid, then $X_{n:n} \leq_{lr} X_{n+1:n+1}$ and $X_{1:n} \geq_{lr} X_{1:n+1}$. Raqab and Amin (1996) strengthened this result and proved that $X_{i:n} \leq_{lr} X_{j:m}$, whenever $i \leq j$ and $n - i \geq m - j$. Using implications (2.1), we get, for $i \leq j$ and $n - i \geq m - j$, $X_{i:n} \leq_{hr} X_{j:m}$ which in turn implies that $X_{i:n} \leq_{st} X_{j:m}$. Removing the identically distributed assumption, it is interesting to investigate the above stochastic inequalities among order statistics. Boland, El-Neweihi and Proschan (1994) showed that if random variables are independent and $X_k \leq_{hr} X_{n+1}$, $k = 1, \ldots, n$, then $X_{i-1:n} \leq_{hr} X_{i:n+1}$, $i = 1, \ldots, n+1$. They also proved that if X_i's are independent and $X_{n+1} \leq_{hr} X_k$, $k = 2, \ldots, n$, then $X_{i:n} \geq_{hr} X_{i:n+1}$, $i = 1, \ldots, n$. The reader may be wondering whether likelihood ratio ordering among order statistics holds for the case when X_i's are independent but not necessarily identically distributed. Assuming $X_1 \leq_{lr} X_2 \leq_{lr} \ldots \leq_{lr} X_n$, Bapat an Kochar (1994) proved that $X_{i:n} \leq_{lr} X_{j:n}$, $i < j$.

We end this section by discussing some results on dispersive ordering of order statistics. David and Groenveld (1982) proved that if X_i's are iid random variables with a common DFR distribution, then $var(X_{i:n}) \leq var(X_{j:n})$, for $i < j$. Kochar (1996) strengthened this result to prove that under the same conditions, $X_{i:n} \leq_{disp} X_{j:n}$, $i < j$. In Theorem 3.2 below, due to Khaledi and Kochar (2000 a), this result has been further extended. It is proved that if X_i's are iid with DFR distribution, then $X_{i:n} \leq_{disp} X_{j:m}$, whenever $i \leq j$ and $n - i \geq m - j$. We will find the following result useful in proving it.

THEOREM **3.1** *(Saunders (1984)). The random variable X satisfies $X \leq_{disp} X + Y$. for any random variable Y independent of X if and only if X has a log-concave density.*

Using Theorem 3.1, first the result is proved for exponential distribution.

LEMMA 3.1 *Let $X_{i:n}$ be the ith order statistic of a random sample of size n from an exponential distribution. Then*

$$X_{i:n} \leq_{disp} X_{j:m} \quad \text{for } i \leq j \text{ and } n - i \geq m - j. \tag{3.1}$$

PROOF : Suppose we have two independent random samples, $X_1, \ldots, X_n$ and $X_1^{'}, \ldots, X_m^{'}$ of sizes n and m from an exponential distribution with failure rate λ. The *ith* order statistic $X_{i:n}$ can be written as a convolutions of the sample spacings as

$$\begin{aligned} X_{i:n} &= (X_{i:n} - X_{i-1:n}) + \cdots + (X_{2:n} - X_{1:n}) + X_{1:n} \\ &\stackrel{dist}{=} \sum_{k=1}^{i} E_{n-i+k} \end{aligned} \tag{3.2}$$

where for $k = 1, \ldots, i$, E_{n-i+k} is an exponential random variable with failure rate $(n - i + k)\lambda$. It is a well known fact that E_{n-i+k}'s are independent. Similarly we can express $X_{j:m}^{'}$ as

$$X_{j:m}^{'} \stackrel{dist}{=} \sum_{k=1}^{j} E_{m-j+k}^{'} \tag{3.3}$$

where again for $k = 1, \ldots, j$, $E_{m-j+k}^{'}$ is an exponential random variable with failure rate $(m - j + k)\lambda$ and $E_{m-j+k}^{'}$'s are independent. It is easy to verify that $E_{n-i+1} \leq_{disp} E_{m-j+1}^{'}$ for $n - i \geq m - j$.

Since the class of distributions with log-concave densities is closed under convolutions (cf. Dharmadhiakri and Joag-dev (1988), p. 17), it follows from the repeated applications of Theorem 3.1 that

$$\sum_{k=1}^{i} E_{n-i+k} \leq_{disp} \sum_{k=1}^{i} E'_{m-j+k}. \tag{3.4}$$

Again since $\sum_{k=i+1}^{j} E'_{m-j+k}$, being the sum of independent exponential random variables has a log-concave density and since it is independent of $\sum_{k=1}^{i} E'_{n-i+k}$, it follows from Theorem 3.1 that the R.H.S of (3.4) is less dispersed than $\sum_{k=1}^{j} E'_{m-j+k}$ for $i \leq j$

That is,

$$X_{i:n} \stackrel{dist}{=} \sum_{k=1}^{i} E_{n-i+k} \leq_{disp} \sum_{k=1}^{j} E'_{m-j+k} \stackrel{dist}{=} X'_{j:m}.$$

Since $X_{j:m}$ and $X'_{j:m}$ are stochastically equivalent, (3.1) follows from this. ■

The proof of the next lemma can be found in Bartoszewicz (1987).

LEMMA 3.2 *Let $\phi : R_+ \to R_+$ be a function such that $\phi(0) = 0$ and $\phi(x) - x$ is increasing. Then for every convex and strictly increasing function $\psi : R_+ \to R_+$ the function $\psi\phi\psi^{-1}(x) - x$ is increasing.*

In the next theorem we extend Lemma 3.1 to the case when F is a DFR distribution.

THEOREM **3.2** *Let $X_{i:n}$ be the ith order statistic of a random sample of size n from a DFR distribution F. Then*

$$X_{i:n} \leq_{disp} X_{j:m} \quad \text{for } i \leq j \text{ and } n - i \geq m - j.$$

PROOF : The distribution function of $X_{j:m}$ is $F_{j:m}(x) = B_{j:m}F(x)$, where $B_{j:m}$ is the distribution function of the beta distribution with parameters $(j, m - j + 1)$.

Let G denote the distribution function of a unit mean exponential random variable. Then $H_{j:m}(x) = B_{j:m}G(x)$ is the distribution function of the jth order statistic in a random sample of size m from a unit mean exponential distribution. We can express $F_{j:m}$ as

$$\begin{aligned} F_{j:m}(x) &= B_{j:m}GG^{-1}F(x) \\ &= H_{j:m}G^{-1}F(x). \end{aligned} \tag{3.5}$$

To prove the required result, we have to show that for $i \leq j$ and $n-i \geq m-j$,

$$\begin{aligned} &F_{j:m}^{-1}F_{i:n}(x) - x \quad \text{is increasing in } x \\ \Leftrightarrow\ &F^{-1}GH_{j:m}^{-1}H_{i:n}G^{-1}F(x) - x \quad \text{is increasing in } x. \end{aligned} \tag{3.6}$$

By Lemma 3.1, $H_{j:m}^{-1}H_{i:n}(x) - x$ is increasing in x for $i \leq j$ and $n-i \geq m-j$. Also the function $\psi(x) = F^{-1}G(x)$ is strictly increasing and it is convex if F is DFR. The required result now follows from Lemma 3.2. ∎

REMARK: A consequence of Theorem 3.2 is that if we have random samples from a DFR distribution, then

$$X_{i:n+1} \leq_{disp} X_{i:n} \leq_{disp} X_{i+1:n+1}, \qquad \text{for } i = 1, \ldots, n.$$

4 Stochastic Comparisons of Order Statistics in two-sample problem

Let $X_1, \ldots, X_n$ be a set of independent random variables and $Y_1, \ldots, Y_n$ be another set of independent random variables. Ross (1983) proved that if $X_i \leq_{st} Y_i$, $i = 1, \ldots, n$, then $(X_1, \ldots, X_n) \leq_{st} (Y_1, \ldots, Y_n)$. A consequence of this result is that $X_{i:n} \leq_{st} Y_{i:n}$ for $i = 1, \ldots, n$. Lynch, Mimmack and Proschan (1987) generalized this result from stochastic ordering to hazard rate ordering. They showed that if $X_i \leq_{hr} Y_j$,

$i, j \in \{1, \ldots, n\}$, then $X_{i:n} \leq_{hr} Y_{i:n}$, $i = 1, \ldots, n$. A similar result for likelihood ratio ordering has been proved by Chan, Proshcan and Sethuraman (1991). They proved that if $X_i \leq_{lr} Y_j$, $i, j \in \{1, \ldots, n\}$, then $X_{i:n} \leq_{lr} Y_{i:n}$, $i = 1, \ldots, n$. Lillo, Nanda and Shaked (2000) strengthened this result to the case when the number of X_i's and Y_i's are possibly different.

THEOREM **4.1** *Let $X_1, \ldots, X_n$ be independent random variables and $Y_1, \ldots, Y_m$ be another set of independent random variables, all having absolutely continuous distributions. Then $X_i \leq_{lr} Y_j$ for all i, j implies $X_{i:n} \leq_{lr} Y_{j:m}$ whenever $i \leq j$ and $n - i \geq m - j$.*

In the next theorem we establish dispersive ordering between order statistics when the random samples are drawn from different distributions.

THEOREM **4.2** *Let $X_1, \ldots, X_n$ be a random sample of size n from a continuous distribution F and let $Y_1 \ldots, Y_m$ be a random sample of size m from another continuous distribution G. If either F or G is DFR, then*

$$X \leq_{disp} Y \Rightarrow X_{i:n} \leq_{disp} Y_{j:m} \quad \text{for } i \leq j \text{ and } n - i \geq m - j. \tag{4.1}$$

PROOF: Let F be a DFR distribution. The proof for the case when G is DFR is similar. By Theorem 3.2, $X_{i:n} \leq_{disp} X_{j:m}$ for $i \leq j$ and $n - i \geq m - j$. Bartoszewicz (1986) proved that if $X \leq_{disp} Y$ then $X_{j:m} \leq_{disp} Y_{j:m}$. Combining these we get the required result. ∎

Since the property $X \leq_{hr} Y$ together with the condition that either F or G is DFR implies that $X \leq_{disp} Y$ (Theorem 2.1), we get the following result from the above theorem.

COROLLARY 4.1 *Let $X_1, \ldots, X_n$ be a random sample of size n from a continuous distribution F and $Y_1 \ldots, Y_m$ be a random sample of size m from another continuous* *distribution G. If either F or G is DFR, then*

$$X \leq_{hr} Y \Rightarrow X_{i:n} \leq_{disp} Y_{j:m}.$$

Stochastic comparisons of order statistics from heterogeneous populations

An assumption often made in reliability models is that the n components have lifetimes with proportional hazards. Let X_i denote the lifetime of the *ith* component of a reliability system with survival function $\overline{F}_i(t), \quad i = 1, \ldots, n$. Then they have proportional hazard rates (PHR) if there exist constants $\lambda_1, \ldots, \lambda_n$ and a (cumulative hazard) function $R(t) \geq 0$ such that $\overline{F}_i(t) = e^{-\lambda_i R(t)}$ for $i = 1, \ldots, n$.Clearly then the hazard rate of X_i is $r_i(t) = \lambda_i R'(t)$ (assuming it exists). An example of such a situation is when the components have independent exponential lifetimes with respective hazard rates $\lambda_1, \ldots, \lambda_n$. Many researchers have investigated the effect on the survival function, the hazard rate function and other characteristics of the time to failure of this system when we switch the vector $(\lambda_1, \ldots, \lambda_n)$ to another vector say $(\lambda_1^*, \ldots, \lambda_n^*)$. Pledger and Proschan (1971), for the first time, studied this problem and proved the following interesting result among many other results.

THEOREM **4.3** *Let* $(X_1, \ldots, X_n)$ *and* $(X_1^*, \ldots, X_n^*)$ *be two random vectors of independent lifetimes with proportional hazards with* $\lambda_1, \ldots, \lambda_n$ *and* $\lambda_1^*, \ldots \lambda_n^*$ *as the constants of proportionality. Suppose that*

$$\boldsymbol{\lambda} \overset{m}{\succeq} \boldsymbol{\lambda}^*.$$

Then

$$X_{i:n} \geq_{st} X_{i:n}^*, \qquad i = 1, \ldots, n. \tag{4.2}$$

Proshcan and Sethuraman (1976) generalized this result from component wise stochastic ordering to multivariate stochastic ordering. That is, under the same assumptions of Theorem 4.3, they showed that

$$(X_{1:n}, \ldots, X_{n:n}) \geq_{st} (X_{1:n}^*, \ldots, X_{n:n}^*).$$

Boland, El-Neweihi and Proschan (1994) proved that for $n = 2$ the above result can be extended from stochastic ordering to hazard rate ordering. They also showed with the help of a counterexample that for $n > 2$, (4.2) cannot be strengthened from stochastic ordering to hazard rate ordering.

Dykstra, Kochar and Rojo (1997) studied the problem of stochastically comparing the largest order statistic of a set of n independent and non-identically distributed exponential random variables with that corresponding to a set of n independent and identically distributed exponential random variables. Let $X_1, \ldots, X_n$ be independent exponential random variables with X_i having hazard rate λ_i, for $i = 1, \ldots, n$. Let $Y_1, \ldots, Y_n$ be a random sample of size n from an exponential distribution with common hazard rate $\bar{\lambda} = \sum_{i=1}^{n} \lambda_i / n$, the arithmetic mean of the λ_i's. They proved that $X_{n:n}$ is greater than $Y_{n:n}$ according to dispersive as well as hazard rate orderings. In Theorem 4.4 below we prove that similar results hold if instead, we assume that for $i = 1, \ldots, n$, the random variable Y_i has exponential distribution with hazard rate $\tilde{\lambda} = (\prod_{i=1}^{n} \lambda_i)^{1/n}$, the geometric mean of the λ_i's. To prove dispersive ordering between $X_{n:n}$ and $Y_{n:n}$ in Theorem 4.4 we shall need the following lemma.

LEMMA 4.1 *For $z > 0$, the functions $g(z) = (1 - e^{-z})/z$ and $h(z) = (z^2 e^{-z})/(1 - e^{-z})^2$ are both decreasing.*

PROOF : The numerator of the derivative of $g(z)$ is $k(z) = (1 + z)e^{-z} - 1$, which is a decreasing function of z. This implies that $k(z) < 0$ for $z > 0$, since $k(0) = 0$.

It is easy to see after some simplifications that

$$\frac{d}{dz}(\log(h(z))) = \frac{2 - 2e^{-z} - z - ze^{-z}}{z(1 - e^{-z})}. \tag{4.3}$$

Using the fact that $k(z)$ is negative, one can verify that the numerator of (4.3) is decreasing, from which the required result follows. ∎

THEOREM **4.4** *Let $X_1, \ldots, X_n$ be independent exponential random variables with X_i having hazard rate λ_i, $i = 1, \ldots, n$. Let $Y_1, \ldots, Y_n$ be a random sample of size n from an exponential distribution with common hazard rate $\tilde{\lambda} = (\prod_{i=1}^n \lambda_i)^{1/n}$. Then*

(a) $X_{n:n} \geq_{disp} Y_{n:n}$;

(b) $X_{n:n} \geq_{hr} Y_{n:n}$.

PROOF : (a) The distribution function of $X_{n:n}$ is

$$F_{X_{n:n}}(x) = \prod_{i=1}^{n} \left(1 - e^{-\lambda_i x}\right),$$

with density function as

$$f_{X_{n:n}}(x) = \sum_{i=1}^{n} \frac{\lambda_i e^{-\lambda_i x}}{1 - e^{-\lambda_i x}} \prod_{i=1}^{n} \left(1 - e^{-\lambda_i x}\right). \tag{4.4}$$

Replacing λ_i with $\tilde{\lambda}$ in (4.4), we see that the distribution function and the density function of $Y_{n:n}$ are

$$F_{Y_{n:n}}(x) = \left(1 - e^{-\tilde{\lambda} x}\right)^n \quad \text{and} \quad f_{Y_{n:n}}(x) = n\tilde{\lambda} e^{-\tilde{\lambda} x} \left(1 - e^{-\tilde{\lambda} x}\right)^{n-1},$$

respectively. It is easy to verify that $F_{Y_{n:n}}^{-1}(x) = -\frac{1}{\tilde{\lambda}} \log\left(1 - x^{1/n}\right)$. Using these observations, it follows that

$$f_{Y_{n:n}}\left(F_{Y_{n:n}}^{-1} F_{X_{n:n}}(x)\right) = n\tilde{\lambda} \left(1 - \prod_{i=1}^{n} (1 - e^{-\lambda_i x})^{1/n}\right) \left(\prod_{i=1}^{n} (1 - e^{-\lambda_i x})^{1/n}\right)^{n-1}. \tag{4.5}$$

To prove that $X_{n:n} \geq_{disp} Y_{n:n}$, it follows from relation (2.2) that it is sufficient to show that

$$f_{X_{n:n}}(x) \leq f_{Y_{n:n}}\left(F_{Y_{n:n}}^{-1} F_{X_{n:n}}(x)\right) \quad \forall x > 0. \tag{4.6}$$

Using expressions (4.4) and (4.5) in (4.6), one can see after some simplifications that (4.6) is equivalent to

$$\sum_{i=1}^{n} \frac{\lambda_i}{1 - e^{-\lambda_i x}} - n \prod_{i=1}^{n} \left(\frac{\lambda_i}{1 - e^{-\lambda_i x}}\right)^{1/n} \leq \sum_{i=1}^{n} \lambda_i - n \prod_{i=1}^{n} (\lambda_i)^{1/n}. \tag{4.7}$$

To prove that (4.7) holds for all $\lambda_i > 0, i = 1, \ldots, n$, it is sufficient to show that the L.H.S. of (4.7) (denoted by $h(x)$) is increasing in x since for $x > 0$,

$$h(x) \leq lim_{x \to +\infty} h(x) = \sum_{i=1}^{n} \lambda_i - n \prod_{i=1}^{n} (\lambda_i)^{1/n} ,$$

the right hand side of (4.7).

The derivative of $h(x)$ is

$$\begin{aligned} h'(x) &= \left(\sum_{i=1}^{n} \frac{\lambda_i e^{-\lambda_i x}}{1 - e^{-\lambda_i x}} \right) \left(\prod_{i=1}^{n} \frac{\lambda_i}{1 - e^{-\lambda_i x}} \right)^{1/n} - \sum_{i=1}^{n} \frac{\lambda_i^2 e^{-\lambda_i x}}{(1 - e^{-\lambda_i x})^2} \\ &\geq \left(\sum_{i=1}^{n} \frac{\lambda_i e^{-\lambda_i x}}{1 - e^{-\lambda_i x}} \right) \left(\frac{n}{\sum_{i=1}^{n} \frac{1 - e^{-\lambda_i x}}{\lambda_i}} \right) - \sum_{i=1}^{n} \frac{\lambda_i^2 e^{-\lambda_i x}}{(1 - e^{-\lambda_i x})^2} , \end{aligned}$$

since the geometric mean of a set of numbers is always greater than or equal to its harmonic mean. Now $h'(x) \geq 0$ if and only if,

$$n \sum_{i=1}^{n} \frac{\lambda_i e^{-\lambda_i x}}{1 - e^{-\lambda_i x}} \geq \left(\sum_{i=1}^{n} \frac{\lambda_i^2 e^{-\lambda_i x}}{(1 - e^{-\lambda_i x})^2} \right) \left(\sum_{i=1}^{n} \frac{1 - e^{-\lambda_i x}}{\lambda_i} \right) . \tag{4.8}$$

Multiplying both sides of (4.8) by $x (> 0)$ and replacing the $\lambda_i x$ with z_i for $i = 1, \ldots, n$, it is enough to prove that

$$n \sum_{i=1}^{n} \frac{z_i e^{-z_i}}{1 - e^{-z_i}} \geq \left(\sum_{i=1}^{n} \frac{z_i^2 e^{-z_i}}{(1 - e^{-z_i})^2} \right) \left(\sum_{i=1}^{n} \frac{1 - e^{-z_i}}{z_i} \right) . \tag{4.9}$$

The inequality in (4.9) follows immediately from Čebyšev's inequality (Theorem 1, p. 36 of Mitrinović, 1970), Lemma 4.1 and by writing

$$\frac{z_i e^{-z_i}}{1 - e^{-z_i}} = \left(\frac{z_i^2 e^{-z_i}}{(1 - e^{-z_i})^2} \right) \left(\frac{1 - e^{-z_i}}{z_i} \right) .$$

This proves that $h(x)$ is increasing in x and hence the result.

(b) It follows from Theorem 5.8 of Barlow and Proschan (1981) that $Y_{n:n}$ is IFR. Using this and part (a), the required result follows from Theorem 2.1. ∎

From the above results, we get the following convenient bounds on the hazard rate and the variance of $X_{n:n}$.

COROLLARY 4.2 *Under the conditions of Theorem 4.4,*

(a) the hazard rate $r_{X_{n:n}}$ *of* $X_{n:n}$ *satisfies*

$$r_{X_{n:n}}(x;\boldsymbol{\lambda}) \le \frac{n\tilde{\lambda}\left(1-exp(-\tilde{\lambda}x)\right)^{n-1} exp(-\tilde{\lambda}x)}{1-\left(1-exp(-\tilde{\lambda}x)\right)^{n}} ,$$

(b)

$$var(X_{n:n};\boldsymbol{\lambda}) \ge \frac{1}{\tilde{\lambda}^2}\sum_{i=1}^{n}\frac{1}{(n-i+1)^2} .$$

Dykstra, Kochar and Rojo (1997) proved a result similar to Theorem 4.4 by assuming that the random variables Y_i's are exponential with common hazard rate $\overline{\lambda} = \sum_{i=1}^{n} \lambda_i/n$ and obtained bounds on the hazard rate and the variance of $X_{n:n}$ in terms of $\overline{\lambda}$. The new bounds given in Corollary 4.2 are better because $r_{Y_{n:n}}$ and $var(Y_{n:n})$ are increasing and decreasing function of $\tilde{\lambda}$, respectively, and the fact that the geometric mean of λ_i's is smaller than their arithmetic mean.

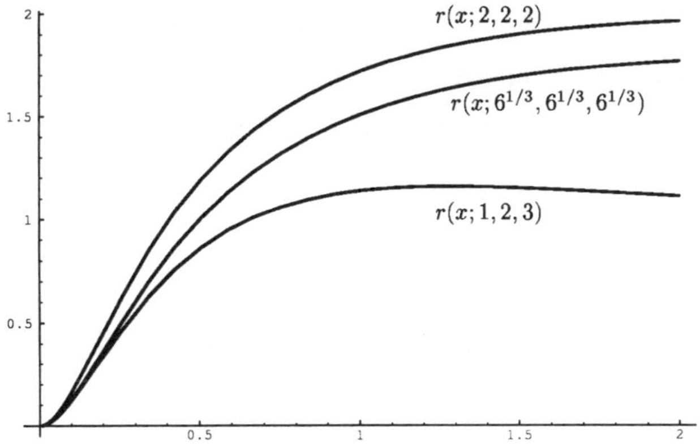

Figure 4.1. Graphs of hazard rates of $X_{3:3}$

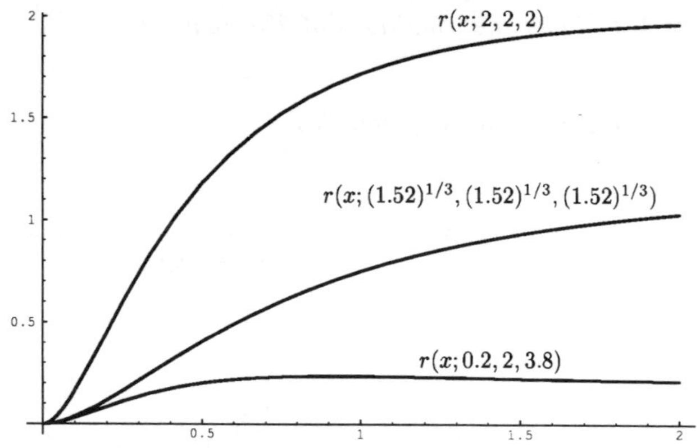

Figure 4.2. Graphs of hazard rates of $X_{3:3}$

In Figures 4.1. and 4.2. above, we plot the hazard rates of parallel systems of three exponential components along with the upper bounds as given by Dykstra, Kochar and Rojo (1997) and the one's given by Corollary 4.2 (a). The vector of parameters in Figure 4.1 is $\boldsymbol{\lambda_1} = (1, 2, 3)$ and that in Figure 4.2 is $\boldsymbol{\lambda_2} = (0.2, 2, 3.8)$. Note that $\boldsymbol{\lambda_2} \overset{m}{\succeq} \boldsymbol{\lambda_1}$. It appears from these figures that the improvements in the bounds are relatively more if λ_i's are more dispersed in the sense of majorization. This is true because the geometric mean is Schur-concave and the hazard rate of a parallel system of i.i.d. exponential components with a common parameter $\tilde{\lambda}$ is increasing in $\tilde{\lambda}$.

Let $\overline{F}$ denote the survival function of a nonnegative random variable X with hazard rate h. According to the PHR model, the random variables $X_1, \ldots, X_n$ are independent with X_i having survival function $\overline{F}^{\lambda_i}(.)$, so that its hazard rate is $\lambda_i h(.)$, $i = 1, \ldots, n$.

Next, we extend Theorem 4.4 from exponential to PHR models. To prove this we need the following theorem due to Rojo and He (1991).

THEOREM **4.5** *Let X and Y be two random variables such that $X \leq_{st} Y$. Then $X \leq_{disp} Y$ implies that $\gamma(X) \leq_{disp} \gamma(Y)$ where γ is a nondecreasing convex function.*

THEOREM **4.6** *Let $X_1, \ldots, X_n$ be independent random variables with X_i having survival function $\overline{F}^{\lambda_i}(x)$, $i = 1, \ldots, n$. Let $Y_1, \ldots, Y_n$ be a random sample of size n from a distribution with survival function $\overline{F}^{\tilde{\lambda}}(x)$, where $\tilde{\lambda} = (\prod_{i=1}^n \lambda_i)^{1/n}$. Then*

(a) $X_{n:n} \geq_{hr} Y_{n:n}$; and

(b) if F is DFR, then $X_{n:n} \geq_{disp} Y_{n:n}$.

PROOF : (a)

Let $H(x) = -\log \overline{F}(x)$ denote the cumulative hazard of F. Let $Z_i = H(X_i)$, $i = 1, \ldots, n$ and $W_i = H(Y_i)$, $i = 1, \ldots, n$. Since X_i's follow the PHR model, then it is easy to show that Z_i is exponential with hazard rate λ_i, $i = 1, \ldots, n$. Similarly, W_i is exponential with hazard rate $\tilde{\lambda}$, $i = 1, \ldots, n$. Theorem 4.4 (b) implies that $Z_{n:n} \geq_{hr} W_{n:n}$. Using this fact, (since H^{-1}, the right inverse of H, is nondecreasing) it is easy to show that $H^{-1}(Z_{n:n}) \geq_{hr} H^{-1}(W_{n:n})$ from which the part (a) follows.

(b) Theorem 4.4 (a) and (b), respectively, imply that $Z_{n:n} \geq_{disp} W_{n:n}$ and $Z_{n:n} \geq_{st} W_{n:n}$. The function $H^{-1}(x)$ is convex, since F is DFR, and is nondecreasing. Using these observations, it follows from Theorem 4.5 that $H^{-1}(Z_{n:n}) \geq_{disp} H^{-1}(W_{n:n})$ which is equivalent to $X_{n:n} \geq_{disp} Y_{n:n}$. ■

In Theorem 4.9 below we prove that for the largest order statistic, the conclusion of Theorem 4.3 holds under the weaker p-larger ordering. The proof of this theorem hinges on the following results.

THEOREM **4.7** *(Marshall and Olkin, 1979, p. 57) Let $I \subset \mathbb{R}$ be an open interval and let $\phi : I^n \to \mathbb{R}$ be continuously differentiable. Necessary and sufficient conditions for ϕ to be Schur-convex on I^n are ϕ is symmetric on I^n and for all $i \neq j$,*

$$(z_i - z_j)[\phi_{(i)}(z_i) - \phi_{(j)}(z_j)] \geq 0 \quad \textit{for all } z \in I^n,$$

where $\phi_{(i)}(z)$ denotes the partial derivative of ϕ with respect to its ith argument.

THEOREM **4.8** *(Marshall and Olkin, 1979, p. 59) A real-valued function ϕ on the set $A \subset \mathbb{R}^n$ satisfies*

$$\mathbf{x} \overset{w}{\succeq} \mathbf{y} \text{ on } A \Longrightarrow \phi(\mathbf{x}) \geq \phi(\mathbf{y})$$

if and only if ϕ is decreasing and Schur-convex on A.

LEMMA 4.2 *The function $\psi : \mathbb{R}^{+^n} \to \mathbb{R}$ satisfies*

$$\mathbf{x} \overset{p}{\succeq} \mathbf{y} \Longrightarrow \psi(\mathbf{x}) \geq \psi(\mathbf{y}) \tag{4.10}$$

if and only if,

(i) $\psi(e^{a_1}, \ldots, e^{a_n})$ is Schur-convex in $(a_1, \ldots, a_n)$

(ii) $\psi(e^{a_1}, \ldots, e^{a_n})$ is decreasing in a_i, for $i = 1, \ldots, n$,

where $a_i = \log x_i$, for $i = 1, \ldots, n$.

PROOF : Using relation (2.3), we see that (4.10) is equivalent to

$$\mathbf{a} \overset{w}{\succeq} \mathbf{b} \Longrightarrow \psi(e^{a_1}, \ldots, e^{a_n}) \geq \psi(e^{b_1}, \ldots, e^{b_n}), \tag{4.11}$$

where $a_i = \log x_i$ and $b_i = \log y_i$, for $i = 1, \ldots, n$.

Taking $\phi(a_1, \ldots, a_n) = \psi(e^{a_1}, \ldots, e^{a_n})$ in Theorem 4.8, we get the required result. ∎

Now we are ready to prove the next theorem.

THEOREM **4.9** *Let $X_1, \ldots, X_n$ be independent random variables with X_i having survival function $\overline{F}^{\lambda_i}(x)$, $i = 1, \ldots, n$. Let $Y_1, \ldots, Y_n$ be another set of random variables with Y_i having survival function $\overline{F}^{\lambda_i^*}(x)$, $i = 1, \ldots, n$. Then*

$$\boldsymbol{\lambda} \overset{p}{\succeq} \boldsymbol{\lambda}^* \Longrightarrow X_{n:n} \geq_{st} Y_{n:n}.$$

PROOF : The survival function of $X_{n:n}$ can be written as

$$\overline{F}_{X_{n:n}}(x) = 1 - \prod_{i=1}^{n}(1 - e^{-e^{a_i}H(x)}), \tag{4.12}$$

where $a_i = \log \lambda_i$, $i = 1, \ldots, n$ and $H(x) = -\log \overline{F}(x)$.

Using Lemma 4.2, we find that it is enough to show that the function $\overline{F}_{X_{n:n}}$ given by (4.12) is Schur-convex and decreasing in a_i's. To prove its Schur-convexity, it follows from Theorem 4.7 that, we have to show that for $i \neq j$, $(a_i - a_j)(\frac{\partial \overline{F}_{X_{n:n}}}{\partial a_i} - \frac{\partial \overline{F}_{X_{n:n}}}{\partial a_j}) \geq 0$. That is,

$$H(x)(a_i - a_j)\left(\prod_{i=1}^{n}(1 - e^{-e^{a_i}H(x)})\right)\left(\frac{e^{a_j}e^{-e^{a_j}H(x)}}{1 - e^{-e^{a_j}H(x)}} - \frac{e^{a_i}e^{-e^{a_i}H(x)}}{1 - e^{-e^{a_i}H(x)}}\right) \geq 0, \text{for } i \neq j \tag{4.13}$$

since

$$\frac{\partial \overline{F}_{X_{n:n}}}{\partial a_i} = -\prod_{i=1}^{n}(1 - e^{-e^{a_i}H(x)})\left(\frac{H(x)e^{a_i}e^{-e^{a_i}H(x)}}{1 - e^{-e^{a_i}H(x)}}\right).$$

It is easy to see that the function $be^{-bH(x)}/(1-e^{-bH(x)})$ is decreasing in b, for each fixed $x > 0$. Replacing b with e^{a_i}, it follows that the function $e^{a_i}e^{-e^{a_i}H(x)}/(1 - e^{-e^{a_i}H(x)})$ is also decreasing in a_i for $i = 1, \ldots, n$. This proves that (4.13) holds. The partial derivative of $\overline{F}_{X_{n:n}}$ with respect to a_i is negative and which in turn implies that the survival function of $X_{n:n}$ is decreasing in a_i for $i = 1, \ldots, n$. This completes the proof.

■

The following result due to Khaledi and Kochar (2000 b) is a special case of Theorem 4.9.

COROLLARY 4.3 *Let $X_1, \ldots, X_n$ be independent exponential random variables with X_i having hazard rate λ_i, $i = 1, \ldots, n$. Let $Y_1, \ldots, Y_n$ be another set of independent exponential random variables with Y_i having hazard rate λ_i^*, $i = 1, \ldots, n$. Then*

$$\boldsymbol{\lambda} \overset{p}{\succeq} \boldsymbol{\lambda}^* \Longrightarrow X_{n:n} \geq_{st} Y_{n:n}.$$

Boland, El-Neweihi and Proschan (1994) pointed out that for $n > 2$, (4.2) cannot be strengthened from stochastic ordering to hazard rate ordering. Since majorization implies p-larger ordering, it follows that, in general, Theorem 4.9 cannot be strengthened to hazard rate ordering.

As shown in the next example, a result similar to Theorem 4.9 may not hold for other order statistics.

EXAMPLE 4.1 : Let X_1, X_2, X_3 be independent exponential random variables with $\boldsymbol{\lambda} = (0.1, 1, 7.9)$ and Y_1, Y_2, Y_3 be independent exponential random variables with $\boldsymbol{\lambda}^* = (1, 2, 5)$. It is easy to see that $\boldsymbol{\lambda} \overset{p}{\succeq} \boldsymbol{\lambda}^*$. The $X_{1:3}$ and $Y_{1:3}$ have exponential distributions with respective hazard rates 9 and 8 and which implies that $Y_{1:3} \geq_{st} X_{1:3}$.

5 Stochastic Comparisons of Sample Spacings

Let $X_1, \ldots, X_n$ be n random variables. The random variables $D_{i:n} = X_{i:n} - X_{i-1:n}$ and $D^*_{i:n} = (n - i + 1)D_{i:n}$, $i = 1, \ldots, n$, with $X_{0:n} \equiv 0$, are respectively called spacings and normalized spacings. They are of great interest in various areas of statistics, in particular, in characterizations of distributions, goodness-of-fit tests, life testing and reliability models. In the reliability context they correspond to times elapsed between successive failures of components in a system. It is well known that the normalized spacings of a random sample from an exponential distribution are i.i.d. random variables having the same exponential distribution. Such a characterization may not hold for other distributions and much of the reliability theory deals with this aspect of spacings. In this section we review stochastic properties of spacings when original random variables are i.i.d. as well as when they are independent but not identically distributed.

Many authors have studied the stochastic properties of spacings from restricted families of distributions. Barlow and Proschan (1966) proved that if $X_1, \ldots, X_n$ is a random sample from a DFR distribution, then the successive normalized spacings are stochastically increasing. Kochar and Kirmani (1995) strengthened this result from stochastic ordering to hazard rate ordering, that is, for $i = 1, \ldots, n-1$,

$$D^*_{i:n} \leq_{hr} D^*_{i+1:n}. \tag{5.1}$$

The corresponding problem when the random variables are not identically distributed, has also been studied by many researchers, including Pledger and Proschan (1971), Shaked and Tong (1984), Kochar and Korwar (1996), Kochar and Rojo (1996), Nappo and Spizzichino (1998), among others. For a review of this topic see Kochar (1998). Here we give some new results obtained recently by the authors.

Kochar and Korwar (1996) conjectured that a result similar to (5.1) holds in the case when $X_1, \ldots, X_n$ are independent exponential random variables with X_i having hazard rate λ_i, for $i = 1, \ldots, n$. Khaledi and Kochar (2001) proved this conjecture when random variables X_i's follow a single outlier model with parameters λ and λ^*, that is when $\lambda_1 = \ldots = \lambda_{n-1} = \lambda$ and $\lambda_n = \lambda^*$. To prove this we shall be using the following results.

The joint density function of the spacings when λ_i's are possibly different is given by (cf. Kochar and Korwar, 1996),

$$f_{D_{1:n},\ldots,D_{n:n}}(x_1, \ldots, x_n) = \sum_{(\mathbf{r})} \frac{\prod_{i=1}^n \lambda_i}{\prod_{i=1}^n \sum_{j=i}^n \lambda(r_j)} \prod_{i=1}^n (\sum_{j=i}^n \lambda(r_j)) exp\{-x_i \sum_{j=i}^n \lambda(r_j)\}, \tag{5.2}$$

for $x_i \geq 0$, $i = 1, \ldots, n$, where $(\mathbf{r}) = (r_1, \ldots, r_n)$ is a permutation of $(1, \ldots, n)$ and $\lambda(i) = \lambda_i$. It is a mixture of products of exponential random variables. From (5.2) it is easy to find that the joint pdf of $(D_{i:n}, D_{j:n})$ for $1 \leq i < j \leq n$, is

$$\begin{aligned} f_{D_{i:n},D_{j:n}}(x, y) &= \sum_{(\mathbf{r})} \frac{\prod_{i=1}^n \lambda_i}{\prod_{i=1}^n \sum_{j=i}^n \lambda(r_j)} \\ &\times (\sum_{m=i}^n \lambda(r_m)) exp\{-x \sum_{m=i}^n \lambda(r_m)\} (\sum_{m=j}^n \lambda(r_m)) exp\{-y \sum_{m=j}^n \lambda(r_m)\}, \end{aligned} \tag{5.3}$$

for x, $y \geq 0$. Now (5.2) can be written as

$$f_{D_{1:n},\dots,D_{n:n}}(x_1,\dots,x_n) = \sum_{\theta=1}^{n} \frac{(n-1)!\lambda^*(\lambda)^{n-1}}{\prod_{i=1}^{\theta}((n-i)\lambda+\lambda^*)\prod_{i=\theta+1}^{n}(n-i+1)\lambda}$$

$$\times \prod_{i=1}^{\theta}((n-i)\lambda+\lambda^*)e^{-((n-i)\lambda+\lambda^*)x_i} \prod_{i=\theta+1}^{n}(n-i+1)\lambda e^{-(n-i+1)\lambda x_i}, \tag{5.4}$$

which can be further expressed as

$$f_{D_{1:n},\dots,D_{n:n}}(x_1,\dots,x_n) = \sum_{\theta=1}^{n} h(\theta) \prod_{i=1}^{\theta} \alpha_i^* e^{-\alpha_i^* x_i} \prod_{i=\theta+1}^{n} \alpha_i e^{-\alpha_i x_i},$$

where $\alpha_i = (n-i+1)\lambda$, $\alpha_i^* = (n-i)\lambda + \lambda^*$, $i = 1,\dots,n$ and using α_i and α_i^*, the function h is given by

$$h(\theta) = \frac{(n-1)!\lambda^{n-1}\lambda^*}{\prod_{i=1}^{\theta}\alpha_i^* \prod_{i=\theta+1}^{n}\alpha_i}, \quad \theta = 1,\dots,n. \tag{5.5}$$

The marginal density function of $D_{i:n}$ can be expressed as

$$f_{D_{i:n}}(x) = H_i \alpha_i e^{-\alpha_i x} + \overline{H}_i \alpha_i^* e^{-\alpha_i^* x}, \tag{5.6}$$

where

$$H_i = \sum_{\theta=1}^{i-1} h(\theta), \quad i = 2,\dots,n \text{ and } H_1 = 0. \tag{5.7}$$

Thus, the density function of $D_{i:n}$ is a mixture of two exponential random variables with parameters α_i and α_i^*. Now we prove the main theorem.

THEOREM **5.1** *Let $X_1,\dots,X_n$ follow the single-outlier exponential model with parameters λ and λ^*. Then*

$$D_{i+1:n}^* \geq_{hr} D_{i:n}^*, \quad i = 1,\dots,n-1.$$

PROOF : We prove the result when $\lambda^* > \lambda$. The proof for the case $\lambda^* < \lambda$ follows using the same kind of arguments. From (5.6) we find that the survival function of

$D^*_{i:n}$ is $\overline{F}_{D^*_{i:n}}(x) = H_i e^{-\lambda x} + \overline{H}_i e^{-\eta_i x}$, where $\eta_i = \frac{(n-i)\lambda+\lambda^*}{n-i+1}$. To prove the theorem we have to show that for any $i \in \{1, \ldots, n-1\}$,

$$g(x) = \frac{\overline{F}_{D^*_{i+1:n}}(x)}{\overline{F}_{D^*_{i:n}}(x)}$$

is increasing in x. The numerator of $g'(x)$, the derivative of $g(x)$ is

$$\begin{aligned}
A(x) &= [H_i e^{-\lambda x} + \overline{H}_i e^{-\eta_i x}][-\lambda H_{i+1} e^{-\lambda x} - \eta_{i+1}\overline{H}_{i+1} e^{-\eta_{i+1} x}] \\
&\quad + [H_{i+1} e^{-\lambda x} + \overline{H}_{i+1} e^{-\eta_{i+1} x}][\lambda H_i e^{-\lambda x} + \eta_i \overline{H}_i e^{-\eta_i x}] \\
&= (\lambda^* - \lambda)\left\{ \frac{\overline{H}_i H_{i+1}}{n-i+1} e^{-(\eta_i+\lambda)x} \right. \\
&\quad \left. - \frac{\overline{H}_{i+1} H_i}{n-i} e^{-(\eta_{i+1}+\lambda)x} - \frac{\overline{H}_i \overline{H}_{i+1}}{(n-i+1)(n-i)} e^{(\eta_i+\eta_{i+1})x} \right\} \\
&\geq (\lambda^* - \lambda)\left\{ \left(\frac{\overline{H}_i H_{i+1}}{n-i+1} - \frac{\overline{H}_{i+1} H_i}{n-i} \right) e^{-(\eta_{i+1}+\lambda)x} \right. \qquad (5.8) \\
&\quad \left. - \frac{\overline{H}_i \overline{H}_{i+1}}{(n-i+1)(n-i)} e^{(\eta_i+\eta_{i+1})x} \right\} \\
&= \frac{(\lambda^* - \lambda)}{(n-i)(n-i+1)} \left\{ \left\{ (n-i)\overline{H}_i - (n-i+1)\overline{H}_{i+1} + \overline{H}_i \overline{H}_{i+1} \right\} \right. \\
&\quad \times \left. e^{-(\eta_{i+1}+\lambda)x} - \overline{H}_i \overline{H}_{i+1} e^{-(\eta_i+\eta_{i+1})x} \right\}. \qquad (5.9)
\end{aligned}$$

The inequality in (5.8) follows, since $\lambda^* > \lambda$ implies $\eta_{i+1} > \eta_i$.

Again $\lambda^* > \lambda$ implies $\lambda < \eta_i$ and which in turn implies $e^{-(\eta_{i+1}+\lambda)x} \geq e^{-(\eta_i+\eta_{i+1})x}$ for every $x \geq 0$. Also for $\lambda^* > \lambda$,

$$\begin{aligned}
\left\{ (n-i)\overline{H}_i - (n-i+1)\overline{H}_{i+1} \right\} &= (n-i)h(i) - \overline{H}_{i+1} \\
&\geq 0, \qquad (5.10)
\end{aligned}$$

since for $\lambda^* > \lambda$, $h(j)$ is a decreasing function of j. Using these results in (5.9) we find that $A(x)$ and hence $g'(x)$ is nonnegative for $x \geq 0$. This proves the required result. ∎

Let $X_1, \ldots, X_n$ be independent exponential random variables with hazard rates $\lambda_1, \ldots, \lambda_n$, respectively. Pledger and Proschan (1971) proved that for $i \in \{1, \ldots, n\}$, $D_{i:n}$ is stochastically larger when the hazard rates are unequal than when they are all equal. Kochar and Rojo (1996) strengthened this result to likelihood ratio ordering. The natural question is to examine whether the survival function of $D_{i:n}$ is Schur-convex in $(\lambda_1, \ldots, \lambda_n)$. Pledger and Proschan (1971) came up with a counterexample to show that this is not true in general. Kochar and Korwar (1996) proved that in the special case of second spacing, whereas the survival function of $D_{2:n}$ is Schur-convex in $(\lambda_1, \ldots, \lambda_n)$, its hazard rate is not Schur-concave. They proved, however, that the hazard rate of $D_{2:2}$ is Schur-concave. We now examine this question when $X_1, \ldots, X_n$ follow the single-outlier exponential model with parameters λ and λ^*. In the rest of this section, we assume that $\lambda^* < \lambda$. We will treat it as a part of the model. It is easy to see that in this case, $(\lambda_1^*, \lambda_1, \ldots, \lambda_1) \overset{m}{\succeq} (\lambda_2^*, \lambda_2, \ldots, \lambda_2)$ if and only if $\lambda_1^* < \lambda_2^* < \lambda_2 < \lambda_1$ and $\lambda_1^* + (n-1)\lambda_1 = \lambda_2^* + (n-1)\lambda_2$. We prove later in this section that for the single-outlier model, for $i \in \{1, \ldots, n\}$, the hazard rate of $D_{i:n}$ is Schur-concave in λ's. To prove it we need the following lemmas.

LEMMA 5.1 *Let $X_1, \ldots, X_n$ follow the single-outlier exponential model with parameters λ and λ^*. Then*

$$\lambda^* < \lambda \Longrightarrow H_i \le \frac{i-1}{n}, \quad \textit{for } i = 1, \ldots, n, \tag{5.11}$$

where H_i is given by (5.7). *The inequality in* (5.11) *is reversed for $\lambda^* > \lambda$.*

PROOF : $\lambda^* < \lambda$ implies that the function $h(j)$ in (5.5) is increasing in j, $j = 1, \ldots, n$. Note that

$$(h(1), h(2), \ldots, h(n)) \overset{m}{\succeq} (1/n, \ldots, 1/n).$$

The required result follows from the definition of majorization.

The proof for the case $\lambda^* > \lambda$ follows from the same kind of arguments. ■

LEMMA 5.2 *Let $X_1, \ldots, X_n$ follow the single-outlier exponential model with parameters λ_1 and λ_1^*. Let $Y_1, \ldots, Y_n$ be another set of random variables following the single-outlier exponential model with parameters λ_2 and λ_2^*. If*

(i) $\lambda_1^* < \lambda_2^* < \lambda_2 < \lambda_1$, then $\Theta_1 \geq_{lr} \Theta_2$,

(ii) $\lambda_1 < \lambda_2 < \lambda_2^* < \lambda_1^*$, then $\Theta_1 \leq_{lr} \Theta_2$,

where Θ_1 and Θ_2 correspond to random variable Θ with probability mass function $h(j)$ in (5.5) for X_i's and Y_i's, respectively.

PROOF : (i) We prove that

$$\frac{h_2(\theta+1)}{h_1(\theta+1)} \leq \frac{h_2(\theta)}{h_1(\theta)},$$

where h_1 and h_2 are probability mass functions of Θ_1 and Θ_2, respectively. This inequality holds if and only if

$$\frac{(n-\theta-1)\lambda_1+\lambda_1^*}{(n-\theta-1)\lambda_2+\lambda_2^*} \leq \frac{\lambda_1}{\lambda_2}. \tag{5.12}$$

Since $\lambda_1^* < \lambda_2^*$ and $\lambda_2 < \lambda_1$, it is easy to see that (5.12) is true.

(ii) In this case the inequality in (5.12) is reversed which in turn implies that $\Theta_1 \leq_{lr} \Theta_2$. This proves the result. ∎

THEOREM **5.2** *Let $X_1, \ldots, X_n$ follow the single-outlier exponential model with parameters λ_1 and λ_1^* with $\lambda_1^* < \lambda_1$. Then for $i \in \{1, \ldots, n\}$, the hazard rate of $D_{i:n}$ is Schur-concave in $\{\lambda_1, \ldots, \lambda_1, \lambda_1^*\}$.*

PROOF: Let $Y_1, \ldots, Y_n$ be another set of random variables following the single-outlier exponential model with parameters λ_2 and λ_2^* ($\lambda_2^* < \lambda_2$) such that $(\lambda_1^*, \lambda_1, \ldots, \lambda_1) \overset{m}{\succeq} (\lambda_2^*, \lambda_2, \ldots, \lambda_2)$. As discussed above this holds if and only if $\lambda_1^* < \lambda_2^* < \lambda_2 < \lambda_1$ and

$\lambda_1^* + (n-1)\lambda_1 = \lambda_2^* + (n-1)\lambda_2$. Without loss of generality, let us assume that $\lambda_1^* + (n-1)\lambda_1 = 1$. We have to prove that under the given conditions for $i = 1, \ldots, n$,

$$D_{i:n}^{(1)} \geq_{hr} D_{i:n}^{(2)},$$

where $D_{i:n}^{(1)}$ ($D_{i:n}^{(2)}$) denotes the ith spacing of X_i's (Y_i's). From (5.6) the survival functions of $D_{i:n}^{(1)}$ and $D_{i:n}^{(2)}$ are

$$\overline{F}_{D_{i:n}^{(1)}}(x) = P_i e^{-\alpha_{i1}x} + \overline{P}_i e^{-\alpha_{i1}^* x},$$

$$\overline{F}_{D_{i:n}^{(2)}}(x) = Q_i e^{-\alpha_{i2}x} + \overline{Q}_i e^{-\alpha_{i2}^* x},$$

where P_i and Q_i correspond to H_i in (5.6) for $D_{i:n}^{(1)}$ and $D_{i:n}^{(2)}$, respectively and $\alpha_{i1} = (n-i+1)\lambda_1$, $\alpha_{i1}^* = (n-i)\lambda_1 + \lambda_1^*$, $\alpha_{i2} = (n-i+1)\lambda_2$ and $\alpha_{i2}^* = (n-i)\lambda_2 + \lambda_2^*$.

We have to show that

$$\phi(x) = \frac{\overline{F}_{D_{i:n}^{(1)}}(x)}{\overline{F}_{D_{i:n}^{(2)}}(x)}$$

is increasing in x. After some simplifications we find that the numerator of $\phi'(x)$, the derivative of $\phi(x)$ is

$$\begin{aligned} g(x) &= -(\alpha_{i1} - \alpha_{i2})P_i Q_i e^{-(\alpha_{i1}+\alpha_{i2})x} + (\alpha_{i2}^* - \alpha_{i1}^*)\overline{P}_i \overline{Q}_i e^{-(\alpha_{i1}^*+\alpha_{i2}^*)x} \\ &- (\alpha_{i1}^* - \alpha_{i2})Q_i \overline{P}_i e^{-(\alpha_{i2}+\alpha_{i1}^*)x} + (\alpha_{i2}^* - \alpha_{i1})\overline{Q}_i P_i e^{-\alpha_{i1}+\alpha_{i2}^*)x}, \end{aligned} \tag{5.13}$$

Using the assumption $\lambda_1^* < \lambda_2^* < \lambda_2 < \lambda_1$ and the fact the $\lambda_i^* + (n-1)\lambda_i = 1$, $i = 1, 2$, it follows, $\alpha_{i1} + \alpha_{i2}^* < \alpha_{i1} + \alpha_{i2}$, $\alpha_{i1} + \alpha_{i2}^* > \alpha_{i1}^* + \alpha_{i2}^*$, $\alpha_{i1} + \alpha_{i2}^* > \alpha_{i1}^* + \alpha_{i2}$ and all $(\alpha_{i1} - \alpha_{i2})$, $(\alpha_{i2}^* - \alpha_{i1}^*)$, $(\alpha_{i2} - \alpha_{i1}^*)$, are nonnegative. Using these observations in (5.13), we see

$$\begin{aligned} g(x) \geq\ & e^{-(\alpha_{i1}+\alpha_{i2}^*)x}\{-(\alpha_{i1} - \alpha_{i2})P_i Q_i + (\alpha_{i2}^* - \alpha_{i1}^*)\overline{P}_i \overline{Q}_i \\ & -(\alpha_{i1} - \alpha_{i2}^*)\overline{Q}_i P_i + (\alpha_{i2} - \alpha_{i1}^*)Q_i \overline{P}_i\} \end{aligned}$$

$$
\begin{aligned}
&= \frac{e^{-(\alpha_{i1}+\alpha_{i2}^*)x}}{n-1}\{Q_i - P_i - (nQ_i - (i-1))\lambda_2^* + (nP_i - (i-1))\lambda_1^*\} \\
&\geq \frac{e^{-(\alpha_{i1}+\alpha_{i2}^*)x}}{n-1}\{Q_i - P_i - n(Q_i - P_i)\lambda_2^*\} \qquad (5.14) \\
&= \frac{e^{-(\alpha_{i1}+\alpha_{i2}^*)x}}{n-1}(Q_i - P_i)(1 - n\lambda_2^*) \\
&\geq 0. \qquad (5.15)
\end{aligned}
$$

The inequality in (5.14) follows, since by Lemma 5.1 $P_i \leq \frac{i-1}{n}$ and $\lambda_1^* < \lambda_2^*$. From Lemma 5.2 it follows that $Q_i \geq P_i$, since it is known the likelihood ratio ordering implies usual stochastic ordering. This observation along with the fact that $\lambda_2^* \leq 1/n$ implies the inequality in (5.15). ■

Remark : The conclusion of Theorem 5.2 holds if instead of $\lambda_1^* < \lambda_1$ and $\lambda_2^* < \lambda_2$ we assume that $\lambda_1^* > \lambda_1$ and $\lambda_2^* > \lambda_2$.

It is known that spacings of independent exponential random variables have DFR distributions (cf. Kochar and Korwar, 1996). Combining this observation with Theorem 2.1, we have proved the following corollary.

COROLLARY 5.1 *Under the assumptions of Theorem 5.2,*

$$D_{i:n}^{(1)} \geq_{disp} D_{i:n}^{(2)}.$$

A consequence of Corollary 5.1 is that $var(D_{i:n}^{(1)}) \geq var(D_{i:n}^{(2)})$, $i = 1, \ldots, n$.

6 Stochastic ordering for sample range

Sample range is one of the criteria for comparing variabilities among distributions and hence it is important to study its stochastic properties. First we study the stochastic properties of the range of a random sample from a continuous distribution. Let $X_1, \ldots, X_n$ be a random sample from F and let $Y_1, \ldots, Y_n$ be an independent random sample from another distribution G. It follows from Lemma 3(c) of Bartoszewic

(1986) that $X \geq_{disp} Y \Rightarrow X_{n:n} - X_{1:n} \geq_{st} Y_{n:n} - Y_{1:n}$. This observation along with Theorem 2.1 (a) leads to the following theorem.

THEOREM **6.1** *Let $X \geq_{hr} Y$ and let either F or G be DFR. Then*

$$X_{n:n} - X_{1:n} \geq_{st} Y_{n:n} - Y_{1:n}. \tag{6.1}$$

Next we consider the case when the parent observations are independent exponentials but with unequal parameters. Let $X_1, \ldots, X_n$ be independent exponential random variables with X_i having hazard rate λ_i, $i = 1, \ldots, n$. Let $Y_1, \ldots, Y_n$ be a random sample of size n from an exponential distribution with common hazard rate $\bar{\lambda}$, the arithmetic mean of the λ_i's. Finally, let $R_X = X_{n:n} - X_{1:n}$ and $R_Y = Y_{n:n} - Y_{1:n}$ denote the sample ranges of X_i's and Y_i's, respectively. Kochar and Rojo (1996) proved that $R_X \geq_{st} R_Y$. Khaledi and Kochar (2000 c) proved the following result which is in terms of $\tilde{\lambda}$, the geometric mean of the λ_i's.

THEOREM **6.2** *Let $X_1, \ldots, X_n$ be independent exponential random variables with X_i having hazard rate λ_i, for $i = 1, \ldots, n$. Let $Y_1, \ldots, Y_n$ be a random sample of size n from an exponential distribution with common hazard rate $\tilde{\lambda}$. Then,*

$$R_X \geq_{st} R_Y.$$

PROOF : The distribution function of R_X (see David, 1981, p. 26) is

$$F_{R_X}(x) = \frac{1}{\sum_{i=1}^n \lambda_i} \sum_{i=1}^n \frac{\lambda_i}{1 - e^{-\lambda_i x}} \prod_{i=1}^n (1 - e^{-\lambda_i x}). \tag{6.2}$$

and that of R_Y is

$$G_{R_Y}(x) = \left(1 - e^{-\tilde{\lambda} x)}\right)^{n-1}. \tag{6.3}$$

Using (6.2) and (6.3), we have to show that

$$\sum_{i=1}^n \frac{\lambda_i}{1 - e^{-\lambda_i x}} \prod_{i=1}^n (1 - e^{-\lambda_i x}) \leq \sum_{i=1}^n \lambda_i \left(1 - e^{-\tilde{\lambda} x)}\right)^{n-1}. \tag{6.4}$$

Multiplying both sides of (6.4) by $x(>0)$, it is sufficient to prove that

$$\sum_{i=1}^{n} \frac{\lambda_i x}{1-e^{-\lambda_i x}} \prod_{i=1}^{n}(1-e^{-\lambda_i x}) \leq \left(\sum_{i=1}^{n} \lambda_i x\right)\left(1-e^{-\tilde{\lambda} x}\right)^{n-1}. \tag{6.5}$$

Dykstra, Kochar and Rojo (1997) proved that

$$\sum_{i=1}^{n} \frac{y_i}{1-e^{-y_i}} \leq \left(\sum_{i=1}^{n} y_i\right) \prod_{i=1}^{n}(1-e^{-y_i})^{-\frac{1}{n}},$$

where $y_i > 0$ for $i = 1, \ldots, n$. Making use of this inequality on the L.H.S. of (6.5), we get

$$\sum_{i=1}^{n} \frac{\lambda_i x}{1-e^{-\lambda_i x}} \prod_{i=1}^{n}(1-e^{-\lambda_i x}) \leq \left(\sum_{i=1}^{n} \lambda_i x\right) \prod_{i=1}^{n}\left(1-e^{-\lambda_i x}\right)^{\frac{n-1}{n}} \tag{6.6}$$

A consequence of Theorem 4.4 (*b*) is that $X_{n:n} \geq_{st} Y_{n:n}$, which is equivalent to $\prod_{i=1}^{n}(1-e^{-\lambda_i x})^{1/n} \leq 1-e^{-\tilde{\lambda} x}$. Using this result, we find that the expression on the R.H.S. of (6.6) is less than or equal to that on the R.H.S. of (6.5) and from which the required result follows. ■

As a consequence of this result we get the following upper bound on the distribution function of R_X in terms $\tilde{\lambda}$.

Corollary 6.1 *Under the conditions of Theorem 6.2, for $x > 0$,*

$$P[X_{n:n} - X_{1:n} \leq x] \leq \left[1-e^{-\tilde{\lambda} x}\right]^{n-1}. \tag{6.7}$$

This bound is better than the one obtained in Kochar and Rojo (1996) in terms of $\overline{\lambda}$, since the expression on the R.H.S. of (6.7) is increasing in $\tilde{\lambda}$ and $\tilde{\lambda} \leq \overline{\lambda}$.

Now we extend Theorem 6.1 to the PHR model. We assume that F is *new worse than used* (NWU), that is,

$$\overline{F}(x+y) \geq \overline{F}(x)\overline{F}(y), \quad \text{for } x, y \geq 0,$$

or equivalently,

$$H(x+y) \leq H(x) + H(y), \quad \text{for } x, y \geq 0,$$

where $H(x) = -\log \overline{F}(x)$ denotes the cumulative hazard of F.

THEOREM **6.3** *Let $X_1, \ldots, X_n$ be independent random variables with X_i having survival function $\overline{F}^{\lambda_i}(x)$, $i = 1, \ldots, n$. Let $Y_1, \ldots, Y_n$ be a random sample of size n from a distribution with survival function $\overline{F}^{\tilde{\lambda}}(x)$, where $\tilde{\lambda} = (\prod_{i=1}^n \lambda_i)^{1/n}$. If F is NWU, then $X_{n:n} - X_{1:n} \geq_{st} Y_{n:n} - Y_{1:n}$.*

PROOF :

The distribution function of the sample range $X_{n:n} - X_{1:n}$ (see David, 1981, p. 26) is

$$\begin{aligned} F_{R_n^X}(x) &= \sum_{i=1}^n \int_0^{+\infty} \lambda_i h(t) e^{-\lambda_i H(t)} \prod_{j \neq i}^n \left(e^{-\lambda_j H(t)} - e^{-\lambda_j H(t+x)}\right) dt \\ &\leq \sum_{i=1}^n \int_0^{+\infty} \lambda_i h(t) e^{-\lambda_i H(t)} \prod_{j \neq i}^n \left(e^{-\lambda_j H(t)} - e^{-\lambda_j H(t)} e^{-\lambda_j H(x)}\right) dt \\ &\quad \text{(since } F \text{ is NWU)} \\ &= \sum_{i=1}^n \lambda_i \prod_{j \neq i} (1 - e^{-\lambda_j H(x)}) \int_0^{+\infty} h(t) \prod_{j=1}^n e^{-\lambda_j H(t)} dt \\ &= \sum_{i=1}^n \lambda_i \prod_{j \neq i} (1 - e^{-\lambda_j H(x)}) \int_0^{+\infty} h(t) e^{-H(t) \sum_{j=1}^n \lambda_j} \\ &= \frac{1}{\sum_{i=1}^n \lambda_i} \sum_{i=1}^n \frac{\lambda_i}{1 - e^{-\lambda_i H(x)}} \prod_{i=1}^n (1 - e^{-\lambda_i H(x)}), \quad x > 0, \end{aligned}$$

Now, replacing x with $H(x)$ in the proof of Theorem 6.2, it is easy to see that

$$F_{R_n^X}(x) \leq F_{R_n^Y}(x). \qquad \blacksquare$$

References

1. Arnold, B. C., Balakrishnan, N., and Nagaraja, H. N. (1992). *A First Course in Order Statistics.* Wiley, New York.
2. Bagai, I. and Kochar, S. C. (1986). On tail ordering and comparison of failure rates. *Comm. Statist. Theory and Methods* **15**, 1377-1388.

3. Balakrishnan, N. and Rao, C. R. (1998a). *Handbook of Statistics* **16** - *Order Statistics : Theory and Methods.* Elsevier, New York.

4. Balakrishnan, N. and Rao, C. R. (1998b). *Handbook of Statistics* **17** - *Order Statistics : Applications.* Elsevier, New York.

5. Bapat, R. B. and Beg, M. I. (1989). Order statistics for nonidentically distributed variables and permanents. *Sankhyā Ser. B* **51**, 79-93.

6. Bapat, R. B. and Kochar, S. C. (1994). On likelihood ratio ordering of order statistics. *Linear Algebra and Its Applications* **199**, 281-291.

7. Barlow, R. E. and Proschan, F. (1966). Inequalities for linear combinations of order statistics from restricted families. *Ann. Math. Statist.* **37**, 1574-1592.

8. Barlow, R. E. and Proschan, F. (1981). *Statistical Theory of Reliability and Life Testing.* To Begin With : Silver Spring, Maryland.

9. Bartoszewicz, J. (1986). Dispersive ordering and the total time on test transformation. *Statist. Probab. Lett.* **4**, 285- 288.

10. Bartoszewicz, J. (1987). A note on dispersive ordering defined by hazard functions. *Statist. Probab. Lett.* **6**, 13-17.

11. Boland, P.J., El-Neweihi, E. and Proschan, F. (1994). Applications of the hazard rate ordering in reliability and order statistics. *J. Appl. Probab.* **31**, 180-192.

12. Boland, P. J., Hollander, M., Joag-Dev, K. and Kochar, S. (1996). Bivariate dependence properties of order statistics. *J. Multivariate Anal.* **56**, 75-89.

13. Boland, P.J., Shaked, M. and Shanthikumar, J.G. (1998). Stochastic ordering of order statistics . In N. Balakrishnan and C. R. Rao, eds, *Handbook of Statistics* **16** - *Order Statistics : Theory and Methods.* Elsevier, New York, 89-103. [Technical report of the University College Dublin, 1995].

14. Bon, J. L. and Paltanea, E. (1999). Ordering properties of convolutions of exponential random variables. *Lifetime Data Anal.* **5**, 185- 192.

15. Chan, W., Proschan, F. and Sethuraman, J. (1991). Convex ordering among functions, with applications to reliability and mathematical statistics. In *Topics in Statistical Dependence*, ed. H. W. Block, A. R. Sampson and T. H. Savits. IMS Lecture Notes **16**, 121-134.

16. David, H. A. (1981). *Order Statistics* (2nd ed.). Wiley, New York.

17. David, H.A. and Groeneveld, R.A. (1982). Measures of local variation in a distribution: Expected length of spacings and variances of order statistics. *Biometrika 69,* 227-232.

18. Deshpande, J. V. and Kochar, S. C. (1983). Dispersive ordering is the same as tail ordering. *Adv. in Appl. Probab.* **15**, 686-687.

19. Dharmadhikari, S. and Joeg-dev, K. (1988). Unimodality, Convexity and Applications. *Academic press, INC.*

20. Dykstra, R., Kochar, S. C. and Rojo, J. (1997). Stochastic comparisons of parallel systems of heterogeneous exponential components. *J. Statist. Plann. Inference* **65**, 203-211.

21. Khaledi, B. and Kochar, S. C. (1999). Stochastic ordering between distributions and their sample spacings. *Statist. Probab. Lett.* **44**, 161-166.

22. Khaledi, B. and Kochar, S. C. (2000 a). On dispersive ordering among order statistics in one-sample and two-sample problems. *Statist. Probab. Lett.* **46**, 257-261.

23. Khaledi, B. and Kochar, S. C. (2000 b). Some new results on stochastic comparisons of parallel systems. *J. Appl. Probab.* **37**,1123-1128.

24. Khaledi, B. and Kochar, S. C. (2000 c). Sample range-some stochastic comparisons results. *Calcutta Statistical Association Bulletin* **50**, 283-291.

25. Khaledi, B. and Kochar, S. C. (2001). Stochastic properties of Spacings in a Single-Outlier Exponential Model. *Probability in Engineering and Information Sciences* **15** (2001), 401-408

26. Kochar, S. C. (1996). Dispersive ordering of order statistics. *Statist. Probab. Lett.* **27**, 271-274.

27. Kochar, S. C. (1998). Stochastic comparisons of spacings and order statistics. *Frontiers in Reliability.* World Scientific : Singapore. 201-216. eds., Basu, A. P., Basu, S. K. and Mukhopadhyay, S.

28. Kochar, S.C. (1999). On stochastic ordering between distributions and their sample spacings. *Statist. Probab. Lett.* **42**, 345-352.

29. Kochar, S. C. and Kirmani, S.N.U.A. (1995). Some results on normalized spacings from restricted families of distributions. *J. Statist. Plann. Inference* **46**, 47-57.

30. Kochar, S. C. and Korwar, R. (1996). Stochastic orders for spacings of heterogeneous exponential random variables. *J. Multivariate Anal.* **57**, 69-83.

31. Kochar, S. C. and Ma, C. (1999). Dispersive ordering of convolutions of exponential random variables. *Statist. Probab. Lett.* **43**, 321-324.

32. Kochar, S. C. and Rojo, J. (1996). Some new results on stochastic comparisons of spacings from heterogeneous exponential distributions. *J. Multivariate Anal.* **59**, 272-281.

33. Lewis, T. and Thompson, J. W. (1981). Dispersive distribution and the connection between dispersivity and strong unimodality. *J. Appl. Probab.* **18**, 76-90.

34. Lillo, R. E., Nanda, A.K. and Shaked, M (2000) Preservation of some likelihood ratio stochastic orders by order statistics. *Statistics and Probability Letters*, To appear.

35. Lynch, J. Mimmack, G. and Proschan, F. (1987). Uniform stochastic orderings and total positivity. *Canad. J. Statist.* **15**,63-69.

36. Marshall, A. W. and Olkin, I. (1979). *Inequalities : Theory of Majorization and Its Applications.* Academic Press, New York.

37. Mitrinovic, D. S. (1970). *Analytic Inequalities.* Springer Verlag, Berlin.

38. Nappo, G. and Spizzichino, F. (1998). Ordering properties of the TTT-plot of lifetimes with Schur joint densities. *Statist. Probab. Lett.* **39**, 195-203.

39. Pledger, P. and Proschan, F. (1971). Comparisons of order statistics and of spacings from heterogeneous distributions. *Optimizing Methods in Statistics.* Academic Press, New York., 89-113. ed. Rustagi, J. S.

40. Proschan, F. and Sethuraman, J. (1976). Stochastic comparisons of order statistics from heterogeneous populations, with applications in reliability. *J. Multivariate Anal.* **6**, 608-616.

41. Raqab, M.Z. and Amin, W.A. (1996). Some ordering results on order statistics and record values. *IAPQR Transactions.* **21, No. 1**, 1-8.

42. Rojo, J. and He, G.Z. (1991). New properties and characterizations of the dispersive ordering. *Statist. Probab. Lett.* **11**, 365-372.

43. Ross, S.M. (1983). *Stochastic Processes.* Wiley, New York.

44. Saunders, D.J. (1984). Dispersive ordering of distributions. *Adv. Appl. Prob.* **16**, 693-694.

45. Saunders, I. W. and Moran, P. A. P. (1978). On quantiles of the gamma and F distributions. *J. Appl. Probab.* **15**, 426-432.

46. Sen, P. K. (1970). A note on order statistics for heterogeneous distributions. *Ann. Math. Statist.* **41**, 2137-2134.

47. Shaked, M. and Shanthikumar, J. G. (1994). *Stochastic Orders and their Applications.* Academic Press, San Diego, CA.

48. Shaked, M. and Tong, Y. L. (1984). Stochastic ordering of spacings from dependent random variables. *Inequalities in Statistics and Probability*, IMS Lecture Notes-Monograph series, **5**, 141-149.

Parametrics to Nonparametrics: Extending Regression Models

Abhinanda Sarkar

IBM India Research Lab

1 Introduction

Central to modern statistics, both in theory and application, lies the notion of a model. While there has been some debate as to what the strict definition of a model should be, the practicing statistician has typically taken the view "I know a model when I use it". This chapter is about statistical models, what they represent and how advances in mathematics and computing have enabled the expansion of what permissible models are and what they can be used for. To fix ideas, we shall restrict ourselves to the case of regression with one predictor variable. This class of models is rich enough and the applications interesting enough to illustrate much.

We shall make a formal distinction between two classes of models, namely parametric and nonparametric. One of the objectives of this chapter is to make the observation that the distinction is not as much as it appears at first sight and that a unified perspective on statistical models is possible and, arguably, desirable. The choice of topics reflects personal preferences. Nonetheless we hope that this view of models in regression serves to cast some light on the unity in the apparent diversity in the area.

1.1 Regression: the basic model

Consider a sample of independent observations denoted by $Y_1, Y_2, \ldots, Y_n$. For example, Y_i can be the selling price the ith car in an auction of n cars. These observations are subject to uncertainty and they can be considered to be independent random variables. As random variables, they have expected values (averages) denoted by $E(Y_i)$. The regression problem arises when there is reason to believe that these expected values are related to other observables. Suppose that there are observations $x_1, x_2, \ldots, x_n$ and a function f such that $E(Y_i) = f(x_i)$. In our example, x_i can be the price at which the auction starts for the ith car. Note that (in regression) we are not interested in the uncertainty in the x_i. Indeed, they need not be considered random at all and it suffices to think of the x_i as fixed or given apriori. For these given values, the random Y_i are observed. The conditions

(i) Y_i are independent and

(ii) $E(Y_i) = f(x_i)$

will be common to all the models we will consider for this scenario. The function f is generally called the *regression function* and modelling and making inferences for it is the subject matter of regression analysis.

2 Parametric regression

In the most common form of classical regression analysis, the bare bones model of Section 1 is fleshed out by making further assumptions. The material in this section is by now considered traditional and is discussed in standard texts such as [4] and [23]. They can be consulted for the mathematical derivations we omit. We present this material as review as well as to set the stage for more recent methods discussed in later sections.

2.1 Polynomial regression

Recall that the Gaussian (or normal) distribution with expectation μ and variance σ^2, denoted by $N(\mu, \sigma^2)$, has the symmetric density function $\frac{1}{\sigma\sqrt{2\pi}}e^{-\frac{(x-\mu)^2}{2\sigma^2}}$.

A simple model for $Y_1, Y_2, \ldots, Y_n$ stipulates that Y_i has $N(\mu_i, \sigma^2)$ distribution where $\mu_i = \beta_0 + \beta_1 x_i + \beta_2 x_i^2 + \ldots + \beta_{p-1} x_i^{p-1}$. Thus, in addition to the basic assumptions in Section 1, we further assume that

(i) Y_i have Gaussian distribution,

(ii) the regression function f is a polynomial in x_i and

(iii) Y_i have the same variance.

This complete specification is called the polynomial regression model. If p is 2, then the ever-popular linear regression emerges. The case of Y_i being independent and identically distributed (iid) Gaussian random variables is captured by setting p to be 1.

Observe that if the values of $\beta_0, \beta_1, \ldots, \beta_{p-1}$ and σ^2 are known, then the distribution of $Y_1, Y_2, \ldots, Y_n$ is completely known. Thus $p+1$ constants, or parameters, can be used to identify the exact distribution of the observed sample. This identification with a finite number of parameters is what allows us to call this model a *parametric* model.

The ability to identify parameters also allows routine inference. For the polynomial regression model, the maximum likelihood (ML) method can be used for estimation. The estimates $(\hat{\beta}_0, \hat{\beta}_1, \ldots, \hat{\beta}_{p-1}, \hat{\sigma}^2)$ are those that maximize the likelihood of observing the sample, namely,

$$\prod_{i=1}^{n} \frac{1}{\sigma\sqrt{2\pi}} \exp\left[-\frac{(Y_i - \beta_0 - \beta_1 x_i - \ldots - \beta_{p-1}x_i^{p-1})^2}{2\sigma^2}\right].$$

It is easy to see that this amounts to minimizing $\sum_{i=1}^{n}(Y_i - \beta_0 - \beta_1 x_i - \ldots - \beta_{p-1}x^{p-1})^2$ in order to obtain $\hat{\beta} = (\hat{\beta}_0, \hat{\beta}_1, \ldots, \hat{\beta}_{p-1})$. This is the celebrated least squares (LS) method of estimation which is thus shown to be equivalent to ML estimation if we assume a Gaussian model. See [21] for a historical account of the central role that LS estimation played in statistics and data analysis.

A principal analytical tool in regression is the collection of fitted values. In our general model of Section 1, the estimate of the regression function f is denoted by $\hat{f}$ and the fitted values are $\hat{Y}_i = \hat{f}(x_i)$. For the polynomial regression model, $\hat{Y}_i = \hat{\beta}_0 + \hat{\beta}_1 x_i + \ldots + \hat{\beta}_{p-1}x^{p-1}$.

To ease notation, let $Y = (Y_1, \ldots, Y_n)$, $\hat{Y} = (\hat{Y}_1, \ldots, \hat{Y}_n)$, and $\beta = (\beta_0, \beta_1, \ldots, \beta_{p-1})$ For the polynomial regression model, define a $n \times p$ covariate matrix X with

$(i,j)^{th}$ element x_i^{j-1}. Then the LS estimators are

$$\hat{\beta} = \text{argmin}_\beta (Y - X\beta)^T(Y - X\beta) = (X'X)^{-1}X'Y$$

and the predicted values are given by $\hat{Y} = X\hat{\beta} = HY$ where H is a function of (non-random) $x = (x_1, \ldots, x_n)$ but not of (random) Y. For obvious reasons, H is called a hat matrix and can be shown to be of rank p. Thus $\hat{Y}$ is a linear function of Y and such a fitting procedure is said to be a linear smoother. For now, smoothing refers to the reduced dimensionality of the fitted values $\hat{Y}$; they lie on a p-dimensional subspace of the n-dimensional Euclidean space containing Y. Note that the dimension of this subspace (and the rank of the smoothing matrix H) is the same as the number of parameters used to specify the regression function. It is also true that H is a projection and $H^2 = H$. From this it follows that the trace of H is also p, which is the number of regression parameters.

Of course, mere point estimation of the regression parameters is inadequate for most applications. We need a measure of uncertainty for our estimates. The standard deviations of the estimates, usually called standard errors, serve as such as a measure. For the Gaussian models above, it can be shown that $se(\hat{\beta}_j) = \sigma\sqrt{(X'X)^{-1}_{jj}}$. If estimates of the standard errors are required, we can replace σ by, for example, the ML estimate $\hat{\sigma} = \left[\frac{1}{n}(Y - X\hat{\beta})^T(Y - X\hat{\beta})\right]^{\frac{1}{2}}$.

2.2 Model selection

While a reasonably complete description of model and inference has been given for polynomial regression, a crucial model-related issue remains to be settled. This is the choice of p or, equivalently, the choice of the degree of the polynomial to be fitted.

It is instructive to think of why the choice of p is important. Too small a p may fail to capture the complexity in the regression function. For example, the rate of a chemical reaction may increase with the concentration of a reagent, but may stabilize beyond a certain concentration. If we restrict ourselves to linear regression ($p=2$), then the increase can be captured, but not the stabilization. The model has too few parameters to capture the features of interest.

On the other hand, playing safe and stipulating a large value of p leads to another kind of shortcoming. If there are too many parameters, each parameter is estimated poorly, i.e. with large standard error. The entire model is thus poorly estimated and is unlikely to be very useful when applied to another data set. In the computer science literature, such models are said to "generalize" badly. An extreme situation arises when we attempt to fit n regression parameters ($p=n$) with n points. The LS fit is then an interpolation with the fitted values being the data points themselves. The model has adapted perfectly to this particular data and is very unlikely to have this high a fidelity to another realization from the same natural source. We will have more to say on this in later sections.

There are various criteria that are used in determining p that balance the above two sources of model misspecification. One of the most useful and popular is the Akaike Information Criteria (AIC) proposed in [1]. Let $\hat{L}_p$ denote the likelihood that has been maximized over p parameters. Then we can define the AIC, a function of p, as $AIC(p) = -2\log\hat{L}_p + 2p$. As model complexity

p increases, the maximized likelihood increases. (This assumes that if $p' < p$, the model with p' parameters is a submodel of the model with p parameters.) In most models used in practice $AIC(p)$ is first decreasing in p and then increases as the complexity term $2p$ begins to dominate. The choice of p that minimizes $AIC(p)$ is considered a choice of p that compromises complexity and generalizability adequately.

2.3 Other parametric models: logistic regression

While the assumption that the random variables have Gaussian distributions is common in regression analysis, it is often clearly unjustifiable. There are other models for other kinds of observables. By way of an example, we illustrate regression with binary responses.

Example: Sarkar and Ananthanarayanan in [19] carried out a study of auctions carried out on the Internet. The data consisted of 250 cars of a specific brand that were available for sale on an auction website in 2000-2001. The start price (decided by the seller), whether the car eventually got sold or not, and the selling price if the car got sold were recorded. One objective of the study was to see how the start price affected the probability of the car finding a buyer and the eventual selling price. Denoting the start price of the ith car that got sold by x_i and the selling price by Y_i, a simple linear regression of the form $E(Y_i) = \alpha_1 + \beta_1 x_i$ was considered. The parameters were estimated by LS, effectively making the assumption that the selling prices are Gaussian. Some results are presented later in this section.

It was also of interest to model the probability that the ith car got sold. Denoting this by P_i, a so-called logistic regression model can be stipulated as $\log\left(\frac{P_i}{1-P_i}\right) = \alpha_2 + \beta_2 x_i$. Let S_i be a binary random variable that takes the value 1 if the ith car got sold and 0 otherwise. The log-likelihood corresponding to the data can be expressed as

$$\log \prod_{i=1}^{n} P_i^{S_i}(1-P_i)^{1-S_i} = -n\log(1+e^{\alpha_2+\beta_2 x_i}) + \alpha_2 \sum_{i=1}^{n} S_i + \beta_2 \sum_{i=1}^{n} S_i x_i$$

The ML estimates of the parameters α_2 and β_2 can now be found by maximizing this log-likelihood. A general treatment of estimation in logistic regression is in [14].

We report some fitted values from both the linear and logistic regressions. As in the linear regression case, the fitted values for P_i are found by plugging-in the LS estimates. The results are as expected. Setting higher selling prices can lead to higher sale prices if the car gets sold, but that eventuality becomes less likely.

Starting price	2000	3000	4000	5000	6000	7000	8000
Fitted selling price	7855	7961	8066	8172	8278	8383	8489
Fitted sale probability	0.81	0.76	0.69	0.62	0.54	0.46	0.38

We could, mechanically, run a polynomial regression of the binary variables S_i on x_i, but that is scientifically flawed. Binary random variables are discrete, as opposed to continuous, and are moreover bounded (between 0 and 1 in this case). The Gaussian assumption that justifies LS is untenable here. Hence the need for alternative models like logistic regression.

It is noteworthy that the likelihood involves the data only through two statistics, namely, the total number of cars sold ($\sum S_i$) and the total starting price of all the cars sold ($\sum S_i x_i$). In statistical inference such statistics that summarize the data to the extent of specifying the likelihood completely are called sufficient statistics. In the logistic regression model, there are two sufficient statistics, however many observations there are. One of the reasons for taking this model for binary data is the consequential availability of such low dimensional sufficient statistics.

We have seen in this section regression with Gaussian as well as binary data. There is a general theory which allows a large class of distributions (called exponential families) to be assumed for regression models. The resulting models are called generalized linear models and the models of this section are special cases. See [15] for other possibilities.

3 Nonparametric regression

In Section 2, we considered a model where Y_i had $N(f(x_i), \sigma^2)$ distribution and the regression function f was completely specified by p parameters. Such a model is called a parametric model as it can be specified in terms of a number of parameters that is (a) finite and (b) independent of the sample size n. We now proceed to drop these restrictions and take a look at models that are not parametric, i.e. *nonparametric* models.

There are, of course, many ways in which conditions (a) and (b) can be violated. For example, we can specify f to be a polynomial, $E(Y_i) = f(x_i)$, and $Var(Y_i) = \sigma^2$ but only stipulate that Y_i has a distribution with finite variance. Thus f and σ^2 do not completely specify the distribution of Y_i. In fact, the class of distributions with finite variance has an infinite number of members, and it is not possible to find a finite set of parameters that suffices for such a complete specification. We shall take a closer look at this type of nonparametric model in Section 4.

A comment on terminology needs to be made here. The term nonparametric does not mean that there are no parameters, but rather that there are so many parameters that it is not useful to think of the model in terms of a parametric representation.

3.1 Kernel smoothing

In this section we shall take a closer look at another class of nonparametric models wherein Y_i still has $N(f(x_i), \sigma^2)$ distribution but f belongs to an infinite class of functions. For example, f can be stipulated to be continuous.

A commonly used estimator for such models are kernel estimators. A kernel is a function K satisfying (for our purposes here) the properties of a symmetric density function; namely

(i) $K(x) \geq 0$ for all x,

(ii) $K(-x) = K(x)$ for all x, and

(iii) $\int_{-\infty}^{\infty} K(u)du = 1$.

A nonparametric estimator (shorthand for an estimator under a nonparametric model) of f is then

$$\hat{f}(x) = \frac{\sum_{i=1}^{n} K(x - x_i) Y_i}{\sum_{i=1}^{n} K(x - x_i)}.$$

This is a weighted average of the Y_i and is often referred to as a kernel smoother (and the operation is then called kernel smoothing). If the kernel is unimodal in the sense that $K(x'') < K(x')$ for $x'' > x' > 0$, then these weights will be the most for the Y_i that correspond to the x_i closest to x. An example of a kernel together with its use will be presented later in this section and more general details can be found in [10].

It can be seen that, in the notation of Section 1, $\hat{Y} = SY$ and the estimator is a linear smoother. However, unlike in the parametric case, S is typically a full rank matrix and, generally, $S^2 \neq S$. However, the trace of S still carries useful information on the number of "parameters" there are, should one care to think in terms of parameters. See [11] for alternatives to the trace.

The degree of smoothness in kernel smoothers in controlled by a bandwidth or window-width specification that determines how sharply the weights decay in the weighted average. The kernel of bandwidth h is defined by $K_h(x) = \frac{1}{h} K\left(\frac{x}{h}\right)$. A very small h corresponds to little smoothing with very local averages and a very large h corresponds to heavy smoothing towards a global average. Let S_h be the corresponding smoothing matrix. The intuition is confirmed by the observations that

$$\lim_{h \to 0} S_h = I \quad \text{and} \quad \lim_{h \to \infty} S_h = \frac{1}{n} 11'$$

where I is the identity matrix and 1 is the column vector or all ones. Thus, if we undersmooth, $\hat{Y}_i$ is too close to Y_i and the regression estimate is essentially an interpolation. If we oversmooth, $\hat{Y}_i$ is essentially $\bar{Y} = \frac{1}{n}\sum_{i=1}^{n} Y_i$ and all structure is lost. Somewhere in between is a desirable fit that captures structure and is yet not wedded to the data at hand. Note the implication that $\lim_{h\to 0} trace(S_h) = n$ (there are n "parameters" in the estimate that interpolates n points) and $\lim_{h\to\infty} trace(S_h) = 1$ (there is 1 "parameter" in the estimate that smooths to the global average). Thus for a desirable smoother $trace(S_h)$ captures the number of "parameters". This does not, however, make the model parametric. As the number of data points n increases, the desirable choice of h will change and will decrease increase with n. Moreover, while we may intuitively agree that this model has a complexity comparable with a parametric model with $trace(S_h)$ "parameters", it is quite another matter to label and extract these "parameters" (even after an integer approximation to the trace, a real number).

Example: To illustrate the delicacy of the problem of bandwidth selection, we take a look at an example from biometry. Reynolds in [18] looks at the body temperature of beavers to study activity levels of the animals over the 24-hour diurnal cycle. Figure 1 below shows the data for one animal with the temperatures as points. There are a hundred observations of the body temperature of an adult beaver taken at ten minute intervals. At the start of the observation sequence, the animal was asleep. She then awoke and became active. For the purposes of this example, we ignore the time series aspects of the data.

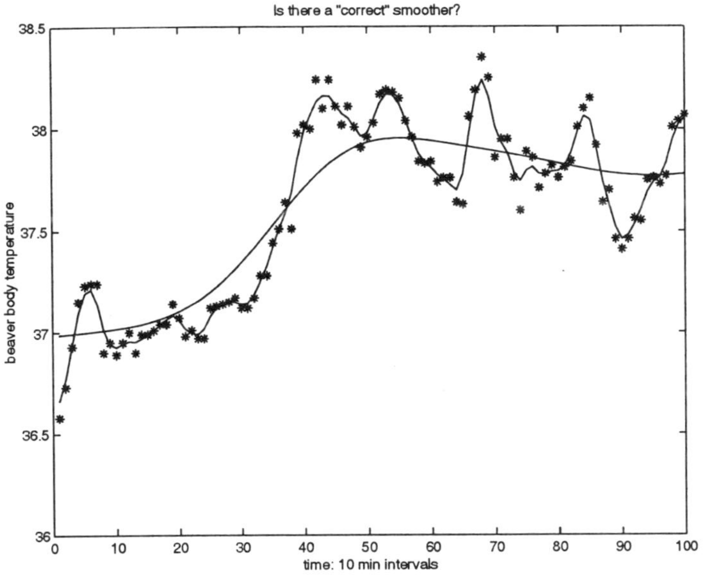

Figure 1: beaver data

The original data is shown as points. We also show two kernel smoothers. The kernel used was a Gaussian kernel with $K(x) = (\sqrt{2\pi})^{-1}e^{-\frac{1}{2}x^2}$. The more variable smoother is one with $h = 1$ and faithfully reproduces most of the fluctuations in body temperature. However, one of the scientific purposes of the study was to analyze the rise in body temperature as activity increases. If so, these fluctuations can be treated as noise and a smoother should not estimate them as part of the regression function. The other smoother plotted with $h = 10$ achieves that end. Thus the choice of bandwidth depends intimately on the use to which the analysis is to be put. There may not be a "correct" bandwidth.

3.2 Cross-validation

To clarify the dependency of the kernel regression estimate on h, we now denote it by $\hat{f}_h$. For a fixed point x, we can see the following expansion for the so-called mean square error (MSE) in estimation

$$E\left[\hat{f}_h(x) - f(x)\right]^2 = E\left[\hat{f}_h(x) - E(\hat{f}_h(x))\right]^2 + \left[E(\hat{f}_h(x)) - f(x)\right]^2$$

The first term is the variance of $\hat{f}_h(x)$ and can be shown (with some restrictions on the kernel) to be approximately $\frac{1}{nh}\sigma^2 \int K_h^2(u)du$. The second term is the square of the bias of $\hat{f}_h(x)$ and the bias can be shown to be approximately $\frac{h^2}{2} f''(x) \int u^2 K_h(u)du$. Derivations can be found in [10]. From the approximation for the bias, it follow that where $f''>0$ ($f''<0$), $\hat{f}_h$ overestimates (underestimates) f. This implies, as Figure 1 also illustrates, that a kernel smoother underestimates peaks and overestimates troughs in the data. This smoothing of features is what gives smoothers their name and is the characteristic of most regression estimates, parametric and nonparametric.

As the bandwidth shrinks there is less averaging, bias is reduced but variance increases. The opposite occurs when the bandwidth increases and there is more averaging. One possible way to optimize this so-called bias-variance trade-off is to choose h so as to minimize the MSE. This, however, requires knowledge of two things: the observational variance σ^2 and the curvature of the regression function f''. Apriori, before any estimation is done, both are unknown.

Cross-validation (CV) proposes an alternative, more intuitive, solution. Consider the problem of predicting the value of Y' corresponding to a new x' and the error in this prediction. This error will be high if h is large as not enough local structure will have been captured due to undersmoothing. This error will also be high if h is small as the function will have adapted too well to the data (that does not include x' and Y'). Thus minimizing prediction error is one natural criterion for optimal choice of h. Given the available data, average prediction error can be estimated by the cross-validation score defined by

$$CV(h) = \frac{1}{n}\sum_{i=1}^{n}[Y_i - \hat{f}_h^{(-i)}(x_i)]^2.$$

Here $\hat{f}_h^{(-i)}$ is the estimate of f obtained by using the kernel K_h on all the data points except x_i and the corresponding Y_i. In principle, this amounts to running n regressions. If this is a burden, an approximation called generalized cross-validation (GCV) can be used:

$$GCV(h) = \frac{1}{n}\sum_{i=1}^{n}\left[\frac{Y_i - \hat{f}_h(x_i)}{1 - n^{-1}trace(S_h)}\right]^2.$$

The strategy then is to compute $\hat{f}_h$ over a reasonable grid of values for h and choose an h that approximately minimizes $CV(h)$ or $GCV(h)$. Note that there are two competing terms in the GCV score: the error-in-fit term $\sum_{i=1}^{n}[Y_i - \hat{f}_h(x_i)]^2$ which increases in h and a model complexity term $trace(S_h)$ which, as we argued in Section 3, represents the number of "parameters" and decreases in h. GCV (and CV) thus trades off fidelity to the particular sample at hand with the number of parameters or degrees of freedom. See [22] for further discussion on cross-validation and its variants.

3.3 Other nonparametric regression estimates: splines

Kernels are not the only way to estimate nonparametric regression models. A book length treatment of a variety of methods is in [7]. Among the other popular methods, we take a brief look at regression with splines and wavelets.

The difficulty with fitting arbitrary functions to data is that the "best" fit is one that fits the data perfectly (in the case that all the x_i are distinct). This, as has been observed, does not generalize well to other data realizations. One approach to fitting functions is to restrict the curvature of the functions fitted, so that they cannot completely adapt to the data. Assuming that f is defined on an interval $[a, b]$ that includes all the x_i, one such estimate ($\lambda > 0$) is

$$\hat{f}_\lambda = \text{argmin}_f \left[\sum_{i=1}^{n} [Y_i - f(x_i)]^2 + \lambda \int_a^b [f''(x)]^2 \, dx \right].$$

It can be shown that the solution to this is a cubic spline, i.e. a function which (assuming that the x_i are labelled in increasing order)

(i) has two continuous derivatives on $[a, b]$ and

(ii) is a cubic polynomial on the intervals $(a, x_1), (x_1, x_2), \ldots, (x_{n-1}, x_n), (x_n, b)$.

As before, the fitted values are $\hat{Y}_i = \hat{f}_\lambda(x_i)$. Like kernel smoothers, smoothing with splines in linear and we can write $\hat{Y} = S_\lambda Y$ for a suitably defined S_λ. For details on splines and smoothing with splines, see [8].

As in the case of kernel smoothers, it is instructive to consider the nature of the spline smoother in limiting cases. In the limit $\lambda \to 0$, the minimization reduces to minimizing the sum of squared errors and $\hat{f}_\lambda$ approaches the interpolating function. Alternatively, we have $\lim_{\lambda \to 0} S_\lambda = I$ where I is the identity matrix. At the other extreme, $\lambda \to \infty$ implies that the integral minimization dominates the minimization problem. In the limit, the second derivative, f'', becomes arbitrarily small and $\hat{f}_\lambda$ approaches the LS fit for linear regression. This can be expressed as $\lim_{\lambda\to\infty} S_\lambda = H$, where H is the hat matrix corresponding to linear regression, i.e. $p = 2$ in the polynomial regression model. Like the bandwidth h in kernel smoothing, the parameter λ controls the effective number of "parameters". This is easily seen from $\lim_{\lambda\to 0} trace(S_\lambda) = n$ and $\lim_{\lambda\to\infty} trace(S_\lambda) = 2$.

If the true, unknown, regression function f has variable smoothness, a single value for h (in kernel smoothing) or λ (in spline smoothing) may be too restrictive. One would like the flexibility of choosing h or λ to be large where f is smooth and to be small where f shows oscillatory behaviour. This will allow us to, so to speak, redistribute parameters with more parameters going towards modelling regions where they are more useful, i.e. where the (unknown) regression is not smooth. The recent developments in the use of wavelets in regression attempt to do just that, optimally and automatically. This chapter cannot do true justice to the elegance of the ideas and methods of wavelet regression and the reader is encouraged to look up the already abundant literature on the subject. Donoho and Johnstone wrote a series of seminal papers (for example, [3]) and [17] is a simpler treatment.

4 Resampling

We now take a closer look at inference in the first nonparametric model we briefly considered in Section 3. Here we state it in a different, but essentially equivalent, form. Let $Y_i = \beta_0 + \beta_1 x_i + \ldots + \beta_{p-1} x_i^{p-1} + \epsilon_i$ where the ϵ_i are independent and identically distributed (iid) errors with expectation zero. Despite the parametric specification of the regression function, this model is nonparametric on account of the infinitely many distributions possible for the errors ϵ_i which are not necessarily Gaussian.

Applied statisticians may, and indeed typically do, still use least squares (LS) to estimate $\beta_0, \beta_1, \ldots, \beta_{p-1}$. But without the Gaussian assumptions these estimates $\hat{\beta}_0, \hat{\beta}_1, \ldots, \hat{\beta}_{p-1}$ are no longer maximum likelihood (ML). Even worse, it is not clear how their sampling distributions are to be determined, given the nonparametric model specification. Thus standard errors and confidence intervals for the regression parameters are not directly available. This is the typical scenario for effective use of resampling methods.

4.1 Nonparametric bootstrap

Conceptually, perhaps the simplest resampling method is the bootstrap, proposed by Efron in [5]. Before discussing the general motivation behind the bootstrap, we first present a bootstrap algorithm for estimating the standard errors of the LS estimates $\hat{\beta}_j$ nonparametrically.

1. Compute the residuals $r_1, r_2, \ldots, r_n$ from the LS fitted model with $r_i =$

$Y_i - \hat{\beta}_0 - \hat{\beta}_1 x_i - \ldots - \hat{\beta}_{p-1} x_i^{p-1}$. Let F_n denote the empirical distribution of the r_i, i.e. the distribution that assigns probability $\frac{1}{n}$ to each of $r_1, \ldots, r_n$.

2. Draw $e_1^*, e_2^*, \ldots, e_n^*$ independently from F_n. Form the resample $Y_1^*, Y_2^*, \ldots, Y_n^*$ with $Y_i^* = \hat{\beta}_0 + \hat{\beta}_1 x_i + \ldots + \hat{\beta}_{p-1} x_i^{p-1} + e_i^*$.

3. Compute the resampled LS estimates using the resample Y^* and the original unperturbed x.

4. Repeat steps 2 and 3 B times to get B resamples of the LS estimates. Denote these by $\hat{\beta}_j^{(b)}$ for $j = 0, 1, \ldots, p-1$ and $b = 1, 2, \ldots, B$.

5. For $j = 0, 1, \ldots, p-1$, the standard error of the LS estimate $\hat{\beta}_j$ is estimated by $\left[\frac{1}{B-1} \sum_{b=1}^{B} (\hat{\beta}_j^{(b)} - \hat{\beta}_j^{(\cdot)})^2\right]^{\frac{1}{2}}$ where $\hat{\beta}_j^{(\cdot)} = \frac{1}{B} \sum_{b=1}^{B} \hat{\beta}_j^{(b)}$.

The general motivation of the bootstrap is illustrated by the following picture adapted from [6] and further discussed in [9].

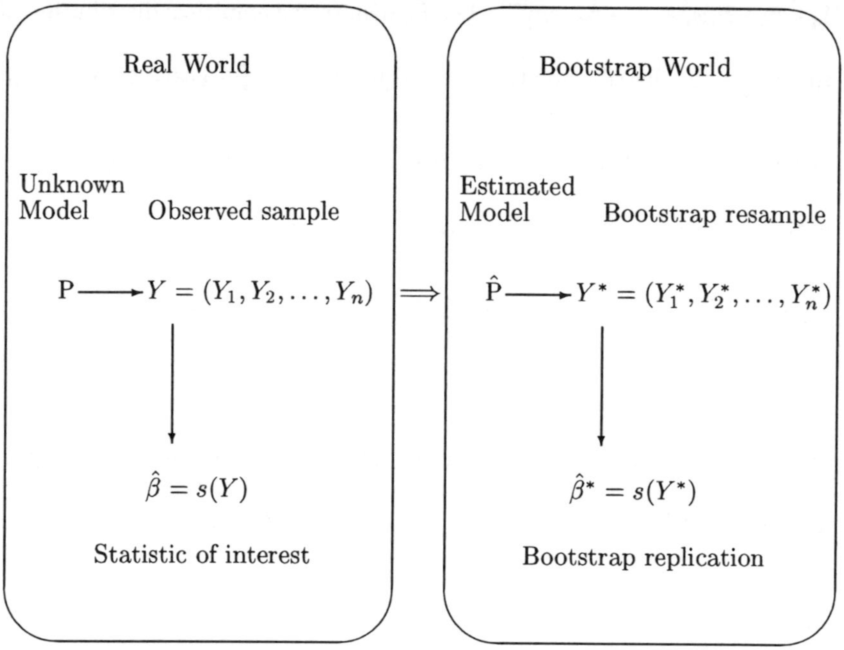

The name of the game in statistics can usually be stated as "if this scenario were to repeat many times, what would be same and what would be different". The (random) differences across such replications lead to sampling distributions and standard errors of estimates. The trouble is that nature gives only one sample and the scenario generally does not so replicate itself. The bootstrap uses the data to estimate a model which can then be used repeatedly to generate resamples. These resamples can then take the place of the unavailable replications and can be used to estimate standard errors and other characteristics of sampling distributions in the usual way.

The bootstrap resamples of the LS estimates can be used for estimating much more than standard errors. For example, an estimate of the bias of $\hat{\beta}_j$, namely $E(\hat{\beta}_j) - \beta_j$, is given by $\hat{\beta}_j^{(\cdot)} - \hat{\beta}_j$.

4.2 Parametric bootstrap

The bootstrap method described above is nonparametric in the sense that it is designed to adapt to arbitrary distributions for the errors. Even if a specific distribution for the errors can be assumed, bootstrap methods are often used for estimating standard errors, bises, etc. A typical scenario involves a nonlinear parametric model.

Example: Consider the celebrated Michaelis-Menten model from chemistry (see, for example, [16]) where reaction rates Y_i are modelled on concentrations x_i as $Y_i = \frac{\alpha x_i}{\beta + x_i} + \epsilon_i$. The positive coefficients α and β are to be estimated. The errors ϵ_i are assumed iid Gaussian and classical LS gives estimates $\hat{\alpha}$ and $\hat{\beta}$. But this is nonlinear LS as the model is not linear in the parameters. As a result the resulting smoother is not a linear smoother and finding standard errors and confidence intervals is no longer an easy problem. While approximations can be made, the parametric bootstrap provides a computational solution.

In order to apply the parametric bootstrap, we need a more complete model specification. With the Gaussian assumption on the errors, Then the LS estimates based on n observations are ML estimates. It can be shown (by considering log-likelihoods) that

$$(\hat{\alpha}, \hat{\beta}, \hat{\sigma}^2) = \text{argmin}_{\alpha,\beta,\sigma^2} \left[n \log \sigma^2 + \frac{1}{2\sigma^2} \sum_{i=1}^{n} \left(Y_i - \frac{\alpha x_i}{\beta + x_i} \right)^2 \right]$$

Steps 1 and 2 of the nonparametric bootstrap algorithm are now be replaced by the step:

Draw $e_1^*, e_2^*, \ldots, e_n^*$ independently from $\mathrm{N}(0, \hat{\sigma}^2)$. Form the resample $Y_1^*, Y_2^*, \ldots, Y_n^*$

with $Y_i^* = \frac{\hat{\alpha} x_i}{\hat{\beta}+x_i} + e_i^*$.

The remaining steps can be carried out as before with the same expressions as in the nonparametric case to yield estimates of standard errors.

Note that the parametric bootstrap adheres to the general philosophy as described in the bootstrap picture. What makes it parametric is that a parametric estimate of the probability model is used to generate the resample. The end benefit is the same; the messy problem of analytic computation or approximation of sampling distributions is avoided by use of resampling.

4.3 Other methods of resampling: the jackknife

Historically, the bootstrap is a relatively new resampling technique made possible by the availability of fast and cheap computing. Theoretically, the method is related to the older idea of the jackknife, which we take a quick look at.

The jackknife considers resamples which differ from the original data only to the extent of deleting single observations. Consider our problem of estimating standard error and bias of the jth regression parameter β_j. Let $\hat{\beta}_j$ denote our estimator of choice (it need not be an LS or ML estimator) and let $\hat{\beta}_j^{(-i)}$ be the same estimator computed for the data deleting the ith observation with $\hat{\beta}_j^{(\cdot)} = \frac{1}{n}\sum_{i=1}^{n} \hat{\beta}_j^{(-i)}$. The jackknife estimates of the standard error and bias of the original estimator are (see [6] for justifications)

$$se(\hat{\beta}_j) = \left[\frac{n-1}{n}\sum_{i=1}^{n}\left(\hat{\beta}_j^{(-i)} - \hat{\beta}_j^{(\cdot)}\right)^2\right]^{\frac{1}{2}} \quad \text{and} \quad bias(\hat{\beta}_j) = (n-1)\left(\hat{\beta}_j^{(\cdot)} - \hat{\beta}_j\right)$$

Note that, unlike the bootstrap, the jackknife is not a simulation based method. However, like the bootstrap, it is nonparametric and estimates characteristics of sampling distributions without making distributional assumptions.

The idea of deleting observations to create resamples can be extended to the delete-d jackknife where d observations are systematically deleted from the original data. The jackknife method proposed above is the case $d=1$.

Resampling methods such as the jackknife and the bootstrap are appealing because of their conceptual and computational simplicity. But there are technical limitations that need to be imposed on their use and such cautionary issues are also discussed in [6].

5 Further reading

Nonparametric statistics has roots that did not encompass regression. The early work of Mosteller, Wilcoxon and others was intended to provide "quick and dirty" methods for inference based on rank tests and order statistics. This classical view of nonparametrics is detailed in texts like [20] and [13]. A principal catalyst for the growth in applied nonparametric regression was the modern computer. In that spirit, [12] is a recent compendium of computational methods and the ideas behind them. We should also mention the special case of regression with survival or lifetime data which led to the development of what is now called semiparametric regression. The proportional hazards regression model proposed by Cox in [2] essentially models survival times nonparametrically, but includes the effect of predictors using parameters. Semiparametric models are of much current interest, notably in econometrics.

6 Exercises

One objective of this chapter has been to emphasize the connections between various methods of estimating regression functions, together with allied issues like model selection and estimation of standard error. Here are a couple of exercises that emphasize the essential unity in statistical modelling - parametric and nonparametric.

1. *AIC and GCV.* The AIC was presented as a model selection criterion for parametric models. We also interpreted the trace of the linear smoother matrix as the equivalent number of parameters for some nonparametric regression models. Assuming Gaussian distributions, devise an AIC for selecting the bandwidth parameter for kernel smoothing. Compare and contrast this with the generalized cross-validation strategy for selecting the same.

2. *Bootstrap for logistic regression.* Consider the problem of estimating bias and standard error of regression coefficients in the logistic regression model for binary data. Devise a parametric bootstrap scheme to do this. Discuss the challenges in devising a corresponding nonparametric bootstrap scheme.

References

[1] Akaike, H. Information theory and an extension of the maximum likelihood principle. In *Second International Symposium on Information Theory* (edited by Petrov and Czaki), Budapest, 267-281, 1973.

[2] Cox, D.R. Regression models and life tables. *J. Roy. Stat. Soc. B*, 34, 187-200, 1972.

[3] Donoho, D.L. and Johnstone, I.M. Adapting to unknown smoothness via wavelet shrinkage. *J. Amer. Statist. Ass.*, 90, 1200-1224, 1995.

[4] Draper, N.R. and Smith, H. *Applied Regression Analysis.* Second edition. Wiley and Sons, New York, 1981.

[5] Efron, B. Bootstrap methods: another look at the jackknife. *Annals of Statistics*, 7, 1-26, 1979.

[6] Efron, B. and Tibshirani, R.J. *An Introduction to the Bootstrap.* Chapman and Hall, New York, 1993.

[7] Fan, J. and Gijbels, I. *Local Polynomial Modelling and its Applications.* Chapman and Hall, London, 1996.

[8] Green, P.J. and Silverman, B.W. *Nonparametric Regression and Generalized Linear Models.* Chapman and Hall, London, 1994.

[9] Hall, P. and Sarkar, A. Bootstrap methods in statistics. *Resonance*, 5, 41-48, Indian Academy of Sciences, 2000.

[10] Hardle, W. *Applied Nonparametric Regression.* Cambridge University Press, Cambridge, 1990.

[11] Hastie, T.J. and Tishirani, R.J. *Generalized Additive Models.* Chapman and Hall, London, 1990.

[12] Hastie, T., Tibshirani, R., and Friedman, J. *The Elements of Statistical Learning.* Springer, New York, 2001.

[13] Hollander, M. and Wolfe, D.A. *Nonparametric Statistical Methods.* Wiley-Interscience, New York, 1999.

[14] Hosmer, D.W. and Lemeshow, S. *Applied Logistic Regression.* Second edition. Wiley and Sons, New York, 2000.

[15] McCullagh, P. and Nelder, J.A. *Generalized Linear Models.* Second edition. Chapman and Hall, London, 1989.

[16] Moran, L.A., Scrimgeour, K.G., Horton, H.R., Ochs, R.S., and Rawn, J.D. *Biochemistry.* Second edition. Prentice Hall, Englewood Cliffs, 1994.

[17] Ogden, R.T. *Essential Wavelets for Statistical Applications and Data Analysis.* Birkhauser, Boston, 1996.

[18] Reynolds, P.S. Time-series analysis of beaver body temperature. In *Case Studies in Biometry* (edited by Lange et. al.), 211-228. Wiley-Interscience, New York, 1994.

[19] Sarkar, A. and Ananthanarayanan, R. Learning from e-commerce: incomplete data and decision support. *Bulletin of the ISI*, 53rd session proceedings, 1, 295-298, 2001.

[20] Siegel, S. and Castellan, N.J. *Nonparametric Statistics for the Behavioral Sciences.* Second edition. McGraw Hill, New York, 1988.

[21] Stigler, S.M. *The History of Statistics.* Belknap Press, Cambridge MA, 1986.

[22] Wahba, G. *Spline Models for Observational Data.* SIAM, Philadelphia, 1990.

[23] Weisberg, S. *Applied Linear Regression.* Second edition. Wiley and Sons, New York, 1985.

TESTING GOODNESS OF FIT OF A SPECIFIC PARAMETRIC PROBABILITY MODEL

J. V. Deshpande, U. V. Naik-Nimbalkar
Department of Statistics,
University of Pune,
Pune - 411 007, India

Sometimes the experimenter has a suspicion or prior belief that the distribution belongs to a particular parametric family, like the normal, exponential, Poisson etc. This could be because the experimental conditions point to a particular distribution as the appropriate one or because of past experience of similar experiments. He then wishes to either confirm or reject this prior belief through a 'test of goodness of fit'. There are three major ways of carrying out such tests : (i) the chi-squared Goodness of fit test of Karl Pearson, (ii) the Kolmogorov - Smirnov goodness of fit test based on the empirical distribution function, and (iii) the Hellinger distance based methods. We shall describe these in successive sections. Then these will be followed by methods developed for testing goodness of fit of two specific popular distributions viz., the exponential and the normal. There are certain graphical procedures used as diagnostic indicators of the family governing the outcomes. These will be discussed in the last section.

Key words : *Chi-square statistic, Kolmogorov - Smirnov statistic; Hellinger distance; tests for exponentiality and normality; graphical diagnostic procedures.*

1 Chi-squared Goodness of Fit Test

The random sample consists of n independent observations $X_1, \cdots, X_n$. The idea is to see whether they occur according to a given common probability distribution F_0. The ideal situation is when we can completely specify the suspected distribution function $F_0(x)$. Often, we can only point to a particular family without being able to specify the values of its parameters. These two cases will be dealt with separately.

(i) Completely specified distribtuion function F_0.

We set up the null hypothesis

$$H_0 : F = F_0$$

for testing. Since F_0 is completely known, we can find the probabilities given by it for any partition $(-\infty, a_1], (a_1, a_2], \cdots, (a_{k-2}, a_{k-1}], (a_{k-1}, \infty]$ in k intervals formed by the $k-1$ numbers

$$-\infty = a_0 < a_1 < \cdots < a_{k-1} < a_k = \infty, \quad k \geq 2.$$

Let these be $p_1, p_2, \cdots, p_k, p_i > 0, i = 1, 2, \cdots, k$ and $\sum_{i=1}^{k} p_i = 1$. Let O_i be the (observed) number of observations in the i-th interval, $\sum_{i=1}^{k} O_i = n$. The probability of the i-th interval is p_i hence the expected number of observations in it is np_i. In [18] Pearson suggested that we should look at the discrepency between the observed and expected frequencies through the chi-squared statistics

$$\chi^2 = \sum_{i=1}^{k} \frac{(O_i - np_i)^2}{np_i}.$$

The denominator is a normalizing factor to make the variances of the terms comparable. If F_0 is indeed the true distribution function then the difference between O_i and np_i is expected to be small, only due to random variations, rather than systematic, which will arise if the probabilities p_i's are not the true probabilities. In fact, let us slightly modify the hypothesis testing problem to :

H_0' : p_i is the probability of interval $(a_{i-1}, a_i], i = 1, 2, \cdots, k$

vs H_1' : q_i (which are not all equal to p_i) are the probabilities of these intervals.

Then the vector $(O_1, O_2, \cdots, O_k)$ will have a multinomial distribution given under H_0' by

$$P_{H_0'}(O_1 = n_1, \cdots, O_k = n_k) = \frac{n!}{\pi_{i=1}^{k} n_i!} \quad \pi_{i=1}^{k} p_i^{n_i}$$

and under H_1'

$$P_{H_1'}(O_1 = n_1, \cdots, O_k = n_k) = \frac{n!}{\pi_{i=1}^{n} n_i!} \quad \pi_{i=1}^{n} q_i^{n_i}.$$

The likelihood ratio test for a simple vs. a composite hypothesis is based on the statistics

$$\begin{aligned} L &= \log \frac{\sup_{H_1'} P_{H_1'}}{P_{H_0'}} \\ &= \log \frac{\pi_{i=1}^{k} \left(\frac{n_i}{n}\right)^{n_i}}{\pi_{i=1}^{k} p_i^{n_i}} \\ &= \log \pi_{i=1}^{k} \left(\frac{n_i}{np_i}\right)^{n_i} \\ &= \sum_{i=1}^{k} n_i (\log \frac{n_i}{n} - \log p_i) \end{aligned}$$

since $\frac{n_i}{n}$ are the maximum likelihood estimators of q_i.

By Taylor expansion, and neglecting terms of order $0(1/n)$ we get

$$2L \approx \sum \frac{(n_i - np_i)^2}{n_i}.$$

Replacing n_i by the quantity np_i in the denominator which it estimates consistently we get Pearson's chi-squared statistic. Hence, asymptotically the chi-squared statistic has the same distribution as the likelihood ratio statistic. The latter, by general principles of likelihood theory is known to have the chi-square distribution with $k - 1$ degrees of freedom ([24], Chapter 13).

As large deviations between O_i and np_i, the observed and expected frequencies, provide evidence against the null hypothesis, we reject it if the observed value of the chi-squared statistic is greater than the upper $\alpha\%$ value of the chi-square distribution with $k-1$ d.f., i.e. the test is to reject H_0 if

$$\chi^2 > \chi^2_{k-1,1-\alpha}.$$

It is clear that if there is a distribution F_1, different from F_0, but specifying the same probabilities p_i for the intervals then the test will not be effective in detecting this alternative. The construction of the intervals is rather arbitrary, it is possible that different decisions may be reached through different such constructions. The number of intervals should not be too small, but at the same time it should be kept in mind that the approximation provided by the asymptotic distribution would not be good if the probability under the null hypotheses for any interval is too small. A rule of thumb which most statisticians recommend and follow is that n and each p_i should be large enough so that no np_i is less than 5 or so and in any case should not be less than 1. This may be achieved by reducing the number of intervals, through pooling.

(ii) **Some parametes of F_0 are unknown.**

We have said at the beginning of this section that we suppose that F_0 is a completely known distribution function. The experimenter sometimes may have an inkling only of the family of the distribution, but not the values of the parameters identifying the exact distribution within the family. For example, the experimenter may suspect, due to the experimental conditions, that the distribution governing the outcomes is normal (μ, σ^2), but may not be able to specify, even as a hypothesis to be tested, the values of the mean μ and the variance σ^2. In such situations, it is usually suggested that the unknown scalar or vector parameter θ be estimated by its minimum chi-square estimator $\hat{\theta}$. Then the estimated value be substituted in F_0 from which the probabilities $\hat{p}_i, i = 1, 2, \cdots, k$ should be obtained for the k intervals. Then the statistic

$$\chi^2 = \sum_{i=1}^{k} \frac{(O_i - n\hat{p}_i)^2}{n\hat{p}_i}$$

be computed as before. The asymptotic distribution of the statistic based on $\hat{p}_i$ has the chi-square with $k-\ell-1$ degrees of freedom where ℓ is the number of parameters (dimensionality of θ) which are estimated from the data. This result again follows from the standard asymptotic theory of likelihood ratio tests. So, the critical points for the test should be chosen from the chi-square distribution with $k-\ell-1$ degrees of freedom. It is thus clear that we may at most estimate $k-2$ parameters from the data while testing goodness of fit.

In [18] the χ^2 test of goodness of fit of a simple (completely specified distribution) null hypotheses is developed and the asymptotic distribution of the statistic is found to be χ^2_{k-1} where k is the number of classes in which the sample space is partitioned. In [11] Fisher dealt with the case when the distribution is not completely

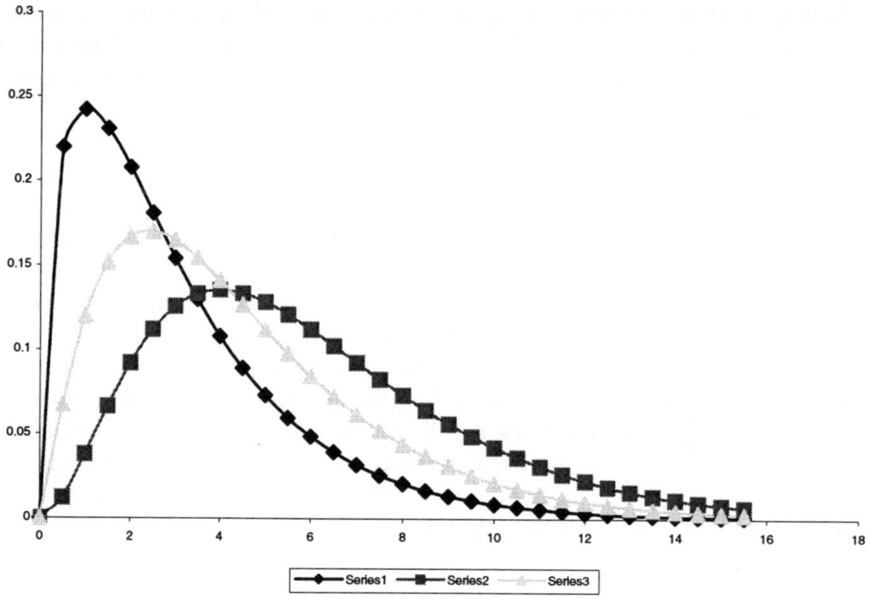

Figure 1: $Series1 : -\chi^2_{k-p-1}$, $Series2 : -\chi^2_{k-1}$, $Series3 : -\chi^2_{k-p-1} + Z^2$

specified but contains p unknown parameters. He also proved that if the estimators obtained by the minimum χ^2 technique are substituted for the unknown parameters then the asymptotic distribution is χ^2_{k-p-1}. Furthermore, in [5] it is shown that if estimators obtained by the more efficient maximum likelihood method of estimation are used then the asymptotic distribution is that of $T = \chi^2_{k-p-1} + Z^2$ where $Z^2 = \sum_{i=1}^{p} \lambda_i X_i^2, X_i$ being independent $N(0,1)$ random variables also independent of the χ^2_{k-p-1} variable and $0 < \lambda_i < 1$. Thus, the asymptotic distribution of T is stochastically bounded between χ^2_{k-1} and χ^2_{k-p-1} random variables. In this situation using the critical points from the χ^2_{k-1} distribution will lead to a conservative test and using those from the χ^2_{k-p-1} distribution will lead to an anticonservative test, i.e., the actual level of significance will be larger than the stated one.

In [16] it is proved that the quadratic form defined below

$$D = \xi'(\hat{\underline{\theta}})\hat{\Sigma}^{+}\xi(\hat{\underline{\theta}})$$

has asymptotically χ^2 distribution with $k - p - 1$ degrees of freedom.

Here

$$\xi(\hat{\underline{\theta}}) = n^{1/2} \begin{pmatrix} \hat{F}(I_1) - p_1(\hat{\underline{\theta}}) \\ \hat{F}(I_2) - p_2(\hat{\underline{\theta}}) \\ \vdots \\ \hat{F}(I_k) - p_k(\hat{\underline{\theta}}) \end{pmatrix},$$

$\hat{F}(I_j)$ are the probabilities given to the intervals $I_j, j = 1, \cdots, k$ by the estimator $\hat{F}$ of $F_{\underline{\hat{\theta}}}$ whether the $H_0 : F_{\underline{\hat{\theta}}}, \underline{\hat{\theta}} \in \Theta$ holds or not with the property $n^{1/2}(\hat{F} - F) \xrightarrow{d} W$, a continuous Gaussian process with a nonsingular correlation structure which can be consistently estimated; $p_j(\underline{\hat{\theta}}), j = 1, \cdots, k$ are the probabilities of the same intervals given by the model F under H_0 where $\underline{\hat{\theta}}$ are estimators of $\underline{\theta}$ obtained by minimizing $\xi'(\underline{\theta})D^2(\underline{\theta})\xi(\underline{\theta})$ under mild conditions of W and D. Here $\hat{\Sigma}^+$ is the Moore - Penrose inverse of the estimated asymptotic variance covariance matrix of $\xi(\underline{\hat{\theta}})$.

This is a general statistic having χ^2_{k-p-1} degrees of freedom. If $\hat{F}$ is the empirical distribution function then it reduces to the Fisher - Pearson χ^2 statistics with p estimated parameters. If the data is randomly censored then $\hat{F}$ may be taken as the Kaplan - Meier product limit estimator and the test can still be carried out.

Example 1. The following are suspected to be 50 values generated from the Poisson distribution with mean 1 using a certain computer programme.

Values	0	1	2	3	4	5
frequency	11	17	10	9	2	1

Thus we wish to test the hypothesis $H_0 : F_0$ is Poisson with mean 1.

i	$p_i = P_1[X = i]$	np_i	np_i	O_i	$(np_i - O_i)^2$	$\frac{(np_i - O_i)^2}{np_i}$
0	0.367879	18.3940	18.3940	11	54.6712	2.9722
1	0.367879	18.3940	18.3940	17	1.9432	0.1056
2	0.183940	9.1970	9.1970	10	0.6448	0.0701
3	0.061313	3.0657	3.9654	12	64.5548	16.2795
4	0.15328	0.7664				
5	0.003066	0.1533				

Since the last two cell values np_i of column 3 and also their sum is less than 1, they have been added to the previous cell value and reported in column 4.

We then get $\chi^2 = 19.427$.

The upper 5% value of the chi-square distribution with $k - 1 = 3$ d.f. is $\chi^2_{3,.95} = 7.815$. Since the calculated $\chi^2 > 7.815$, we reject H_0.

The p-value in this case is less than 0.001.

Example 2. The data is taken from [12]. (Original Source : Lieblein J. and Zelen M. [17]).

The number of cycles to failure of 22 ball bearings are given. The data is already in the ordered form.

17.88	28.92	33.00	41.52	52.12
45.60	48.48	51.84	51.96	54.12
55.56	67.40	68.64	68.88	84.12
93.12	98.64	105.12	105.84	127.92
128.04	173.40			

The aim is to test $H_0 : F_0(x) = 1 - e^{-\lambda x}, x > 0$ that is exponential with mean $1/\lambda, \lambda$ unknown.

The maximum likelihood estimator of λ is

$$\hat{\lambda} = \frac{n}{\sum_{i=1}^{n} X_i} = \frac{1}{72.3873} = 0.0138.$$

Partition	$\hat{p}_i = e^{-\hat{\lambda} a_{i-1}} - e^{-\hat{\lambda} a_i}$	$n\hat{p}_i$	O_i	$\frac{(n\hat{p}_i - O_i)^2}{(n\hat{p}_i)}$
(0, 40]	0.424540	9.33988	3	4.30349
(40, 80]	0.244306	5.37473	11	5.88748
(80, 120]	0.140588	3.09294	5	1.17587
(120, 160]	0.080903	1.77987	2	0.02723
(160, 200]	0.046556	1.02423	1	0.00057

Thus $\chi^2 = 11.3946$.

Note that in this case χ^2 is stochastically bounded between χ_4^2 and χ_3^2. The .05 level critical points of the respective distributions are $\chi^2_{4,.95} = 9.488$ and $\chi^2_{3,.95} = 7.815$.

The null hypothesis is rejected using the larger critical point, thus at a level somewhat less than 0.05.

2 The Kolmogorov - Smirnov Goodness of Fit Test

This test is based directly on the difference between the distribution function specified by the null hypothesis F_0 and its estimator, the empirical distribution function F_n.

Again, let the null hypothesis H_0 completely specify the distribution function:

$$H_0 : F = F_0$$

assumed to be a continuous distribution function.

The random sample $X_1, X_2, \cdots, X_n$ is used to construct the empirical distribution function $F_n(x)$ defined as $F_n(x) = \frac{i}{n}$, if exactly i of the n X's are less than or equal to x. Calculate the Kolmogorov - Smirnov statistics

$$\begin{aligned} D_n &= \sup_{-\infty < x < \infty} |F_n(x) - F_0(x)| \\ &= \max_{1 \le i \le n} \{max\{|F_n(x_{(i)}) - F_0(x_i)|, |F_n(x_{(i)}-) - F_0(x_{(i)}-)|\}\}, \\ &= \max_{1 \le i \le n} \{\max\{|\frac{i}{n} - F_0(x_{(i)})|, |\frac{i-1}{n} - F_0(x_{(i)})|\}\}, \end{aligned}$$

the maximum of $2n$ positive quantities, where $x_{(i)}$ is the i-th order statistic of the random sample.

We should reject H_0 if D_n appears to be too large.

The test is based on the fact that if F_0 is indeed the true distribution function then D_n will have a sampling distribution which does not depend upon F_0, due to the probability integral transformation. The exact distribution for small sample size n is rather complicated. It has been however tabulated and exact critical points for use in the test are available. When n is large the asymptotic distribution of $\sqrt{n}D_n$

as $n \to \infty$, is given by Kolmogorov and well tabulated. Hence in either case (n small or large) the test is

Reject H_0 if $D_n > d_{n,1-\alpha}$

where $d_{n,1-\alpha}$ is the upper 100 $\alpha\%$ critical point from either the exact or the asymptotic distribution of the statistic.

If the distribution function under the null hypothesis is not completely known, but known only upto the family with values of parameters still unknown, then one may use values of consistent estimators of the parameters to completely specify $\hat{F}_0$ and compare it with F_n through D_n. But neither the exact nor the asymptotic distributions used above would hold in this case. If we still use the upper $100\alpha\%$ critical points of these distributions to carry out the test, we would be performing a conservative test, i.e., the actual level of significance of the test would be smaller than (or at most equal to) the stated level of significance α. This is so because when some parameters are estimated from the data, to specify F_0, the difference between it and the empirical distribution function F_n, which is totally based on the data, would be stochastically smaller than what it would have been in case F_0 were totally specified.

In case the experimenter knows that the distribution from which the data has been realized, if not F_0, falls entirely above F_0 then it is more efficient to compute

$$D_n^+ = \sup_{-\infty<x<\infty} \{F_n(x) - F_0(x)\}.$$

In the opposite case, when the data, if not from F_0, is expected to be from a distribution lying entirely below F_0, one may calculate

$$D_n^- = \sup_{-\infty<x<\infty} \{F_0(x) - F_n(x)\}.$$

The exact and asymptotic null distributions of D_n^+ and D_n^- are somewhat easier to handle. These too are well tabulated and level α tests will reject H_0 if the value of D_n^+ is larger than its $(1-\alpha)100\%$ percentile. The same distributions and hence the same critical points apply to D_n^- as well.

It can be easily seen that

$$D_n = \max\{D_n^+, D_n^-\}.$$

The comments made above about estimation of unknown parameters apply here also.

Comparison of the chi-square and Kolmogorov tests for the goodness of fit hypotheses.

The distribution of the chi-squared statistics, under the null hypothesis is known only asymptotically so we do not have any exact critical points for small sample sizes. Also, the test cannot distinguish the null hypothesis from another distribution which gives the same probabilities for the system of intervals. However, there is a well defined method to deal with null hypotheses which leave values of some parameters unspecified and can be applied with equal ease to continuous or discrete distributions.

In case of the Kolmogorov test, exact critical points are available for small samples also. The test is able to distinguish any distribution which is at all different from the distribution under the null hypothesis. However, if certain parameters are unspecified and estimated from the data then we do not know much about the error rates of the test except that it behaves in a conservative manner. Also, the distribution of the test statistic when the null hypothesis specifies a discrete distribution cannot be provided.

Hence in case of discrete distributions the chi-square test is recommended.

Example 3. Data from Example 2 is used to test the hypothesis $H_0 : F_0(x) = 1 - e^{-\lambda x}, x > 0$, λ unknown.

The maximum likelihood estimator of λ is given by $\hat{\lambda} = 0.0138$. In the following Table $\hat{F}_0(x) = 1 - e^{-\hat{\lambda} x}$ (Mean $= 1/\hat{\lambda} = 72.3873$).

i	$x_{(i)}$	$\hat{F}_0(x_{(i)})$	i/n	D_n^+	D_n^-
1	17.88	0.218864	0.04545	0.000000	0.218864
2	28.92	0.329358	0.09091	0.000000	0.283903
3	33.00	0.366112	0.13636	0.000000	0.275203
4	41.52	0.436498	0.18182	0.000000	0.300134
5	42.12	0.441149	0.22727	0.00000	0.259331
6	45,60	0.467380	0.27273	0.000000	0.240107
7	48.48	0.488155	0.31818	0.000000	0.215428
8	51.84	0.511370	0.36364	0.000000	0.193189
9	51.96	0.512180	0.40909	0.000000	0.148543
10	54.12	0.526521	0.45455	0.000000	0.117430
11	55.56	0.535487	0.50000	0.000000	0.081302
12	67.80	0.608054	0.54545	0.000000	0.108054
13	68.64	0.612576	0.59091	0.000000	0.067122
14	68.88	0.613859	0.63636	0.022505	0.022950
15	84.12	0.687167	0.68182	0.000000	0.050804
16	93.12	0.723742	0.72727	0.003531	0.041923
17	98.64	0.744025	0.77273	0.028702	0.016752
18	105.12	0.765944	0.81818	0.052238	0.000000
19	105.84	0.768260	0.86364	0.095376	0.000000
20	127.92	0.829184	0.90909	0.079907	0.000000
21	128.04	0.829467	0.95455	0.125079	0.000000
22	173.40	0.908869	1.00000	0.091131	0.000000

Thus $D_{22} = 0.3001$.

From the table for critical values of the Kolmogorov - Smirnov one sample test statistic we get $d_{22,0.95} = 0.281$ for the two sided test. Since $D_{22} > 0.281$ we reject H_0. The p - value in this case is 0.038. Note that since a parameter is estimated, we have actually a conservative test.

3 Testing Goodness of Fit with Censored Data

It is known that in case of randomly censored data, when the variables lifetime and censoring time are independent, the Kaplan - Meier (K-M) product limit estimator is consistent for the true distribution function of the life time, (See [14]). A test of goodness of fit then can be carried out on the basis of the difference between the two. Again let the null hypothesis be

$$H_0 : F = F_0.$$

The K-M product limit estimator is defined as

$$\hat{F}(t) = 1 - \prod_{\{j:t_j \leq t\}} (1 - \frac{d_j}{n_j})^{\delta_j}$$

where t_j are distinct values of the lifetimes / censoring times, d_j, the number of deaths at t_j (excluding the censorings at t_j) and $\delta_j = 1$ if $d_j > 0$ and zero otherwise, and n_j is the number of observation in the risk set just before t_j. Form the statistic

$$\hat{D}_{n,T} = \sup_{0<t<T} |\hat{Z}_n(t)|$$

where T is an upper bound upto which the K-M product limit estimator provides a consistent estimator (i.e. a number no more than the largest uncensored observation), and

$$\hat{Z}_n(t) = \frac{\hat{Y}_n(t)}{[1 + a_n(t)](1 - F_0(t))}$$

$$\hat{Y}_n(t) = n^{1/2}\{\hat{F}(t) - F_0(t)\}$$

$$a_n(t) = n \sum_{j:t_j \leq t} \frac{\delta_j}{n_j(n_j + 1)}.$$

The additional terms in the formula like $a_n, 1 - F_0$ have become necessary because the observations are subject to censoring by an unknown censoring distribution.

The asymptotic distribution of the statistic depends upon T, the point of truncation and is rather complicated. The critical points are available in [15]. One may also use the critical points from the Kolmogorov distribution which are used for the statistic D_n in the uncensored data case. The test will again be a conservative one.

A general method for testing goodness of fit of a specific family of distributions $\{F_\theta, \theta \in \Theta\}$ with unknown values of parameters is to calculate. $\hat{F}_n(x)$ and $F_{\hat{\theta}}(x)$, where $\hat{F}_n$ is the K - M product limit estimator of the distribution function based on the data and $\hat{\theta}$ is the maximum likelihood estimator of θ in the above family. Then calculate the Kolmogorov distance

$$\hat{D}_n = \sup_x |F_n(x) - F_{\hat{\theta}}(x)|.$$

The exact (or asymptotic) null distribution of the statistic will not be free of the function F or the true value θ_0 of θ. Hence critical values from the actual distribution are impractical.

Conservative tests, usually loose power for relevant alternatives as is well demonstrated by various Monte Carlo studies. See [8] in the context of the normal family. Hence for testing goodness of fit of families which are commonly used as models such as normal or exponential, it is suggested that more specific tests based on statistics sensitive to departures from certain prime features of the family, e.g. the values $\beta_1 = 0$ and $\beta_2 = 3$ for the coefficients of skewness and kurtosis of the normal distribution or lack of memory property of the exponential distribution. These generally have more power for detecting departures from such features at the cost of generality.

4 Goodness-of-fit Tests based on Hellinger Distance

Goodness-of-fit tests may also be based on distances or disparities between the probability density functions (p.d.f.). The squared Hellinger distance $HD(f,g)$ between two p.d.f.s f and g is defined as

$$HD(f,g) = \int (f^{1/2}(x) - g^{1/2}(x))^2 dx.$$

Let $X_1, X_2, \cdots, X_n$ be a random sample from the p.d.f. g. The aim is to test the null hypothesis that g belongs to a specified parametric family $\mathcal{F} = \{f_\theta, \theta \in \Theta\}$ of p.d.f.s. Minimum Hellinger distance estimator (MHDE) $\hat{\theta}_n$ of θ is the value that minimizes $HD(f_\theta, \hat{g}_n)$, where $\hat{g}_n$ is some nonparametric density estimator of g. That is

$$\hat{\theta}_n = argmin_\theta HD(f_\theta, \hat{g}_n).$$

The minimized distance $HD(f_{\hat{\theta}_n}, \hat{g}_n)$ then provides a natural goodness-of-fit statistic.

For continuous models, asymptotic properties of $\hat{\theta}_n$ and $HD(f_{\hat{\theta}_n}, \hat{g}_n)$ are obtained in [4] by Beran when $\hat{g}_n$ is a kernel density estimator.

Let

$$\hat{g}_n(x) = \frac{1}{nC_nS_n} \sum_{i=1}^{n} w\left(\frac{x - X_i}{C_nS_n}\right)$$

where $h_n = C_nS_n$ is called the bandwidth, $\{C_n\}$ is a sequence of positive constants, $S_n = S_n(X_1, X_2, \cdots, X_n)$ is a robust scale estimator and the kernel $w(\cdot)$ is a density function.

Let

$$R_n = \max_{1 \le i \le n} X_i - \min_{1 \le i \le n} X_i,$$

$$\mu_n = \frac{1}{4} R_n \int w^2(x) dx$$

and

$$\sigma_n^2 = \frac{1}{8} C_n R_n \int (w * w)^2(x) dx$$

where

$$w * w(x) = \int w(x-t) w(t) dt.$$

Then under certain mild assumptions

$$H_n = \sigma_n^{-1}[nC_n HD(f_{\hat{\theta}_n}, \hat{g}_n) - \mu_n]$$

converges in distribution to a standard normal variable $N(0,1)$ under f_θ as $n \to \infty$.

Thus the α- level test is to reject H_0 if $|H_n| > z_{1-\alpha}$, where $z_{1-\alpha}$ is the upper $\alpha\%$ value of the standard normal distribution.

We note that under these assumptions, the limiting distribution of $\sqrt{n}(\hat{\theta}_n - \theta)$ is normal with mean 0 and variance $\frac{1}{4}[\int \dot{h}_\theta(x)\dot{h}_\theta^T(x)dx]^{-1}$ under f_θ, where $\dot{h}_\theta(x) = (\dot{h}_\theta^{(1)}(x), \cdots, \dot{h}_\theta^{(p)}(x))^T$ with $\dot{h}_\theta^{(j)}(x); 1 \leq j \leq p$ denoting the first order partial derivatives of $f_\theta^{1/2}$ with respect to θ and T denoting the transpose.

Thus, for example, the goodness of fit of any location scale family $\{\sigma^{-1} f(\sigma^{-1}(x-\mu)); \sigma > 0, -\infty < \mu < \infty\}$ where f is continuous can be tested.

The most popular choice of the kernel function in density estimation is the Epanechnikov kernel given by

$$w(x) = .75(1-x^2) \quad \text{for} \quad |x| \leq 1,$$

for which $\int_{-1}^{1} w(x)^2 dx = 3/5$ and $\int_{-1}^{1} (w * w(x))^2 dx = \frac{167}{355}$.

The following numerical example is reproduced from [4] to illustrate the feasibility of the procedure.

Example 4. A random sample of size 40 was drawn from a standard normal distribution. The 40 realized sample values were:

−0.706781	0.143266	0.123015	−0.745385	2.16105
0.654191	1.14438	−0.118696	0.258899	−0.154302
0.352057	−1.28269	0.885335	2.51841	−1.09603
2.04580	0.402274	0.0431284	−0.456585	−2.07226
−1.64175	−0.0192038	1.70932	0.929303	0.144781
−0.885728	−0.588767	−0.169394	0.699988	−0.162130
0.0621123	0.729453	0.655040	1.67987	−0.194017
1.01924	−0.927988	−0.524994	0.133760	−0.412047

The aim here is to test $H_0 : f_\theta$ belongs to the family $\{N(\mu, \sigma^2), -\infty < \mu < \infty, 0 < \sigma^2 < \infty\}$.

The MHDEs of μ and σ^2 were obtained by using an iterative algorithm with initial estimates as $\hat{\mu}^{(0)} = \text{median } \{x_i\}$ and

$$\hat{\sigma}^{(0)} = (0.674)^{-1} \text{ median } \{|x_i - \hat{\mu}^{(0)}|\}.$$

The density estimator $\hat{g}_n(x)$ was based upon the Epanechnikov kernel, with the scale statisic $S_n = \hat{\sigma}^{(0)}$. The value of C_n was taken to be 0.7 as the corresponding MDHEs of μ and σ were both close to the sample mean (0.158) and the sample standard deviation (= 1.012). The following Table gives the MHDEs of μ, σ, the goodness-of-fit statistic $H_n(f_{\hat{\theta}_n}, \hat{g}_n)$ and the asymptotic upper 0.10 critical value $h_{.90} = (z_{.90}\sigma_n + \mu_n)/nC_n$, where $z_{.90}$ is the upper .10 critical value of the standard normal distribution, $\mu_n = R_n(3/5)1/4$, and $\sigma_n = .7 \times R_n \times 167/355 \times 1/8$. The Table reports the effects on the estimators of changing the value nearest to zero in the data set, namely $x_{22} = -0.0192038$, by a series of increasing positive values. The 0.10 upper critical values from the asymptotic distribution of $HD(\hat{f}_{\theta_n}, \hat{g}_n)$ are all larger than the corresponding observed values of the statistic, suggesting that the fitted normal distribution gives a good fit. This is as it should be, as changing one observation (or having one outlier) out of the 40 should not affect the bulk of the sample and hence the decision based on it.

Table

μ $x_{22} \rightarrow$	original value	1	2	3	4	5	10	15
$\hat{\mu}$	0.143	0.173	0.191	0.218	0.194	0.156	0.150	0.151
Sample mean	0.158	0.184	0.209	0.234	0.259	0.284	0.409	0.534
$\hat{\sigma}$	1.007	1.019	1.044	1.091	1.080	1.032	1.020	1.018
sample standard deviation	1.012	1.020	1.052	1.106	1.179	1.268	1.855	2.555
$HD(\hat{f}_{\hat{\theta}_n}, \hat{g}_n)$	0.0176	0.0134	0.0198	0.0219	0.0322	0.0401	0.0418	0.0424
asymptotic upper .10 critical value $h_{.90}$	0.0437	0.0437	0.0437	0.0473	0.0545	0.0616	0.0957	0.128

For the discrete models, goodness-of-fit tests based on power divergence statistics have been introduced in [6] and [19]. The power divergence I^λ between densities f and g is defined by

$$I^\lambda(g, f) = \frac{1}{\lambda(\lambda+1)} \int g(x) \left[\left(\frac{g(x)}{f(x)} \right)^\lambda - 1 \right] dx.$$

The power divergence statistics of [5] is of the form

$$I_n^\lambda = \frac{1}{n\lambda(\lambda+1)} \sum_{i=1}^{k} O_i \left\{ \left(\frac{O_i}{np_i} \right)^\lambda - 1 \right\}, \lambda \in R$$

where O_i are the observed frequencies and np_i the expected frequencies. The Pearson's χ^2 $(\lambda = 1)$, log likelihood ratio statistic $(\lambda \rightarrow 0)$, Freeman - Tukey statistic

($\lambda = -\frac{1}{2}$) are all special cases of the above. The statistic for $\lambda = 2/3$ is shown to be a good alternative to the χ^2 test.

For the discrete models, goodness-of-fit tests based on the blended weight Hellinger distance methods have been introduced and their comparisons given in [3] and [22].

5 Tests of Exponentiality

The two most important continuous probability distributions from the modelling point of view are the exponential and the normal distributions. The exponential distribution is the single most important distribution used for modelling lifetimes. It is the only continuous distribution with the **memoryless property** (i.e. $P(X > x+t|X > t) = P(X > x) \;\; \forall \;\; x, t \geq 0$) hence it is the proper model for the lifetimes of electronic and other non-ageing components. Also, it plays a central role in life testing as a norm, deviations from which have to be noted and studied. So it is extremely important to test goodness-of-fit of the exponential distribution to collected sets of data on lifetimes. Besides, the experimenter wishes to understand what other types of models may be the true models, if not the exponential. The omnibus tests like the Pearson chi-square or Kolmogorov goodness of fit tests do not provide this further information, after rejection. Hence certain tests are devised which reject the H_0 of exponentiality if certain relevant types of alternatives hold.

As mentioned above the exponential distribution uniquely possesses the memoryless or no ageing property. But there are components which are subject to wear and tear or those which deteriorate with age. This phenomenon is known as positive ageing. One type of **positive ageing** is defined as

$$P(X > x+t|X > t] \leq P[X > x], \;\; \forall \;\; x, t \geq 0,$$

with strict inequality for some x and t. In words we may say that a unit which has already been used for t units of time has smaller probability of surviving another x units of time than a new (unused) unit $\forall \;\; x, t \geq 0$. A random variable X, or its c.d.f. F, which possesses this property is said to possess **New Better than Used (NBU)** property. A finer positive ageing property is the Increasing Failure Rate (IFR) property in which the above inequality is changed to

$$F(X > x+t_2|X > t_2] \leq P[X > x+t_1|X > t_1], \;\; \forall \;\; x, 0 < t_1 \leq t_2 < \infty.$$

There are many other classes of distributions including the **Increasing Failure Rate Average (IFRA)** and **Decreasing Mean Residual Life (DMRL)** classes. A reference to any standard book of Reliability Theory, say [1] will give detailed discriptions of and interrelationships between these and such classes of distributions.

(i) **The Hollander - Proschan Test** (see [13]).
The testing problem considered here is
$H_0 : F(x) = 1 - e^{-\lambda x}, x \geq 0, \lambda > 0$, unspecified versus
$H_1 : \overline{F}(s+t) < \overline{F}(s)\overline{F}(t)$, i.e. F belongs to the NBU class. Here $\overline{F} = 1 - F$.

Let $X_1, X_2, \cdots, X_n$ be a random sample from the distribution F. Then the Hollander - Proschan test is based on the U-statistic estimator of the parameter

$$\gamma = \int_0^\infty \int_0^\infty \overline{F}(s+t)dF(s)dF(t)$$
$$= P[X_1 > X_2 + X_3].$$

Define a kernel function

$$\psi(X_1, X_2, X_3) = \begin{cases} 1 \,, & \text{if} \quad X_1 > X_2 + X_3 \\ 0 \,, & \text{otherwise} \end{cases}$$

and let h^* be its symmetrized version. Then

$$U = \frac{1}{\binom{n}{3}} \sum^{*} h^*(X_{i_1}, X_{i_2}, X_{i_3})$$

where $\sum^*$ is the sum over all the $\binom{n}{3}$ combinations of the indices (i_1, i_2, i_3) from the integers $(1, 2, \cdots, n)$.

It is seen that

$E(U) = \gamma$ which is $1/4$ under H_0 and strictly greater than $1/4$ under H_1. Also, the null asymptotic variance of $\sqrt{n}U$ is seen to be $5/432$. Hence the asymptotic distribution of

$$Z = \frac{\sqrt{n}(U - 1/4)}{\sqrt{5/432}}$$

is $N(0,1)$. The test is

Reject H_0 if

$$Z > Z_{1-\alpha}$$

where $Z_{1-\alpha}$ is the $(1-\alpha)$-th quantile of either the exact distribution of Z or its asymptotic $(N(0,1))$ distribution. Hollander and Proschan have shown that the test is consistent for the entire NBU class of distributions and has good efficiency for several common models belonging to this class.

(ii) **The Deshpande Test** (see [9]).

The class of Increasing Failure Rate Average (IFRA) distributions is often encountered in reliability as it is the smallest class containing the exponential distribution and closed under the formation of coherent systems. The IFRA class may be characterized by the property

$$[\overline{F}(x)]^b \leq \overline{F}(bx), \text{`}\ 0 \leq b \leq 1, \quad 0 \leq x < \infty$$

with strict inequality for some b and x. So the testing problem is formed as

$H_0 : F(x) = 1 - e^{-\lambda x}, x > 0, \lambda > 0, \lambda$ unknown, versus

$H_1 : (\overline{F}(x))^b \leq [\overline{F}(bx)], \quad 0 \leq b \leq 1, 0 \leq x < \infty$ and F is not exponential.

To test the null hypothesis the U-statistic estimator of the parameter

$$M = \int_0^\infty \overline{F}(bx)dF(x)$$

is used.

It is easily seen that $M = \frac{1}{(b+1)}$ under H_0 and strictly greater than $\frac{1}{(b+1)}$ under H_1. Hence the U-statistics

$$J_b = \frac{1}{\binom{n}{2}} \sum^{*} h^*(X_{i_1}, X_{i_2})$$

where h^* is the symmetric version of the kernel

$$\begin{aligned} h(X_1, X_2) &= 1, \quad \text{if} \quad X_1 > bX_2 \\ &= 0, \quad \text{otherwise,} \end{aligned}$$

and $\sum^*$ is the sum over all the $nchoose2$ combinations of (i_1, i_2) from the integers $(1, 2, \cdots, n)$.

The asymptotic variance of $\sqrt{n} J_b$ is

$$\xi = \left\{1 + \frac{b}{2+b} + \frac{1}{2b+2} + \frac{2(b-1)}{1+b} - \frac{2b}{1+b+b^2} - \frac{4}{(b+1)^2}\right\}.$$

Then by the U-statistics theorem we know that under H_0

$$Z = \frac{\sqrt{n}(J_b - \frac{1}{b+1})}{\sqrt{\xi}}$$

has $N(0,1)$ distribution. Hence the test is to reject H_0 if $Z > Z_{1-\alpha}$ where $Z_{1-\alpha}$ is again the $(1-\alpha)$-th quantile of the exact null distribution or the asymptotic $(N(0,1))$ distribution of Z. There is the question of choosing b for defining the statistic. Generally $b = 0.5$ or 0.9 is recommended. Test based on $J_{0.5}$ is consistent against the larger NBU class and $J_{0.9}$ seems to have somewhat larger power for many common IFRA distributions.

The statistics J_b is simple to compute. Multiply each observation by the chosen value of b. Arrange $X_1, X_2, \cdots, X_n$ and $bX_1, bX_2, \cdots, bX_n$ together in increasing order of magnitude. Let R_i be the rank of X_i in the combined order and let

$$S = \sum_{i=1}^{n} R_i - \frac{n(n+3)}{2}.$$

Then it is seen that

$$J_b = \{n(n-1)\}^{-1} S.$$

It may be noted that it is essentially the Wilcoxon rank sum statistic for the data of $X_1, X_2, \cdots, X_n$; and $bX_1, bX_2, \cdots, bX_n$.

6 Tests for Normality

The normal distribution is the single most commonly used model for describing the occurrence of outcomes of random experiments and phenomena. Ever since the days of Gauss and Laplace in the eighteenth and nineteenth century it has been recognized as a very useful model. For a considerable time it was believed that most of random phenomena actually give rise to normally distributed data, at least after appropriate transformations. Theory of errors as developed for application in Physics and Astronomy, basically makes the normality assumption. However, by and by it came to be recognized that there are many situations where other models are much more realistically discriptive of real data. Hence there arose the need for testing whether a given set of data, i.e. realizations of independent, identically distributed random variables is described well by the normal distribution or not. Probability plotting as explained later is a useful graphical tool in this respect. Here we describe a formal test based on the quantities involved in probability plotting.

The Shapiro - Wilk - Francia - D'Agostino Tests

Let $X_1, X_2, \cdots, X_n$ be the random sample and $X_{(1)}, X_{(2)}, \cdots, X_{(n)}$ be the corresponding order statistic. Then the test is based on the statistic

$$W = \frac{\left(\sum_{i=1}^{n} a_{i,n} X_{(i)}\right)^2}{\sum_{i=1}^{n}(X_i - \overline{X})^2}$$

which is the ratio of the slope of the normal probability plot, or the square of the weighted least squares estimator of the standard deviation, to the usual estimation of the variance. The values of $a_{i,n}$ for $i = 1, 2, \cdots, n, n = 2, \cdots, 50$ have been tabulated. If the sample size is large, say, greater than 50, the following modified statistic has been proposed in [21].

$$W' = \frac{\left(\sum_{i=1}^{n} b_{i,n} X_{(i)}\right)^2}{\sum_{i=1}^{n}(X_i - \overline{X})^2 \sum_{i=1}^{n} b_{i,n}^2}$$

where $b_{i,n} = \Phi^{-1}\left(\frac{i}{n+1}\right)$ and Φ is the standard normal distribution function. Exact critical values of W (for $n \le 50$) and for $W'(n \le 100)$ are available. For even larger sample sizes in [7], the test statistic

$$D = \frac{\sum_{i=1}^{n}(i - \frac{1}{2}(n+1))X_{(i)}}{n^2\sqrt{\Sigma(X_i - \overline{X})^2}}.$$

has been proposed and its exact critical values for values of n upto 1000 have been provided.

These Shapiro - Wilk - Francia - D'Agostino tests are considered to be omnibus tests as they are able to detect departures from normality in all directions.

7 Diagnostic Methods for Identifying the Family of Distribution Functions

The goodness-of-fit tests described earlier in this chapter provide the means of carrying out formal statistical inference, with known probability of first type of error, about the respected distribution function governing the data. The methods described in this section are less formal. They provide indications to the true distribution functions through graphical procedures. A suspected distribution is at the back of our mind and we compare its shape (or that of some releted functions) with graphs obtained from the data.

(i) **The Q-Q Plot**

The Q - Q or quantile - quantile plot compares the theoretical quantiles of a distribution with the corresponding sample quantiles represented by the order statistics. These plots have been discussed in detail in [23].

Suppose that the suspected distribution function F belongs to a scale-location family $F_0\left(\frac{t-\mu}{\sigma}\right)$ where standard values of μ and σ, say 0 and 1 give a completely known standardized distribution F_0 in this family. For example, F may represent the normal family with mean μ and variance σ^2, with $\mu = 0$ and $\sigma^2 = 1$ giving the standard normal distribution.

Let $F_n^*(t)$ be a slightly modified version of the empirical distribution function given by $F_n^*(t_{(i)}) = \frac{i-\frac{1}{2}}{n}$ rather than i/n. This has been done as generally a theoretical distribution may give $-\infty$ and ∞ as the values of $F^{-1}(z)$ at $z = 0$ and 1. We then compare $F *_n^{-1}\left(\frac{i-1/2}{n}\right)$ of $F^{-1}(z)$ at $z = 0$ and 1. We then compare $F_n^{*-1}\left(\frac{i-1/2}{n}\right) = t_{(i)}$ and $F_0^{-1}\left(\frac{i-1/2}{n}\right)$ by plotting the points $\left(t_{(i)}, F_0^{-1}\left(\frac{i-1/2}{n}\right)\right)$ in a graph. If the true distribution F is belongs to the scale - location family based on F_0 then we expect that this graph called the Q - Q plot will be situated on or near a straight line. This is because

$$F_0^{-1}F(t) = F_0^{-1}F_0\left(\frac{t-\mu}{\sigma}\right) = \frac{t-\mu}{\sigma}$$

is a straight line with slope $\frac{1}{\sigma}$ and intercept $\frac{\mu}{\sigma}$. A straight line is easy for the eye to comprehend and departures from it can be quickly recognized. While not proposing a formal test, the plot does give an indication whether the proposed scale - location family is the appropriate model or not. The slope and the intercept would provide very rough estimates of the parameters which could be useful as initial values in an iterative scheme to find, say, the maximum likelihood estimators or other more formal estimators.

The values of the inverse function F_0^{-1} at the points $\frac{i-\frac{1}{2}}{n}$ for $i = 1, 2, \cdots, n$, are sometimes easy to obtain by direct calculations, someties they are available in wellknown tables (e.g. Φ^{-1}, the inverse of the standard normal distribution). For many distributions they can be obtained by numerical integration or other computer based calculations or packages.

(ii) **The log Q - Q plot**

This is a modification of the Q - Q chart. For some positive valued random variables the distributions of its logarithm belong to a scale - location family. For example the lognormal or the Weibull distributions have this property. Therefore arguing as before we plot the points $\left\{\log t_{(i)}, F^{-1}\left(\frac{i-\frac{1}{2}}{n}\right)\right\}$. For example, in the Weibull case

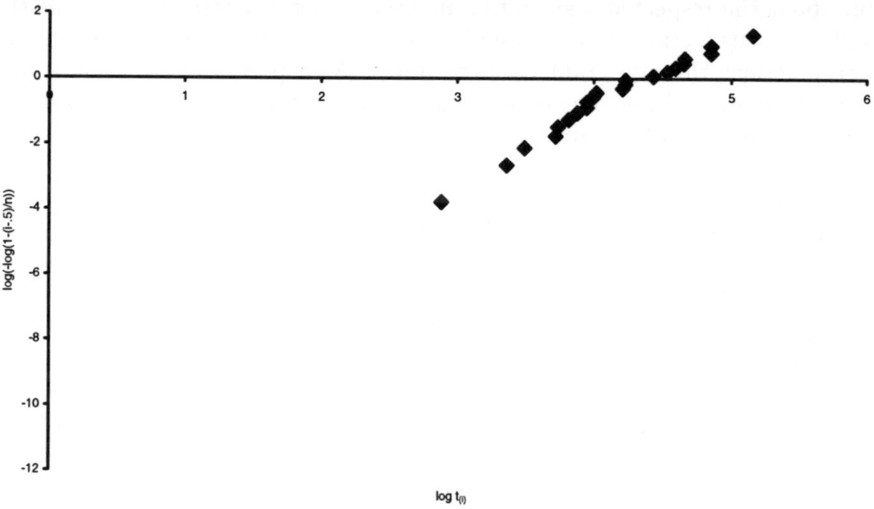

Figure 2: log Q-Q Plot

$[F(t) = 1 - e^{-\lambda t^{\nu}},\ t > 0]$. Hence $[F^{-1}(y) = \frac{\log[-\log(1-y)] - \log\lambda}{\nu},\ \ 0 < y < 1]$, and $\lambda = \nu = 1$ leads to the standard exponential distributions with distribution function $F_0(t) = 1 - e^{-t}, t > 0$ in this family.

Hence if we plot the points $(\log t_{(i)}, \log(-\log\left(1 - \frac{i-\frac{1}{2}}{n}\right)))$ they are expected to lie on a straight line with slope ν and intercept $\log\lambda$. Thus the fact that the points look like being on a straight line will indicate that the distribution is Weibull, with the slope and intercept leading to preliminary estimation of the parameters. The log Q - Q plot for the data of Example 2 is given in Figure 2. It can be seen that the intercept and the slope are approximately -10 and 2.2, respectively, leading to preliminarly estimates $\hat{\lambda}_0 = e^{-10}$ and $\hat{\nu}_0 = 2.2$.

(iii) **The P-P Plot**

The P - P (Probability - probability) plot charts the points $(F_n(t_{(j)}),$ $F(t_{(j)}, \hat{\theta}))$ where F is the proposed family of distributions, possibly dependent upon parameter θ (see [23]). The parameter θ may be estimated by some method suitable for this family, like the method of maximum likelihood and the estimate substituted for the true value. As before $F_n(t_{(j)}) = \frac{j-1/2}{n}$. This plot is restricted to the square $(0,1) \times (0,1)$ and the points are expected to lie on the diagonal from $(0,0)$ to $(1,1)$ if the model holds.

In Figure 3 it will be hard to say that the points do not lie on or near the diagonal, whereas in Figure 4 the plot seems to be concave in nature rather than the

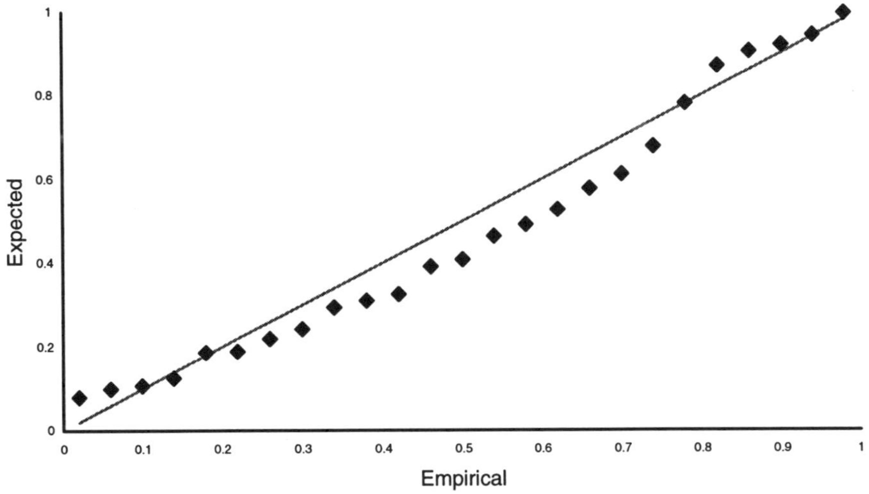

Figure 3: P-P Plot

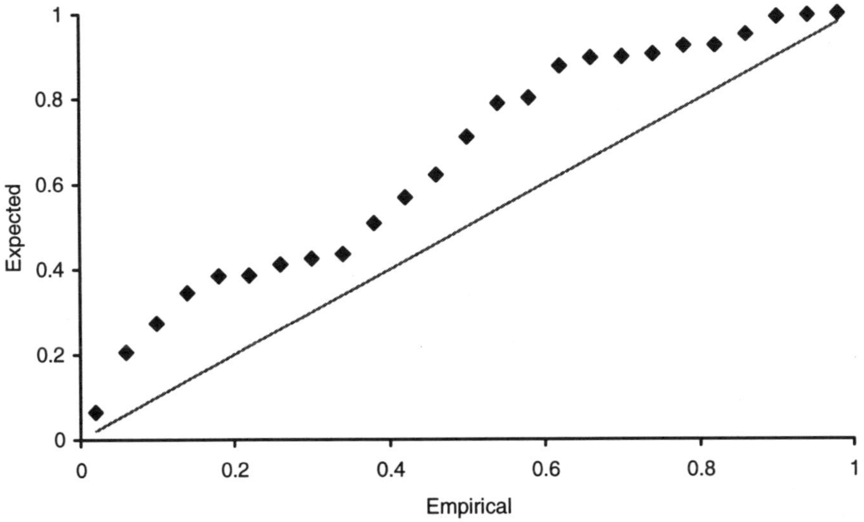

Figure 4: P-P Plot

straight line of the diagonal. The shape of the graph of these points when it is not a straight line also gives some indications regarding the true distribution vis-a-vis the suspected distribution. In particular, if the graph is concave as in Figure 4, it is indicated that the ratio $h_{F_{true}}/h_{F_0}$ of the failure rate of the true distribution with that of the suspected distribution is increasing. This in turn can be interpreted to mean the data comes from a distribution which is aging faster than the suspected distribution. These considerations helps us in selecting appropriate models from the point of view of survival theory.

(iv) **The T-T-T Plot**

The total time on test (T-T-T) plot is very useful in adherence to the exponential model and also departures form it. In specific departures which are of interest in lifetime studies the basis is the scaled T-T-T transform of distribution function defined by

$$T_F(u) = \frac{\int_0^{F^{-1}(u)} \overline{F}(t)dt}{\int_0^\infty \overline{F}(t)dt} \quad , \quad 0 < u < 1.$$

Like other transforms, this is also in 1:1 correspondence with probability distributions. It is easy to see that for the exponential distribution ($\overline{F}(t) = e^{-\lambda t}, t > 0$) it is the straight line segment (diagonal) joining $(0,0)$ with $(1,1)$. Hence the technique is to define the sample version of the scaled T-T-T transform as the T-T-T statistic given by

$$\begin{aligned} T_{F_n}(i/n) &= \frac{\int_0^{F_n^{-1}(i/n)} \overline{F}_n(t)dt}{\int_0^\infty \overline{F}_n(t)dt} \\ &= \frac{\sum_{j=1}^{i}(n-j+1)(X_{(j)} - X_{(j-1)})}{n\overline{X}} \end{aligned}$$

where $\overline{X}$ is the sample mean and $0 = X_{(0)} \leq X_{(1)} \leq \cdots \leq X_{(n)}$ are the order statistics of the random sample. The numerator of $T_{F_n}(i/n)$ is the total time on test (or under operation) of all the n items put on test, simultaneously, upto the i-th failure. Hence the name of the statistic and the transform. The points $\left\{\frac{i}{n}, T_{F_n}(i/n)\right\}, i = 1, 2, \cdots, n$ are plotted in the square $(0,1) \times (0,1)$. If they lie on the diagonal or near it and not systematically on one side, then the exponential distribution is indicated. If a systematic pattern, apart from the diagonal, is discernible then certain alternative models may be more appealing.

The TTT-statistic was first introduced in [10] but its applications in analysis of failure data began with a paper by Barlow and Campo [2].

In the Figure 5 the exponential distribution is indicated, whereas in the Figure 6 the jumps in the values of the sample scaled T-T-T transform seem to become larger and larger indicating a distribution in which failures occur progressively less frequently in time compared to the exponential distribution. If the graph appears to be convex then a DFR distribution and if it is only below the diagonal without being convex then some other NWU distribution is expected to fit better to the data.

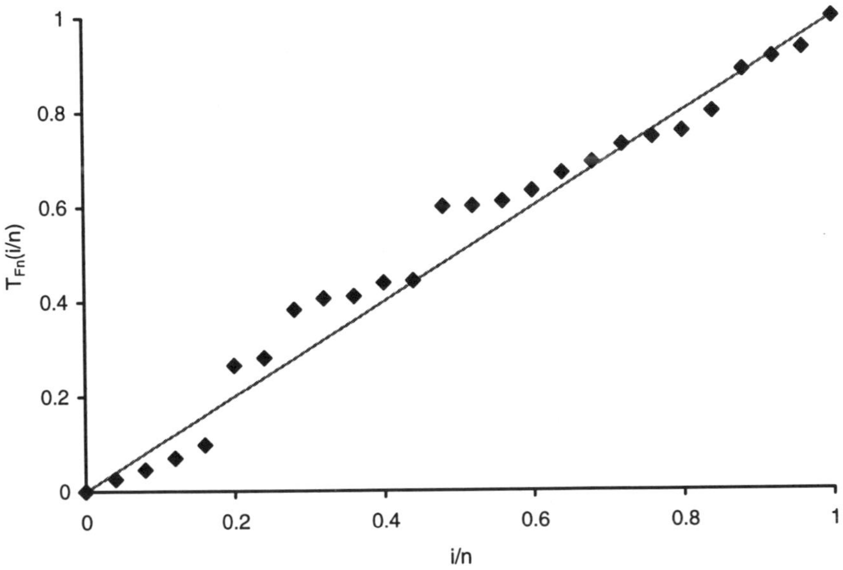

Figure 5: TTT Transform

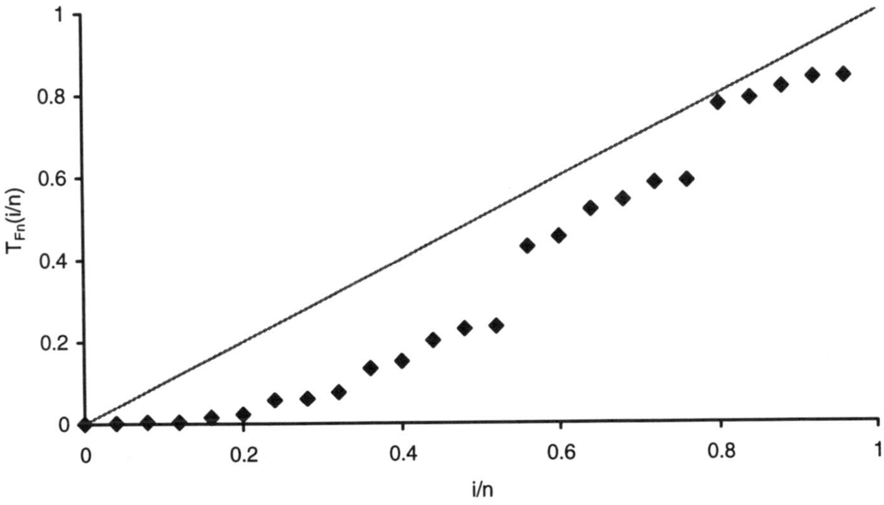

Figure 6: TTT Transform

References

[1] Barlow, R. E. and Proschan, F. Statistical Theory of Reliability and Life Testing : Probability Models. To Begin With, Silver Spring, Maryland (Second edition; first edition published by Holt, Reinhart and Winston 1975) (1981).

[2] Barlow, R. E. and Campo, R. Total time on test processes and applications to failure data analysis. In Reliability and Fault Tree Analysis edited by Barlow, R. E., Fussell, J. and Singpurwalla, N. D., SIAM, Philadelphia, (1975).

[3] Basu, A. and Sarkar, S. On disparity based goodness-of-fit tests for multinomial models. Stat. and Prob. Lett., 19, 307-312 (1994).

[4] Beran, R. J. Minimum Hellinger distance estimates for parametric models. Ann. Statist., 5, 445-463 (1977).

[5] Chernoff, H. and Lehmann E. L. The use of maximum likelihood estimates in χ^2 test for goodness of fit. Ann. Math. Statist., 25, 579-586 (1954).

[6] Cressie, N. and Read, T. R. C. Multinomial goodness-of-fit tests. J. Roy. Statist. Soc., B 46, 440-464 (1984).

[7] D'Agostino, R. B. An omnibus test of normality for moderate and large sample sizes. Biometrika, 58, 341-348 (1971).

[8] D'Agostino, R. B. and Stephens, M. A. Goodness-of-fit Techniques. Marcel Dekker, Inc. (1986).

[9] Deshpande, J. V. A class of tests for exponentiality against increasing failure rate average altenatives. Biometrika, 70, 514-518 (1983).

[10] Epstein, L. and Sobel, M. Life Testing. J. Amer. Statist. Assoc., 48, 486-502 (1953).

[11] Fisher, R. A. The conditions under which χ^2 measures the discrepancy between observation and hypothesis. J. Royal Statist. Soc., 87, 442-450 (1924).

[12] Hand, D. J., Daly, F., Lunn, A. D., McConway, K. J. and Ostrowski, E. (editors). A Handbook of Small Data Sets. Chapman and Hall (1994).

[13] Hollander, M. and Proschan, F. Tests for the mean residual life. Biometrika, 62, 585-594 (1972).

[14] Koziol, J. A. Goodness-of-fit tests for randomly censored data. Biometrika, 67, 693-696 (1980).

[15] Koziol, J. A. and Byar, D. P. Percentage points of the asymptotic distributions of one and two sample $k-s$ statistics for truncated or censored data. Technometrics, 17, 507-510 (1975).

[16] Li, G. and Doss, H. Generalized Pearson - Fisher chi-square goodness-of-fit tests with applications to models with life history data. Ann. Statist., 21, 772-797 (1993).

[17] Lieblein, J. and Zelen, M. Statistical investigation of the fatigue life of deep groove ball bearing. J. Res. Nat. Bur. Stand., 57, 273-316 (1956).

[18] Pearson, K. On the criterion that a given system of deviations from the probable in the case of a correlated system of variables is such that it can be reasonably supposed to have arisen from random sampling. Philosophical Magazine, Ser. 5, 50, 157-172 (1900).

[19] Read, T. R. C. and Cressie, N. Goodness-of-fit Statistics for Discrete Multivariate Data, New York, Springer Verlag. (1988).

[20] Shapiro, S. S. and Wilk, M. B. An analysis of variance test for normality. J. Amer. Statist. Assoc., 67, 215-216 (1965).

[21] Shapiro, S. S. and Francia, R. S. Approximate analysis of variance test for normality. J. Amer. Statist. Assoc. 67, 215-216 (1972).

[22] Shin, D. W., Basu, A. and Sarkar, S. Comparisons of the blended weight Hellinger distance based goodness-of-fit test statistics. Sankhya, B57, 365-376.

[23] Wilk, M. B. and Ganadesikan, R. Probability plotting methods for the analysis of data. biometrika, 55, 1-17 (1968).

[24] Wilks, S. S. Mathematical Statistics, John Wiley and Sons, Inc. (1962).

U STATISTICS AND M_m ESTIMATES

ARUP BOSE
Indian Statistical Institute, Kolkata

Abstract. After a quick introduction to some basic properties of U statistics with examples, we discuss M_m estimators and their asymptotic properties under easily verifiable conditions. In particular, these estimators are approximately U statistics and as a consequence, a huge collection of commonly used estimators are consistent and asymptotically normal. We also establish some higher order asymptotic properties of these estimates. The material is more or less self contained.

Key Words and Phrases: U statistics, M_m estimates, strong consistency, asymptotic normality, almost sure representation, U quantiles, multivariate medians.

1. Introduction

This article is broadly divided into two parts. In the first part we deal with U statistics. We concentrate on some results which are useful from a statistician's point of view. As applications, we establish the asymptotic distribution of many common statistics which are either U statistics or simple functions of U statistics. This material is fairly standard but our concise treatment makes the article reasonably self contained.

In the second part, we deal with M_m estimators and their asymptotic properties. A huge class of common and also not so common estimators fall in this category. The asymptotic properties of these estimates have been treated under different sets of conditions in the literature. The most general results for these estimators require very sophisticated treatment using techniques from the theory of empirical processes. But here we strive for a simple approach. The conditions we assume are few and simple but general enough to be applicable widely. We demonstrate how one can check the necessary conditions of the general theory in particular cases.

2. U statistics and its basic properties

2.1 Definition and first examples. Let $X_1, X_2, \cdots, X_n$ be observations, not necessarily real valued. We shall assume throughout that they are independent and identically distributed (iid). Suppose $h(x_1, \cdots, x_m)$ is a real valued function which is *symmetric in its arguments.*

Definition 1. The U statistics of *order* or *degree* m, with *kernel* h is:

$$U_n = \binom{n}{m}^{-1} \sum_{1 \leq i_1 < \cdots < i_k \leq n} h(X_{i_1}, \cdots, X_{i_m}) \tag{2.1}$$

The systematic study of U statistics began with Hoeffding (1948). Many of the basic properties of U statistics are due to him. In this section we will cover a few basic properties of U statistics and provide a few examples. Further material on U statistics are provided by Lee (1990) and Koroljuk and Borovskich (1993). Some results on U statistics that we specifically need for the study of M_m estimates are given in the next section.

Example 1. (Sample mean) Letting $m = 1$, $h(x) = x$, we obtain $U_n = n^{-1} \sum_{i=1}^n X_i$.

Example 2. (Sample variance) Letting $m = 2$, $h(x_1, x_2) = \frac{(x_1 - x_2)^2}{2}$, we get

$$U_n = \binom{n}{2}^{-1} \sum_{1 \leq i_1 < i_2 \leq n} [X_{i_1} - X_{i_2}]^2/2.$$

It is easily seen that $U_n = (n-1)^{-1} \sum_{i=1}^n (X_i - \overline{X})^2$, the *sample variance.*

Example 3. (Sample covariance) Suppose (X_i, Y_i), $1 \leq i \leq n$ are the observations, $m = 2$ and $h((x_1, y_1), (x_2, y_2)) = \frac{1}{2}(x_1 - x_2)(y_1 - y_2)$. Then U_n is the sample covariance between $\{X_i\}$ and $\{Y_i\}$.

Example 4. (Kendall's tau) Suppose (X_i, Y_i), $1 \leq i < n$ are bivariate observations. A measure of *discordance* is Kendall's tau, defined by

$$t_n = \binom{n}{2}^{-1} \sum_{1 \leq i < j \leq n} \text{sign}(X_i - X_j)(Y_i - Y_j). \tag{2.2}$$

This is a U statistic with $h((x_1, x_2), (y_1, y_2)) = \text{sign}(x_1 - x_2)(y_1 - y_2)$.

Example 5. *Gini's mean difference*, a measure of inequality is defined as

$$U_n = \binom{n}{2}^{-1} \sum_{1 \leq i < j \leq n} | X_i - X_j | .$$

If the observations are $N(0, \sigma^2)$, $E(U_n) = (2/\pi)^{1/2}\sigma$. Thus U_n is a measure of dispersion. This is a U statistic with $h(x_1, x_2) =| x_1 - x_2 |$.

Example 6. (Wilcoxon's one sample rank statistic) Suppose X_i, $1 \leq i \leq n$ are continuous observations. Let $R_i =$ Rank $(| X_i |)$, $1 \leq i \leq n$. Wilcoxon one sample rank statistic is defined as $T^+ = \sum_{i=1}^n R_i \, I(X_i > 0)$. T^+ can be written as a linear combination of two U statistics with kernels of size 1 and 2. To do this, note that for $i \neq j$,

$$I\{X_i + X_j > 0\} = I\{X_i > 0\} I\{| X_j |< X_i\} + I\{X_j > 0\} I\{| X_i |< X_j\}$$

Hence

$$\begin{aligned}
\sum_{1 \leq i \leq j \leq n} I\{X_j + X_j > 0\} &= \sum_{1 \leq i < j \leq n} I\{X_i > 0\} I\{| X_j |< X_i\} \\
&+ \sum_{1 \leq i < j \leq n} I\{X_j > 0\} I\{| X_i |< X_j\} \\
&+ \sum_{i=1}^{n} I\{X_i > 0\} \\
&= \sum_{i=1}^{n} \sum_{j=1}^{n} I\{X_i > 0\} I\{| X_j |\leq X_i\} \\
&= \sum_{i=1}^{n} I\{X_i > 0\} R_i = T^+
\end{aligned}$$

Define two kernels, $h_1(x_1) = I\{x_1 > 0\}$, and $h_2(x_1, x_2) = I\{x_1 + x_2 > 0\}$. Define two U statistics as

$$U_n(h_1) = \frac{1}{n}\sum_{i=1}^{n} h_1(X_i), \text{ and } U_n(h_2) = \binom{n}{2}^{-1} \sum_{1 \le i < j \le n} h_2(X_i, X_j).$$

Then

$$T^+ = \binom{n}{2} U_n(h_2) + nU_n(h_1). \tag{2.3}$$

2.2. Some properties of U statistics

(i) **Variance of U_n.** Computing the variance needs computing the covariances between $h(X_{j_1}, \cdots, X_{jm})$ and $h(X_{i_1} \cdots X_{i_m})$ which depends on the number of common indices. Let $\delta_c = \text{cov}[h(X_{i_1}, \cdots, X_{i_m}), h(X_{j1}, \cdots, X_{jm})]$ when the number of common indices is c. It is easy to see that $\delta_c \geq 0$ for all c. By a simple combinatorial argument

$$V(U_n) = \binom{n}{m}^{-2} \sum_{c=1}^{m} \binom{n}{m}\binom{m}{c}\binom{n-m}{m-c}\delta_c = \binom{n}{m}^{-1} \sum_{c=1}^{m} \binom{m}{c}\binom{n-m}{m-c}\delta_c.$$

As a consequence of this, we also have

$$V(U_n) = \frac{m^2\delta_1}{n} + 0(n^{-2}) \text{ and } V(n^{1/2}(U_n - \theta)) \to m^2\delta_1. \tag{2.4}$$

Example 7. Suppose $U_n = s_n^2 = (n-1)^{-1}\sum_{i=j}^{n}(X_i - \overline{X}_n)^2$. Here the kernel is $h(x_1, x_2) = \frac{(x_1 - x_2)^2}{2}$. It can be verified directly that, if $\sigma^2 = V(X_1)$, then $\delta_1 = \frac{E(X - E(X))^4 - \sigma^4}{4}$, $\delta_2 = \sigma^4$ and $V(U_n) = \frac{4\,\delta_1}{n} + \frac{2}{n(n-1)}\delta_2$.

(ii) **First projection of U_n.** Note that the leading term in the asymptotic variance of $n^{1/2}U_n$ is given by $m^2\delta_1$. It is also not hard to check that δ_1 is given by $\delta_1 = \text{cov}[h(X_1, \cdots, X_m), h(X_1, X_{m+1}, \cdots, X_{2m})] = Var(h_1(X_1))$ where $h_1(x_1) = E\ h(x_1, X_2, \cdots, X_m)$ is the *conditional expectation* of h given *one* of the coordinates. This function h_1 is called the *first projection* of U_n. Let $\tilde{h}_1$ be its centered version:

$$\tilde{h}_1(x_1) = h_1(x_1) - \theta$$

so that $E\,\tilde{h}_1(X_1) = E[h(X_1, \cdots, X_m)] - \theta = 0$. Let

$$R_n = U_n - \theta - \frac{m}{n}\sum_{i=1}^{n} \tilde{h}_1(X_i). \tag{2.5}$$

By explicit calculations it can be easily seen that the decomposition (2.5) is an *orthogonal decomposition* in the following sense :

$$\text{cov}[\tilde{h}_1(X_i), R_n] = 0 \quad \forall\, i = 1, \cdots, n. \tag{2.6}$$

(iii) **Convergence of U statistics**. From the above it immediately follows that if $V[h(X_1, \ldots, X_m)] < \infty$, then $U_n - \theta \xrightarrow{P} 0$ as $n \to \infty$. In fact a much stronger statement is true: If $E[|h(X_1, \ldots, X_m)|] < \infty$, then $U_n - \theta \xrightarrow{a.e.} 0$. This can be proved by using SLLN for either reverse martingales or forward martingales. A rate result when higher moment exists is given in Lemma 3 in Section 3.

(iii) **Asymptotic normality of U_n**. From the relations (2.5) and (2.6), we get $V(U_n) = \frac{m^2}{n}\delta_1 + V(R_n)$. But on the other hand, from (2.4) $V(U_n) = \frac{m^2}{n}\delta_1 + 0(n^{-2})$. This shows that $V(n^{1/2}R_n) \to 0$ and hence $n^{1/2}R_n \to 0$ in probability. Now appealing to the decomposition given in (2.5), and the usual central limit theorem, we immediately obtain,

Theorem 1. *If* $V[h(X_1, \ldots X_m)] < \infty$, *then*

$$n^{1/2}(U_n - \theta) \xrightarrow{\mathcal{D}} N(0, \sigma^2) \;\text{ where }\; \sigma^2 = m^2 V(\tilde{h}_1(X_1)) = m^2\delta_1.$$

Remark 1. By using the Cramer-Wold device, it is easy to see that the *multivariate* version of Theorem 1 holds. This is useful in applications where more than one U statistics is involved.

Example 8. Consider the U statistic s_n^2. In Example 7, we have calculated $\delta_1 = \frac{\mu_4 - \sigma^4}{4}$, where $\mu_4 = E(X - (EX))^4$. Thus if $\mu_4 < \infty$,

$$n^{1/2}(s_n^2 - \sigma^2) \xrightarrow{\mathcal{D}} N(0,\ \mu_4 - \sigma^4).$$

Example 9. Suppose $h(x_1, x_2) = x_1 x_2$. Let $\mu = E(X_1)$. Then

$$\begin{aligned} \delta_1 &= \operatorname{cov}(X_1X_2, X_1X_3) \\ &= EX_1^2X_2X_3 - (EX_1X_2)(EX_1X_3) \\ &= \mu^2 EX_1^2 - \mu^4 = \mu^2 V(X_1). \end{aligned}$$

Hence if $V(X_1) < \infty$, then

$$n^{1/2}\left(\binom{n}{2}^{-1} \sum_{1 \le i < j \le n} X_iX_j - \mu^2\right) \xrightarrow{\mathcal{D}} N(0, 4\,\mu^2\sigma^2).$$

Example 10. Consider Kendall's tau defined in Example 4. It is used to test the *null* hypothesis that X and Y are independent. We shall derive the asymptotic null distribution of this statistic. Let F, F_1, F_2 denote the distributions of (X, Y), X and Y respectively, Then

$$\begin{aligned} h_1(x, y) &= E\, h((x, y), (X_2, Y_2)) \\ &= P\{(x - X_2)(y - Y_2) > 0\} - P\{(x - X_2)(y - Y_2) < 0\} \\ &= P\{(X_2 > x, Y_2 > y),\ \text{or}\ (X_2 < x, Y_2 < y)\} \\ &\quad - P\{(X_2 > x, Y_2 < y)\ \text{or}\ (X_2 < x, Y_2 > y)\} \\ &= 1 - 2F(x, \infty) - 2F(\infty, y) + 4\,F(x, y) \\ &= (1 - 2F_1(x))(1 - 2F_2(y)) + 4(F(x, y) - F_1(x)F_2(y)). \end{aligned}$$

Under the null hypothesis X and Y are independent and hence for all x and y, $F(x, y) = F_1(x)F_2(y)$. Hence in that case, $h_1(x, y) = (1 - 2F_2(x))(1 - 2F_2(y))$. To compute its variance, note that $U = 1 - 2F_1(X_1)$ and $V = 1 - 2F_2(Y_1)$ are independent $U(-1,\ 1)$ random variables. Hence

$$V[h_1(X, Y)] = V(UV) = EU^2EV^2 = (\frac{1}{2}\int_{-1}^{1} u^2 du)^2 = 1/9$$

Moreover, under independence, $\theta = E[\text{ sign } (X_1 - X_2)(Y_1 - Y_2)] = 0$.

Hence *under independence*, $n^{1/2}U_n \xrightarrow{\mathcal{D}} N(0, \sigma^2)$ where $\sigma^2 = 2^2/9 = 4/9$.

Example 11. Wilcoxon's statistic defined in Example 6 is used for testing the null hypothesis that the distribution F of X_1 is symmetric about 0. Recall the expression for T^+ in (2.3). We concentrate first on $U_n(h_1)$. $E\, U_n(h_1) = P(X_1 + X_2 > 0) = \theta$, say. Under the null hypothesis, $\theta = 1/2$. Further,

$$\begin{aligned} V(\tilde{h}_1) &= \text{Cov } [I(X_1 + X_2 > 0), \;\; I(X_1 + X_2 > 0] \\ &= P(X_1 + X_2 > 0, \; X_1 + X_3 > 0) - (1/2)^2. \end{aligned}$$

Now assume that F *is continuous.* Then under null hypothesis, by symmetry, $P(X_1 + X_2, \;\; X_1 + X_3 > 0) = 1/3$. Thus $n^{1/2}(U_n(h_2) - 1/2) \;\; \xrightarrow{\mathcal{D}} N(0, \sigma^2)$ where $\sigma^2 = 2^2(1/3 - 1/4) = 1/3$. It also follows easily that

$$n^{-3/2}(nU_n(h_1)) \to 0 \text{ in probability.}$$

Hence we have, after algebraic adjustments,

$$n^{-3/2}\sqrt{(12)}(T^+ - n^2/4) \xrightarrow{\mathcal{D}} N(0, 1).$$

3. U statistics and M_m estimators

M estimators and their general versions M_m estimators were introduced by Huber (1964) from robustness considerations. The literature on these estimators is very rich. There are a variety of conditions under which the asymptotic properties of these estimators have been studied. It is known that under suitable conditions these estimates are consistent and asymptotically normal and satisfy appropriate almost sure representations. The goal of this section is to offer easily verifiable conditions to derive some of the asymptotic properties of these estimators. A huge class of M_m estimators turn out to be approximate U statistics. Hence the theory of U statistics plays a crucial role in this approach. We give several examples to show how the general results can be applied to many estimators. In particular, several multivariate estimates of location are discussed in details.

3.1 Basic definitions and examples.

M_m **parameter**. Let $f(x_1, \cdots, x_m, \theta)$ be a real valued function. The argument θ is assumed to belong to $\mathcal{R}^d$. Let $X_1, \cdots, X_m$ be i.i.d. random variables. Define

$$Q(\theta) = E\, f(X_1, \cdots, X_m, \theta).$$

Let θ_0 be the **unique** minimizer of $Q(\theta)$. We consider θ_0 to be the unknown parameter. It is usually called the *M_m parameter*. The special case when $m = 1$ is the one that is most commonly studied and in that case θ_0 is traditionally called the *M parameter*.

M_m **estimator**. Suppose $\{X_1, \cdots, X_n\}$ is a sequence of independent and identically distributed observations. Under the absence of any further information on the distribution of X, a natural estimate of the M_m parameter θ_0 is the minimiser of the sample analogue Q_n of Q,

$$Q_n(\theta) = \binom{n}{m}^{-1} \sum_{1 \le i_1 < i_2 \cdots < i_m \le n} f(X_{i_1}, \cdots, X_{i_m}, \theta).$$

Definition 2. Any (measurable) value θ_n which minimizes $Q_n(\theta)$ is called an *M_m estimator* of θ_0.

Example 12. Let $f(x, \theta) = (x - \theta)^2 - x^2$. Clearly $Q(\theta) = \theta^2 - 2E(X)\theta$ which is minimized uniquely at $\theta_0 = E(X)$. Its M estimator is the sample mean.

Example 13. (Sample median and quantiles). For $0 < p < 1$, the population pth quantile is the point where the distribution function exceeds p for the first time. To see these as M parameters, let $f(x, \theta) = \mid x - \theta \mid - \mid x \mid -(2p - 1)\theta$. It is easy to check that,

$$f(x, \theta) = \theta\,[2I(x \le 0) - 1] + 2\int_0^\theta [I(x \le s) - I(x \le 0)]ds - (2p - 1)\theta.$$

Hence if F_X denotes the distribution function of X then

$$Q(\theta) = 2\int_0^\theta F_X(s)ds - 2p\theta.$$

$Q(\theta)$ is minimized at

$$\theta_0 = F_X^{-1}(p) = \inf\{x : F_X(x) \geq p\}.$$

Then θ_0 is a population pth quantile. It is unique if F_X is *strictly increasing* at θ_0. If $p = 1/2$, it is called the population median. The M estimator of θ_0 is called the sample pth quantile. Unlike the previous example, this estimator is *not* necessarily unique. If $p = 1/2$, we get the sample median.

Example 14. (L_1 median) There are several reasonable definitions of the "median" when the observations are multivariate. The reader may consult Small (1990) for a first exposure to the various notions of median/location for multivariate observations. One such median is the L_1 median. Suppose X is a d dimensional random vector $= (Z_1 \cdots Z_d)$. Let

$$f(x, \theta) = [\sum_{i=1}^{d}(x_i - \theta_i)^2]^{1/2} - [\sum_{i=1}^{d} x_i^2]^{1/2}.$$

It can be shown that if F_X does not put all its mass on a hyperplane (that is, if $P\{\sum_{i=1}^d C_i\, Z_i = \text{Constant }\} \neq 1$ for any choice of real number $(C_1 \cdots C_d)$), then $Q(\theta)$ is minimized at a *unique* θ_0. This θ_0 is called the L_1 median. The corresponding M estimator is called the (sample) L_1 median. It is unique if all the sample values do not lie in a lower dimensional hyperplane. If $d = 1$, the L_1 median reduces to the usual median discussed in Example 13.

We now give examples of M_m estimates where $m > 1$. Traditionally, M estimators are thought of as measures of location. However, M_m estimators encompass both measures of location and of scale.

Example 15. For a function $h(x_1, \ldots, x_m)$ which is symmetric in its arguments let

$$f(x_1 \ldots, x_m, \theta) = [\theta - h(x_1, \ldots, x_m)]^2 - [h(x_1, \ldots, x_m)]^2.$$

Then $\theta_0 = E\ h(X_1, \ldots, X_m)$ and θ_n is the U statistic with kernel h. So all U statistics are M_m estimators. In particular, the sample variance is an M_2 estimator.

Example 16. (Oja median) A multivariate median due to Oja (1983) is defined as follows. Suppose $X_1, \cdots, X_d$ are d-dimensional i.i.d. random vectors. Let $\Delta(X_1, \cdots, X_d, \theta)$ denote the (absolute) volume of the simplex formed by the $(d+1)$ points $\{X_1, \cdots, X_d, \theta\}$ in $\mathcal{R}^d$. Let

$$f(x_1, \cdots, x_d, \theta) = \Delta(x_1, \cdots, x_d, \theta) - \Delta(x_1, \cdots, x_d, 0).$$

If $E|X_1| < \infty$, then $Q(\theta) = E\ f(X_1, \cdots, X_d, \theta)$ exists. It is also known that if the distribution of X_1 does not concentrate on a hyperplane of a lower dimension, then θ_0 is unique and is called the *Oja median*. The corresponding M_m estimate is called the (sample) Oja median.

Example 17. (Hodges-Lehmann measure of location) Suppose $X_1, \cdots, X_n$ are i.i.d. observations. Instead of the usual mean as a measure of location, one may consider the *median* of $\{\frac{X_i+X_j}{2},\ \ 1 \leq i < j \leq n\}$ as the sample measure of location. Here $m = 2$ and

$$f(x_1, x_2, \theta) = \left| \frac{x_1 + x_2}{2} - \theta \right| - \left| \frac{x_1 + x_2}{2} \right|.$$

The parameter θ_0 is the *median* of G where G is the distribution of $\frac{X_1+X_2}{2}$.

Example 18. (Robust measure of scale/dispersion) The variance as a measure of dispersion is influenced by extreme observations. To address this problem, Bickel and Lehmann (1979) considered the distribution of $| X_1 - X_2 |$ and took its median to be a measure of dispersion. Here $m = 2$ and

$$f(x_1, x_2, \theta) = || x_1 - x_2 | - \theta | - | x_1 - x_2 |.$$

Example 19. (U quantiles) The ideas of the previous two examples can be extended to define U quantiles (Choudhury and Serfling (1988)). Let $h(x_1, \cdots, x_m)$ be a symmetric kernel. Define

$$f(x_1, \cdots, x_m, \theta) = | h(x_1, \cdots, x_m) - \theta | - | h(x_1, \cdots, x_m) |.$$

Then θ_0, the minimizer of $E[f(X_1, \cdots, X_m, \theta)]$ is called a U-median. Other U quantiles can be defined in a way similar to the sample quantiles in Example 13. Note that just like the sample quantiles in Example 13, these estimates, in general, are not unique. Multivariate versions of these U-quantiles defined by Helmers and Huskova (1994) are also M_m estimates.

Many researchers have studied the asymptotic properties of M estimators and M_m estimators. Early works on the asymptotic properties of M_1 estimators and M_2 estimators are Huber (1967) and Maritz et. al. (1977). Oja (1984) proved the *consistency and asymptotic normality* of M_m estimators under conditions similar to Huber (1967). His results apply to some of the estimators above.

We emphasize that all examples of f we have considered so far have a common feature. They are all *convex* functions of θ. Statisticians prefer to work with convex loss functions for various reasons. We shall make this blanket assumption here. This does entail some loss of generality. But convexity leads to a significant simplification in the study of M_m estimators while at the same time, still encompassing a huge class of estimators.

As Examples 13 and 19 showed, an M_m estimator is not necessarily unique. However, it can be shown that by using the convexity assumption, a *measurable minimiser* can always be chosen. This can be done by Corollary 1 in the Appendix of Niemiro (1992). The asymptotic results that we will discuss hold for any measurable sequence of minimizers of $Q_n(\theta)$.

Several works have assumed and exploited this convexity in similar contexts. Perhaps the earliest use of this convexity was by Heiler and Willers(1988) in linear regression models. See also Hjort and Pollard(1993). For example, for $m = 1$, Habermann (1989) established the consistency and asymptotic normality of θ_n and Niemiro (1992) established a Bahadur type representation $\theta_n = \theta_0 + S_n/n + R_n$ where R_n is of suitable order *almost surely* and S_n is the partial sum of a sequence if iid random variables. In the next subsections, we shall exploit the convexity heavily and establish some large sample properties of M_m estimates.

Remark 2. Even though our set up covers a lot of interesting multivariate location and scale estimators, it does *not* cover several other estimators such as the medians of Liu (1990), Tukey (1975) and Rousseeuw (1986) etc since the convexity condition is not satisfied. One general approach in the absence of convexity is provided by Jureckova (1977)). See also de la Pena and Gine (1999, page 279).

3.2 Strong consistency. The immediate consequence of convexity is the strong consistency of M_m estimates. Assume that:

(I) $f(x_1, \ldots, x_m, \theta)$ is convex in θ for every $(x_1, \ldots, x_m)$.
(II) $Q(\theta)$ is finite for all θ.
(III) θ_0 exists and is unique.

Remark 3. Often the parameter space is restricted. If (I) and (II) are satisfied for a subset of $\mathcal{R}^d$, then all the results we give below remain valid if θ_0 is an interior point of this subset.

Theorem 2. (Strong consistency) Under Assumptions I, II and III,

$$\theta_n \to \theta_0 \quad \text{almost surely, as } n \to \infty.$$

Remark 4. The above theorem, in particular implies that *all* the estimators introduced so far in our examples are strongly consistent under minimal assumptions, (I)–(III).

To prove the Theorem, we need the following Lemma. Recall that convex functions converge if they converge on a dense set and the convergence is uniform over compact sets. Using this and a diagonalisation argument, the Lemma can be easily proved. Details can be found in Niemiro (1992).

Lemma 1. Suppose that $h_n(\alpha)$, $\alpha \in \mathcal{R}^d$ is a sequence of random convex functions which converge to $h(\alpha)$ for every fixed α either in probability or almost surely. Then this convergence is uniformly on any compact set of α, respectively in probability or almost surely.

Proof of Theorem 2. Note that by the strong law for U statistics, $Q_n(\alpha)$ converges to $Q(\alpha)$ for each α almost surely. By Lemma 1, this convergence is uniform on any compact set almost surely.

Let B be a ball of arbitrary radius ϵ around θ_0. If θ_n is not consistent, then there is a set S in the probability space such that $P(S) > 0$ and for each sample point in S, there is a subsequence of θ_n that lies outside this ball. We assume without loss that for each point in this set, the convergence of Q_n to Q also holds. For a fixed sample point, we continue to denote such a sequence by $\{n\}$.

Consider the point θ_n^* which is the intersection of the line joining θ_0 and θ_n with the ball B. Then for some sequence $0 < \gamma_n < 1$, $\theta_n^* = \gamma_n\theta_0 + (1-\gamma_n)\theta_n^*$. By convexity of Q_n and the fact that θ_n is a minimiser of Q_n,

$$Q_n(\theta_n^*) \leq \gamma_n Q_n(\theta_0) + (1-\gamma_n)Q_n(\theta_n) \leq \gamma_n Q_n(\theta_0) + (1-\gamma_n)Q_n(\theta_0) \leq Q_n(\theta_0).$$

Note that the right side converges to $Q(\theta_0)$. Pick a subsequence of θ_n^* which converges to, say θ_1. Since the convergence of Q_n to Q is uniform, the left side of the above equation converges to $Q(\theta_1)$. Hence, $Q(\theta_1) \leq Q(\theta_0)$. This is a contradiction to the uniqueness of θ_0. This proves the theorem.

3.3 Asymptotic normality. We now give an *in probability* representation result for M_m estimators. This representation implies the asymptotic normality of M_m estimators. To state the result, first note that since f is convex, it has a *subgradient* $g(x, \theta)$. This subgradient has the property that for all α, β, x,

$$f(x,\alpha) + (\beta - \alpha)'g(x,\alpha) \leq f(x,\beta). \tag{3.1}$$

If f is differentiable, then this subgradient is simply the ordinary derivative. Further g is measurable in x for each α. This is possible by an appropriate selection theorem, such as Corollary 2 in the Appendix of Niemiro (1992), or see Castaing and Valadier (1977).

It is easy to see that by using (3.1), under assumption (II), the expectation of g is finite. Moreover, the gradient vector $\nabla Q(\theta)$ of Q at θ exists and

$$\nabla Q(\theta) = E[g(X_1, \ldots, X_m, \theta)] < \infty.$$

Denote the matrix of second derivatives of Q at θ, whenever it exists, by $\nabla^2 Q(\theta)$. We also define

$$H = \nabla^2 Q(\theta_0)$$

and

$$U_n = \binom{n}{m}^{-1} \sum_{1 \le i_1 \cdots < i_m \le n} g(X_{i_1} \cdots, X_{i_m}, \theta_0).$$

Let N be an appropriate neighbourhood of θ_0. We list the following additional assumptions to derive asymptotic normality.

(IV) $E|g(X_1, \ldots, X_m, \theta)|^2 < \infty \ \forall \ \theta \in N$.
(V) $H = \nabla^2 Q(\theta_0)$ exists and is positive definite.

The following theorem is a consequence of the works of Habermann (1989) and Niemiro (1992) for $m = 1$, and Bose (1998) for general m.

Theorem 3. *Suppose Assumptions (I)–(V) hold. Then,*

(i) $\theta_n - \theta_0 = -H^{-1}U_n + o_P(n^{-1/2})$

(ii) $n^{1/2}(\theta_n - \theta_0) \xrightarrow{\mathcal{D}} N(0, k^2 H^{-1} K H^{-1})$ *where K is the variance covariance of the first projection of the gradient vector $g(X_1, \ldots, X_m, \theta_0)$.*

Remark 5. The M_m estimators given in section 3.1 all satisfy the conditions of Theorem 3 under suitable conditions on F. Thus Theorem 3 implies the asymptotic normality of a huge collection of estimators. After we give the proof of the theorem, we illustrate its use through a discussion of the appropriate conditions required in some specific cases.

For the proof of this theorem as well for those theorems given later, assume without loss that $\theta_0 = 0$ and $Q(\theta_0) = 0$. Also let S denote the set of all m element increasingly ordered subsets of $\{1, \ldots, n\}$. For any $s = \{i_1, \ldots, i_m\} \in S$, let Y_s denote the random vector $(X_{i_1}, \ldots, X_{i_m})$ and $X(s, \alpha) = Q(Y_s, \alpha)$.

Proof of Theorem 3. For any fixed α, and $s \in S$ let $X_{ns} = f(Y_s, n^{-1/2}\alpha) - f(Y_s, 0) - n^{-1/2}\alpha^T g(Y_s, 0)$.

Note that $V_n = \binom{n}{m}^{-1} \sum_{s\in S} X_{ns}$ is a U statistic. From the decomposition given in (2.5) and (2.6), using (3.1), it follows that

$$\begin{aligned} V(\binom{n}{m}^{-1} \sum X_{ns}) &\le \frac{m}{n} E[(X_{ns} - EX_{ns})]^2 \\ &\le K\frac{m}{n} EX_{ns}^2 \\ &\le K\frac{m}{n^2} E[\alpha'\{g(X_{ns}, n^{-1/2}\alpha) - g(X_{ns}, 0)\}]^2 \end{aligned}$$

Let Y be identically distributed as any Y_s. Let $Y_n = \alpha'\{g(Y, n^{-1/2}\alpha) - g(Y, 0)\}$. Note that $Y_n \ge 0$ and Y_n is nonincreasing. Let $\lim Y_n = Y_0 \ge 0$. Thus $E(Y_n) \uparrow E(Y_0)$. But $EY_n \to 0$. Hence $Y_0 = 0$ a.s.. This implies that $EY_n^2 \to 0$. Noting that $E\ X_{ns} = Q(n^{-1/2}\alpha)$, it follows that for each fixed α,

$$n\binom{n}{m}^{-1} \sum (X_{ns} - EX_{ns}) = nQ_n(\frac{\alpha}{\sqrt{n}}) - nQ_n(0) - n^{1/2}\alpha' U_n - nQ(\frac{\alpha}{\sqrt{n}}) \to 0$$

in probability. By Assumption (VII),

$$nQ(\alpha/\sqrt{n}) \to \alpha' H\alpha/2$$

and due to convexity, both the above convergences are uniform on any compact set by Lemma 1. Thus for every small $\epsilon > 0$ and every $M > 0$, the inequality

$$\sup_{|\alpha|\le M} |nQ_n(\alpha/\sqrt{n}) - nQ_n(0) - \alpha' n^{1/2} U_n - \alpha' H\alpha/2| < \epsilon \tag{3.2a}$$

holds with probability at least $(1 - \epsilon/2)$ for large n.

Define the quadratic form $B_n(\alpha) = \alpha' n^{1/2} U_n + \alpha' H\alpha/2$. Its minimiser is $\alpha_n = -H^{-1} n^{1/2} U_n$ which converges in distribution to $N(0, m^2 H^{-1} K H^{-1})$. The minimum value of the quadratic form is $-n^{1/2} U_n' H^{-1} n^{1/2} U_n'/2$. Further, from the U statistics central limit theorem, $n^{1/2} U_n$ is bounded in probability. So we can select an M such that

$$P\{|-H^{-1} n^{1/2} U_n| < M - 1\} \ge 1 - \epsilon/2. \tag{3.2b}$$

The rest of the argument is on the intersection of the two sets in (3.2a) and (3.2b), which has probability at least $1-\epsilon$.

Consider the convex function $A_n(\alpha) = nQ_n(\alpha/\sqrt{n}) - nQ_n(0)$. From (3.2a), its value at α_n is bounded above by

$$\epsilon - n^{1/2}U_n' H^{-1} n^{1/2} U_n'/2. \tag{3.2c}$$

Now consider the value of A_n on the sphere $\{\alpha : |\alpha - \alpha_n| = K\epsilon^{1/2}\}$ where K will be chosen. Again by using (3.2a), on the sphere, its value is *at least*

$$B_n(\alpha) - \epsilon. \tag{3.2d}$$

Comparing the two bounds in (3.2c) and (3.2d), and using the condition that α lies on the sphere, it can be shown that the bound in (3.2d) is always larger than the one in (3.2c) once we choose $K = 2[\lambda_{min}(H)]^{-1/2}$ where λ_{min} denotes the minimum eigen value.

On the other hand A_n has the minimiser $n^{1/2}\theta_n$. So using the fact that A_n is convex, it follows that its minimiser satisfies $|n^{1/2}\theta_n - \alpha_n| < K\epsilon^{1/2}$. Since this holds with probability at least $(1-\epsilon)$ where ϵ is arbitrary, this proves the first part. The second part now follows from Theorem 1.

Example 20. Under suitable conditions, the maximum likelihood estimator (mle) is consistent and asymptotic normal. See van der Vaart (1999) for sets of conditions under which this is true. If we are ready to assume that the loglikelihood function is concave in the parameter, then these claims follow from the above theorem.

Example 21. (Sample quantiles) From Example 13, it follows that if the distribution of X has a positive density f_X at the population pth quantile θ_0, then

$$H = Q''(\theta_0) = 2\, f(\theta_0) > 0.$$

Further,

$$g(x,\theta) = I(\theta \geq x) - I(x \leq \theta) - (2p-1) = 2I(\theta \geq x) - 2p.$$

Since g is bounded, Assumption (IV) is trivially satisfied. Thus all the conditions (I)–(V) are satisfied. Moreover

$$K = V[2I(\theta_0 \geq X)] = 4\,V[I(X \leq \theta_0)] = 4p(1-p).$$

Hence if $f_X(\theta_0) > 0$, then the sample pth quantile θ_n satisfies

$$n^{1/2}(\theta_n - \theta_0) \xrightarrow{\mathcal{D}} N(0,\ p(1-p)(f^2(\theta_0))^{-1}).$$

Example 22. If the assumptions of Theorem 3 are not satisfied, the limiting distribution of the M estimate need not be normal. Smirnov (1952) had studied the sample quantiles in such nonregular situations in complete details, identifying the class of distributions possible. Jureckova (1983) considered general M estimates in nonregular situations. See also Bose and Chatterjee (2001a).

Example 23. (U Quantiles) The arguments of Example 21 apply without change to U quantiles introduced in Example 19. Let $X_1, \ldots, X_n$ be i.i.d. random variables with distribution F, h be a function from $\mathcal{R}^m$ to $\mathcal{R}$ which is symmetric in its arguments. Let H_F denote the distribution function of $h(X_1, \ldots, X_m)$ and let $H_F^{-1}(p)$ be the pth quantile of H_F. Then $\theta_0 = H_F^{-1}(p)$, which is unique if H_F has a positive density at θ_0.

As in Example 21, if H_F is differentiable at θ_0 with a positive density $h_F(\theta_0)$, then Assumption (V) holds with $H = 2h_F(\theta_0)$. The gradient vector is given by

$$g(x, \theta) = 2I[\theta \geq h(X_1, \ldots X_m)] - 2p.$$

This is bounded and hence (IV) holds trivially.

Let

$$H_n(y) = \binom{n}{m}^{-1} \sum_{1 \leq i_1 < \ldots < i_m \leq n} I(h(X_{i_1}, \ldots, X_{i_m}) \leq y)$$

be the *empirical distribution*. The M_m estimate is then $H_n^{-1}(p)$, the pth quantile of $H_n(\cdot)$.

By application of Theorem 3,

$$n^{1/2}(H_n^{-1}(p) - H^{-1}(p)) \xrightarrow{\mathcal{D}} N(0,\ (p(1-p)(h_F(\theta_0))^{-1}).$$

Particular examples of this result are the following four estimates.

(i) *Univariate Hodges-Lehmann estimator* where

$$h(X_1, \ldots X_m) = m^{-1}(X_1 + \ldots, X_m),$$

(ii) *Dispersion estimator* of Bickel and Lehmann (1979) where

$$h(X_i, X_j) = |X_i - X_j|)$$

(iii) *Regression coefficient estimator* introduced by Theil (see Hollander and Wolfe (1973, pp. 205-206) where $(X_i,\ Y_i)$ are bivariate i.i.d. random variables and

$$h((X_i, Y_i),\ (X_j, Y_j)) \ = (Y_i - Y_j)/(X_i - X_j).$$

(iv) *A location estimate of Maritz (1977)* can also be treated in this way. Let β be any fixed number between 0 and 1. Let $L(\theta, x_1, x_2) = |\beta x_1 + (1-\beta)x_2 - \theta| \ + |\beta x_2 + (1-\beta)x_1 - \theta|$. The minimizer of $E\ [L(\theta, X_1, X_2) - L(0, X_1, X_2)]$ is a measure of location of X_i (Maritz (1977)) and its estimate is the median of $\beta X_i + (1-\beta)X_j,\ i \neq j$ ($\beta = 1/2$ yields the Hodges-Lehmann estimator of order 2). Conditions similar to above guarantee asymptotic normality for this estimator.

Example 24. The L_1 median was defined in Example 14. If the dimension $d = 1$, then the L_1 median is the usual median whose asymptotic normality was discussed in Example 21. So we assume that $d \geq 2$. The gradient vector is

$$g(\alpha, x) = \begin{cases} \dfrac{\alpha - x}{\mid \alpha - x \mid} & \text{if} \quad \alpha \neq x \\ 0 & \text{if} \quad \alpha = x \end{cases}$$

Thus g is a bounded function and Assumption (IV) is satisfied. Note that g is differentiable (except when $x = \theta$). Define

$$h(x, \theta) = \frac{1}{|\theta - x|}\left(I - \frac{(\theta - x)(\theta - x)}{|\theta - x|^2}\right), x \neq \theta.$$

Note that $E|X-\theta_0|^{-1}<\infty$ implies $E|h(X_1,\theta)|<\infty$.

Recall that $\nabla Q(\theta)=E[g(X_1,\theta)]$. By simple algebra, for $|x|\le|\theta|$,

$$|g(x,\theta)-g(x,0)|\le 2|\theta|/|x|.$$

Similarly, for $|x|>|\theta|$,

$$|g(x,\theta)-g(x,0)-h(x,0)\theta|\le 5\frac{|\theta|^2}{|x|^2}+\frac{|\theta|^3}{|x|^3}.$$

Using these two inequalities, and the inverse moment condition, it is easy to check that, the matrix H exists and can be evaluated as

$$H=E[h(X,\theta)].$$

Example 25. (Oja median) Recall the Oja median defined in Example 16. Let X denote the $d\times d$ random matrix whose ith column is $X_i=(X_{1i},\dots X_{di})'$ $1\le i\le d$. Let $X(i)$ be the $d\times d$ matrix obtained from X by deleting its ith row and replacing it by a row of 1's at the end. Finally let $M(\theta)$ be the $(d{+}1)\times(d{+}1)$ matrix obtained by augmenting the column vector $\theta=(\theta_1,\dots,\theta_d)'$ and a $(d{+}1)$ row vector of 1's respectively to the first column and last row of X. Note that $f(X_1,\dots X_d,\theta)$ equals $||M(\theta)||-||M(0)||$ where $||\cdot||$ denotes the absolute determinant. It is easily seen that

$$f(X_1,\dots X_d,\theta)=||M(\theta)||-||M(0)||=|\theta'Y-Z|-|Z| \tag{3.3}$$

where

$$Y=(Y_1,\dots Y_d)'$$

and

$$Y_i=(-1)^{i+1}|X(i)|,\quad Z=(-1)^d|X|.$$

Hence Q is well defined if $E\ |X_1|<\infty$. Further, the ith element of the gradient vector of f is given by

$$g_i=Y_i\cdot\text{sign}(\theta'Y-Z),\quad i=1,\dots,d$$

and is similar to the gradient in Examples 21 and 23. Note that $E|X_1|^2 < \infty$ implies $E|Y|^2 < \infty$ which in turn implies $E|g_i|^2 < \infty$ and thus Assumption (IV) is satisfied.

To obtain condition (V), first assume that F is continuous. Note that by arguments similar to those in Example 13,

$$\begin{aligned} Q(\theta) - Q(\theta_0) &= 2E[\theta' Y\ I(Z \leq \theta' Y) - \theta_0' Y\ I(Z \leq \theta_0' Y)] \\ &\quad +2E[Z\ I(Z \leq \theta' Y) - \ Z\ I(Z \leq \theta_0' Y)]. \end{aligned}$$

It easily follows that the ith element of the gradient vector of $Q(\theta)$ is given by

$$Q_i(\theta) = 2E[Y_i\ I(Z \leq \theta' Y)].$$

If F has a density, it follows that the derivative of $Q_i(\theta)$ with respect to θ_j is given by

$$Q_{ij}(\theta) = 2E[Y_i Y_j f_{Z|Y}(\theta' Y)]$$

where $f_{Z|Y}(\cdot)$ denotes the conditional density of Z given Y. Thus

$$H = ((Q_{ij}(\theta_0))).$$

Clearly then (V) will be satisfied if we assume that, the density of F exists and the H defined above exists and is positive definite. This condition is satisfied by many common densities.

The pth *order Oja median* for $1 < p < 2$ is defined by minimizing

$$Q(\theta) = E[\Delta^p(X_1, \ldots, X_d, \theta) - \Delta^p(X_1, \ldots, X_d, 0)].$$

The quantities g_i and H are now given by

$$g_i(\theta) = pY_i|\theta' Y - Z|^{p-1}\text{sign } (\theta' Y - Z),\ i = 1, \ldots, d,$$

$$H = ((h_{ij})) = p(p-1)((E[Y_i Y_j |\theta_0' Y - Z|^{p-2}])).$$

Now it is easy to formulate conditions for the asymptotic normality of the pth order Oja median.

3.4 Finer asymptotic properties. In this section we demonstrate that if some of the conditions assumed so far are strengthened, then the results on consistency and asymptotic normality can be sharpened considerably. Below, N is an appropriate neighbourhood of θ_0. $r > 1$ and $0 \le s < 1$ with further restrictions on them in the theorems.

We first state the assumptions to strengthen the strong consistency.

(VIa) $E[\exp(t|g(X_1, \ldots, X_m, \theta)|)] < \infty \ \forall\ \theta \in N$ and some $t = t(\theta) > 0$.
(VIb) $E|g(X_1, \ldots, X_m, \theta)|^r < \infty \ \forall\ \theta \in N$.

Theorem 4.

(a) Suppose (VIa) holds. Then for every $\delta > 0$, there exists an $\alpha > 0$ such that,

$$P(\sup_{k \ge n} |\theta_k - \theta_0| > \delta) = O(\exp(-\alpha n)).$$

(b) Suppose (VIb) holds with some $r > 1$. Then for every $\delta > 0$,

$$P(\sup_{k \ge n} |\theta_k - \theta_0| > \delta) = o(n^{1-r}) \text{ as } n \to \infty.$$

Remark 6.

(a) Part (a) of the theorem says that the rate of convergence is exponentially fast. This implies that $\theta_n \to \theta_0$ *completely*: for every $\delta > 0$,

$$\sum_{n=1}^{\infty} P\{|\theta_n - \theta_0| > \delta\} < \infty.$$

Note that if $r < 2$, then Assumption (VIb) is *weaker* than Assumption (IV) needed for the asymptotic normality. If $r > 2$, then Assumption (VIb) is stronger than Assumption (IV) but still implies complete convergence.

(b) The *last time* that the estimator is ϵ distance away from the parameter is of interest as ϵ approaches zero. See Bose and Chatterjee (2001b) and the references there for information on this problem.

To prove Theorem 4, we need two Lemmae, but first a definition.

Definition 3. Let A_0 and B be sets in $\mathcal{R}^d$. We say that B is a δ *triangulation* of A_0 if every $\alpha \in A_0$ is a finite linear combination of points $\beta_i \in B$ such that $|\beta_i - \alpha| < \delta$ for all i.

Lemma 2. Let $A \subset A_0$ be convex sets in $\mathcal{R}^d$ such that $|\alpha - \beta| > 2\delta$ whenever $\alpha \in A$ and $\beta \notin A_0$. Assume that B is a δ-triangulation of A_0. If Q is a function on A_0 satisfying $|Q(\alpha) - Q(\beta)| < L|\alpha - \beta|$ and h is a convex function on A_0 then

$$\sup_{\beta \in B} |Q(\beta) - h(\beta)| < \epsilon \text{ implies } \sup_{\alpha \in A} |Q(\alpha) - h(\alpha)| < 5\delta L + 3\epsilon.$$

Remark 7. The next lemma is on U statistics and supplements the law large numbers for U statistics mentioned in Section 2. Parts (ii) and (iii) will also be useful in the proof of Theorems 5 and 6.

Lemma 3. Let h be a real valued function on $\mathcal{R}^m$ which is symmetric in its arguments. Let $U_n(h)$ be the corresponding U statistic. Let $\mu = Eh(X_1, \ldots, X_m)$.

(i) If $E|h(X_1, \ldots, X_m)|^r < \infty$ for some $r > 1$, then for every $\epsilon > 0$,

$$P\left(\sup_{k \geq n} |U_k(h) - \mu| > \epsilon\right) = o(n^{1-r}).$$

(ii) If $\psi(s) = E\{s \exp[|h(X_1, \ldots, X_m)|]\} < \infty$ for some $0 < s \leq s_0$, then for $k = [n/m]$, and $0 < s \leq s_0 k$,

$$E[\exp(sU_n)] \leq [\psi(s/k)]^k.$$

(iii) Under the same assumption as (ii), for every $\epsilon > 0$, there exist constants C and $\delta < 1$, such that

$$P\{\sup_{k \geq n} |U_k - \mu| > \epsilon\} \leq C\delta^n.$$

Proof of Lemma 3. The proofs of (ii) and (iii) can be found in Serfling (1980, page 200–202). Here we give a sketch of the proof of (i).

If $m = 1$, $U_n(h)$ reduces to a sample mean and a proof is given in Petrov (1975, Chapter 9, Theorem 2.8).

If $m > 1$, use the decomposition given in section 2 on U statistics and write

$$U_n(h) - \theta = \frac{m}{n}\sum_{i=1}^{n} \tilde{h}_1(X_i) + R_n.$$

Since the result is already established for $m = 1$, it is now enough to prove that

$$P\left\{\sup_{k\geq n} |R_k| \geq \epsilon\right\} \leq= o(n^{1-r}). \tag{3.4}$$

Note that R_n is a U statistic. Every U statistic is a *reverse martingale* and from the well known reverse martingale inequality, it follows that

$$P\left(\sup_{k\geq n} |R_k| \geq \epsilon\right) \leq CP\{|R_n| \geq \epsilon\}. \tag{3.5}$$

Further, R_n is a *degenerate* U statistic, that is, it is a U statistic whose first projection is zero. Hence using Theorem 2.1.3 of Koroljuk and Borovskich (1993, page 72), for $1 < r < 2$ and Theorem 2.1.4 of Koroljuk and Borovskich (1993, page 73), for $r \geq 2$, it follows that

$$E|R_n| = O(n^{2(1-r)}) = o(n^{1-r}). \tag{3.6}$$

Using this in (3.5) verifies (3.4) and proves Lemma 3 (i) completely.

Proof of Theorem 4. We first prove part (b). Fix $\delta > 0$. Note that Q is convex and hence continuous. It is also Lipschitz (with Lipschitz constant L say) in a neighbourhood of 0. Hence there exists an $\epsilon > 0$ such that $Q(\alpha) > 2\epsilon$ for all $|\alpha| = \delta$.

Fix α. By Assumption (VIb) and Lemma 3 (i),

$$P(\sup_{k\geq n} |Q_k(\alpha) - Q_k(0) - Q(\alpha)| > \epsilon) = o(n^{1-r}) \tag{3.7}$$

Now choose ϵ' and δ' both positive such that $5\delta' L + 3\epsilon' < \epsilon$. Let $A = \{\alpha : |\alpha| \le \delta\}$ and $A_0 = \{\alpha : |\alpha| \le \delta + 2\delta'\}$. Let B be a *finite* δ' triangulation of A_0. From (3.7),

$$P(\sup_{k \ge n} \sup_{\alpha \in B} |Q_k(\alpha) - Q_k(0) - Q(\alpha)| > \epsilon) = o(n^{1-r}). \tag{3.8}$$

Since $Q_k(\cdot)$ is convex, using Lemma 2 (with $h = Q_k$) and (3.8),

$$P(\sup_{k \ge n} \sup_{|\alpha| \le \delta} |Q_k(\alpha) - Q_k(0) - Q(\alpha)| < 5\delta' L + 3\epsilon' < \epsilon) = 1 - o(n^{1-r}) \tag{3.9}$$

Suppose that the event in (3.9) occurs. Using the fact that $f_k(\alpha) = Q_k(\alpha) - Q_k(0)$ is convex, $f_k(0) = 0$, $f_k(\alpha) > \epsilon$ for all $|\alpha| = \delta$, we conclude that $f_k(\alpha)$ attains its minimum on the set $|\alpha| \le \delta$. This proves part (b) of the theorem.

To prove part (a), follow the argument given in the proof of part (b) but use Lemma 3 (iii) to obtain the required exponential rate. The rest of the proof remains unchanged. We omit the details.

Example 26. Whenever the gradient is uniformly bounded, Assumption (VIa) is trivially satisfied. In particular, this is the case for U quantiles and the L_1 median.

Example 27. Recall the Oja median defined in Example 16 and discussed further in Example 25. Note that the rth moment of the gradient g is finite if the rth moment of Y is finite which in turn is true if the rth moment of X_1 is finite.

We now proceed to strengthen the asymptotic normality theorem by imposing further assumptions. As before, let N be an appropriate neighbourhood of θ_0 while $r > 1$ and $0 \le s < 1$ are numbers. Suppose that as as $\theta \to \theta_0$.

(VII) $|\nabla Q(\theta) - \nabla^2 Q(\theta_0)(\theta - \theta_0)| = O(|\theta - \theta_0|^{(3+s)/2})$.
(VIII) $E|g(X_1, \ldots, X_m, \theta) - g(X_1, \ldots, X_m, \theta_0)|^2 = O(|\theta - \theta_0|^{(1+s)})$.
(IX) $E|g(X_1, \ldots, X_m, \theta)|^r = O(1)$.

Theorem 5. Suppose the above assumptions hold for some $0 \le s < 1$ and $r > (8 + d(1+s))/(1-s)$. Then almost surely as $n \to \infty$,

$$n^{1/2}(\theta_n - \theta_0) = -H^{-1}n^{1/2}U_n + O(n^{-(1+s)/4}(\log n)^{1/2}(\log\log n)^{(1+s)/4}).$$

Theorem 5 holds for $s = 1$, with the interpretation $r = \infty$ and g is bounded. This is of special interest and we state this separately in the next theorem.

Theorem 6. Assume that g is bounded and (VII) - (VIII) hold with $s = 1$. Then almost surely as $n \to \infty$,

$$n^{1/2}(\theta_n - \theta_0) = -H^{-1}n^{1/2}U_n + O(n^{-1/2}(\log n)^{1/2}(\log\log n)^{1/2}).$$

Remark 7. The almost sure result obtained in Theorems 5 and 6 are by no means exact. We shall discuss this issue in details in Remark 8 later.

To prove the theorems, we need two Lemmae. The first is a refinement of Lemma 2 on convex functions to the *gradient* of convex functions. A proof may be found in Niemiro (1992).

Lemma 4. Let $A \subset A_0$ be convex sets in $\mathcal{R}^d$ such that $|\alpha - \beta| > 2\delta$ whenever $\alpha \in A$ and $\beta \notin A_0$. Assume that B is a δ-triangulation of A_0. Let k be an $\mathcal{R}^d$ valued function on A_0, satisfying $|k(\alpha) - k(\beta)| < L|\alpha - \beta|$. Let g be a subgradient of some convex function on A_0. Then

$$\sup_{\beta \in B} |k(\beta) - g(\beta)| < \epsilon \text{ implies } \sup_{\alpha \in A} |k(\alpha) - g(\alpha)| < 4\delta L + 2\epsilon$$

The second Lemma is a result on probability of deviations for U statistics.

Lemma 5. Let $\{h_n\}$ be a sequence of (symmetric) kernels of order m and let $\{X_{ni}, 1 \le i \le n\}$ be i.i.d. real valued random variables for each n. Let

$$U_n(h_n) = \binom{n}{m}^{-1} \sum_{1 \le i_1 < \ldots < i_m \le n} U_n(h_n(X_{ni_1}, \ldots, X_{ni_m})).$$

Further suppose that for some $\delta > 0$, and some $v_n \leq n^\delta$,

$$E\ U_n(h_n(X_n, ..., X_{nm})) = 0,$$

$$E\ |h_n(X_{ni}, ..., X_{nm})|^2 \leq v_n^2 \quad \text{and}$$

$$E\ |h_n(X_{ni}, ..., X_{nm})|^r \leq b < \infty \quad \text{for some } r > 2.$$

Then for all large K,

$$P(n^{1/2}|U_n(h_n)| > Kv_n\ (\log n)^{1/2})\ \leq\ Dn^{1-r/2}v_n^{-r}\ (\log n)^{r/2}.$$

Proof of Lemma 5. Let $\widetilde{h_n}= h_n I(|h_n| \leq m_n)$, $h_{n1} = \widetilde{h}_n - E\widetilde{h}_n, h_{n2} = h_n - h_{n1}$ where $\{m_n\}$ will be chosen. Note that $\{h_{n1}\}$ and $\{h_{n2}\}$ are mean zero kernels and have the same properties as $\{h_n\}$. Further $U_n(h_n) = U_n(h_{n1}) + U_n(h_{n2})$. Let $a_n = K\ (\log n)^{1/2}/2$ and $\Psi_n(t) = E[\exp\{tU_n(h_{n1}(X_{n1}, \ldots, X_{n,m}))\}$. Note that $\Psi_n(t)$ is finite for each t since h_{n1} is bounded. Letting $k = [n/m]$, and using Lemma 3, part (ii),

$$\begin{aligned} A_1 &= P(n^{1/2}U_n(h_{n1}) \geq v_n a_n) = P(tn^{1/2}U_n(h_{n1})/v_n \geq t\ a_n) \\ &= \exp(-ta_n)\ [\Psi_n(n^{1/2}t/v_n k)]^k = \exp(-ta_n)\ [E\exp(n^{1/2}t/v_n kY)]^k, \quad \text{say.} \end{aligned}$$

Using the fact that $|Y| \leq m_n$, $EY = 0$, and $EY^2 \leq v_n^2$, we get

$$\begin{aligned} E\exp\left(\frac{n^{1/2}t}{kv_n}Y\right) &\leq 1 + E\sum_{j=2}^{\infty} Y^2 \left(\frac{n^{1/2}t}{kv_n}\right)^j m_n^{j-2}/j! \\ &\leq 1 + \frac{t^2 n}{2k^2 v_n^2}(EY^2)\sum_{j=0}^{\infty}\left(\frac{n^{1/2}t}{kv_n}m_n\right)^j /j! \leq 1 + \frac{t^2 n}{k^2} \end{aligned}$$

provided $t \leq n^{-1/2}kv_n/_2m_n$. With such a choice of t,

$$A_1 \leq \exp\left(-ta_n + \frac{t^2 n}{k}\right) \tag{3.10}$$

Let $t = K(\log n)^{1/2}/4(2m-1)$. Then for all large n, the exponent in (3.10) equals

$$-K^2(\log n)/8(2m-1) + nK^2(\log n)/16(2m-1)k \leq -K^2(\log n)/16(2m-1).$$

Thus we have shown that

$$P(|n^{1/2}U_n(h_{n1})| > Kv_n(\log n)^{1/2}/2) \leq n^{-K^2/16(2m-1)}. \tag{3.11}$$

To tackle $U_n(h_{n2})$, we proceed as follows.

$$\begin{aligned} P(|n^{1/2}U_n(h_{n2})| \geq a_n v_n/2) &\leq 4v_n^{-1}a_n^{-1}n^{1/2}E|h_{n2}(X_{n1},\ldots,X_{nm})| \\ &\leq 8v_n^{-1}a_n^{-1}n^{1/2}[E|h_n|^r]^{1/r}[P(|h_n| \geq m_n)]^{1-1/r} \\ &\leq 8v_n^{-1}a_n^{-1}n^{1/2}b^{1/r}(m_n^{-r})^{1-1/r}b^{1-1/r}. \end{aligned}$$

Choosing $m_n = n^{1/2}v_n \;/\; K(\log n)^{1/2}$,

$$P(|n^{1/2}U_n(h_{n2})| \geq a_n v_n/2) \leq 8bv_n^{-r}K^{r-2}n^{1-r/2}(\log n)^{(r-1)/2}. \tag{3.12}$$

Note that the choice of m_n, K and t are indeed compatible. The Lemma follows by using (3.11) and (3.12) and the given condition on v_n.

Proof of Theorem 5. Recall the notations S, s, Y_s introduced before the beginning of proof of Theorem 3. Define $G(\alpha) = DQ(\alpha)$, $G_n(\alpha) = \binom{n}{m}^{-1}\sum_{s\epsilon S} g(Y_s,\alpha)$ and $X_{ns} = g(Y_s, \frac{\alpha}{\sqrt{n}}) - g(Y_s, 0)$. Note that $E(X_{ns}) = G(\frac{\alpha}{\sqrt{n}})$, and $\binom{n}{m}^{-1}\sum_{s\epsilon S} X_{ns} = [G_n(\frac{\alpha}{\sqrt{n}}) - U_n]$. By (VIII), $E|X_{ns}|^2 = O((n^{-1/2}l_n)^{1+s})$ uniformly for $|\alpha| \leq Ml_n = M(\log\log n)^{1/2}$.

By applying Lemma 5 with $v_n^2 = C^2 n^{-(1+s)/2} l_n^{1+s}$,

$$\begin{aligned} \sup_{|\alpha|\leq Ml_n} P\Big(n^{1/2}|G_n(\frac{\alpha}{\sqrt{n}}) - U_n - G(\frac{\alpha}{\sqrt{n}})| > KCn^{-(1+s)/4}l_n^{(1+s)/2}(\log n)^{1/2}\Big) \\ \leq Dn^{1-r/2}Cn^{r(1+s)/4}l_n^{r(1+s)/2}(\log n)^{r/2} \\ = Dn^{1-r(1-s)/4}(\log n)^{r/2}(\log\log n)^{-r(1+s)/4}. \end{aligned}$$

This is the main probability inequality required to establish the Theorem. The rest of the proof is similar to that of Theorem 3. The refinements needed now are provided by the triangulation Lemma 6 and the law of iterated logarithm for U_n.

Condition (VII) implies that for each $M > 0$,

$$\sup_{|\alpha|\le Ml_n}\left|H\alpha - n^{1/2}G(\frac{\alpha}{\sqrt{n}})\right| = O(n^{-(1+s)/4}(\log\log n)^{(3+s)/4}).$$

and so in the left side of above probability inequality, we can replace $n^{1/2}G(\frac{\alpha}{\sqrt{n}})$ by $H\theta$.

Let

$$\epsilon_n = n^{-(1+s)/4} l_n^{(1+s)/2}(\log n)^{1/2}.$$

Consider a $\delta_n = n^{-(1+s)/4}(\log n)^{1/2}$ triangulation of the ball $B = \{\alpha : |\alpha| \le Ml_n + 1\}$. We can select such a triangulation consisting of $O(n^{d(1+s)/4})$ points. From the probability inequality above it follows that

$$|n^{1/2}G_n(\frac{\alpha}{\sqrt{n}}) - n^{1/2}U_n - H\alpha| \le KC\epsilon_n$$

holds simultaneously for all α belonging to the triangulation with probability $1 - O(n^{d(1+s)/4+1-r(1-s)/4}(\log n)^{r/2})$. Now use Lemma 6 to extend this inequality to all points α in the ball. Letting $K_1 = KC(2|H|+1)$, we obtain

$$P\{\sup_{|\alpha|\le Ml_n} n^{1/2}|G_n(\frac{\alpha}{\sqrt{n}}) - U_n - n^{-1/2}H\alpha| > K_1\epsilon_n\}$$
$$= O(n^{d(1+s)/4+1-r(1-s)/4}(\log n)^{r/2}.$$

Since $r > [8 + d(1+s)]/(1-s)$, the right side is summable and hence we can apply the Borel-Cantelli Lemma to conclude that almost surely, for large n,

$$\sup_{|\alpha|\le Ml_n} |n^{1/2}G_n(\frac{\alpha}{\sqrt{n}}) - n^{1/2}U_n - H\alpha| \le K_1\epsilon_n.$$

Using the *law of iterated logarithm* for U statistics which implies that $n^{1/2}U_n(\log\log n)^{-1/2}$ is bounded almost surely as $n \to \infty$, we can choose M so that $|n^{1/2}H^{-1}U_n| \le Ml_n - 1$ almost surely for large n. To conclude the proof, we consider the convex function $nQ_n(n^{-1/2}\alpha) - nQ_n(0)$ on the sphere $S = \{\alpha : |\alpha - H^{-1}n^{1/2}U_n| = K_2\epsilon_n\}$ where $K_2 = 2K_1[\inf_{|e|=1} e'He]^{-1}$. Clearly,

$e'n^{1/2}G_n(-H^{-1}n^{1/2}U_n + K\epsilon_n e) \geq e'HeK\epsilon_n - K_1\epsilon_n > 0$, and so the radial directional derivatives of the function are positive. This shows that

$$|n^{1/2}\theta_n + H^{-1}n^{1/2}U_n| \leq K\epsilon_n$$

with probability one for large n, proving Theorem 5.

Proof of Theorem 6. Let v_n and X_{ns} be as in the proof of Theorem 5. Let U_n be the U statistic with kernel $X_{ns} - EX_{ns}$ which is now bounded since g is bounded. By the arguments similar to those given in the proof of Lemma 5 for the kernel h_{n1},

$$P\{|n^{1/2}U_n| \geq v_n(\log n)^{1/2}\} \leq \exp\{-Kt(\log n)^{1/2} + t^2 n/k\},$$

provided $t \leq n^{-1/2}kv_n/2m_n$, where $k = [n/m]$ and m_n is bounded by C_0 say. Letting $t = K_0(\log n)^{1/2}$, it easily follows that the right side of the above inequality is bounded by $\exp(-Cn)$ for some c. The rest of the proof is same as the proof of Theorem 5.

Example 28. U quantiles were defined in Example 19. Chaudhury and Serfling (1988) proved a representation for them by using the approach of Bahadur (1966). Such a result now follows directly from Theorem 6. The gradient vector given in Example 23 is bounded. Suppose that

(VIII)$'$ H_F has a density h_F which is continuous around θ_0.

It may then be easily checked that

$$E|g(\theta, x) - g(\theta_0, x)|^2 \leq 4|H_F(\theta) - H_F(\theta_0)| = O(|\theta - \theta_0|).$$

Thus (VIII) holds with $s = 0$.

It is also easily checked (see Example 13) that

$$\nabla Q(\theta) \;=\; Eg(X, \theta) \;=\; 2H_F(\theta) - 2p.$$

Assume that

(VII)′ $H_F(\theta) - H_F(\theta_0) - (\theta - \theta_0)h_F(\theta_0) = O(|\theta - \theta_0|^{\frac{3}{2}})$ as $\theta \to \theta_0$.

Then $Q(\theta)$ is twice differentiable at $\theta = \theta_0$ with $H = \nabla^2 Q(\theta_0) = 2h_F(\theta_0)$.

Thus, under the assumptions (VII)′ and (VIII)′, Theorem 5 holds for U quantiles. The same arguments also show that the location measure of Maritz et.al. (1977) also satisfies Theorem 6 under conditions similar to above.

Example 29. (Oja median) Recall the notations of Examples 16, 25 and 27. The ith element of the gradient vector of f is given by $g_i = Y_i \cdot \text{sign}(\theta' Y - Z)$, $i = 1, \ldots, d$.

Condition (VIII) is satisfied if

$$E||Y||^2[I(\theta' Y \le Z \le \theta_0' Y) + I(\theta_0' Y \le Z \le \theta Y)] = O(|\theta - \theta_0|^{1+s}) \quad (3.13)$$

If F has a density then so does the conditional distribution of Z given Y. By conditioning on Y it is easy to see that (VIII) holds with $s = 0$ if this conditional density is bounded uniformly in $\theta' Y$ for θ in a neighbourhood of θ_0 and $E||Y||^2 < \infty$. For the case $d = 1$, this is exactly condition (VIII)′ in Example 28.

To obtain condition (VII), recall that if F has a density, derivative Q_{ij} of $Q_i(\theta)$ with respect to θ_j and the matrix H are given by

$$Q_{ij}(\theta) = 2E[Y_i Y_j f_{Z|Y}(\theta' Y)] \text{ and } H = ((Q_{ij}(\theta_0)))$$

where $f_{Z|Y}(\cdot)$ denotes the conditional density of Z given Y.

Hence (VII) will be satisfied if we assume that for each i, as $\theta \to \theta_0$,

$$E[|Y_i\{F_{Z|Y}(\theta' Y) - F_{Z|Y}(\theta_0' Y) - f_{Z|Y}(\theta_0' Y)(\theta - \theta_0)')Y\}|] = O(|\theta - \theta_0|^{(3+s)/2}) \quad (3.14)$$

This condition is satisfied by many common densities. The other required condition (IX) is satisfied by direct moment conditions on Y or X.

By a similar approach, it is easy to formulate conditions under which Theorem 6 holds for pth Oja median for $1 < p < 2$.

Example 30. (L_1 median, mth order Hodges - Lehmann estimate, geometric quantiles in dimension $d \geq 2$). Suppose $X, X_1, X_2, \ldots, X_n$ are i.i.d. d dimensional random variables.

(i) (L_1 median). Since results for the univariate median (and quantiles) are very well known (see for example Bahadur (1966) Kiefer (1967)), we confine our attention to the case $d \geq 2$.

Proposition 1. Suppose θ_0 is unique. If for some $0 \leq s \leq 1$,

$$E|X - \theta_0|^{-(3+s)/2} < \infty,$$

then according as $s < 1$ or $s = 1$, the representation of Theorem 5 or 6 holds for the L_1 median with $S_n = \sum_{i=1}^{d}(X_i - \theta_0)/|X_i - \theta_0|$ and H defined in Example 24 earlier.

To establish the proposition, we verify the appropriate conditions. Conditions (I) and (II) are trivially satisfied. Recall the gradient vector given in Example 24 which is bounded. Hence Assumptions (I)–(V) are trivially satisfied. Let F be the distribution of X_1.

To verify (VIII), without loss of generality assume that $\theta_0 = 0$. Noting that g is bounded by 1 and $|g(x, \theta) - g(x, 0)| \leq 2|\theta|/|x|$, we have

$$\begin{aligned} E|g(X,\theta) - g(X,0)|^2 &\leq 4|\theta|^2 \int_{|x|>|\theta|} |x|^{-2} dF(x) + \int_{|x|<|\theta|} dF(x) \\ &\leq 4|\theta|^{1+s} \int_{|x|>|\theta|} |x|^{-(1+s)} dF(x) \\ &\quad + |\theta|^{1+s} \int_{|x|<|\theta|} |x|^{-(1+s)} dF(x) \\ &\leq 4|\theta|^{1+s} E|X|^{-(1+s)}. \end{aligned}$$

The moment assumption assures that (VIII) is satisfied since $(1 + s) \leq (3 + s)/2$. Recall the function $h(\theta, x)$ and H defined in Example 24. Note that under our assumptions H is positive definite. By using arguments similar to those given in Example 24, it is easily seen that for $|x| \leq |\theta|$,

$$|g(x, \theta) - g(x, 0) - h(x, 0)\theta| \leq 4|\theta|/|x|.$$

Similarly, for $|x| > |\theta|$,

$$|g(x,\theta) - g(x,0) - h(x,0)\theta| \le 6\frac{|\theta|^2}{|x|^2}.$$

Using these two inequalities, and taking expectation,

$$|\nabla Q(\theta) - \nabla Q(0) - H\theta| \le I_1 + I_2$$

where

$$I_1 \le 4|\theta| \int_{|x|\le|\theta|} |x|^{-1} dF(x) \le 2|\theta|^{(3+s)/2} \int_{|x|\le|\theta|} |x|^{-(3+s)/2} dF(x).$$

and using the fact that $0 \le s \le 1$,

$$I_2 \le 6|\theta|^2 \int_{|x|>|\theta|} |x|^{-2} dF(x) \le 6|\theta|^{(3+s)/2} \int_{|x|\ge|\theta|} |x|^{-(3+s)/2} dF(x).$$

The moment condition assures that (VII) holds with $\nabla^2 Q(\theta_0) = H$. Thus we have verified all the the conditions needed.

Let us investigate the nature of the inverse moment condition that we assumed. If X has a density f bounded on every compact subset of $\mathcal{R}^d$ then $E|X-\theta|^{-2} < \infty$ if $d \ge 3$ and E $|X - \theta_0|^{-(1+s)} < \infty$ for any $0 \le s < 1$ if $d = 2$ and Theorem 5 is applicable. However, this boundedness or even the *existence* of a density as such is *not* needed if $d \ge 2$. This is in marked contrast with the situation for $d = 1$ where the existence of the density is required since it appears in the leading term of the representation. For most common distributions the representation holds with $s = 1$ from dimension $d \ge 3$ and with some $s < 1$ for dimension $d = 2$.

The weakest representation corresponds to $s = 0$ and gives a remainder $O(n^{-1/4}(\log n)^{1/2}(\log\log n)^{1/4})$ if $E|X-\theta|^{-3/2} < \infty$.

The strongest representation corresponds to $s = 1$ and gives a remainder $O(n^{-1/2}(\log n)^{1/2}(\log\log n)^{1/2})$ if $E\ |X - \theta|^{-2} < \infty$.

The moment condition forces F to necessarily assign zero mass at the median. Curiously, if F assigns zero mass to an entire neighbourhood of the median, then the moment condition is automatically satisfied.

Now assume that the L_1 median is zero and X is dominated in the neighbourhood of zero by a variable Y which has a radially symmetric density $f_Y(|x|)$. Transforming to polar coordinates, note that the moment condition is satisfied if the integral of $g(r) = r^{-(3+s)/2+d-1} f_Y(r)$ is finite. If $d = 2$ and f is bounded in a neighbourhood of zero then the integral is finite for all $s < 1$. If $f_Y(r) = O(r^{-\beta})$, $(\beta > 0)$, then the integral is finite if $s < 2d - 3 - 2\beta$. In particular, if f is bounded $(\beta = 0)$, then any $s < 1$ is feasible for $d = 2$ and $s = 1$ for $d = 3$.

(ii) (Hodges-Lehmann estimate) The above arguments also show that if the moment condition is changed to $E|m^{-1}(X_1 + \cdots + X_m) - \theta_0|^{-(3+s)/2} < \infty$, Proposition 1 holds for the Hodges - Lehmann estimator with

$$S_n = \sum_{1 \leq i_1 < i_2 < \ldots < i_m \leq n} g(m^{-1}(X_{i_1} + \cdots + X_{i_m}),\ \theta_0)).$$

(iii) (Geometric quantiles) For any u such that $|u| < 1$, the uth geometric quantile of Chaudhuri (1996) is defined by taking $f(\theta, x) = |x - \theta| - |x| - u'\theta$. Note that $u = 0$ corresponds to the L_1 median. The arguments given in the proof of Proposition 1 remain valid and the representation of Theorem 5 or 6 holds for these estimates. One can also define the Hodges - Lehmann version of these quantiles and the representations would still hold.

Remark 8. Obtaining the exact order is a delicate and hard problem.

The higher order asymptotic properties of the sample median was extensively studied with suitable conditions on the density by Bahadur (1966) and Keifer (1967) via the fluctuations of the sample distribution function which puts mass n^{-1} at the sample values.

This approach has been used by several authors in other similar situations. For example, a representation for U quantiles was proved by Chowdhury and Serfling (1988) by studying the fluctuations of the distribution function which puts equal mass at all the $\binom{n}{m}$ points $h(X_{i_1}, \ldots, X_{i_m})$, $1 \leq i_1 \leq i_2 \leq \ldots \leq i_m \leq n$. Chaudhuri (1992) proved a representation for the L_1 median and its Hodges-Lehmann version in higher dimensions by the same approach.

Results from the theory of *empirical processes* is a very valuable tool in the study of properties of estimators. For instance, Arcones (1996) derives some *exact* almost sure rates for U quantiles under certain "local variance conditions" by using empirical processes.

Generally speaking, the exact rate depends on the nature of the function f. See Arcones and Mason (1997) for some refined almost sure results in general M estimation problems. As an example, consider the L_1 median when $d = 2$. If the density of the observations exists in a neighbourhood of the median, is continuous at the median and $E\ g(X_1, \theta)$ has a second order expansion at the median, then the *exact* almost sure order of the remainder is $O(n^{-1/2}(\log n)^{1/2}(\log\log n))$.

References

Arcones, M. A. (1996). The Bahadur-Kiefer representation for U-quantiles. *Ann. Statist.*, **24**, 1400–1422.

Arcones, M. A. (1998). The Bahadur-Kiefer representation of two dimensional spatial medians. it Ann. Inst. Stat. Maths. **50**, 71–86.

Arcones, M. A., Chen, Z. and Gine, E. (1994). Estimators related to U-processes with applications to multivariate medians: asymptotic normality. *Ann. Statist.* **22**, 1460–1477.

Arcones, M. A. and Mason, D.M. (1997). A general approach to Bahadur-Kiefer representation for M-estimators. *Math. Meth. Statist.*, **6**, 267–292.

Bahadur, R.R. (1966). A note on quantiles in large samples. *Ann. Math. Statist.* **37**, 577 - 580.

Bickel, P.J. and Lehmann, E.L. (1979) Descriptive statistics for nonparametric models, IV. Spread. In *Contributions to Statistics* Ed. J. Jureckova. pp. 33-40. Academia, Prague.

Bose, Arup (1998) Bahadur representation of M_m estimates, *Ann. Statist.*, **26**, 771-777.

Bose, Arup and Chatterjee, S. (2001a). Generalised bootstrap in nonregular M estimation problems. *Statist. Prob. Letters*, **55**, 319-328.

Bose, Arup and Chatterjee, S. (2001b). Last passage times of minimum contrast estimators. *Jour. of Aus. Math. Soc.*, **71**, 1, 1-10.

Castaing, C. and Valadier, M. (1977) *Convex Analysis and measurable Multifunctions. Lecture Notes in Maths.*, 580. Springer, New York.

Chaudhuri, P. (1992) Multivariate location estimation using extension of R - estimates through U-statistics type approach, *Ann. Statist.*, **20**, 897 - 916.

Chaudhuri, P. (1996) On a geometric notion of quantiles for multivariate data, *J. Amer. Statist. Assoc.*, **91**, 862-872.

Chen, X. (1980). On limiting properties of U-statistics and von Mises statistics. *Sci. Sinica*, **23**, 1079–1091.

Choudhury, J. and Serfling, R.J. (1988) Generalized order statistics, Bahadur representations, and sequential nonparametric fixed width confidence intervals, *Jour. Statist. Plann. Inference*, **19**, 269 - 282.

de la Pena, V. H. and Gine, E. (1999). *Decoupling.* Springer-Verlag, New York.

Habermann, S. J. (1989) Concavity and estimation, *Ann. Statist.*, **17**, 1631-1661.

Hjort, N.L. and Pollard, D. (1993) Asymptotics for minimizers of convex processes, Statistical Research report, Univ. of Oslo.

Hoeffding, W. (1948). A class of statistics with asymptotically normal distribution. *Ann. Math. Statist.* 19, 293-325.

Hollander, M. and Wolfe, D.A. (1973) *Nonparametrical Statistical Methods*, John Wiley and Sons, New York.

Huber, P. J. (1964) Robust estimation of a location parameter, *Ann. Math. Statist.*, **35**, 73-101.

Jureckova, J. (1983) Asymptotics behavior of M-estimators of location in nonregular cases. *Statistics and Decisions* **1**, 323-340.

Koroljuk, V. S. and Borovskich, Yu. V. (1993). *Theory of U statistics*. Kluwer Academic Publishers.

Kiefer, J. (1967). On Bahadur's representation of sample quantiles. *Ann. Math. Statist.* **38**, 1323 - 1342.

Lee, A.J. (1990). *U - Statistics Theory and Practice.* Marcel Dekker, Inc., New York.

Liu, R.Y. (1990). On a notion of data depth based on random simplices. *Ann. Statist.* **18**, 405 - 414.

Maritz, J.S., Wu, M. and Staudte, R.G. (1977) A location estimator based on U statistic, *Ann. Statist.*, **5**, 779-786.

Niemiro, W. (1992) Asymptotics for M-estimators defined by convex minimization, *Ann. Statist.*, **20**, 1514-1533.

Oja, H. (1983) Descriptive statistics for multivariate distribution, *Statist. Prob. Letters*, **1**, 327 - 333.

Oja, H. (1984). Asymptotical properties of estimators based on U statistics. Preprint, Dept. of Appl Math. Univ. of Oulu, Finland.

Oja, H. and Niimimaa, A. (1985). Asymptotic properties of the generalized median in the case of multivariate normality. *J.R. Statist. Soc. B.* **47**, 2, 372 - 377.

Petrov V.V. (1975). *Sums of Independent Random Variables.* Springer, Berlin.

Rousseeuw, P.J. (1986). Multivariate estimation with high breakdown point. In *Mathematical Statistics and Applications* (W. Grossman, G. Pflng, I. Vinage and W. Wertz, eds.) 283 - 297. Riedel, Dordrecht.

Serfling, R.J. (1980). *Approximation Theorems of Mathematical Statistics. Wiley*, New York.

Small, Christopher, G. (1990). A survey of multidimensional medians. *Int. Stat. Rev.*, 58, 263-277.

Smirnov, N.V. (1952) Limit distributions for the terms of a variational series. *American Mathematical Society Translation Series (1)*, 11, 82-143.

Tukey, J.W. (1975). Mathematics and the filtering of data. In *Proc. International Congress of Mathematicians*, Vancouver, 1974, **2**, 523 - 531.

van der Vaart, A. W. and Wellner, J. A. (1996) *Weak convergence and empirical processes. With applications to statistics*, Springer Series in Statistics. Springer-Verlag, New York.

PARAMETRIC INFERENCE WITH GENERALIZED RANKED SET DATA

BARRY C. ARNOLD

ROBERT J. BEAVER

Department of Statistics, University of California, Riverside CA 92521-0138, USA

Assume that observations have a common distribution function $F_{\underline{\theta}}$, which belongs to a family of distributions indexed by $\underline{\theta} \in \Theta$. We are interested in making inferences about the unknown parameter vector $\underline{\theta} \in \Theta$ based upon generalized rank set data, i.e. J independent order statistics $X_{i_j:n_j}$, j = 1, 2, …,J with a common parent distribution $F_{\underline{\theta}}(\cdot)$. We will discuss (i) the problem of estimating $\underline{\theta}$ or deriving probability bounds for $\underline{\theta}$ in the Bayesian sense, (ii) testing composite hypotheses concerning $\underline{\theta}$, and (iii) testing goodness of fit to the model $F_{\underline{\theta}} : \underline{\theta} \in \Theta$.

1 Introduction

Ranked set sampling, proposed by McIntyre [1952] results when n samples, each consisting of n observations, are drawn and the n units of each sample are ranked, usually by visual inspection with respect to the magnitude of the characteristic being studied. The unit quantified as the smallest from the first sample is selected, the unit having the second smallest rank is selected from the second sample, and so on, with the unit with the largest rank selected from the nth sample. The resulting n observations are independent order statistics with a common parent distribution $F_{\underline{\theta}}(\cdot)$. The potential n^2 observations after ordering are given in the following array.

Sample	Smallest	Second smallest	…	Largest
1	$X_{1:1}$	$X_{2:1}$	…	$X_{n:1}$
2	$X_{1:2}$	$X_{2:2}$	…	$X_{n:2}$
⋮	⋮	⋮	⋱	⋮
n	$X_{n:1}$	$X_{n:2}$	…	$X_{n:n}$

The actual observations taken under rank set sampling are given in the next array.

Sample	Smallest	…	Largest
1	$X_{1:1}$		
2		$X_{2:2}$	
⋮			⋮
n			$X_{n:n}$

This process is repeated m times so that the total number of items drawn from the population is mn^2 and the total number of units upon which observations are taken is mn. Ranked set samples are generally useful when the sampling units can be easily drawn from the population, the exact measurement of the characteristic to be studied is costly monetarily, or in time or effort required to obtain the measurement, and when the units within a sample can be readily ordered by visual inspection or by other rough gauging methods not requiring an assessment of actual values. For this to be easily accomplished, the value of n is usually small, and in order to have a reasonable total sample size, m is usually large. McIntyre's procedure will be called balanced ranked set sampling (BRSS).

Generalized ranked set sampling (GRSS), introduced by Kim and Arnold [1999], relaxes the condition that the measurements consist of a smallest, a second smallest, …, and finally a largest order statistic, and that all order statistics are selected from independent samples of the same size n. In their scheme an arbitrary order statistic is selected from the first sample consisting of n_1 observations, a second order statistic is selected from the second sample of n_2 observations, and so on, until the last order statistic is selected from a sample of size n_J.

Set		Measurement	
1	…	$X_{i_1:n_1}$	…
2	…	$X_{i_2:n_2}$	…
∂	…	⋮	…
J	…	$X_{i_J:n_J}$	…

Hence for generalized ranked set sampling (GRSS) the data consists of J independent order statistics, $X_{i_1:n_1}, X_{i_2:n_2}, \ldots, X_{i_J:n_J}$. Typically, the sample sizes n_j's will be small, but J will be large.

We wish to make inferences about $F_{\underline{\theta}}$ based upon GRSS by considering the observations not taken as missing data. Several approaches can be employed in this context. Let $\underline{X}$ represent the GRSS data, and let $\underline{Z}$ represent the missing observations. We propose to use (i.) the E-M algorithm approach in estimation and in some generalized goodness of fit tests and (ii.) a Bayesian approach whereby the missing observations can be generated using the conditional distribution $\underline{Z} \mid \underline{X}, \underline{\theta}$, via the Gibbs sampler.

2 Estimation

For parametric inference we assume that $F_{\underline{\theta}}$ is absolutely continuous with respect to a convenient measure, and that *f*, the common parent density of the observations belongs to the parametric family of densities $\{f(x/\underline{\theta}) : \underline{\theta} \in \Theta\}$. We wish to estimate $\underline{\theta}$ based on our GRSS. The joint density corresponding to the order statistics $\underline{X}' = (X_{i_1:n_1}, X_{i_2:n_2}, \ldots, X_{i_J:n_J})$ is given by

$$f_{\underline{X}|\underline{\theta}}(\underline{x}|\underline{\theta}) = \prod_{j=1}^{J} \frac{n_j!}{(i_j-1)!(n_j-1)!} [F_{\underline{\theta}}(x_{i_j:n_j})]^{i_j-1} \times [1 - F_{\underline{\theta}}(x_{i_j:n_j})]^{n_j-1} f_{\underline{\theta}}(x_{i_j:n_j}). \tag{2.1}$$

Maximum likelihood estimation may be feasible if the specific form of (2.1) is tractable, or if an efficient optimization program is available. However, this will typically not be the case. An alternative is to consider the GRSS as a missing data problem. In this scenario we have one measurement $X_{i_1:n_1}$ from the first set of observations from which n_1 - 1 of the observations are missing. Similarly, $X_{i_2:n_2}$ has n_2 - 1 missing observations associated with it. With $N = \sum_{j=1}^{J} n_j$ we have J observations and (N - J) missing values. Let $\underline{X}$ represent the vector of measurements resulting from the GRSS, and let $\underline{Z}$ represent the vector of (N - J) missing observations. For $j = 1, 2, \ldots, J$, let us denote by $\underline{Z}_{(j)}$ the vector of missing observations from the set of n_j observations associated with $X_{i_j:n_j}$. Thus $\underline{Z}_j$ consists of all of the n_j order statistics except $X_{i_j:n_j}$, the one that was observed. Note that $\underline{Z}$ consists of all the coordinate random variables in

$\underline{Z}_{(1)}, \underline{Z}_{(2)}, \ldots, \underline{Z}_J$ concatenated into a single vector. Using this notation we can write the joint likelihood of the observed and unobserved data as:

$$f_{\underline{X},\underline{Z}|\underline{\theta}}(\underline{x}, \underline{z} \mid \underline{\theta}) = \prod_{j=1}^{J} f_{X_{i_j:n_j}|\underline{\theta}}(x_{i_j:n_j} \mid \underline{\theta}) \prod_{j=1}^{J} f_{\underline{Z}_j, X_{i_j:n_j}, \underline{\theta}}(\underline{z}_j \mid x_{i_j:n_j}, \underline{\theta}) \qquad (2.2)$$

2.1 The E-M Algorithm

The E-M algorithm is an interative computational procedure for obtaining maximum likelihood estimates when the observations can be viewed as incomplete data. Viewing GRSS in this light, we have "incomplete data" whereby we have J observed measurements, $\underline{X}$, and N - J unobserved measurements, $\underline{Z}$, with the complete data vector given by $(\underline{X}', \underline{Z}')'$. The E-M algorithm consists of two steps that are iterated until convergence to the MLE is obtained. Let the likelihood of the complete data be given by

$$f_{\underline{X},\underline{Z}|\underline{\theta}}(\underline{x}, \underline{z} \mid \underline{\theta}) \qquad (2.3)$$

and the conditional likelihood of $(\underline{Z}/\, \underline{x}, \underline{\theta})$ by

$$f_{\underline{Z}|\underline{X},\underline{\theta}}(\underline{z} \mid \underline{x}, \underline{\theta}) = \frac{f_{\underline{X},\underline{Z}|\underline{\theta}}(\underline{x}, \underline{z} \mid \underline{\theta})}{f_{\underline{X}|\underline{\theta}}(\underline{x} \mid \underline{\theta})}, \qquad (2.4)$$

where

$$f_{\underline{X}|\underline{\theta}}(\underline{x} \mid \underline{\theta}) = \int_{\underline{Z}} f_{\underline{X},\underline{Z}|\underline{\theta}}(\underline{x}, \underline{z} \mid \underline{\theta}) d\underline{z}. \qquad (2.5)$$

Let p denote the current iteration step. If $\underline{\theta}^{(0)}$ represents the initial or starting value of the unknown parameter vector, then the E-step consists of finding

$$\underline{Z}^{(1)} = E_{\underline{\theta}^{(0)}}(\underline{Z} \mid \underline{X} = \underline{x}). \qquad (2.6)$$

The value of $\underline{Z}^{(1)}$ is used in the maximization step whereby we find $\underline{\theta}^{(1)}$ satisfying

$$\underline{\theta}^{(1)} = \arg\max_{\underline{\theta}} f_{\underline{X},\underline{Z}^{(1)}|\underline{\theta}}(\underline{x}, \underline{z}^{(1)} \mid \underline{\theta}).$$

Therefore the E-M algorithm consists of iterating these two steps, denoted by

$$\underline{Z}^{(p+1)} = E_{\underline{\theta}^{(p)}}(\underline{Z} \mid \underline{X} = \underline{x}) \tag{2.7}$$

and

$$\underline{\theta}^{(p+1)} = \arg\max_{\underline{\theta}} f_{\underline{X},\underline{Z}^{(1)}|\underline{\theta}}(\underline{x}, \underline{z}^{(p+1)} \mid \underline{\theta}) \tag{2.8}$$

When the density in (2.1) belongs to a one parameter exponential family, (2.7) and (2.8) are simplified considerably. The E-step consists of estimating the missing values that would comprise the sufficient statistics, $t(\underline{x}, \underline{z})$ for a the parameter vector, $\underline{\theta}$ by solving the following equation for $\underline{z}^{(p+1)}$,

$$\underline{z}^{(p+1)} = E[\underline{Z} \mid \underline{x}, \underline{\theta}^{(p+1)}], \tag{2.9}$$

while the M-step would consist of solving the following conditional likelihood equation for $\underline{\theta}^{(p+1)}$,

$$E[t(\underline{X}, \underline{Z}) \mid \underline{\theta}^{(p+1)}] = t(\underline{x}, \underline{z}^{(p+1)}). \tag{2.10}$$

See Dempster et al.[1977] for further details. In the Section 3 we apply these techniques when the sampled population is exponential or normal.

2.2 The Gibbs Sampler

Consider the density given in (2.3). Suppose that a convenient conjugate prior family for $\underline{\theta}$, say $g(\underline{\theta} \mid \underline{\eta})$ is available. Then the posterior density of $\underline{\theta}$, $f(\underline{\theta} \mid \underline{x}, \underline{z}, \underline{\eta})$ will be nice. If $\underline{\theta}$ is known it is relatively easy to simulate the missing data since $f(\underline{z} \mid \underline{x}, \underline{\theta})$ will also be nice. This can be accomplished as follows. Consider the observation $X_{i_j:n_j} = x_{i_j:n_j}$, the i_j-th order statistic from sample j (j = 1, 2, ..., J). Since $f_{X|\underline{\theta}}(x \mid \underline{\theta})$ and $F_{X|\underline{\theta}}(x \mid \underline{\theta})$, the density and distribution function of the underlying random variable are known, so is $F_{\underline{\theta}}^{-1}(u)$. Hence to generate the $(n_j - 1)$ missing observations from sample j, generate $(i_j - 1)$ uniform $(0, u_{i_j:n_j})$ variables and $(n_j - i_j)$

uniform $(u_{i_j:n_j}, 1)$ variables, where $u_{i_j:n_j} = F_{\underline{\theta}}(x_{i_j:n_j} \mid \underline{\theta})$. These random uniform variates are transformed to the space of the x's using $F_{\underline{\theta}}^{-1}(u)$. This procedure continues for each sample, $j = 1, 2\ldots, J$ until all of the missing units have been simulated. An easier but less efficient approach to completing the samples is to simulate data from $f_{X|\underline{\theta}}(x \mid \underline{\theta})$ using $F_{\underline{\theta}}^{-1}(u)$ where the u's are uniform (0, 1) variables, keeping $(i_j - 1)$ of those that are less than $x_{i_j:n_j}$ and $(n_j - 1)$ of those greater than $x_{i_j:n_j}$. This approach has the potential to generate excess data whereby, say for sample j, more than $(i_j - 1)$ values less than $x_{i_j:n_j}$ may be generated before generating the requisite number of values greater than $x_{i_j:n_j}$, or visa versa, but it is slightly easier to program.

Our goal is to learn about the posterior density of $\underline{\theta}$ given $\underline{X} = \underline{x}$, i.e. given the observed data. We can use a conditional Gibbs sampler (given $\underline{X} = \underline{x}$) to simulate realizations from $f_{\underline{\theta}|\underline{X}}(\underline{\theta} \mid \underline{x})$ as follows. Begin with an initial value of $\underline{\theta}$, say $\underline{\theta}^{(0)}$. Use this to generate the missing data, say $\underline{z}^{(0)}$ using the procedure just described. Now treating $(\underline{x}, \underline{z}^{(0)})$ as a sample of size N from $f_{X|\underline{\theta}}(x \mid \underline{\theta})$, we can readily evaluate the posterior conditional density $f_{\underline{\theta}|\underline{X},\underline{Z}}(\underline{\theta} \mid \underline{x}, \underline{z}^{(0)})$ (recall that we assumed a conjugate prior was available). From this posterior density we can simulate a realization, say $\underline{\theta}^{(1)}$. This will then be used to generate a new set of missing data, say $\underline{z}^{(1)}$. The process is continued, say K times, obtaining in this fashion $\underline{\theta}^{(1)}, \underline{\theta}^{(2)}, \ldots, \underline{\theta}^{K}$. After discarding the first k of these, for burn in, the remaining ones can be viewed as a simulated sample of size K - k from $f_{\underline{\theta}|\underline{X}}(\underline{\theta} \mid \underline{x})$. Of course a good initial choice for $\underline{\theta}^{(0)}$ will accelerate the process and reduce the time required. Similarly, in the E-M approach, a good initial estimate of $\underline{\theta}$ will accelerate the estimation process.

In many situations we are dealing with location and scale families so that $\underline{\theta}' = (\mu, \sigma)$. To implement the missing data approaches for these families, we can use the crude initial estimates of μ and σ that were given by David [1981] based upon the observed order statistics but ignoring variance heterogeneity. The estimate of μ is given by

$$\hat{\mu} = \sum_{j=1}^{J} b_j X_{i_j:n_j} \qquad (2.11)$$

for

$$b_j = \left(\frac{i}{n_j} - \frac{\overline{\alpha}(\alpha_j - \overline{\alpha})}{\sum_{i=1}^{J} (\alpha_i - \overline{\alpha})^2} \right), \tag{2.12}$$

with

$$\alpha_j = E(X_{i_j:n_j}), \tag{2.13}$$

while the estimate of σ is

$$\hat{\sigma} = \sum_{j=1}^{J} c_j X_{i_j:n_j}, \tag{2.14}$$

with

$$c_j = \frac{(\alpha_j - \overline{\alpha})}{\sum_{i=1}^{J} (\alpha_i - \overline{\alpha})}. \tag{2.15}$$

3 Applications of the E-M Algorithm and the Gibbs Sampler

3.1 The Exponential Distribution.

Let us consider the two parameter exponential density with location μ and precision λ given by

$$f_{X|\mu,\lambda}(x|\mu,\lambda) = \lambda \exp(-\lambda(x-\mu)) \quad x > \mu, \lambda > 0. \tag{3.1}$$

3.1.3 The E-M Algorithm

We consider the joint likelihood of $\underline{Y}' = (\underline{X}', \underline{Z}')$ given by

$$L(\underline{x}, \underline{z}|\mu,\lambda) \propto \prod_{j=1}^{J} \{\lambda \exp(-\lambda(x_{i_j:n_j} - \mu)) \prod_{i \neq i_j}^{n_j} \lambda \exp(-\lambda(z_{i:n_j} - \mu))\} \tag{3.2}$$

and the log-likelihood given by

$$\ln L(\underline{x}, \underline{z} \mid \mu, \lambda) = N\lambda - \lambda\left(\sum_{j=1}^{J} (x_{i_j:n_j} + \sum_{i \neq i_j}^{n_j} z_{i:n_j}) - N\mu \right) + c. \tag{3.3}$$

(The crude estimators given by David [1981] can be used as the initial estimators of μ and λ.) The E-step consists of estimating the missing observations, given the observed vector $\underline{x}$ and the current estimate of $\underline{\theta}' = (\mu, \lambda)$. This is accomplished by solving

$$\underline{z}^{(p+1)} = E_{\underline{\theta}^{(p)}} (\underline{Z} \mid \underline{X} = \underline{x}) . \tag{3.4}$$

The M-step consists of maximizing (3.3), subject to the constraints $x_{i_j:n_j} > \mu$ and $z_{i:n_j} > \mu$. The well-known maximum likelihood estimators for μ and λ are

$$\mu^{(p+1)} = \min\{\underline{x}, \underline{z}^{(p+1)}\} \tag{3.5}$$

and

$$\lambda^{(p+1)} = \frac{N}{\sum_{j=1}^{J} (x_{i_j:n_j} + \sum_{i \neq i_j}^{n_j} z_{i:n_j}) - N\mu} . \tag{3.6}$$

3.1.2 The Gibbs Sampler

The prior and posterior distributions associated with an exponential distribution with unknown scale and intensity parameters, although tractable, are not as easy to work with, as are those associated with a Pareto distribution. Let us assume that the distribution of the random variable X is given by

$$f_{X|\mu,\lambda}(x \mid \mu, \lambda) = \lambda \exp(-\lambda(x - \mu)) , \qquad x > \mu, \tag{3.7}$$

where $\mu \in \Re$ and $\lambda > 0$. Then $Y = e^X$ has a Pareto distribution with shape or inequality parameter λ and precision parameter $\tau = e^{-\mu}$ ($\mu = -\log\tau$) with $y > \tau$. Hence the distribution of Y is given by

$$f_{Y|\lambda,\tau}(y\,|\,\lambda,\tau) = \tau\lambda(\tau y)^{-(\lambda+1)}, \qquad y > 1/\tau. \tag{3.8}$$

If λ were known, a natural conjugate family of priors for τ would be the gamma family, while if τ were known, a natural conjugate family of priors for λ would be the Pareto family. However, with both τ and λ unknown, we follow Arnold et al. [1998] and consider a conjugate prior for which $\tau|\lambda$ has a gamma distribution for each λ, and λ/τ has a Pareto distribution for each τ. This general class of such priors is of the form

$$\begin{aligned} f(\lambda,\tau) \propto \exp[b\log\tau + m_{12}\log\lambda\log\tau]\times \\ \exp[a_1\lambda + a_2\log\lambda + m_{11}\lambda\log\tau], \end{aligned} \qquad \tau c > 1 \tag{3.9}$$

(The first factor on the right hand side consists of hyperparameters whose values are unaffected by the data.) The conditional density of τ / λ is gamma with shape parameter $\gamma(\tau) = (1 + a_1 + m_{12}\log\tau)$ and intensity parameter $\lambda(\tau) = -(a_1 + m_{11}\log\tau)$. The conditional density of λ / τ is Pareto with shape (or inequality) parameter $\delta(\lambda) = -(1 + b + m_{11}\lambda + m_{12}\log\lambda)$ and precision parameter $v(\lambda) = c$, hence the condition $\tau c > 1$ in (3.9). The classical conjugate prior family introduced by Lwin [1972] corresponds to setting $b = m_{12} = 0$. The independent gamma and Pareto priors suggested by Arnold and Press [1989] correspond to the choice of $m_{11} = m_{12} = 0$. Such independent priors have hyperparameters that may be easier to assess.

Using initial values for λ and $\tau = e^{\mu}$, say λ_0 and τ_0, the GRSS is transformed first to have a Pareto distribution using $Y = e^X$, and then to have a uniform distribution using $u = 1-(\tau_0 y)^{-\lambda_0}$. The missing values are generated for each sample, as considered in Section 2.2. These uniform variates are then transformed to have a Pareto distribution using the inverse probability transformation $z = (1-u)^{\lambda_0}/\tau_0$. The next step is to use the complete sample, $(\underline{Y},\underline{Z})$, to update the prior distribution for (λ,τ). Table 1 gives the relationship between the prior and posterior values of these parameters. (See Arnold et al., 1998, p. 237.)

Table1. Prior and posterior values of the parameters in (3.9)

Parameter	Prior value	Posterior value
a_1	a_1^*	$a_1^* - \sum_j (x_{i_j:n_j} - \sum_{i \neq i_j} z_{i:n_j})$
a_2	a_2^*	$a_2^* + N$
b	b^*	b^*
m_{11}	m_{11}^*	m_{11}^* - N
m_{12}	m_{12}^*	m_{12}^*
c	c^*	$\min(x_{i_j:n_j}, z_{j:n_j}, c^*)$

Using the full joint distribution for λ and τ, or either of the Lwin or Arnold and Press specified prior distributions, we can now simulate values for λ and τ from the posterior distributions, and repeat the process of simulating the missing values, and using these values plus the original data to update the conditional distributions of λ and τ. This is continued for a large number of times. This whole process is continued, say K times, generating an empirical distribution for (λ, τ).

3.2 The normal distribution

3.2.1 The E-M Algorithm

Since the normal distribution is a member of the regular exponential family with jointly sufficient statistics

$$t_1(\underline{x}, \underline{z}) = \sum_{j=1}^{J} (x_{i_j:n_j} + \sum_{i \neq i_j}^{n_j} z_{i:n_j})$$

and

$$t_2(\underline{x}, \underline{z}) = \sum_{j=1}^{J} (x_{i_j:n_j}^2 + \sum_{i \neq i_j}^{n_j} z_{i:n_j}^2),$$

the p-th iteration of the E-step consists of solving the following equations for $\underline{z}$

$$\underline{z} = E_{\underline{\theta}^{(p)}}(\underline{Z} \mid \underline{X} = \underline{x}), \tag{3.10}$$

while the p-th iteration of the M-step consists of solving the following set of equations for $\underline{\theta}$

$$E\begin{bmatrix} t_1(\underline{X},\underline{Z}) \mid \underline{\theta}^{(p+1)} \\ t_2(\underline{X},\underline{Z}) \mid \underline{\theta}^{(p+1)} \end{bmatrix} = \underline{t}^{(p)}(\underline{x},\underline{z}), \tag{3.11}$$

can be rewritten as

$$\begin{aligned} J\mu^{(p+1)} + (N-J)\mu^{(p+1)} &= \sum_{j=1}^{J}(x_{i_j:n_j} + \sum_{i\neq i_j}^{n_j} z_{i:n_j}) \\ J\left((\sigma^{(p+1)})^2 + (\mu^{(p+1)})^2\right) + (N-J)\left((\sigma^{(p+1)})^2 + (\mu^{(p+1)})^2\right) &= \sum_{j=1}^{J}(x^2_{i_j:n_j} + \sum_{i\neq i_j}^{n_j} z^2_{i:n_j}) \end{aligned} \tag{3.12}$$

Therefore step $(p+1)$ has the solution

$$\begin{aligned} \mu^{(p+1)} &= \left(\sum_{j=1}^{J}(x_{i_j:n_j} + \sum_{i\neq i_j}^{n_j} z_{i:n_j})\right) / N \\ (\sigma^2)^{(p+1)} &= \left(\sum_{j=1}^{J}(x^2_{i_j:n_j} + \sum_{i\neq i_j}^{n_j} z^2_{i:n_j})\right) / N - (\mu^{(p+1)})^2 \end{aligned} \tag{3.13}$$

3.2.2 The Gibbs Sampler

We are interested in estimating the mean μ and the precision $\tau = 1/\sigma^2$ by using informative prior distributions in the face of incomplete data. If μ were known, a natural conjugate prior for τ would be the gamma family, while if τ were known, a natural conjugate prior for μ would be the normal family. Hence, we would like to have an appropriate prior distribution for μ and τ whereby the distribution of $\mu \mid \tau$ is normally distributed and the distribution of $\tau \mid \mu$ has a gamma distribution. The class of such gamma-normal distributions constituting an eight parameter exponential family is discussed in Castillo and Galambos [1987]. Arnold et al. [1998] use the following parameterization for the joint conditionally specified prior distribution for μ and τ.

$$\begin{aligned} f(\mu,\tau) \propto {} & \exp(a_1\mu + a_2\mu^2 + m_{12}\mu\log\tau + m_{22}\mu^2\log\tau)\times \\ & \exp(b_1\tau + b_2\log\tau + m_{11}\mu\tau + m_{21}\mu^2\tau \end{aligned} . \tag{3.14}$$

With this prior distribution the conditional density of μ given τ is normal with mean

$$E(\mu \mid \tau) = \frac{-(a_1 + m_{11}\tau + m_{12}\log\tau)}{2(a_2 + m_{21}\tau + m_{22}\log\tau)} \tag{3.15}$$

and precision

$$1/Var(\mu \mid \tau) = -2(a_2 + m_{21}\tau + m_{22}\log\tau). \tag{3.16}$$

The conditional density of τ given μ is gamma with shape parameter $\alpha(\mu)$ and intensity parameter $\lambda(\mu)$. The conditional mean and variance of τ given μ are

$$E(\tau \mid \mu) = \frac{1 + b_2 + m_{12}\mu + m_{22}\mu^2}{-(b_1 + m_{11}\mu + m_{21}\mu^2)}, \tag{3.17}$$

$$Var(\tau \mid \mu) = \frac{1 + b_2 + m_{12}\mu + m_{22}\mu^2}{(b_1 + m_{11}\mu + m_{21}\mu^2)^2}. \tag{3.18}$$

DeGroot [1970] has postulated a joint prior for μ and σ^2 in which the precision $\tau = 1/\sigma^2$ has a marginal gamma distribution with shape parameters α and intensity parameter λ, and μ/τ has a normal distribution with mean θ and precision $a\tau$, a scalar multiple of τ. This is equivalent to setting

$$a_1 = a_2 = m_{12} = m_{22} = 0 \tag{3.19}$$

in (3.14). The posterior conditional mean and variance of μ using DeGroot's formulation are

$$E(\mu \mid \tau, \underline{x}, \underline{z}) = \frac{(a\theta + \sum_{j=1}^{J}(x_{i_j:n_j} + \sum_{i \neq i_j}^{n_j} z_{i:n_j}))}{a + N}, \tag{3.20}$$

$$V(\mu \mid \tau, \underline{x}, \underline{z}) = \frac{1}{a + N}. \tag{3.21}$$

while the posterior conditional mean and variance of τ are

$E(\tau \mid \mu, \underline{x}, \underline{z}) =$

$$\frac{\alpha + 1/2 + N/2}{\lambda - a/2 + \left(\sum_{j=1}^{J} (x_{i_j:n_j} + \sum_{i \neq i_j}^{n_j} z_{i:n_j}) \right) /2 - \left(a\theta + \sum_{j=1}^{J} (x_{i_j:n_j} + \sum_{i \neq i_j}^{n_j} z_{i:n_j}) \right) \mu + \left(\frac{a+N}{2} \right) \mu^2} \tag{3.22}$$

and.

$Var(\tau \mid \mu, \underline{x}, \underline{z}) =$

$$\frac{\alpha + 1/2 + N/2}{\left(\lambda - a/2 + \left(\sum_{j=1}^{J} (x_{i_j:n_j} + \sum_{i \neq i_j}^{n_j} z_{i:n_j}) \right) /2 - \left(a\theta + \sum_{j=1}^{J} (x_{i_j:n_j} + \sum_{i \neq i_j}^{n_j} z_{i:n_j}) \right) \mu + \left(\frac{a+N}{2} \right) \mu^2 \right)^2} . \tag{3.23}$$

Press [1982] approached this problem using an independent normal $(\theta, 1/\delta)$ prior for μ and an independent gamma prior distribution for τ with shape parameter α and intensity parameter λ in order to more easily assess the values of the hyperparameters. This corresponds to initially setting

$$m_{11} = m_{12} = m_{21} = m_{22} = 0 \tag{3.24}$$

in (3.14). Although the prior distributions are independent, the posterior distributions for μ and τ are not. The posterior conditional mean and variance of μ are given by

$$E(\mu \mid \tau, \underline{x}, \underline{z}) = \frac{(\delta\theta + (\sum_{j=1}^{J} (x_{i_j:n_j} + \sum_{i \neq i_j}^{n_j} z_{i:n_j}))\tau}{\delta + N\tau} \tag{3.25}$$

and

$$V(\mu \mid \tau, \underline{x}, \underline{z}) = 1/(\delta + N\tau) . \tag{3.26}$$

The posterior distribution of τ *given* μ has a gamma distribution with shape parameter $(\alpha + N/2)$ and intensity parameter

$$\lambda + \left(\sum_{j=1}^{J} (x^2_{i_j:n_j} + \sum_{i \neq i_j}^{n_j} z^2_{i:n_j}) \right) / 2 - \left(\sum_{j=1}^{J} (x_{i_j:n_j} + \sum_{i \neq i_j}^{n_j} z_{i:n_j}) \right) \mu + N\mu^2 / 2 . \tag{3.27}$$

The unrestricted prior and posterior values of the parameters (3.14) are given in Table 2.

Table 2. The prior and posterior values of the parameters in (3.14)

Parameter	Prior Value	Posterior Value
a_1	a_1^*	a_1^*
a_2	a_2^*	a_2^*
b_1	b_1^*	$b_1^* - (\sum_{j=1}^{J} (x_{i_j:n_j} + \sum_{i \neq i_j}^{n_j} z_{i:n_j})$
b_2	b_2^*	$b_2^* + \frac{N}{2}$
m_{11}	m_{11}^*	$m_{11}^* + \sum_{j=1}^{J} (x^2_{i_j:n_j} + \sum_{i \neq i_j}^{n_j} z^2_{i:n_j})$
m_{12}	m_{12}^*	m_{12}^*
m_{21}	m_{21}^*	$m_{21}^* - \frac{N}{2}$
m_{22}	m_{22}^*	m_{22}^*

Notice that only b_1, b_2, m_{11} and m_{21} are updated by the data. To use the Gibbs sampler, initial estimates of μ and $\tau = 1/\sigma^2$ can be obtained using the crude estimates given by David (1981). With $u_{i_j:n_j} = \Phi(\tau(x_{i_j:n_j} - \mu))$, generate $(i_j - 1)$ uniform (0, $u_{i_j:n_j}$) random variables, and $(n_j - 1)$ uniform ($u_{i_j:n_j}$, 1) random variables, $j = 1, 2, \ldots, J$. The values of $\underline{Z}$ are obtained using the inverse transformation $z = \mu + \tau^{-1}\Phi^{-1}(u)$. The Gibbs sampler proceeds by using the updated values of the hyperparameters in Table 2 to simulate new values of μ and τ which are in turn used to transform the values of the GRSS to uniform order statistics, and then by simulating the missing data as before. The inverse probability transformation is used to find the new values of $\underline{Z}$. The complete sample is then used in updating the posterior distribution, from which new values of μ and τ are selected. This procedure is continued for a fixed, but large number of iterations. This whole iteration procedure is run a large number of times, say K, at which time the resulting empirical distributions for μ and τ (given $\underline{x}$) can be used to find estimates for μ and τ together with measures of precision for the estimates.

4 Testing of Hypotheses

In this section we shall examine the topic of testing hypotheses from several different viewpoints, each involving GRSS data. We will begin by considering the use of goodness of fit tests in situations when an hypothesis completely specifies a distribution, and in situations when the hypothesis only specifies a parametric family of distributions. Another approach is to use maximum likelihood using the E-M algorithm in conjunction with generalized likelihood ratios. Yet another is the Bayesian approach of calculating posterior odds if the hypothesis is not sharp, perhaps using diffuse priors.

4.1 Goodness of Fit Tests

The classic problem of goodness of fit involves determining whether a set of i.i.d. observations can be reasonably supposed to have common distribution function F_0, a completely specified distribution. It is often assumed, and is assumed here, that F_0 is continuous. Thus, via a straightforward transformation, we reduce the problem to one of testing goodness of fit to either a uniform or an exponential distribution, whichever is deemed convenient. We assume that our data will consist of independent order statistics with common parent distribution F constituting a GRSS (Kim and Arnold [1999]). In both cases we will wish to test $H: F = F_0$. It is natural to also consider the problem of testing a composite hypothesis $H: F \in \{F_{\underline{\theta}} : \underline{\theta} \in \Theta\}$ using ranked set data configurations. In such a situation the first step will be to use the data to estimate $\underline{\theta}$.

The data consist of J independent order statistics $X_{i_1:n_1}, X_{i_2:n_2}, \ldots, X_{i_J:n_J}$ from a common parent distribution, F. To test $H: F = F_0$, we consider

$$Y_{i_j:n_j} = F_0(X_{i_j:n_j}) \tag{4.1}$$

and ask whether these can be reasonably supposed to be uniform order statistics. A goodness of fit statistic in this case could be of the form

$$T = \sum_{j=1}^{J} \frac{\left(Y_{i_j:n_j} - i_j/(n_j+1)\right)^2}{i\ (n_j - i_j + 1)(n_j+1)^{-2}(n_j+2)^{-1}}. \tag{4.2}$$

Note that

$$T = \sum_{j=1}^{J} \frac{(Y_{i_j:n_j} - E(Y_{i_j:n_j}))^2}{Var(Y_{i_j:n_j})} \tag{4.3}$$

where the moments of $Y_{i_j} : n_j$ are computed under the hypothesis that H is true, i.e. that the $Y_{i_j:n_j}$'s are order statistics from a uniform(0,1) distribution. Large values of T will be cause for rejection of H. The null distribution of T would be expected to be approximately χ_J^2 if J is large, if $n_1, n_2, \ldots, n_J$ are large and if the ratios i_j/n_j are not too extreme. In practice however, the n_j's will be small. If J is large a χ_J^2 approximation may be adequate. If J is small then a more accurate evaluation of the null distribution of T will be needed. A balanced rank set sample (BRSS) is most commonly used. These consist of m independent replicates of a complete set of n independent order statistics $X_{1:n}, X_{2:n}, \ldots, X_{n:n}$ where n is small and m is generally not small. Simulation based upper 90, 95 and 99th percentiles of the statistic T for such balanced ranked set samples are provided by Arnold et al. [2001]

Of course, one could instead have transformed to get exponential order statistics instead of uniform ones using the transformation

$$Y'_{i_j:n_j} = -\log(1 - F_0(X_{i_j:n_j}). \tag{4.4}$$

The test statistic in this case, say $\tilde{T}$, defined as

$$\tilde{T} = \sum_{j=1}^{J} \frac{\left(Y'_{i_j:n_j} - E(Y'_{i_j:n_j})\right)^2}{Var(Y'_{i_j:n_j})}, \tag{4.5}$$

with the mean and variance of $Y'_{i_j:n_j}$ given by

$$E(Y'_{i_j:n_j}) = \sum_{k=0}^{i_j-1} \frac{1}{(n_j - k)} \tag{4.6}$$

and

$$Var(Y'_{i_j:n_j}) = \sum_{k=0}^{i_j-1} \frac{1}{(n_j - k)^2}\,. \tag{4.7}$$

Tabled values of the statistic $\tilde{T}$ can be found in Arnold et al. [2001]. For both T and $\tilde{T}$, the χ^2 approximation underestimates the simulated percentage points in all but a few cases studied and hence, tables of critical values for both of these statistics are required.

When $F_{\underline{\theta}}$ is not completely specified, estimates of the parameters must be found before testing for goodness of fit. The estimation techniques involving missing data, specifically the E-M algorithm approach presented in Section 3, can be used to advantage here. This has the effect of simplifying the problem in that we no longer have to work with the joint distribution of independent order statistics, but rather with the parent distribution directly. Arnold et al. [2001] used the Gibbs sampler approach to estimate both the observations that were not recorded as well as the values of unknown parameters used in the simulation process. However, the statistic used was Stephen's [1974] modified version of the Watson [1961] U^2 statistic based upon uniform order statistics. (See Agostino and Stephens [1986].) All $N = \sum_{j=1}^{J} n_j$ actual and simulated observations are transformed to uniform order statistics and the resulting reordered statistics are denoted by $Y_{(1)}, Y_{(2)}, \ldots, Y_{(N)}$. The U^2 statistic is

$$U^2 = \frac{1}{12N} + \sum_{i=1}^{N} \left(\frac{2i-1}{2N} - Y_{(i)} \right)^2 - N(\bar{Y} - 0.5)^2 \tag{4.8}$$

with the modified statistic given by

$$U^2_{MOD} = \{U^2 - \frac{0.1}{N} + \frac{0.1}{N^2}\}\{1 + \frac{0.8}{N}\}. \tag{4.9}$$

For N greater than 10, the critical values given by Stephens are: 90th percentile = 0.152, 95th percentile = 0.187, and 99th percentile = 0.267. Simulated values for these same percentiles when $N < 10$ differ only slightly from the values given here.

Power studies at the .05 level of significance involving BRSS for testing the null hypothesis that F was a standard normal distribution when the true distribution was either: Normal (0, 4), Normal (2, 1), Logistic (0, 1), and Logistic (0, 4) revealed that almost uniformly over the range of values of m and n the test based upon $\tilde{T}$ was more powerful than the T test, which was more powerful than the U^2_{MOD} test. This study also showed that all three tests had very little power in discriminating between a Normal (0,1) and a Logistic (0, 1) distribution.

4.2 An Alternative Goodness of Fit Test for GRSS.

Suppose that we have a GRSS from a distribution F and we wish to test whether our sample could have come from a specified distribution F_0. Let $U_{i_j:n_j} = F_0(X_{i_j:n_j})$, $j = 1, 2, \ldots, J$. Estimators of the unobserved order statistics from each sample are found in the following simplistic way. Suppose we had the third order statistic $U_{3:7}$ based on a sample of 7 observations drawn from a uniform (0, 1) distribution. Estimators of the remaining 6 order statistics are given as:

$$\begin{aligned} &\hat{U}_{1:7} = U_{3:7}/3,\ \hat{U}_{2:7} = 2U_{3:7}/3,\ \hat{U}_{3:7} = U_{3:7}, \\ &\hat{U}_{4:7} = U_{3:7} + (1 - U_{3:7})/5,\ \hat{U}_{5:7} = U_{3:7} + 2(1 - U_{3:7})/5, \\ &\hat{U}_{6:7} = U_{3:7} + 3\left(1 - U_{3:7}\right)/5,\ and\ \hat{U}_{7:7} = U_{3:7} + 4(1 - U_{3:7})/5. \end{aligned} \tag{4.10}$$

Notice that we are providing unbiased estimators of the missing order statistics based on the value of the one order statistic observed. Of course, this is done for each of the samples in the data set. The relationships in (4.10) can be summarized as:

$$\hat{U}_{r:n_j} = \begin{cases} \dfrac{r}{i_j} U_{i_j:n_j} & 1 \le r \le i_j \\ U_{i_j:n_j} + \left(\dfrac{r - i_j}{n_j - i_j + 1} \right)(1 - U_{i_j:n_j}) & i_j \le r+1 \le n_j \end{cases} . \tag{4.11}$$

These estimates are now used in place of the missing data, and the test statistic is that given in (4.2) with the obvious modifications to allow for all $N = \sum_{j=1}^{J} n_j$ observations to be included in the calculations. The statistic in (4.2) becomes

$$T = \sum_{j=1}^{J} \sum_{r=1}^{n_j} \frac{\left(\hat{U}_{r:n_j} - r/(n_j+1)\right)^2}{r(n_j - r + 1)(n_j+1)^{-2}(n_j+2)^{-1}} . \tag{4.12}$$

When the number of repetitions, *m*, is greater than one, T in (4.12) is also summed over *m*. Simulation based percentiles of the distribution of T in (4.12) are given in Table 3 for different values of *m* and *n*, based upon 100,000 runs for each combination. Notice that the given percentiles do not appear to be approximately distributed as a χ^2 variable with n^2m degrees of freedom as one might expect, and therefore separate tabled values are needed. More extensive tables are available in Arnold et al. [2001].

Table 3. Simulation based percentage points of the statistic *T* for different values of *m* and *n* when missing data is imputed using (4.11).

n	m	$T_{.90}$	$T_{.95}$	$T_{.99}$
1	1	1.7085	4.0132	12.8382
3	1	8.4715	12.7374	26.5639
5	1	17.9390	24.0911	42.1798
10	1	55.0919	66.5028	95.5670
1	3	6.7501	10.9304	24.2815
3	3	23.0126	29.7781	49.3253
5	3	46.5705	55.7791	80.1884
10	3	141.3330	157.8699	197.0167

n	m	$T_{.90}$	$T_{.95}$	$T_{.99}$
1	5	10.9071	16.2701	31.5959
3	5	35.4255	43.5205	65.4460
5	5	72.0565	83.1354	111.9622
10	5	222.3327	242.5671	288.5463
1	10	20.1603	26.9115	45.8466
3	10	64.2386	74.5578	100.4973
5	10	132.6031	147.2900	180.8142
10	10	414.6961	441.1265	497.7994
1	25	42.7897	52.0217	74.6665
3	25	143.3099	157.9101	191.0295
5	25	301.7058	322.0480	366.6617
10	25	972.5086	1010.5220	1091.2590
1	50	75.4745	87.2622	114.1974
3	50	266.9689	285.7319	327.7236
5	50	572.7384	599.0957	655.3024
10	50	1879.2019	1929.6000	2032.3724
1	100	136.9693	151.9760	184.2377
3	100	505.1316	529.6553	582.0591
5	100	1102.6717	1138.0480	1209.1400
10	100	3663.7339	3733.8757	3868.9519

Alternatively, the uniform order statistics can transformed to exponential order statistics and analyzed using the test based upon $\tilde{T}$ defined in (4.12) using the imputation technique given in (4.11). Table 4 gives the simulation based percentiles of the statistic $\tilde{T}$ for different combinations of m and n based upon 100,000 runs for each combination.

Again one might expect the empirical percentage points in Table 4 to approximate those of a χ^2 distribution with n^2m degrees of freedom. As in Table 3, the tabled values are substantially smaller than those for a χ^2 with the appropriate degrees of freedom.

Table 4. Simulation based percentage points of the statistic $\tilde{T}$ for different values of m and n when missing data is imputed using (4.11).

n	m	$\tilde{T}_{.90}$	$\tilde{T}_{.95}$	$\tilde{T}_{.99}$
1	1	1.7085	4.0132	12.8382
3	1	8.4715	12.7374	26.5639
5	1	17.9390	24.0911	42.1798
10	1	55.0919	66.5028	95.5670
1	3	6.7501	10.9304	24.2815
3	3	23.0126	29.7781	49.3253
5	3	46.5705	55.7791	80.1884
10	3	141.3330	157.8699	197.0167
1	5	10.9071	16.2701	31.5959
3	5	35.4255	43.5205	65.4460
5	5	72.0565	83.1354	111.9622
10	5	222.3327	242.5671	288.5463
1	10	20.1603	26.9115	45.8466
3	10	64.2386	74.5578	100.4973
5	10	132.6031	147.2900	180.8142
10	10	414.6961	441.1265	497.7994
1	25	42.7897	52.0217	74.6665
3	25	143.3099	157.9101	191.0295
5	25	301.7058	322.0480	366.6617
10	25	972.5086	1010.5220	1091.2590
1	50	75.4745	87.2622	114.1974
3	50	266.9689	285.7319	327.7236
5	50	572.7384	599.0957	655.3024
10	50	1879.2019	1929.6000	2032.3724
1	100	136.9693	151.9760	184.2377
3	100	505.1316	529.6553	582.0591
5	100	1102.6717	1138.0480	1209.1400
10	100	3663.7339	3733.8757	3868.9519

Earlier power studies have shown that when no imputation of the missing values in the order statistics is used, but rather only the observed BRSS values, the statistic $\tilde{T}$ is more powerful than T when testing H: N(0,1) versus the alternatives N(0, 4), N(2, 1), and Logistic (0, 4), while both $\tilde{T}$ and T have very little power in distinguishing a standard logistic from a N(0, 1). Simulation studies to determine how these two statistics, (using both observed and imputed values for the order statistics not measured) perform against these same alternatives are underway.

4.3 Likelihood Ratio Tests

Consider testing the hypothesis $H: \underline{\theta} \in \Theta_0$ versus the alternative $A: \underline{\theta} \in \Theta_1$ when the data consists of GRSS data. A straightforward testing procedure in this case would be to use the general likelihood ratio procedure in which the E-M algorithm is used to estimate both the missing observations and the values of the parameters under test. In keeping with earlier sections, we will restrict attention to the families of distributions studied earlier in this report.

4.3.1 Normal Distribution.

We consider the case in which the GRSS resulted from sampling a normal population, and we wish to test a hypothesis concerning $\underline{\theta}' = (\mu, \sigma^2)$. Suppose that we wish to test $H: (\mu, \sigma^2) = (\mu_0, \sigma^2)$, that is, that H: $\mu = \mu_0$ with σ^2 unspecified, against the alternative $A: \mu \neq \mu_0$. Using the initial estimate for σ in (2.14) and (2.15) we can invoke the E-M algorithm to generate pseudo data in place of the missing measurements. Hence we can begin by transforming the observed GRSS, $X_{i_1:n_1}, X_{i_2:n_2}, \ldots, X_{i_J:n_J}$, to $U_{i_1:n_1}, U_{i_2:n_2}, \ldots, U_{i_J:n_J}$ using $u_{i_j:n} = \Phi(\frac{x_{i_j:n} - \mu_0}{\hat{\sigma}})$, $j = 1, \ldots, J$, where $\Phi(\cdot)$ denotes the standard normal distribution function. Next the missing observations are estimated by generating $(i_j - 1)$ uniform $(0, u_{i_j:n_j})$ and $(n_j - 1)$ uniform $(u_{i_j:n_j}, 1)$ variates $j = 1. \ldots, J$, which are transformed to normal variates. This procedure continues until the estimate of σ converges. Hence, under H the likelihood becomes

$$L(\underline{x}\mid \mu_0,\hat{\sigma}^2) = \prod_{j=1}^{J}\frac{n_j!}{(i_j-1)!(n_j-1)!}[\Phi(\frac{x_{i_j:n_j}-\mu_0}{\hat{\sigma}}\mid \mu_0,\hat{\sigma}^2)]^{i_j-1}\times \\ [1-\Phi(\frac{x_{i_j:n_j}-\mu_0}{\hat{\sigma}}\mid \mu_0,\hat{\sigma}^2)]^{n_j-1}\phi(\frac{x_{i_j:n_j}-\mu_0}{\hat{\sigma}}\mid \mu_0,\hat{\sigma}^2) \quad . \tag{4.13}$$

To evaluate the likelihood under A, initial estimates of both μ and σ can be found using (2.11) through (2.15) which in turn can be used to begin the E-M algorithm whereby the missing data are replaced by pseudo data and μ and σ^2 are estimated using this augmented data set. This cycling continues until the estimates of μ and σ^2 converge to $\tilde{\mu}$ and $\tilde{\sigma}^2$, at which point the likelihood is evaluated as

$$L(\underline{x}\mid \tilde{\mu},\tilde{\sigma}^2) = \prod_{j=1}^{J}\frac{n_j!}{(i_j-1)!(n_j-1)!}[\Phi(\frac{x_{i_j:n_j}-\tilde{\mu}}{\tilde{\sigma}}\mid \tilde{\mu},\tilde{\sigma}^2)]^{i_j-1}\times \\ [1-\Phi(\frac{x_{i_j:n_j}-\tilde{\mu}}{\tilde{\sigma}}\mid \tilde{\mu},\tilde{\sigma}^2)]^{n_j-1}\phi(\frac{x_{i_j:n_j}-\tilde{\mu}}{\tilde{\sigma}}\mid \tilde{\mu},\tilde{\sigma}^2). \tag{4.14}$$

With $R = L(\underline{x}\mid \mu_0,\hat{\sigma}^2)/L(\underline{x}\mid \tilde{\mu},\tilde{\sigma}^2)$, the likelihood ratio statistic is given by $-2\ln R$ which has an approximate χ^2-distribution with one degree of freedom. Another case of interest may be to test H: $\sigma^2 = \sigma_0^2$ with μ not specified. A procedure analogous to that outlined above would be implemented by estimating μ rather than σ^2 under H.

4.3.2 Exponential Distribution.

In the same fashion, a likelihood ratio test for location and/or scale parameters in an exponential distribution could be implemented using the E-M algorithm. For example, in testing H: $\lambda = \lambda_0$ with μ unspecified, an initial crude estimate of μ could be taken to be $\hat{\mu} = \min_{i_j}(X_{i_1:n_1}, X_{i_2:n_2}, \ldots, X_{i_J:n_J})$. With λ_0 and $\hat{\mu}$, the E-M algorithm can be implemented to generate the pseudo data used to re-estimate μ, cycling until the estimate of μ converges. In this case

$$L(\underline{x}\mid \hat{\mu},\lambda_0) = \prod_{j=1}^{J}\frac{n_j!}{(i_j-1)!(n_j-1)!}[1-\exp\{-\lambda_0(x_{i_j:n_j}-\hat{\mu})\}]^{i_j-1}\times \\ [\exp\{-\lambda_0(x_{i_j:n_j}-\hat{\mu})\}]^{n_j-1}\lambda_0\exp\{-\lambda_0(x_{i_j:n_j}-\hat{\mu})\}. \tag{4.15}$$

Using $\tilde{\lambda}$ and $\tilde{\mu}$, the converged values of λ and μ under A, in place of λ_0 and $\hat{\mu}$ in (4.14), the likelihood ratio statistic, $-2\ln R$ has an approximate χ^2-distribution with one degree of freedom.

4.3.3 Bayesian approach.

When the hypotheses tested are not sharp, that is when the hypotheses define subsets of the parameter space Θ that are not of measure zero, calculation of an odds ratio, for both the prior and posterior distributions produces a measure that can be used to accept or reject the hypotheses under test. When working with the Pareto distribution via a transformation from an exponential distribution, the prior distribution of λ and τ is given by (3.9), while the posterior distribution is given by (3.9) with the hyperparameters replaced by those in Table 1. The prior odds ratio in this case would be calculated as the ratio π_1/π_2 where

$$\pi_1 = \int_{\Theta_H} f(\lambda,\tau)\,d\lambda d\tau \text{ and } \pi_2 = \int_{\Theta_A} f(\lambda,\tau)\,d\lambda d\tau. \tag{4.16}$$

The posterior odds would be found as the ratio $\pi_1(\underline{x})/\pi_2(\underline{x})$, where

$$\pi_1(\underline{x}) = \int_{\Theta_H} f(\lambda,\tau \mid \underline{x})\,d\lambda d\tau \text{ and } \pi_2(\underline{x}) = \int_{\Theta_A} f(\lambda,\tau \mid \underline{x})\,d\lambda d\tau. \tag{4.17}$$

If the data have come from a normal distribution, then the prior distribution for μ and σ^2 is given by (3.14) while the posterior distribution is given by (3.14) using the updated parameters in Table 2. The prior and posterior odds are approximated by Gibbs sampler based calculations for the corresponding joint density for (λ, τ).

Although the prior odds measure the analyst's beliefs with respect to the distribution of the parameters under investigation, the posterior odds have been updated with sample information and should be a more credible measure of the strength of the hypotheses under test. Obviously, if the odds ratio is larger than one, there is support for the null hypothesis, while values smaller than one provide support for the alternative hypothesis. In examining the joint prior and posterior distributions under both hypotheses, it is clear that (4.14) and (4.15) would need to be evaluated numerically.

5 Summary

The analysis of GRSS data classically uses the distribution of independent order statistics. This distribution is not easy to work with even in classic cases when the data has been sampled from a normal distribution. The suggested analysis based upon the use of the E-M algorithm or the Gibbs sampler to augment the data and estimate underlying parameters reduces the problems to ones that are easily handled using standard techniques.

References

1. Arnold, B. C., Beaver, R. J., Castillo, E., and Sarabia, J. M. (2001). Goodness of fit tests based on record and generalized ranked set data. Technical Report 268, Department of Statistics, University of California, Riverside, CA.
2. Arnold, B. C., Castillo, E., and Sarabia, J. S. (1998). Bayesian analysis for classical distributions using conditionally specified priors. *Sankhya B* 60-2, 228-245.
3. Arnold, B. C. and Press, S. J. (1989). Bayesian estimation and prediction for Pareto data. *Journal of the American Statistical Association,* 84, 1079-1084.
4. Castillo, E. and Galambos, J. (1987. Bivariate distributions with normal conditionals. In **Proceedings of the International Symposium on Simulation, Modeling and Development**. Acta Press, Anaheim, 59-62
5. D'Agostino, R. B. and Stephens, M. A. (1986). **Goodness-of-Fit Techniques.** Marcel Dekker: New York.
6. David, H. A. (1981). **Order Statistics, 2nd Ed.** John Wiley and Sons: New York. 131-132.
7. DeGroot, M. H. (1970). **Optimal Statistical Decisions**, McGraw-Hill, New York.
8. Dempster, A. P., Laird, N. M., and Rubin, D. B. (1977). Maximum Likelihood from incomplete data via the EM algorithm. *Journal of the Royal Statistical Society,* Series B, 39 (1), 1-38.
9. Kim, Y.-H. and Arnold, B. C. (1999). Parameter estimation under generalized ranked set sampling. *Statistics and Probability Letters,* 42, 353-360.

10. Lwin, T. (1972). Estimating the tail of the Paretian law. *Skand. Aktuarietidskrift,* 55, 170-178.
11. McIntyre, G. A. (1952). A method of unbiased selective sampling, using ranked set. *Australian Journal of Agricultural Research,* 3, 385-390.
12. Press, S. J. (1982). **Applied Multivariate Analysis: Using Bayesian and Frequentist Methods of Inference**. Kreiger, Melbourne, FL.
13. Stephens, M. A. (1974). Use of the Kolmogorov-Smirnov, Cramer-von Mises and related statistics without extensive tables. *Journal of the Royal Statistical Society,* B32, 115-122.
14. Watson, G. S. (1961). Goodness-of-fit tests on a circle. *Biometrika,* 48, 109-114.

FISHER INFORMATION IN THE FARLIE-GUMBEL-MORGENSTERN TYPE BIVARIATE EXPONENTIAL DISTRIBUTION

H. N. Nagaraja * and Z. A. Abo-Eleneen

Ohio State University, USA and Zagazig University, Egypt

Abstract

We obtain expressions for the elements of the Fisher information matrix (FIM) for the three parameters of the Gumbel Type II bivariate exponential (G_2BVE) distribution. This distribution belongs to the Farlie-Gumbel-Morgenstern family and has exponential marginals. We evaluate the FIM for various values of the dependence parameter and discuss implications to finite-sample and asymptotic inference from the G_2BVE parent. We also conduct a similar study for the Marshall-Olkin bivariate exponential distribution and compare the results.

Key Words: *Gumbel Type II bivariate exponential distribution; Marshall-Olkin bivariate exponential distribution; Cramér-Rao bound; maximum likelihood estimator; Asymptotic relative efficiency.*

1 Introduction

Suppose X is a continuous random variable with cumulative distribution function (cdf) $F_1(x;\theta)$ and probability density function (pdf) $f_1(x;\theta)$, where θ is a real or a

*Corresponding author; hnn@stat.ohio-state.edu; Department of Statistics, Ohio State University, Columbus OH 43210-1247, USA.

vector valued parameter. The Fisher information (FI) about the real parameter θ contained in X is defined by $I(X;\theta) = E\left(\frac{\partial \log f_1(X;\theta)}{\partial \theta}\right)^2 = -E\left(\frac{\partial^2 \log f_1(X;\theta)}{\partial \theta^2}\right)$, under certain regularity conditions (see, for example, [13], p. 329). When θ is a vector $\boldsymbol{\theta} = (\theta_1, \cdots, \theta_k)$ the Fisher Information Matix (FIM), $\mathbf{I}(X; \theta_1, \cdots, \theta_k)$, is a $k \times k$ matrix whose (i,j)th element is

$$I_{ij} = E\left(\frac{\partial \log f_1(X;\boldsymbol{\theta})}{\partial \theta_i} \frac{\partial \log f_1(X;\boldsymbol{\theta})}{\partial \theta_j}\right) = -E\left(\frac{\partial^2 \log f_1(X;\boldsymbol{\theta})}{\partial \theta_i \partial \theta_j}\right). \tag{1.1}$$

Now suppose (X,Y) is absolutely continuous with joint cdf $F(x,y;\theta)$ and pdf $f(x,y;\theta)$. The FI in (X,Y) is similarly defined. The FI plays an important role in statistical inference through the information (Cramér-Rao) inequality and its association with the asymptotic properties of the maximum likelihood estimators (MLE). For a compact introduction to FI and some historical notes, see [10] (Sec. 2.5, 2.6, 2.8, 6.3 and 6.5).

Beginning with the work of Gumbel in the 1950's and 60's, several bivariate versions of the univariate exponential distribution have appeared. For some of these, the marginal distributions are not even exponential! While a univariate exponential distribution is absolutely continuous and has the lack of memory property, the bivariate extensions with dependent marginals cannot satisfy these requirements and at the same time have exponential marginals ([6], Chapter 5). Quite a few of the bivariate exponential (BVE) distributions are motivated by either the operational models within a reliability set up, or the generalizations of the univariate exponential distribution based on characterizations. For a classification of BVE distributions based on such themes, see the review paper by Barnett [3]. Since his review, numerous other new BVE's have appeared and the most recent comprehensive overview is provided in Chapter 47 of [9].

We investigate the behavior of the FIM of the Gumbel Type II bivariate exponential (G_2BVE) distribution (proposed in [7]). It is a special member of the Farlie-Gumbel-Morgenstern (FGM) family of absolutely continuous bivariate distributions (see [9], p. 51-52, 353). For the FGM family, the cdf is given by

$$F(x,y) = F_1(x)F_2(y)[1 + \alpha(1 - F_1(x))(1 - F_2(y))], \tag{1.2}$$

and the associated pdf is given by

$$f(x,y) = f_1(x)f_2(y)[1 + \alpha(1 - 2\ F_1(x))(1 - 2\ F_2(y))] \tag{1.3}$$

where $-1 < \alpha < 1$, F_2 is the marginal cdf of Y, and f_2 is the pdf of F_2. The parameter α serves as the dependence parameter, and X and Y are independent when it is 0. The marginal distributions F_1 and F_2 can be arbitrary with additional associated parameters. For the FGM family with normal, exponential, and logistic marginals, the correlation coefficient is a scalar multiple of α. Hutchinson and Lai ([8], Sec. 5.2) provide an excellent introduction to the FGM family and discuss its properties and applications to a variety of situations. See also [9], Chapter 44 and 47. Recently Abo-Eleneen and Nagaraja [1] have investigated the FI content about α in a collection of X-order statistics and their Y concomitants from the FGM distribution. Smith and Moffatt [14] have investigated FI about α in FGM type bivariate logistic models with some special sampling schemes.

For the G_2BVE the joint cdf takes the form

$$F(x, y; \boldsymbol{\theta}) = \left\{1 - e^{-(x/\theta_1)}\right\}\left\{1 - e^{-(y/\theta_2)}\right\}\left\{1 + \theta_3\, e^{-[(x/\theta_1)+(y/\theta_2)]}\right\}, \tag{1.4}$$

where $x > 0$, $y > 0$, $\theta_1, \theta_2 > 0$, and $-1 < \theta_3 < 1$, where, for convenience we denote the dependence parameter α by θ_3. Here, the marginal distributions are exponential, the correlation coefficient is $\theta_3/4$, and θ_i is the mean of F_i, $i = 1, 2$. The joint pdf associated with (1.4) is

$$f(x, y; \boldsymbol{\theta}) = \frac{1}{\theta_1} e^{-(x/\theta_1)} \frac{1}{\theta_2} e^{-(y/\theta_2)} \left\{1 + \theta_3 \left[2e^{-(x/\theta_1)} - 1\right]\left[2e^{-(y/\theta_2)} - 1\right]\right\}. \tag{1.5}$$

This pdf is the second of the two bivariate exponential distributions introduced by Gumbel in [7]. He notes that for this bivariate distribution, the conditional expectation of Y given X increases or decreases with X according as the dependence parameter θ_3 is positive or negative (and X and Y are independent when it is 0). Thus θ_3 provides a measure of relationship through the regression function. Here, in Section 2, we obtain explicit expressions for the elements of $\mathbf{I}(X, Y; \boldsymbol{\theta})$, and in Section 3.1, we evaluate it for selected values of θ_3, and discuss some implications to inference.

Similar investigations for other bivariate distributions have been undertaken in the literature. Arnold [2] has obtained the FIM for the BVE distribution proposed by Marshall and Olkin in [11]. Other examples include the work of Oakes and Manatunga [12] who discuss the case of a bivariate extreme value distribution and the recent work of Bjarnason and Hougard [4] who have obtained the FIM for two gamma frailty bivariate Weibull models. In Section 3.2 we revisit Arnold's work and compare the behavior of the asymptotic variances of the MLE's under the G_2BVE and Marshall-Olkin BVE (MOBVE) models. Both these BVE's have exponential marginals and three parameters that identify the bivariate distributions. While the former is absolutely continuous, the latter has the bivariate lack of memory property.

2 The Fisher Information Matrix

We now obtain the FIM for the parameter $\boldsymbol{\theta} = (\theta_1, \theta_2, \theta_3)$, for the pdf given by (1.5). The FIM is $\mathbf{I}(X, Y; \boldsymbol{\theta}) = (I_{ij})_{3\times 3}$, where the elements are computed using one of the expressions given in (1.1). In our derivations we use the transformations $U = X/\theta_1$, $V = Y/\theta_2$ to simplify the expressions. These random variables correspond to the G_2BVE distribution with standard exponential marginals. First we present the diagonal elements.

Since

$$\frac{\partial \log f(x,y;\boldsymbol{\theta})}{\partial \theta_1} = \frac{1}{\theta_1}\left\{\frac{x}{\theta_1} - 1 + \frac{2\theta_3\, x}{\theta_1}\,\frac{\exp\{\frac{-x}{\theta_1}\}(2\,\exp\{\frac{-y}{\theta_2}\} - 1)}{[1+\theta_3(2\,\exp\{\frac{-x}{\theta_1}\} - 1)(2\,\exp\{\frac{-y}{\theta_2}\} - 1)]}\right\},$$

we can write

$$\begin{aligned}\left(\frac{\partial \log f(x,y;\boldsymbol{\theta})}{\partial \theta_1}\right)^2 &= \frac{1}{\theta_1^2}\left\{(u-1)^2 + \frac{4\,\theta_3(u^2-u)\,\exp\{-u\}\,(2\,\exp\{-u\}-1)}{[1+\theta_3\,(2\,\exp\{-u\}-1)\,(2\,\exp\{-v\}-1)]}\right.\\ &\quad + \left.\frac{4\,\theta_3^2 u^2\,\exp\{-2u\}\,(2\,\exp\{-v\}-1)^2)}{[1+\theta_3\,(2\,\exp\{-u\}-1)\,(2\,\exp\{-v\}-1)]^2}\right\}. \qquad (2.1)\end{aligned}$$

Hence

$$\begin{aligned} I_{11} &= E\left(\frac{\partial \log f(X,Y;\boldsymbol{\theta})}{\partial \theta_1}\right)^2 \\ &= \frac{1}{\theta_1^2}\left\{1 + 4\,\theta_3^2 \int_0^\infty\int_0^\infty \frac{u^2\,\exp\{-3u\}\,(2\,\exp\{-v\}-1)^2\,\exp\{-v\}}{[1+\theta_3\,(2\,\exp\{-u\}-1)\,(2\,\exp\{-v\}-1)]}\,du\,dv\right\}. \qquad (2.2)\end{aligned}$$

The denominator of the integrand of the inside integral in (2.2) can be expanded as a power series given by

$$\sum_{j=0}^{\infty}(-\theta_3)^j(2\,\exp\{-v\}-1)^j(2\,\exp\{-u\}-1)^j,$$

and, since $|\theta_3\ (2\ \exp\{-v\}-1)^j(2\ \exp\{-u\}-1)^j| < 1$ for all real x and y, this representation is uniformly convergent. So it is permissible to integrate term by term. Hence we get

$$I_{11} = \frac{1}{\theta_1^2}\left\{ 1 + 4\sum_{j=0}^{\infty}\frac{\theta_3^{2j+2}}{2j+3}\int_0^{\infty} u^2\exp\{-3u\}\,(2\ \exp\{-u\}-1)^{2j}\ du\right\}. \tag{2.3}$$

We obtain I_{22} upon replacing θ_1 by θ_2 in (2.3) above. Now

$$I_{33} = E\left(\frac{\partial \log f(X,Y;\boldsymbol{\theta})}{\partial\theta_3}\right)^2,$$

and we use (1.3) to obtain the element of the FIM corresponding to the dependence parameter for the general FGM distribution. Thus we begin with

$$\frac{\partial \log f(x,y;\boldsymbol{\theta})}{\partial\theta_3} = \frac{(1-2\ F_1(x))(1-2\ F_2(x))}{[1+\theta_3\ (1-2\ F_1(x))(1-2\ F_2(x))]}, \tag{2.4}$$

to obtain

$$I_{33} = \int_0^{\infty}\int_0^{\infty}\frac{[(1-2\ F_1(x))(1-2\ F_2(x))]^2\ f_1(x)f_2(y)}{[1+\theta_3\ (1-2\ F_1(x))(1-2\ F_2(x))]}\ dx\ dy. \tag{2.5}$$

As done in the derivation of I_{11}, we now observe that the denominator of the integrand of the inside integral on the RHS of (2.5) can be expanded as a power series that is uniformly convergent. Upon summation and term by term integration, we obtain

$$I_{33} = \sum_{j=0}^{\infty}(-\theta_3)^j\int_0^{\infty}\int_0^{\infty}[(1-2\ F_1(x))]^2[(1-2\ F_2(y))]^2\ f_1(x)f_2(y)dx\ dy$$

which simplifies to

$$I_{33} = \sum_{j=0}^{\infty}\frac{4\ \theta_3^{2j}}{(2j+3)^2}\ . \tag{2.6}$$

This indicates that I_{33} is independent of the marginal pdfs of X and Y. The expression in (2.6) is available in [1] for the general FGM distribution, and, in [14] in the context of bivariate logistic distribution.

We compute the off-diagonal entries next where the second representation in (1.1) is more convenient. Note that, upon simplification, we obtain

$$\frac{\partial^2 \log f(x,y;\boldsymbol{\theta})}{\partial\theta_1\partial\theta_2} = \frac{4xy\theta_3}{\theta_1^2\theta_2^2} \frac{\exp\{\frac{-x}{\theta_1}\}\exp\{\frac{-y}{\theta_2}\}}{[1+\theta_3(2\exp\{\frac{-x}{\theta_1}\}-1)(2\exp\{\frac{-y}{\theta_2}\}-1)]^2}$$

and consequently

$$\begin{aligned} I_{12} &= -\frac{4\theta_3}{\theta_1\theta_2}\int_0^\infty\int_0^\infty \frac{u\ v\ \exp\{-2u\}\exp\{-2v\}}{[1+\theta_3\,(2\ \exp\{-v\}-1)\,(2\ \exp\{-u\}-1)]}\,du\,dv, \\ &= -\frac{4\theta_3}{\theta_1\theta_2}\sum_{j=0}^{\infty}(-\theta_3)^j\left\{\int_0^\infty u\exp\{-2u\}\,(2\ \exp\{-u\}-1)^j\ du\right\}^2, \qquad (2.7) \end{aligned}$$

upon expanding the integrand as a power series and performing term by term integration. Further,

$$\frac{\partial^2 \log f(x,y;\boldsymbol{\theta})}{\partial\theta_1\partial\theta_3} = \frac{2x}{\theta_1^2} \frac{\exp\{\frac{-x}{\theta_1}\}(2\ \exp\{\frac{-y}{\theta_2}\}-1)}{[1+\theta_3(2\ \exp\{\frac{-x}{\theta_1}\}-1)(2\ \exp\{\frac{-y}{\theta_2}\}-1)]^2}$$

and consequently

$$\begin{aligned} I_{13} &= -E\left(\frac{\partial^2 \log f(X,Y;\boldsymbol{\theta})}{\partial\theta_1\partial\theta_3}\right) \\ &= \frac{2}{\theta_1}\sum_{j=0}^{\infty}\frac{\theta_3^{2j+1}}{2j+3}\int_0^\infty u\exp\{-2u\}\,(2\ \exp\{-u\}-1)^{2j+1}\ du, \qquad (2.8) \end{aligned}$$

upon simplification. We now note that, by symmetry, $\theta_2 I_{23} = \theta_1 I_{13}$ where I_{13} is given by (2.8). Also I_{11} and I_{33} are even functions of θ_3 while I_{13} is an odd function. Thus, one needs to know only I_{11}, I_{33}, I_{12}, and I_{13} to determine all the elements of the FIM for a given θ_3. For the elements $\mathbf{I}$ that correspond to $-\theta_3$, we now need to evaluate only the corresponding I_{12}. Note that when θ_3 is 0, X and Y are independent exponentials and consequently I_{12} is 0, $I_{11} = \theta_1^{-2} = I(X;\theta_1)$, and $I_{22} = \theta_2^{-2} = I(Y;\theta_2)$. Further, in that case, from (2.6) and (2.8) we see that $I_{33} = 4/9$, and $I_{13} = 0$, respectively.

Table 1 provides the values of the above elements that are needed to evaluate all the elements of $\mathbf{I}(X,Y;\boldsymbol{\theta})$ for $\theta_3 = 0, 0.25, 0.5, 0.75, 0.99$, when $\theta_1 = \theta_2 = 1$. These were computed using IMSL routines in FORTRAN. The table indicates that empirically $I_{12}(\theta_3)$ and $-I_{12}(-\theta_3)$ are very close and thus the former can be used

θ_3	0	0.25	0.5	0.75	0.99
$I_{11}(=I_{22})$	1	1.0062	1.0254	1.0595	1.1119
I_{33}	0.4444	0.4548	0.4905	0.5747	0.8743
$I_{13}(=I_{23})$	0	-0.0047	-0.0100	-0.0169	-0.0292
I_{12}	0	-0.0625	-0.1258	-0.1915	-0.2594
$I_{12}(-\theta_3)$	0	0.0629	0.1275	0.1955	0.2679

Table 1: *Essential elements of the FIM for the parameter* $\boldsymbol{\theta} = (\theta_1, \theta_2, \theta_3)$ *for the* G_2BVE *distribution when* $\theta_1 = \theta_2 = 1$.

as a quick approximation to the latter. It also shows the effect of the changes in the value of θ_3 on the FI content of (X, Y) about θ_1 (or θ_2) as well as about θ_3 itself. The I_{11} values there can be used to gauge the contribution of the covariate Y in the increase in the FI in X about its mean θ_1. This relative improvement, $[I(X, Y; \theta_1) - I(X; \theta_1)]/I(X; \theta_1)$, increases with $|\theta_3|$, but never exceeds 12%. This suggests that the knowledge of the covariate increases the information content of the univariate data only to a limited extent.

3 Discussion

3.1 Efficiency and Asymptotic Variance of MLE

Table 1 entries can be used to compute the Cramér-Rao (lower) bound for the variance of unbiased estimators of the parameter θ_i. More importantly they can be used to obtain the variance of the limiting normal distribution of $T_n^i = \sqrt{n}(\hat{\theta}_i - \theta_i)$ $(i = 1, 2, 3)$ as $n \to \infty$ where $\hat{\theta}_i$ is the MLE of θ_i based on a random sample of size n from $f(x, y)$ given in (1.5). Table 2 provides the values of $UB_i = 1/I_{ii}$, and $MB_i = (\mathbf{I}^{-1})_{ii}$ as a function of θ_3, for $i = 1, 3$, assuming $\theta_1 = \theta_2 = 1$. The inverse of the FIM, $\mathbf{I}^{-1}$, was obtained using S-PLUS. The quantity UB_i represents the Cramér-Rao bound as well as the variance of the limit distribution of T_n^i when the other parameters are known, and MB_i corresponds to these quantities when the other parameters are unknown. When θ_3 is 0, X and Y are independent exponentials with mean 1, and thus UB_1 is 1. In this case UB_1 also represents the limiting variance

	θ_1				θ_3	
θ_3	UB_1	MB_1	$MARE_1$	$UMARE_1$	UB_3	MB_3
-0.99	0.8994	0.9553	1.0468	1.0622	1.1438	1.1454
-0.75	0.9438	0.9774	1.0231	1.0356	1.7400	1.7414
-0.50	0.9752	0.9907	1.0094	1.0159	2.0388	2.0396
-0.25	0.9938	0.9978	1.0022	1.0040	2.1989	2.1991
0	1	1			2.25	2.25
0.25		0.9980	1.0020	1.0042		2.1990
0.50		0.9904	1.0097	1.0156		2.3098
0.75		0.9760	1.0246	1.0341		1.7420
0.99		0.9524	1.0500	1.0589		1.1464

Table 2: *Asymptotic variance of the MLE (Cramér-Rao lower bound on the unbiased estimators) of* θ_1 *and* θ_3 *for the* G_2BVE *distribution when* $\theta_1 = \theta_2 = 1$.

of the MLE of θ_1 based on only the X sample values. Further, as the univariate bounds are symmetric functions of θ_3 the cells corresponding to its positive values are left blank.

From Table 2 one can compare the limiting variances of $\hat{\theta}_1$ and $\hat{\theta}_3$ as θ_3 changes. One useful comparison is that of the limiting variances of the MLE's based on the univariate and bivariate samples. For example, the ratio $UARE_1 = UB_1(0)/UB_1(\theta_3)$ would provide the Asymptotic Relative Efficiency (ARE) of the MLE of θ_1 based on the (X, Y) data when compared to the X data alone (which corresponds to $\theta_3 = 0$), when the other parameters are known. This comparison is essentially the ratio of the I_{11} values discussed just above in Section 2. The ratio $MARE_1 = MB_1(0)/MB_1(\theta_3)$ provides such a comparison when the other parameters are unknown. These values are included in Table 2 and they indicate that the improvement is at most 5%. Had the nuisance parameters been known, as observed earlier, the improvement would be under 12%.

One can also compute $UMARE_i = MB_i(\theta_3)/UB_i(\theta_3)$ to examine the effect of the knowledge of the other parameters on the limiting variance of $\hat{\theta}_i$ based on the bivariate data. From the $UMARE_1$ column in Table 2 it follows that for $\hat{\theta}_1$, the ARE increases in a nonsymmetric manner as θ_3 moves away from 0, and is 1.06

when θ_3 is -0.99. This implies that while estimating θ_1, the efficiency gained by the knowledge of the other parameters is at most 6%. For $\hat{\theta}_3$, the MARE values are not shown, but it is easily seen from Table 2 that they are barely above 1.00. This indicates that while estimating the dependence parameter, the knowledge of the parameters of the marginal distributions has hardly any effect in terms of the asymptotic variance of the MLE. We are tempted to suggest that this conclusion would hold for other commonly used FGM type distributions.

Remark: Let $X_{r:n}$ be the rth X-order statistic $(1 \leq r \leq n)$ and $Y_{[r:n]}$ be its concomitant obtained from a random sample of size n from an absolutely continuous pdf $f(x, y)$. (See [5] for a recent review of the area of concomitants of order statistics.) For the FGM density (1.3), Abo-Eleneen and Nagaraja [1] evaluate $I(X_{r:n}, Y_{[r:n]}; \alpha)$ for selected n and r and discuss its properties. While the $(X_{r:n}, Y_{[r:n]})$ are dependent, the FI in $(X_{r:n}, Y_{[r:n]})$ turns out to be additive in r. For the G_2BVE pdf given by (1.5), they evaluate $I(Y_{[r:n]}; \theta_3)$ and compare it with $I(Y_{r:n}; \theta_3)$.

3.2 Marshall-Olkin Bivariate Exponential Distribution

Another BVE distribution with exponential marginals is due to Marshall and Olkin [11]. They introduced a BVE distribution to model the component lifetimes in the context of a shock model. We say (X, Y) has MOBVE with parameters $\lambda_1 > 0$, $\lambda_2 > 0$, and $\lambda_3 \geq 0$, if

$$P(X > x, Y > y) = e^{-\lambda_1 x - \lambda_2 y - \lambda_3 \max(x,y)}, \quad x, y > 0. \tag{3.1}$$

Here X and Y are marginally exponential with means $(\lambda_1 + \lambda_3)^{-1}$ and $(\lambda_2 + \lambda_3)^{-1}$, respectively, and the correlation $\rho = \lambda_3/\lambda$, where $\lambda = \lambda_1 + \lambda_2 + \lambda_3$. This distribution has a singular component and consequently the joint pdf does not exist. In the context of parameter estimation for MOBVE distribution, Arnold [2] has given an explicit expression for the FIM for the parameter vector $\boldsymbol{\lambda} = (\lambda_1, \lambda_2, \lambda_3)$. He shows that

$$\mathbf{I}(X, Y; \boldsymbol{\lambda}) = \frac{1}{\lambda} \begin{pmatrix} a+c & 0 & a \\ 0 & b+d & b \\ a & b & a+b+e \end{pmatrix} \tag{3.2}$$

where

$$a = \lambda_2(\lambda_1 + \lambda_3)^{-2}, b = \lambda_1(\lambda_2 + \lambda_3)^{-1}, c = \lambda_1^{-1}, d = \lambda_2^{-1}, \text{ and } e = \lambda_3^{-1}. \tag{3.3}$$

We now consider the parametric transformation $\eta_1 = \lambda_1 + \lambda_2$, $\eta_2 = \lambda_2 + \lambda_3$ and $\eta_3 = \lambda_3$ so that the first two parameters correspond to the marginal distributions of X and Y, respectively, and the last one is related to the dependence structure. With $\boldsymbol{\eta} = (\eta_1, \eta_2, \eta_3)$, using (2.6.15) in [10] (p. 125), the (i,j)th element of the FIM for the parameter $\boldsymbol{\eta}$ can be expressed as

$$I_{ij}(\boldsymbol{\eta}) = \sum_{k=1}^{3}\sum_{l=1}^{3} I_{kl}(\boldsymbol{\lambda}) \frac{\partial \lambda_k}{\partial \eta_i}\frac{\partial \lambda_l}{\partial \eta_j}.$$

Thus we obtain

$$\mathbf{I}(X, Y; \boldsymbol{\eta}) = \frac{1}{\eta_1 + \eta_2 - \eta_3} \begin{pmatrix} a+c & 0 & -c \\ 0 & b+d & -d \\ -c & -d & c+d+e \end{pmatrix}, \tag{3.4}$$

where $a, \cdots, e$ are given in (3.3). In terms of the η's we may write $a = (\eta_2 - \eta_3)/\eta_1^2$, $b = (\eta_1 - \eta_3)/\eta_2^2$, $c = 1/(\eta_1 - \eta_3)$, $d = 1/(\eta_2 - \eta_3)$ and $e = 1/\eta_3$. Using the reciprocal of the diagonal entries in (3.3) and the diagonal entries of $\mathbf{I}^{-1}(X, Y; \boldsymbol{\eta})$, we can carry out a discussion of the Cramér-Rao lower bound on the unbiased estimators of η_1 and the variance of the limiting distribution of the MLE's as done above in Section 3.1. To fix ideas, we take $\eta_1 = \eta_2 = 1$. Then, $\rho = \eta_3/(2 - \eta_3)$, or equivalently, $\eta_3 = \rho/(1+\rho)$. We compute the bounds corresponding to η_1 as functions of ρ as it varies in $[0, 1)$. The bounds for selected values of ρ are given in Table 3 and these were computed using MAPLE 5.1.

The changes in the values of UB_1 and MB_1 in Table 3 indicate the rapid improvement in the limiting variance of $\hat{\eta}_1$, the MLE of η_1, as ρ increases. Thus, the improvement in its efficiency due to the bivariate data is substantial. The last column in Table 3 provides $UMARE_1(\rho) = MB_1(\rho)/UB_1(\rho)$ to facilitate the examination of the impact of the knowledge of η_2 and η_3 on the limiting variance of $\hat{\eta}_1$. It is clear that the effect is substantial.

From the above discussion we conclude that the improvement in the efficiency of the MLE of the mean of X due to the availability of the covariate values as well

ρ	UB_1	MB_1	$UMARE_1$
0	1	1	1
0.1	.8597	.9029	1.0503
0.2	.7377	.8125	1.1014
0.3	.6320	.7294	1.1541
0.4	.5405	.6532	1.2085
0.5	.4615	.5837	1.2648
0.6	.3933	.5201	1.3224
0.7	.3342	.4620	1.3824
0.8	.2830	.4086	1.4438
0.9	.2386	.3596	1.5071
0.99	.2036	.3187	1.5649

Table 3: *Asymptotic variance of the MLE (Cramér-Rao lower bound on the unbiased estimators) of* η_1 *in terms of* ρ *for the MOBVE distribution when* $\eta_1 = \eta_2 = 1$.

as the knowledge of the nuisance parameters is limited for the G_2BVE distribution whereas, in both these circumstances, the improvement is quite substantial for the MOBVE distribution.

Acknowledgements

The first author's research was supported in part by National Institutes of Health, USA, Grant # M01 RR00034 and the second author's research was supported by a training grant from the Egyptian government.

References

1. Abo-Eleneen Z. A. and Nagaraja H. N., Fisher information in an order statistic and its concomitant. *Ann. Institute of Statistical Mathematics*, to appear.

2. Arnold B. C., Parameter estimation for a multivariate exponential distribution. *J. Amer. Statist. Assoc.* **63**, 848-852 (1968).

3. Barnett V., The bivariate exponential distribution; A review and some new results. *Statistica Neerlandica* **39**, 343-356 (1985).

4. Bjarnason, H. and Hougaard P., Fisher information for two Gamma frailty bivariate Weibull models. *Life Data Analysis* **6**, 59- 71 (2000).

5. David H. A. and Nagaraja H. N., Concomitants of order statistics. In *Handbook of Statistics, Vol. 16*, (Edited by N. Balakrishnan and C. R. Rao), pp. 487-513, Elsevier, Amsterdam (1998).

6. Galambos J. and Kotz S., *Characterizations of Probability Distributions*, Lecture Notes in Mathematics, 675, Springer-Verlag, Berlin (1978).

7. Gumbel E. J., Bivariate exponential distributions. *Jour. Amer. Statist. Assoc.* **55**, 698-707 (1960).

8. Hutchinson T. P. and Lai C. D., *Continuous Bivariate Distributions, Emphasising Applications.* Rumsby Scientific Publishing, Adelaide, Australia (1990).

9. Kotz S., Balakrishnan N. and Johnson N., *Continuous Multivariate Distributions, Vol. 1: Models and Applications, Second Edition.* John Wiley, New York (2000).

10. Lehmann E. L. and Casella G., *Theory of Point Estimation, Second Edition.* Springer-Verlag , New York (1998).

11. Marshall A. W. and Olkin I., A multivariate exponential distribution. *J. Amer. Statist. Assoc.* **62**, 30-44 (1967).

12. Oakes D. and Manatunga A., Fisher information for a bivariate extreme value distribution. *Biometrika* **79**, 827-832 (1992).

13. Rao C. R., *Linear Statistical Inference and Its Applications, Second Edition.* John Wiley, New York (1973).

14. Smith M. D. and Moffatt P. G., Fisher's information on the correlation coefficient in bivariate logistic models. *Austral. & New Zealand J. Statist.* **41**, 315-330 (1999).

CLASSIFICATION INVARIANCE IN DATA ENVELOPMENT ANALYSIS

Lawrence M. Seiford and Joe Zhu

Department of Industrial and Operations Engineering, University of Michigan, Ann Arbor, MI 48109-2117 USA

Department of Management, Worcester Polytechnic Institute, 100 Institute Road, Worcester, MA 01609 USA

Invariance property in data envelopment analysis (DEA) allows negative data in efficiency analysis. In general, there are three cases of invariance under data transformation in DEA. The first case is the "classification invariance" where the classifications of efficiencies and inefficiencies are invariant to the data transformation. The second case is the "ordering invariance" of the inefficient decision making units (DMUs). The last case is the "solution invariance" in which the new DEA model (after data translation) must be equivalent to the old one. The current paper indicates that DMUs with negative output values may be classified as efficient when we use the classification invariance. Although such classification is mathematically correct, it may not be managerially acceptable. The method of finding well-defined facet is suggested to re-evaluate the performance of DMUs with negative values. The paper illustrates the approach with an application to textile firms where negative profit is present.

Key words: Data Envelopment Analysis (DEA); classification invariance; efficiency.

1. Introduction

Since the original data envelopment analysis (DEA) model by Charnes, Cooper and Rhodes (CCR, [3]), many theoretical extensions and empirical studies have appeared in the literature [8]. One research issue which has received widespread attention in the rapidly growing field of DEA is the invariance property. Ali and Seiford [1] discover the translation invariance property in the additive model [4] and the BCC model [2] that does not require positivity of any inputs or outputs. Pastor [7] find out that by the translation invariance property in DEA, input (output) values can be not only zero but also negative. However, the use of DEA models is restricted.

This paper is concerned only with the classification invariance. (For other cases of invariance, see [6, 7].) Note that the key to classification invariance in DEA lies in the convexity constraint. Therefore, we consider only the BCC model. The term *classification invariance* here means that the BCC efficient frontier or the BCC efficiency classification is invariant to data transformation.

The paper indicates that situations when DMUs with negative output values are classified as efficient may need a careful analysis. For example, in a performance study of textile firms, negative profits were found in some firms [9]. In the original study of [9], those firms with negative profits were deleted from the analysis. By the classification invariance, we can evaluate those firms with others. However, some loss firms are classified as efficient. Although such efficiency classification is mathematically correct, it may not be preferred by the management. The current paper discusses this issue and develop an approach to revise the efficiency results.

2. Background

A DEA data domain can be characterized by a data matrix

$$\mathbf{P} = \begin{bmatrix} \mathbf{Y} \\ -\mathbf{X} \end{bmatrix} = [\mathbf{P}_1, \ldots, \mathbf{P}_n]$$

with $s+m$ rows and n columns. Each column corresponds to one of the DMUs. The jth column

$$\mathbf{P}_j = \begin{bmatrix} \mathbf{y}_j \\ -\mathbf{x}_j \end{bmatrix}$$

is composed of an input vector $\mathbf{x}_j$ whose ith component x_{ij} is the amount of input i used by DMU_j and an output vector $\mathbf{y}_j$ whose rth component y_{rj} is the amount of output r produced by DMU_j.

The BCC efficiency can be obtained by calculating the following linear programming problem

$$\begin{aligned} &\max \eta \\ \text{s.t. } &\sum_{j=1}^{n} \lambda_j \mathbf{x}_j + \mathbf{s}^- = \mathbf{x}_o \\ &\sum_{j=1}^{n} \lambda_j \mathbf{y}_j - \mathbf{s}^+ = \eta \mathbf{y}_o \\ &\sum_{j=1}^{n} \lambda_j = 1 \\ &\lambda_j \geq 0 \quad j=1,\ldots,n. \end{aligned} \tag{1}$$

where $\mathbf{x}_o$ and $\mathbf{y}_o$ represent input and output vectors of DMU_o, respectively. This model is an output-based (or output-oriented) program. Similarly, one can have an input-based BCC model. On the basis of optimal solutions of η^*, λ_j^*, $\mathbf{s}^{+*}$ and $\mathbf{s}^{-*}$, DMU_o can be classified as one of the four efficiency classifications [5]: class E (consists of extreme points, i.e., $\eta^* = 1$ as well as $\mathbf{s}^{+*} = \mathbf{s}^{-*} = 0$ with unique solution of λ_j^*), class E′ (linear combination of DMUs in class E), class F ($\eta^* = 1$ with nonzero $\mathbf{s}^{+*}$ and (or) nonzero $\mathbf{s}^{-*}$), class N ($\eta^* > 1$), where the first two classes consist of efficient DMUs and the last two consist of inefficient DMUs.

Next suppose the input vector is displaced by the m rowed vector $\boldsymbol{u}$ and the output vector is displaced by the s rowed vector $\boldsymbol{v}$. That is $\bar{\mathbf{x}}_j = \mathbf{x}_j + \mathbf{u}$ and $\bar{\mathbf{y}}_j = \mathbf{y}_j + \mathbf{v}$ (j =1, 2, ..., n).

We have the following result with respect to the translation invariance in DEA when all input and output data are nonnegative [1].

Classification Invariance: DMU_o is efficient for (1) if and only if DMU_o is efficient for (1) under translated data; DMU_o is inefficient for (1) if and only if DMU_o is inefficient for (1) under translated data.

We can generalize this result by relaxing the nonnegativity condition. However, the type of the BCC model is restricted in use with respect to the following solution invariance when negative input/output values are present [7].

Solution Invariance: The input-based BCC model is output translation invariant and the output-based BCC model is input translation invariant.

3. Negative input/output values in DEA

Although the classification invariance is discovered under the nonnegativity assumption, it can be applied to the situation where negative data are present, since the relative position of DMUs is invariant to the data changes.

Consider the three DMUs used in [7], each with a single negative input and a single positive output. The input-output vector for the three DMUs are: DMU1 = (-4, 1), DMU2 = (-2, 3) and DMU3 = (-1, 2).

By a translation vector of (5,0), we have the corresponding translated DMUs, (1, 1), (3, 3) and (4, 2). The input-based BCC model now classifies DMUs 1 and 2 as efficient (class E) and DMU3 as inefficient (class N).

Thus, the input-based BCC model can be employed when negative inputs are present, if the negative inputs are translated into positive values. (A similar result can be obtained for the output-based BCC model with negative outputs. That is, the output-based BCC model can be used when negative outputs are present, if the negative outputs are translated into positive values). This gives a different result in contrast to what is found in [7]. This is due to the fact that [7] focuses on optimal solutions to the translated and untranslated BCC models, i.e., solution invariance; whereas we here focus on how to determine the efficient frontier, and particularly, the efficiency classifications, rather than the efficiency scores.

This result is useful when compared to that of [7], since the choice of orientation of BCC model is a choice between exogenous and endogenous variables. For instance, if a DMU is able to vary the quantity of all the outputs and is not able to, in short term, act on the inputs, then the output-based BCC model should be selected even some outputs may have negative values. Thus, the choice of orientation depends on the nature of the problem and not the value range of a variable. By using classification invariance, either input-based or output-based BCC model can be employed in the presence of negative input or output values. Note that in this situation, the efficiency scores may not be independent of the selected translation vector. However, the efficiency classification is independent of the translation vector we select.

Next, we observe what will happen if we have both positive and negative outputs. Consider a simple case where the data domain is given by

$$\begin{bmatrix} \mathbf{Y} \\ -\mathbf{X} \end{bmatrix} = \begin{bmatrix} \mathbf{Y}_1^+ & \mathbf{Y}_2^- \\ & -\mathbf{X}^+ \end{bmatrix}$$

where $\mathbf{Y}^-$ represents negative outputs. $\mathbf{X}^+$ and $\mathbf{Y}^+$ stand for positive inputs and positive outputs, respectively.

Table 1: Seven sample DMUs

DMU	1	2	3	4	5	6	7
x (input)	2	3	4	2	1	1	4
y (output)	2	4	5	-1	-1	-3	2
$\bar{y} = y + 4$	6	8	9	3	3	1	6
Efficiency							
Untranslated							
Input-based	E	E	E				N (1/2)*
Output-based	E	E	E				N (5/2)
Translated							
Input-based	E	E	E	N (1/2)	E	F	N (1/2)
Output-based	E	E	E	N (2)	E	N (3)	N (3/2)

*The number in parenthesis represents efficiency score.

If the DMUs with negative output are not considered, then DMUs 1, 2 and 3 are efficient (class E) under the BCC model. Associated with an output displacement of $v = 4$, DMUs 1, 2, 3 and 5 are efficient (class E) (see Table 1). When the output stands for profit, it may be unsatisfactory that DMU5 with negative output is classified as efficient. Thus, the efficiency of DMU5 should be reestimated. We assume that those DMUs, particularly the efficient ones, with positive output values outperform those DMUs with negative output values. Note that for any variable it may be advised to define a range of values considered as admissible. Here the zero is not necessarily a limit for this interval. Nevertheless by the classification invariance, any limit values can be transformed into zero. Thus, this assumption can be made to any situation of limit values.

In order to solve this problem, the current study suggests the following: First, let the data domain D be partitioned into H subdomains, D_h, $h = 1, \ldots,$ H. The first subdomain D_1 consists of all those DMUs with positive input and positive output values. For other subdomain D_h, $h \neq 1$, all DMUs in D_h produce the same type of negative outputs. For example,

$$\begin{matrix} \mathrm{D} \\ \begin{bmatrix} \mathbf{Y} \\ -\mathbf{X} \end{bmatrix} \end{matrix} = \begin{matrix} \mathrm{D}_1 \quad \mathrm{D}_2 \quad \mathrm{D}_3 \\ \begin{bmatrix} \mathbf{Y}_1^+ & \mathbf{Y}_2^+ & \mathbf{Y}_3^- \\ \mathbf{Y}_4^+ & \mathbf{Y}_5^- & \mathbf{Y}_6^- \\ -\mathbf{X}_1^+ & -\mathbf{X}_2^+ & -\mathbf{X}_3^+ \end{bmatrix} \end{matrix}$$

Second, after output data translation (i.e., force all outputs to be positive), we find all well-defined positive multiplier efficient facets in subdomain D_1, and all efficient DMUs in each subdomain D_h, $h \neq 1$. Then, we assign each efficient DMU in D_h, $h \neq 1$ to a proper well-defined efficient facet in D_1 which gives the highest efficiency score. After obtaining the re-estimated efficiency scores for all the efficient DMUs in D_h, $h \neq 1$, we assign each corresponding inefficient DMU in D_h, $h \neq 1$ to the facet to which the referent D_h, $h \neq 1$ efficient DMU is assigned.

The second stage, in fact, can be carried out by the following binary linear programming problem

$$\begin{aligned} &\min \mathbf{v}^{\mathbf{T}}\overline{\mathbf{X}}_{\mathbf{o}} + u_o \\ &s.t.\ \mathbf{u}^{\mathbf{T}}\overline{\mathbf{Y}}_{\mathbf{j}} - \mathbf{v}^{\mathbf{T}}\overline{\mathbf{X}}_{\mathbf{j}} - u_o + z_j = 0,\ j \in \mathrm{E} \\ &\quad \mathbf{u}^{\mathbf{T}}\overline{\mathbf{Y}}_o = 1 \\ &\quad z_j - b_j M \le 0\ \ j \in \mathrm{E} \\ &\quad \sum_{j \in \mathrm{E}} b_j = |\mathrm{E}| - k \\ &\quad b_j \in \{0,1\}, z_j, \mathbf{u}^{\mathbf{T}}\mathbf{v}^{\mathbf{T}} \ge 0,\ u_o\ \ free. \end{aligned} \tag{2}$$

where $\overline{\mathbf{X}}$ and $\overline{\mathbf{Y}}$ represent translated data, E stands for the set of extreme-efficient DMUs in D_1. Model (2) is an output-based BCC-type model in which k determines the dimension of a specific reference facet for an efficient DMU_o in D_h, $h \neq 1$. For example, if set k = m + s, the number of inputs plus outputs, then (2) assigns a specific DMU_o onto a full dimensional efficient facet. The constraint $\sum_{j \in \mathrm{E}} b_j = |\mathrm{E}| - k$ = | E | - m- s ensures that m + s extreme-efficient DMUs in D_1 determine the (full dimensional) reference facet. If there does not exist any full dimensional efficient facet in D_1, then we specify the dimension of the reference facet by k.

Table 2: Reestimated efficiency scores for DMUs in Table 1

DMU	4	5	6
Efficiency score under facet-1: y = 2x + 2	2	4/3	4
Efficiency score under facet-2: y = x + 5	7/3	2	6

Consider the example in Table 1. Table 2 gives the two efficient facets composed by the DMUs having positive input/output values and the corresponding reestimated efficiency scores. In Table 1, DMU5 outperforms DMU7 whereas by model (2), the opposite result is obtained. Also, the nonzero slack on the negative output of DMU6 is suppressed.

In a study of Chinese textile performance, negative profit is present in some textile firms [9]. As a result, firms with negative profit were excluded from the DEA analysis. By means of the BCC classification invariance, we re-evaluate the performance of those textile firms by including firms with negative profits.

We select three inputs: (i) labor which represents the number of staff and workers, (ii) working fund (WF) and investment (INV) which represents the total investment for building and purchasing of fixed assets. We use two outputs: (i) gross industrial output value (GIOV) which represents the general achievement of each firm, and (ii) profit & taxes (P & T) which measures the net contribution of each firm. (For more information on these inputs/outputs, please refer to [9].) Table 3 presents the 33 textile firms and their inputs and outputs in the annual period of 1989 where "rmb" is the Chinese monetary unit. Note that the last five firms had negative profit & taxes values.

Managers of firms in China are rewarded primarily based upon their success in meeting physical output targets, such as, profit and gross industrial output value, set by the local government. Firm itself also pays more attentions to the profit and taxes, therefore an output-based BCC model is employed. Using an output translation vector of (0, 50) and model (1), we have the efficiency results shown in the last column of Table 3. Twelve DMUs were efficient (class E). Among them, DMU32, was a loss firm (i.e., it had negative profit). This result is unsatisfactory in a decision maker's view. Because a firm's manager should be penalized for the firm's deficit in profit and for the firm's inability to pay taxes. Therefore we need to modify the efficiency score of DMU32.

Table 3: Data and Efficiency Scores for 33 textile firms

	Inputs			Outputs		Efficiency
	(person)	(10,000rmb)		(10,000rmb)		scores
DMU No.	Labor	WF	INV	GIOV	P & T	
1	4063	4650.9	6663.5	11867.8	1787.0	1.00000
2	481	638.5	405.4	1621.6	128.7	1.00000
3	4762	2606.6	2154.8	8838.7	1107.0	1.00000
4	1365	1083.2	1441.9	4508.4	280.5	1.12476
5	1267	648.7	259.0	1667.3	66.3	1.03125
6	1342	627.1	573.5	1800.4	40.0	1.83539
7	1185	554.2	1055.2	2042.1	133.5	1.42735
8	534	238.9	100.3	307.5	41.4	1.64481
9	1083	1034.6	1703.7	8249.3	27.0	1.00000
10	837	873.1	1072.1	5804.4	110.6	1.00000
11	843	356.0	487.9	1458.4	142.5	1.15063
12	1594	941.5	729.2	980.7	39.2	3.75105
13	475	278.9	255.9	936.7	81.0	1.21964
14	558	309.6	367.1	715.3	69.1	1.67461
15	454	223.3	33.9	717.5	70.3	1.00000
16	476	231.9	196.6	1024.9	99.4	1.01990
17	395	202.6	87.3	593.3	3.2	1.43593
18	453	150.2	161.6	570	46.9	1.44838
19	252	76.5	52.4	470.6	15.7	1.00000
20	213	110.2	31.1	229.5	12.0	1.00000
21	538	271.3	236.2	756.9	30.3	1.90190
22	706	144.7	236.0	1503.4	50.6	1.00000
23	390	229.2	161.8	424.5	73.6	1.22981
24	448	207.2	156.8	538.3	71.7	1.27760
25	381	133.2	113.4	295.2	101.8	1.00000
26	983	554.8	358.4	1147.9	151.4	1.36252
27	1426	477.4	516.8	1070.2	64.9	2.28246
28	498	175.0	181.2	1259.0	89.9	1.00000
29	1066	300.6	530.9	766.9	-5.2	3.14486
30	1349	1015.1	2003.0	2268.8	-8.4	3.31258
31	1140	357.2	716.4	1705.6	-31.0	1.82590
32	481	187.9	683.4	1601.8	-41.3	1.00000
33	717	174.8	201.9	386.4	-23.5	4.47580

We first partition the (translated) data domain into two subdomains, D_1 (composed of DMUs from 1 to 28) and D_2 (composed of DMUs from 28 to 33), and then find a proper well-defined positive multiplier efficient facet in subdomain D_1 to modify DMU32's current score. (All of the loss firms with negative profit & taxes are in class E under subdomain D_2, i.e., DMU32 is not a referent DMU by other loss firms. Therefore the efficiency scores for other loss firms do not need to be reestimated.) We set k = m + s = 5, and find a full dimensional efficient facet composed by DMUs 9, 10, 19, 22 and 25 which yields the highest new efficiency score of 1.44625 for DMU32. Note that with this reestimated score, DMU32 still outperformed other four DMUs having negative profit & taxes values.

4. Conclusions

It has been shown that, by the classification invariance in the BCC model, both input-based BCC and output-based BCC models can be used to characterize the efficiencies and inefficiencies of DMUs when either negative output or input values occur. Output (input) values are no longer restricted to be positive and the choice of orientation of the BCC model is also unrestricted. This broadens the application of the DEA methodology. The empirical study of textile firms has shown that the technique of finding a well-defined envelopment facet should be used for a specific observation with negative output values to reestimate the efficiency score. The same technique can be applied to situations when negative input values are present.

References

1. Ali, A.I. and Seiford, L.M., Translation invariance in data envelopment analysis. *Operations Research Letters* **9**, 403-405 (1990).

2. Banker, R.D., Charnes, A. and Cooper, W.W., Some models for estimating technical and scale inefficiencies in data envelopment analysis. *Management Science.* **30**, 1078-1092 (1984).
3. Charnes, A., Cooper, W.W. and Rhodes, E., Measuring the efficiency of decision making units. *European Journal of Operational Research* **2**, 429-444 (1978).
4. Charnes, A, Cooper, W.W., Golany, B., Seiford, L.M. and Stutz, J., Foundations of data envelopment analysis for Pareto-Koopmans efficient empirical production functions. *J. Econom.* **30**, 91-107 (1985).
5. Charnes, A., Cooper, W.W. and Thrall, R.M., A structure for classifying and characterizing efficiencies and inefficiencies in DEA. *J. of Productivity Analysis* **2**, 197-237 (1991).
6. Lovell, C.A.K. and Pastor, J.T., Units invariant and translation invariant DEA models. *Operations Research Letters* **18**, 147-151 (1995).
7. Pastor, T., Translation invariance in DEA: a generalization. *Annals of Operations Research* **66**, 93-102 (1996).
8. Seiford, L.M., Data Envelopment Analysis: the evolution of the state of the art (1978-1995). *Journal of Productivity Analysis* **7**, 99-137 (1995).
9. Zhu, J., DEA/AR analysis of 1988-1989 performance of Nanjing Textile Corporation. *Annals of Operations Research* **66**, 311-335 (1996).

A Useful Isometry for Time Reversible Markov Chains, with Applications to Reliability

by

Mark Brown
Department of Mathematics
The City College, CUNY

Abstract

Inequalities and error bounds are derived for finite state, irreducible, time reversible Markov chains in continuous time. The results are illustrated in a reliability example involving a 2 out of 4 repairable system. The inequalities are derived via an isometry between two inner product spaces, one corresponding to the chain of interest, and the other to its companion star chain. This connection between the two Markov chains was originally exploited in Aldous and Brown (1992).

Sponsored by the Office of Equal Opportunity Employment of the National Security Agency.

Section 1. Introduction

In this paper inequalities and error bounds are derived for finite state, irreducible, continuous time, time reversible Markov chains. To motivate the results we will illustrate their use in a reliability example involving a repairable system. The same example can arise from the viewpoint of a weighted random walk on the unit cube. Consider a system of 4 independent components. Component i alternates between working or up periods, exponentially distributed with parameter α_i and repair or down periods, exponentially distributed with parameter β_i. These up and down periods are independent both within and between components.

Assume that the values $\alpha_i, \beta_i, i = 1, 2, 3, 4$ are given by

component		1	2	3	4
failure rate	α_i	1.1	1.8	1.2	.9
repair rate	β_i	19	21	22	18

As the parameters vary with component, the number of down components does not form a birth and death process. Rather, the process of component states,

$$\mathbf{X}(t) = \{(X_1(t), X_2(t), X_3(t), X_4(t)), t \geq 0\}$$

with,

$$X_i(t) = \begin{cases} 1 & \text{if component } i \text{ is up at time } t \\ 0 & \text{if component } i \text{ is down at time } t \end{cases}$$

forms a time reversible Markov chain with state space,

$$I = \Pi_1^4\{0, 1\}_i.$$

Assume that the system is a 2 out of 4 system, meaning that the system is up at time t if and only if at least 2 components are up. The set of states corresponding to system failure is thus,

$$A = \{(0,0,0,0), (1,0,0,0), (0,1,0,0), (0,0,1,0), (0,0,0,1)\}.$$

If we start at time 0 with initial distribution w on I, then the distribution of the waiting time until system failure is denoted by $\mathcal{L}_w T_A, T_A$ being the first passage time to A. The steady state distribution on I, is denoted by π, and thus $\mathcal{L}_\pi T_A$ is the time to system failure starting in steady state.

Define Q to be the transition intensity matrix of the Markov process, Q_A to be the restriction of Q to $A^c \times A^c$, and γ_1 to be the smallest eigenvalue of $-Q_A$, which in this case equals .02131. Employing error bounds found in Aldous and Brown [1992], and Brown [1999], we can well quantify that $\mathcal{L}_\pi T_A$ is approximately exponential with parameter $\gamma_1 = .02131$.

More specifically,

$$.99887e^{-\gamma_1 t} \leq P_\pi(T_A > t) \leq .99959e^{-\gamma_1 t} \tag{1.1}$$

$$\text{(mean)} \quad 46.8734 \leq E_\pi T_A \leq 46.9071$$

$$\text{(standard derivation)} \quad 46.92929 \leq SD_\pi T_A \leq 46.92933$$

$$\text{(skewness)} \quad 2 \leq skew_\pi T_A \leq 2.000004$$

$$\text{(kurtosis)} \quad 6.000002 \leq kur_\pi T_A \leq 6.000015.$$

Thus the distribution of $\mathcal{L}_\pi T_A$ is well understood. Of, at least equal interest, in this reliability example is the distribution of the time to system failure starting in the perfect state, $\mathbf{1} = (1,1,1,1)$. As $\mathbf{1}$ is the best state and stochastic monotonicity is present, $\mathcal{L}_\mathbf{1} T_A$ is stochastically larger than an exponential distribution with parameter γ_1, which is the time to first failure starting from the quasi-stationary distribution on I (see Section 2.2 for a discussion of quasi-stationarity). In Section 4 (4.36), we derive an upper bound for $P_w(T_A > t)$, for an arbitrary initial distribution w. Applying it to our reliability example with $w(\mathbf{1}) = 1$ we obtain,

$$e^{-\gamma_1 t} \leq P_\mathbf{1}(T_A < t) \leq (1.05125)e^{-\gamma_1 t}. \tag{1.2}$$

Thus for example, the 99th percentile of $\mathcal{L}_1(T_A)$ falls between 216.1 and 218.5.

For our next variation, define $T_A(t)$ to be the waiting time starting at t, for the first visit to A. We anticipate that for, t, a moderate multiple of the relation time (which in this case equals $(18.9)^{-1}$, that $\mathcal{L}_w(T_A(t))$ should be close to $\mathcal{L}_\pi T_A$. Two error bounds are derived ((4.6) and (4.25)) which treat different variations of this theme. For approximation of the survival function, for the reliability example with $w(\mathbf{1}) = 1$,

$$\sup_{s \geq 0} \frac{|P_{\mathbf{1}}(T_A(t) > s) - P_\pi(T_A > s)|}{P_\pi(T_A > s)} \leq .23265e^{-18.9t}. \tag{1.3}$$

For example for $t = .5$ (which is 9.45 times the relaxation time), the bound on the righthand side of (1.3) is smaller than 1.831×10^{-5}.

Similarly, in our reliability example,

$$\sup_{\alpha \geq 0} \frac{|E_{\mathbf{1}} T_A^\alpha(t) - E_\pi T_A^\alpha|}{E_\pi T_A^\alpha} \leq .0221e^{-18.9t}. \tag{1.4}$$

The choice $t = .5$ yields a bound which is smaller than 1.74×10^{-6}. Thus, for $t \geq .5$, the difference between $\mathcal{L}_{\mathbf{1}}(T_A(t))$ and $\mathcal{L}_\pi(T_A)$ is negligible. In an amount of time which is a small fraction of $E_\pi T_A$, the chain has for practical purposes "forgotten" its initial state.

As a final illustration, start the chain in steady state and consider the conditional distribution of $\mathbf{X}(t)$ given that $T_A > t$. Aldous (1982), reasoned that if the relaxation time is small compared to $E_\pi T_A$, then after a passage of time, t, which is small compared to $E_\pi T_A$, the chain is unlikely to have hit A, and if not, $\mathbf{X}(t)$ should be close in distribution to the quasi-stationary distribution on I, (defined and discussed in Section 2.2). Consequently the conditional distribution of $\mathcal{L}_\pi T_A$ given $T_A > t$ (and thus the unconditional distribution) should be well approximated by the quasi-stationary distribution, $\mathcal{L}_\alpha T_A$, an exponential distribution with parameter γ_1. This intuition was quantified in Aldous-Brown (1992), and leads to (1.1) above. To quantify the idea that $\mathbf{X}(t)|T_A > t$ converges rapidly to $\mathcal{L}_\alpha T_A$ when the relaxation time is small compared to $E_\pi T_A$ (equivalently to $E_\alpha T_A$), we derive an error bound, (4.24), which when applied to the reliability example yields,

$$\sup_B |P_\pi(\mathbf{X}(t)\epsilon B|T_A > t) - \alpha(B)| \le .01342e^{-18.87869t}. \tag{1.5}$$

For $t = .5$, the bound is smaller than 10^{-6}. The above inequalities are particular cases of results derived in Section 3 and 4. In potential applications, to convert the inequalities to specific numerical bounds we need to compute or approximate two eigenvalues. One is λ_1, the second smallest eigenvalue of $-Q$ (where Q is the transition intensity matrix), and the other is γ_1, the smallest eigenvalue of $-Q_A$, the restriction of $-Q$ to $A^c \times A^c$. For the Markov chain corresponding to repairable systems, we show in Section 5.1 that,

$$\lambda_1 = \min_i(\alpha_i + \beta_i).$$

In the running example of the 2 out of 4 system, λ_1 thus equals 18.9. For many Markov chains, λ_1 will be not be analytically available. For small to moderate size matrices there are a variety of numerical analysis methods for computing eigenvalues. For large matrices, several authors have studied techniques for bounding λ_1. Diaconis and Stroock (1991) review some of this methodology, introduce a new method, and give several illustrative examples. The value $\gamma_1 = .02131$ used in the reliability example, was computed using Mathematica for the 11×11 matrix, $-Q_A$. When not readily computable the quantity γ_1 is easy to upper bound, as the extremal characterization of eigenvalues represents γ_1 as an infimum. (See Aldous and Brown (1992), p8; for further comments). For many of our bounds, only an upper bound on $\gamma_1|\lambda_1$ is needed to obtain numerical values. In practice, for large matrices, we would need to upper bound γ_1 (perhaps using the extremal characterization), and lower bound λ_1 (for example with the Diaconis-Stroock (1991) approach).

Approximations in reliability models has been a topic of considerable interest. Gertsbakh [(1984) and (1989), Chapter 3], surveys work in this area, including many contributions from authors in the former Soviet Union. Some notable works are Gnedenko, et al (1969) and Solovyev [(1971) and (1972)]. The model for a repairable system discussed in the introduction was of major interest to nuclear engineers in the early 1970's in their use of fault tree analysis to help assess the safety of nuclear power plants. A conference volume, Barlow et al (1975), contains several articles by nuclear engineers as well as mathematical articles on time to first failure.

Keilson (1975), (1979)) suggested that $\mathcal{L}_\pi T_A$ and $\mathcal{L}_1 T_A$ could be approximated through the use of the spectral representation for reversible chains, and the resultant complete monotonicity for $\mathcal{L}_\pi T_A$. He anticipated bounds and inequalities based on a parameter, ρ, which reflects departure from exponentiality based on the behavior of the first two moments of $\mathcal{L}_\pi T_A$ (see example (ii), Section 4). Aldous (1982) pointed out the importance of the ratio of the relaxation time to $E_\pi T_A$ (equivalently of $\gamma_1|\lambda_1$), as a quantity for bounding distance to exponentiality. Keilson's approach is further explored in Brown (1983), and the approach of Aldous in Aldous and Brown ((1992) and (1993)), Brown (1999), and the current paper.

Keilson was motivated by models in reliability and queues. More recently there has been a great deal of interest in random walks on graphs, which are also examples of time reversible chains. In this context first passage times are of less interest than ergodicity, but there are connections between the two topics. Diaconis and Fill (1990) develop and explore a notion of duality in which ergodicity can be studied via first passage times. The current paper uses contributions to the study of distance to ergodicity by Diaconis and Stroock (1991), and Fill (1991), in developing inequalities for first passage times.

A forthcoming book by Aldous and Fill (2002), presents an elegant treatment of random walks on graphs.

Background material on the spectral representation, as well as various definitions and properties for reversible chains are found in Section 2. The isometry, which leads to our inequalities, is developed in Section 3. It is based on the relationship between the underlying chain and its companion star chain, the star chain being discussed in Section

2.4. Various bounds and inequalities, which follow from the isometry, and derived in Section 4. Section 5 obtains the eigenvalues of $-Q$, for the Markov chain corresponding to a repairable system.

Section 2. Background and Definitions

We review some results for time reversible chains. Keilson (1979) is an excellent reference.

2.1 Spectral representation. Consider a Markov chain $\{X(t), t \geq 0\}$ with finite state space I and transition rate matrix Q. The chain is assumed irreducible with stationary distribution π, and time reversible $(\pi(i)q_{ij} = \pi(j)q_{ji})$. For a non-empty proper subset of I let Q_A denote the restriction of Q to $A^c \times A^c$, and have D_π denote a diagonal matrix with diagonal entries $\{\pi(i), i \in A^c\}$. By reversibility $D_\pi Q_A = Q'_A D_\pi$. It follows that $D_\pi^{1/2} Q_A D_\pi^{-1/2} = D_\pi^{-1/2} Q'_A D_\pi^{1/2}$, thus the matrix $M_A \stackrel{\text{def}}{=} D_\pi^{1/2} Q_A D_\pi^{-1/2}$ is symmetric and M_A and Q_A are similar. As M_A is a real symmetric matrix it has real eigenvalues and we can choose a complete orthonormal system for $R^{A^c}, \varphi_1 \ldots \varphi_n$, consisting of eigenvectors of M_A. The eigenvalues of M_A (which by similarity coincide with those of Q_A) are denoted by $-\nu_1, \ldots, -\nu_n$ where $0 < \nu_1 \leq \nu_2 \ldots \leq \nu_n$.

Consider the matrix $R_t = e^{M_A t} = \sum_0^\infty \frac{t^k M_A^k}{k!}$. By standard methods it follows that,

$$R_t = D_\pi^{1/2} e^{Q_A t} D_\pi^{-1/2} = D_\pi^{1/2} \tilde{P}_t D_\pi^{-1/2}, \qquad \text{where} \tag{2.1}$$

$$\tilde{P}_t(i,j) = P_i[X(t) = j, T_A > t], \tag{2.2}$$

where T_A is waiting time to reach A, with $T_A = 0$ if $X(0) \in A$. The matrices R_t and $\tilde{P}_t$ have eigenvalues $\{e^{-\nu_j t}, j = 1, \ldots, n\}$, and φ_j is an eigenvector of R_t with eigenvalue $e^{-\nu_j t}, j = 1, \ldots, n$.

From (2.1), (2.2) and the complete orthonormality of $\varphi_1, \dots, \varphi_n$,

$$\sqrt{\pi(i)} P_i[X(t) = j, T_A > t] = \sqrt{\pi(j)} \sum_{k=1}^{n} \varphi_k(i)\varphi_k(j) e^{-\nu_k t}. \tag{2.3}$$

Denote by $\sqrt{\pi}$ the vector $D_\pi^{1/2} 1$; its entries are $\{\sqrt{\pi(i)}, i \in A^c\}$. From (2.3),

$$\sqrt{\pi(i)} P_i(T_A > t) = \sum_{k=1}^{n} (\sqrt{\pi}, \varphi_k)\varphi_k(i) e^{-\nu_k t}. \tag{2.4}$$

Denote by $\gamma_1 < \gamma_2 \dots < \gamma_{\tilde{m}}$ the distinct values among $\nu_1 \dots \nu_n$. Define $S_r = \{k : \nu_k = \gamma_r\}$, and define $\mathcal{S}_r$ to be the subspace generated by $\{\varphi_k, k \in S_r\}$; $\mathcal{S}_r$ is thus the eigenmanifold corresponding to γ_r. Finally, delete from $\gamma_1 \dots \gamma_{\tilde{m}}$ any eigenvalues for which $\sum_{s_r} (\sqrt{\pi}, \varphi_k)^2 = 0$, and relabel the resulting set as $\gamma_1 < \gamma_2 \dots < \gamma_m$. Then (2.4) can be rewritten as,

$$\sqrt{\pi(i)} P_i(T_A > t) = \sum_{r=1}^{m} (P_{\mathcal{S}_r}(\sqrt{\pi}))(i) e^{-\gamma_r t} \tag{2.5}$$

where $P_{\mathcal{S}_r}(\sqrt{\pi})$ is the projection of $\sqrt{\pi}$ on $\mathcal{S}_r$.

Define $p(r) = ||P_{\mathcal{S}_r}(\sqrt{\pi})||^2, r = 1, \dots, m$. From the definition (of $\gamma_1 \dots \gamma_m$), $p(r) > 0, r = 1, \dots, m$, and,

$$\sum_1^m p(r) = ||\sum_1^m P_{\mathcal{S}_r}(\sqrt{\pi})||^2 = ||\sqrt{\pi}||^2 = \pi(A^c) = 1 - \pi(A). \tag{2.6}$$

Next, from (2.3),

$$\begin{aligned} P_\pi(X(t) = j, T_A > t) &= \sqrt{\pi(j)} \sum_{k=1}^{n} (\sqrt{\pi}, \varphi_k)\varphi_k(j) e^{-\nu_k t} \\ &= \sqrt{\pi(j)} \sum_{r=1}^{m} (P_{\mathcal{S}_r}(\sqrt{\pi}))(j) e^{-\gamma_r t}. \end{aligned} \tag{2.7}$$

From either (2.5) or (2.7),

$$P_\pi(T_A > t) = \sum_{r=1}^{m} p(r) e^{-\gamma_r t}. \tag{2.8}$$

Finally we reference for later use an inequality (Aldous and Brown (1992 p.12)),

$$p(1) \geq 1 - \frac{\gamma_1}{\lambda_1}. \tag{2.9}$$

where λ_1 is the second smallest eigenvalue of $-Q$. This leads directly to (Aldous and Brown (1992 p.2))

$$(1 - \frac{\gamma_1}{\lambda_1})e^{-\gamma_1 t} \leq P_\pi(T_A > t) \leq (1 - \pi(A))e^{-\gamma_1 t} \tag{2.10}$$

Inequality (1.1) is a direct application of (2.10).

2.2 Quasi-stationary distribution.

Darroch and Seneta (1965) formalized the concept of a quasi-stationary distribution. In our context, define Q_A to be irreducible if for each pair $x, y \in A^c$, there exists an m, and $\{z_i \in A^c, i = 1, \ldots, m\}$ such that,

$$q_{x,z_1} q_{z_m,y} \prod_{i=1}^{m-1} q_{z_i,z_{i+1}} > 0.$$

Thus, it is possible for the chain to go from state x to state y, without passing through A.

For Q_A irreducible, Darroch and Seneta show that for all $x, j \in A^c, t > 0$

$$\lim_{t\to\infty} P_x(X(t) = j | T_A > t) = \beta(j), \quad \text{and} \tag{2.11}$$

$$\lim_{t\to\infty} P_x(T_A > t + s | T_A > t) = e^{-\gamma_1 s}, \tag{2.12}$$

where,

$$\beta(j) = \frac{\sqrt{\pi(j)}\phi_1(j)}{(\sqrt{\pi}, \phi_1)}, \tag{2.13}$$

and ϕ_1 is the unique eigenvector (up to a constant multiple) of $-Q_A$ corresponding to the eigenvalue γ_1.

If Q_A is not irreducible (but recall that Q is irreducible), then it is still the case that,

$$\lim_{t\to\infty} P_\pi(X(t)=j|T_A>t) = \sqrt{\pi(j)}P_{S_1}(\sqrt{\pi})(j)/p(1), \quad \text{and} \tag{2.14}$$

$$\lim_{t\to\infty} P_\pi(T_A>s+t|T_A>t) = e^{-\gamma_1 s}, \tag{2.15}$$

(2.14) and (2.15) following from (2.7) and (2.8).

Since we will not require that Q_A be irreducible, we will refer to,

$$\alpha(j) \stackrel{\text{def}}{=} \frac{\sqrt{\pi(j)}P_{S_1}(\sqrt{\pi})(j)}{p(1)}, j \in A^c$$

as the quasi-stationary distribution on Q^c, and $\mathcal{L}_\alpha(T_A)$, an exponential distribution with failure rate γ_1, as the quasi-stationary distribution of T_A. Of course when Q_A is irreducible, α and β coincide, and the stronger properties, (2.11) and (2.12) hold.

2.3 Post-recovery Distribution. The post recovery distribution on A^c is defined by,

$$\begin{aligned}\sigma(i) &= \lim_{t\to\infty} P_\pi[X(t)=i|X(t^-)\in A, X(t)\in A^c] \\ &= \sum_{\alpha\in A}\pi(\alpha)q_{\alpha i}/\sum_{\alpha\in A}\pi(\alpha)q_{\alpha A^c} = \pi(i)q_{iA}/\sum_{k\in A^c}\pi(k)q_{kA},\ i\in A^c\end{aligned} \tag{2.16}$$

where $q_{iA}=\sum_{j\in A}q_{ij}$ for $i\in A^c$. The post recovery exit time distribution of T_A is defined by $\mathcal{L}_\sigma T_A$, the distribution of T_A under $X(0)\sim\sigma$. Now,

$$q_{iA} = -\sum_{j\in A^c} q_{ij} = \sum_{j\in A^c,k}\sqrt{\frac{\pi(j)}{\pi(i)}}\varphi_k(i)\varphi_k(j)\nu_k. \tag{2.17}$$

From (2.15),

$$\pi_i q_{iA} = \sqrt{\pi(i)}\sum_k(\sqrt{\pi},\varphi_k)\varphi_k(i)\nu_k = \sqrt{\pi(i)}\sum_{r=1}^m\gamma_r(P_{S_r}(\sqrt{\pi}))(i) \tag{2.18}$$

and,

$$\sum_{i\in A^c}\pi_i q_{iA} = \sum_{r=1}^m \gamma_r p(r) = (\gamma,p), \qquad \text{so that} \tag{2.19}$$

$$\sigma(i) = \sqrt{\pi(i)} \sum_{r=1}^{m} \gamma_r (P_{S_r}(\sqrt{\pi}))(i)/(\gamma, p). \tag{2.20}$$

From (2.5), (2.8) and (2.20),

$$\begin{aligned} P_\sigma(T_A > t) &= \sum_i \sigma(i) P_i(T_A > t) = \sum_{r=1}^{m} \gamma_r p(r) e^{-\gamma_r t}/(\gamma, p) \\ &= f_\pi(t)/f_\pi(0) \end{aligned} \tag{2.21}$$

where f_π is the pdf corresponding to the absolutely continuous component of $\mathcal{L}_\pi T_A$ (all but the atom at $\{0\}$). Thus, as Keilson (1979) observed, the stationary renewal distribution corresponding to $\mathcal{L}_\sigma T_A$ is $\mathcal{L}_{\pi|A^c} T_A$ where $\pi|A^c$ is the restriction of π to $A^c((\pi|A^c)(j) = \pi(j)|\pi(A^c)), j \in A^c$, 0 elsewhere).

2.4 Star chain. The star chain $\{X^*(t), t \geq 0\}$ corresponding to $\{X(t), t \geq 0\}$ and A, is a time reversible Markov chain with state space $\{0, 1, \ldots, m\}$ and transition intensity matrix,

$$Q^*(i,j) = \begin{cases} \gamma_i, i = 1, \ldots, m, \ j = 0 \\ -\gamma_i, j = i \neq 0 \\ \frac{p_j \gamma_j}{\pi(A)}, i = 0, j \neq 0 \\ -\frac{(p,\gamma)}{\pi(A)}, i = j = 0. \end{cases} \tag{2.22}$$

In (2.22) the quantities $\gamma_1 \ldots \gamma_m, p_1 \ldots p_m$ are the same as those appearing in (2.8). It easily follows that the stationary distribution. π^*, of X^* has $\pi^*(i) = p(i), \ i = 0, \ldots, m$ with $p(0) = \pi(A)$. Defining $A^* = \{0\}$ and T_{A^*} as the waiting time to reach $\{0\}$, it easily follows that $\mathcal{L}_{\pi^*} T_{A^*} = \mathcal{L}_\pi T_A$, $\mathcal{L}_{\sigma^*} T_{A^*} = \mathcal{L}_\sigma T_A$ and $\mathcal{L}_{\alpha^*} T_{A^*} = \mathcal{L}_\alpha T_A$ where $\pi^*, \sigma^*, \alpha^*$ are the stationary, post-recovery and quasi-stationary distributions for the star chain.

Note that X and X^* may have an unequal number of states, and π and π^* may be quite different in character. For example we may have π uniform and π^* assigning most of its mass to a single point. Furthermore $Q^*_{A^*}$ is diagonal, while Q_A in most interesting examples will be irreducible. Thus, the resemblance between X and X^* appears to be quite superficial. Nevertheless the results of section 3 show that there are interesting similarities of behavior.

Section 3. Main results.

Consider the m dimensional subspace of R^{A^c} (the set of real-valued functions on A^c), consisting of all linear combinations of $D_\pi^{1/2}P_{S_r}((\sqrt{\pi})), r = 1,\ldots,m$. Call this subspace $\mathcal{S}$. A typical member looks like,

$$x_i = \sqrt{\pi(i)}\sum_{r=1}^{m}\tilde{x}(r)[P_{S_r}(\sqrt{\pi})](i), \quad i \in A^c \tag{3.1}$$

Define a mapping V from R^m to $\mathcal{S}$ by,

$$Vy(i) = \sqrt{\pi(i)}\sum_{r=1}^{m}\frac{y(r)}{p(r)}(P_{S_r}(\sqrt{\pi}))(i), \quad i \in A^c \tag{3.2}$$

where as before $p(r) = ||P_{S_r}(\sqrt{\pi})||^2$.

Note that $D_\pi^{-1/2}(Vy)$ is chosen to have the same Fourier coefficients with respect to $P_{S_r}(\sqrt{\pi})/\sqrt{p(r)}, r = 1,\ldots,m$ as $D_p^{-1/2}y$ has with respect to $e_r, r = 1,\ldots,m$ where $e_r(j) = \delta(r,j), j = 1,\ldots,m$.

Several properties of V are found in Lemma (3.1) below.

Theorem 3.1 The map V from R^m to $\mathcal{S}$, defined above, has the following properties:

(i) $(Vy, Vz)_{\pi^{-1}} \stackrel{\text{def}}{=} \sum_{A^c}\frac{(Vy)_i(Vz)_i}{\pi(i)} = \sum_1^m \frac{y(r)z(r)}{p(r)} \stackrel{\text{def}}{=} (y,z)_{p^{-1}}$

(ii) V is linear, one to one and onto.

(iii) $\sum_{A^c}(Vy)_i = \sum_1^m y(r)$

(iv) $\sum_{A^c}(Vy)_iP_i(T_A > t) = \sum_1^m y(r)e^{-\gamma_r t}$; Vy is the unique element in $\mathcal{S}$ with this property.

(v) If $x \in R^{A^c}$ satisfies $\sum_{A^c} x_i P_i(T_A > t) = \sum_1^m \tilde{x}(r) e^{-\gamma_r t}$, then $||x||_{\pi^{-1}} \geq ||V\tilde{x}||_{\pi^{-1}} = ||\tilde{x}||_{p^{-1}}$, with equality if and only if $x = V\tilde{x}$.

(vi) If w and Vw are probability vectors (over A^{*c} and A^c respectively) then $\mathcal{L}_{Vw} T_A = \mathcal{L}_w T_{A^*}$ and,

$$\chi_1^2(Vw, \pi) \stackrel{\text{def}}{=} \sum_I \frac{((Vw)_i - \pi(i))^2}{\pi(i)} = \Big(\sum_{A^c} \frac{(Vw)_i^2}{\pi(i)}\Big) - 1 = \Big(\sum_1^m \frac{w^2(r)}{p(r)}\Big) - 1$$
$$= \sum_0^m \frac{(w(r) - p(r))^2}{p(r)} \stackrel{\text{def}}{=} \chi_2^2(w, p).$$

Proof. (i) From (3.2),

$$(Vy, Vz)_{\pi^{-1}} = \Big(\sum \frac{y(r)}{p(r)} P_{\mathcal{S}_r}(\sqrt{\pi}),\ \sum \frac{z(r)}{p(r)} P_{\mathcal{S}_r}(\sqrt{\pi})\Big)$$
$$= \sum \frac{y(r)z(r)}{p(r)} = (y, z)_{p^{-1}}.$$

(ii) V is obviously linear. If $x \in \mathcal{S}$ is given by (3.1) then $x = Vy$ with $y(r) = \tilde{x}(r)p(r)$, thus V is onto. If $Vx_1 = Vx_2$ then $||V(x_1 - x_2)||_{\pi^{-1}} = ||x_1 - x_2||_{p^{-1}} = 0$ (by (i)), thus $x_1 = x_2$ and V is one to one.

(iii) From (3.2),

$$\sum (Vy)_i = \sum \frac{y(r)}{p(r)} (\sqrt{\pi}, P_{\mathcal{S}_r}(\sqrt{\pi})) = \sum \frac{y(r)}{p(r)} p(r) = \sum y(r)$$

(iv) From (2.5) and (3.2),

$$\sum (Vy)_i P_i(T_A > t) = \Big(\sum \frac{y(r)}{p(r)} P_{\mathcal{S}_r}(\sqrt{\pi}), \sum_r P_{\mathcal{S}_r}(\sqrt{\pi}) e^{-\gamma_r t}\Big)$$
$$= \sum_r y(r) e^{-\gamma_r t}.$$

Uniqueness follows since if $z \in \mathcal{S}$, then by (ii), $z = Vy'$ for a unique $y' \in R^m$. Thus if $\sum z_i P_i(T_A > t) = \sum (Vy)_i P_i(T_A > t)$, then $\sum (y(r) - y'(r)) e^{-\gamma_r t} = 0$ for all $t \geq 0$, thus $y = y'$ and $z = Vy$.

(v) From (2.5),

$$\sum x_i P_i(T_A > t) = \sum (D_\pi^{-1/2} x, P_{S_r}(\sqrt{\pi}))e^{-\gamma_r t}.$$

Thus $\tilde{x}(r) = (P_{S_r}(D_\pi^{-1/2}x), P_{S_r}(\sqrt{\pi}))$. By Cauchy-Schwartz,

$$\tilde{x}^2(r) \le ||P_{S_r}(D_\pi^{-1/2}x)||^2 \; ||P_{S_r}(\sqrt{\pi})||^2 = p(r)||P_{S_r}(D_\pi^{-1/2}x)||^2 \tag{3.3}$$

with equality if and only if $P_{S_r}(D_\pi^{-1/2}x) = \frac{\tilde{x}(r)}{p(r)} P_{S_r}(\sqrt{\pi})$. Thus, from (3.3),

$$\sum_r \frac{(\tilde{x}(r))^2}{p(r)} \le \sum_r ||P_{S_r}(D_\pi^{-1/2}x)||^2 \le ||D_\pi^{-1/2}x||^2 = \sum_i \frac{x_i^2}{\pi(i)},$$

with equality if and only if $D_\pi^{-1/2}x = \sum \frac{\tilde{x}(r)}{p(r)} P_{S_r}(\sqrt{\pi})$, thus if and only if $x = V\tilde{x}$.

(vi) $\mathcal{L}_{Vw}T_A = \mathcal{L}_w T_{A^*}$ by (iv). The chi-square equality follows from (i). ◇

We next consider product moment identities for certain conditional expectations.

Corollary 3.2 (i) Let g_1, g_2 be functions satisfying $\int |g_i(x)|e^{-\gamma_1 x}dx < \infty, i = 1, 2$. Define $\beta_{g_j}(i) = E_i(g_j(T_A)), i \in I, j = 1, 2$, and $\beta^*_{g_j}(r) = E_r(g_j(T_{A^*})), r = 0, \ldots, m, j = 1, 2$. Then

$$E_\pi[\beta_{g_1}(X(0))\beta_{g_2}(X(0))] = E_p[\beta^*_{g_1}(X^*(0))\beta^*_{g_2}(X^*(0))]$$

(ii) If, in addition, g_1, g_2 are distribution functions of measures μ_1, μ_2 on $[0, \infty)$ with $\mu_j\{0\} = 0, j = 1, 2$, then,

$$E_\pi[\beta_{g_1}(X(0))\beta_{g_2}(X(0))] = E_\pi[(g_1 * g_2)(T_A)] = E_p[(g_1 * g_2)(T_{A^*})]$$

where,

$$g_1 * g_2(t) = \int_0^t g_1(t - x)dg_2(x),$$

the distribution function of $\mu_1 * \mu_2$.

Proof. (i) Define $w_j(r) = p(r)\beta^*_{g_j}(r) = p(r)\int_0^\infty g_j(t)\gamma_r e^{-\gamma_r t}dt, r = 1, \ldots, m, j = 1, 2.$

From (2.5) and (3.2),

$$\begin{aligned}\pi(i)\beta_{g_j}(i) &= \sqrt{\pi(i)}\sum_{r=1}^{m}(P_{S_r}(\sqrt{\pi}))(i)\beta^*_{g_j}(r) \\ &= (Vw_j)(i), \quad i \in A^c\end{aligned} \tag{3.4}$$

Thus by Theorem (3.1) (i) and (3.4),

$$\begin{aligned}E_\pi[\beta_{g_1}(X(0))\beta_{g_2}(X(0))] &= \pi(A)g_1(0)g_2(0) + (Vw_1, Vw_2)_{\pi^{-1}} \\ &= p(0)g_1(0)g_2(0) + (w_1, w_2)_{p^{-1}} = E_p[\beta^*_{g_1}(X^*(0))\beta^*_{g_2}(X^*(0))].\end{aligned} \tag{3.5}$$

(ii) Since $g_1(0) = g_2(0) = 0$ and $g(t)$ is a distribution function,

$$\begin{aligned}\beta^*_{g_j}(r) &= E_r\int_0^{T_{A^*}} dg_j(t) = E_r\int_0^\infty I_{T_{A^*}>t}\, dg_j(t) \\ &= \int_0^\infty e^{-\gamma_r t}dg_j(t) \stackrel{\text{def}}{=} \psi_{g_j}(\gamma_r), \quad j = 1, 2.\end{aligned} \tag{3.6}$$

Thus,

$$(w_1, w_2)_{p^{-1}} = \sum_{r=1}^{m} p(r)\psi_{g_1}(\gamma_r)\psi_{g_2}(\gamma_r) = \sum_{r=1}^{m} p(r)\psi_{g_1 * g_2}(\gamma_r). \tag{3.7}$$

Next,

$$\begin{aligned}E_\pi[(g_1 * g_2)T_A] &= E_p[(g_1 * g_2)(T_{A^*})] = E_p[\beta^*_{g_1 * g_2}(X^*(0))] \\ &= \sum_{r=1}^{m} p(r)\psi_{g_1 * g_2}(\gamma_r) \quad \text{(by (3.6))}.\end{aligned} \tag{3.8}$$

The result follows from (3.5), (3.7) and (3.8).

Corollary 3.3. Suppose that w is a probability distribution on $\{1, \ldots, m\}$, and Vw a probability distribution on A^c. Define $\mathcal{L}_{Vw}[X(t)|T_A > t](\mathcal{L}_w([X^*(t)|T_{A^*} > t])$ to be the conditional distribution of $X(t)(X^*(t))$ given $T_A > t$ $(T^*_A > t)$ under $X(0) \sim Vw(X^*(0)) \sim w)$. Then,

$$V(\mathcal{L}_w(X^*(t)|T_{A^*} > t)) = \mathcal{L}_{Vw}(X(t)|T_A > t) \tag{i}$$

(ii) $$\chi_1^2(\mathcal{L}_{Vw}(X(t)|T_A > t), \pi) = \chi_2^2(\mathcal{L}_w(X^*(t)|T_{A^*} > t), p)$$

(iii)
$$\begin{aligned}
&\sum_{i\in A^c} (\pi_i)^{-1} P_{Vw}(X(t) = i|T_A > t) P_{Vw}(X(s) = i|T_A > s) \\
&= \sum_{r=1}^{m} (p(r))^{-1} P_w(X^*(t) = r|T_{A^*} > t) P_w(X^*(s) = r|T_{A^*} > s) \\
&= \sum_{r=1}^{m} \frac{(w(r))^2}{p(r)} e^{-\gamma_r(t+s)} / P_{Vw}(T_A > t) P_{Vw}(T_A > s)
\end{aligned}$$

Proof (i). From (2.3) and (3.2),

$$\begin{aligned}
P_{Vw}[X(t) = j, T_A > t] &= \sum_{i\in A^c} (Vw)_i P_i[X(t) = j, T_A > t] \\
&= \sqrt{\pi(j)} \sum_{r=1}^{m} e^{-\gamma_r t} \frac{w(r)}{p(r)} \sum_{k\in S_r} \varphi_k(j)(P_{S_r}(\sqrt{\pi}), \varphi_k) \qquad (3.9)\\
&= \sqrt{\pi(j)} \sum_{r=1}^{m} \frac{w(r)}{p(r)} (P_{S_r}(\sqrt{\pi}))(j) e^{-\gamma_r t}.
\end{aligned}$$

Furthermore,

$$P_w[X^*(t) = r, T_{A^*} > t] = w(r) e^{-\gamma_r t}. \qquad (3.10)$$

The result now follows from (3.2), (3.9) and (3.10).

(ii) Follows from (i) and Theorem (3.1), (vi).

(iii) Follows from (i), (3.10) and Theorem (3.1), (i).

Section 4. Examples.

Example (i) Choose $g_1(t) = g_2(t) = t^\alpha, \alpha > 0$.

Then $\psi_g(s) = \int e^{-st} \alpha t^{\alpha-1} dt = \Gamma(\alpha+1)/s^\alpha$. Applying Corollary (3.2) we obtain,

$$E_\pi[E^2(T_A^\alpha|X(0))] = \Gamma^2(\alpha+1) \sum \frac{p(r)}{\gamma_r^{2\alpha}} = \frac{\Gamma^2(\alpha+1)}{\Gamma(2\alpha+1)} E_\pi(T_A^{2\alpha}) \qquad (4.1)$$

Define $c_\alpha = E_\pi T_A^\alpha / \Gamma(\alpha+1)$, then from (4.1),

$$\text{Var}_\pi[E(T_A^\alpha | X(0))]/(E_\pi T_A^\alpha)^2 = \frac{c_{2\alpha}}{c_\alpha^2} - 1. \tag{4.2}$$

Assume $\gamma_1 < \lambda_1$. Multiply all three components of (2.10) by $\alpha t^{\alpha-1}$ and integrate t from zero to ∞, obtaining

$$(1 - \frac{\gamma_1}{\lambda_1})\gamma_1^{-\alpha} \leq c_\alpha \leq (1 - \pi(A))\gamma_1^{-\alpha}. \tag{4.3}$$

From (4.3),

$$\frac{c_{2\alpha}}{c_\alpha^2} - 1 \leq \frac{2\frac{\gamma_1}{\lambda_1} - (\frac{\gamma_1}{\lambda_1})^2 - \pi(A)}{(1 - \frac{\gamma_1}{\lambda_1})^2} \leq 2\frac{\gamma_1}{\lambda_1}(1 - \frac{\gamma_1}{\lambda_1})^{-2}. \tag{4.4}$$

Thus when γ_1/λ_1 is small, so is the squared coefficient of variation of $E(T_A^\alpha | X(0))$.

Define $T_A(t)$ to be the waiting time starting at t to reach A. Set $h(i) = E_i(T_A^\alpha)$, then $E_i h(X(t)) = E[T_A^\alpha(t) | X(0) = i]$. Aldous and Brown (1992 p.7) derive for general h,

$$|E_i h(X(t)) - E_\pi h(X(0))| \leq \sqrt{\frac{1 - \pi(i)}{\pi(i)} \text{Var}_\pi h(X(0))}\; e^{-\lambda_1 t}. \tag{4.5}$$

Applying (4.2) - (4.5) with $h(i) = E_i T_A^\alpha$ we obtain,

$$\sup_{\alpha \geq 0} \frac{|E_i T_A^\alpha(t) - E_\pi T_A^\alpha|}{E_\pi T_A^\alpha} \leq (1 - \frac{\gamma_1}{\lambda_1})^{-1} \sqrt{(2\frac{\gamma_1}{\lambda_1})\frac{1 - \pi(i)}{\pi(i)}}\; e^{-\lambda_1 t}. \tag{4.6}$$

Moreover, for general w, we can replace $\sqrt{\frac{1-\pi(i)}{\pi(i)}}$ on the righthand side of (4.6) by $x_1(w, \pi)$.

Inequality (1.4) is the specialization of (4.6) to the reliability example with $i = \mathbf{1}$ and,

$$\pi(\mathbf{1}) = \prod_1^4 [\beta_i | \alpha_i + \beta_i] = .822367.$$

Example (ii). For X completely monotone, Keilson (1979) suggested $\rho = [(EX^2/2(EX)^2) - 1]$ as a measure of departure between X and an exponential distribution with mean EX. The author (Brown (1983)) showed that small ρ indeed implies small sup norm distance between the survival function of X and that of the approximating exponential distribution.

A further property of ρ follows from example (i). Setting $\alpha = 1$ in (4.2) yields,

$$\text{Var}_\pi E(T_A|X(0))/(E_\pi T_A)^2 = \rho \tag{4.7}$$

Thus ρ is the squared coefficient of variation of $E[T_A|X(0)]$. A small value of ρ indicates that the variance of $E(T_A|X(0))/E_\pi T_A$ is small under $X(0) \sim \pi$. For chains with π uniform and ρ small, (4.7) can be interpreted to mean that most of the quantities $\{E_i T_A/E_\pi T_A, i \in A^c\}$ are close to 1.

Some consequences of a small coefficient of variation for $E(T_A|X(0))$ are explored in Aldous and Brown (1992).

Example (iii). For $s, t > 0$ define $\mu_1(\mu_2)$ to be a one point probability distribution concentrated at $s(t)$. Then $\mu_1 * \mu_2$ is a one point probability distribution concentrated at $s + t$. From Corollary (3.2), (ii),

$$P_\pi(T_A > s + t) = \sum_{i \in A^c} \pi_i P_i(T_A > t) P_i(T_A > s). \tag{4.8}$$

Equivalently,

$$\sum_{i \in A^c} \pi_i [P_i(T_A > s + t) - P_i(T_A > s) P_i(T_A > t)] = 0. \tag{4.9}$$

Thus "on average" T_A behaves as if it were conditionally exponential given $X(0)$ although typically $T_A|X(0) = i$ will not be exponential for any i.

Integrate both sides of (4.8) with respect to s from 0 to ∞ to obtain,

$$\sum_{i\in A^c} \pi(i) E_i T_A[\bar{G}_i(t) - P_i(T_A > t)] = 0 \tag{4.10}$$

where $\bar{G}_i(t) = \int_t^\infty P_i(T_A > s)ds | E_i T_A$, the stationary renewal distribution corresponding to $\mathcal{L}_i T_A$. Thus once again T_A behaves in an average sense as if it were conditionally exponential given $X(0)$.

David Aldous (personal communication) points out that (4.8) follows from the observation that for the stationary version of $\{X(t), -\infty < t < \infty\}$ on the whole real line, that $T_A = \inf\{t \geq 0 : X(t) \in A\}$ and $\tilde{T}_A = -\sup\{t \leq 0 : X(t) \in A\}$ are conditionally i.i.d. given $X(0)$.

Example (iv). Recall the quasi-stationary distribution discussed in Section 2.2. For the star chain the quasi-stationary distribution, α^*, is a one point probability distribution at $\{1\}$. Thus from (2.13) and (3.2),

$$V\alpha^*(i) = \sqrt{\pi(i)} P_{S_1}(\sqrt{\pi})(i)/p(1) = \alpha(i)$$

thus $V\alpha^* = \alpha$ and Theorem 3.1, (vi) yields,

$$\chi_1^2(\alpha, \pi) = \sum_{i\in A^c} \frac{\alpha_i^2}{\pi_i} - 1 = \chi_2^2(\alpha^*, p) = \frac{1}{p_1} - 1. \tag{4.11}$$

When $\gamma_1 < \lambda_1$, the bound (2.9) combines with (4.11) to give,

$$\chi_1^2(\alpha, \pi) \leq \frac{\gamma_1}{\lambda_1 - \gamma_1}. \tag{4.12}$$

which improves slightly upon a bound given in Aldous and Brown ((1992) p.9).

Example (v). Recall the post-recovery distribution of Section (2.3). For the star chain,

$$\sigma^*(r) = \frac{p(r)\gamma(r)}{(p, \gamma)}, \quad r = 1, \ldots, m. \tag{4.13}$$

From (2.16) and (4.13) we see that $V\sigma^* = \sigma$. It thus follows from Theorem 3.1 that,

$$\chi_1^2(\sigma,\pi) = \chi_2^2(\sigma^*,p) = \frac{(p,\gamma^2)}{(p,\gamma)^2} - 1$$
$$= \frac{f_\sigma(0)}{f_\pi(0)} - 1$$

where f_σ is the pdf of $\mathcal{L}_\sigma T_A$.

Since $||\sigma||^2_{\pi^{-1}} = ||\sigma^*||^2_{p^{-1}}$ (Theorem (3.1)) it also holds that,

(4.15)
$$\sum_{i\in A^c} \pi_i q_{iA}^2 = \sum_1^m p(r)\gamma^2(r).$$

Combining (2.6), (2.17) and (4.15) we have,

(4.16)
$$\sum_{i\in A^c} \pi_i q_{iA}^j = \sum_{r=1}^m p(r)\gamma^j(r), \quad j = 0, 1, 2.$$

In (4.16), the identities for $j = 0$ and 1 are well known but $j = 2$ (4.15) appears to be new.

Example (vi). Observe that $Vp(i) = \sqrt{\pi}(i)\sum_{r=1}^m (P_{S_r}(\sqrt{\pi}))(i) = \pi(i)$, $i \in A^c$, thus $\pi = Vp$. Define $\pi_t(j) = P_\pi(X(t) = j|T_A > t)$. From Corollary (3.3), (iii),

(4.18)
$$\sum_{j\in A^c} \frac{\pi_t(j)\pi_s(j)}{\pi(j)} = \sum_{r=1}^m \frac{\pi_t^*(r)\pi_s^*(r)}{p(r)} = \frac{\sum_{r=1}^m p(r)e^{-\gamma_r(t+s)}}{P_\pi(T_A > t)P_\pi(T_A > s)}$$
$$= \frac{P_\pi(T_A > s+t)}{P_\pi(T_A > t)P_\pi(T_A > s)}.$$

As $t \to \infty$, $\pi_t \to \alpha$, the quasi-stationary distribution. Now,

(4.19)
$$||\pi_t - \alpha||^2_{\pi^{-1}} = ||\pi_t^* - \alpha^*||^2_{p^{-1}} = (P_\pi(T_A > t))^{-2}[P_\pi(T_A > 2t) - 2e^{-\gamma_1 t}P_\pi(T_A > t) + \frac{1}{p_1}P_\pi^2(T_A > t)]$$
$$= (P_\pi(T_A > t))^{-2}R(t)$$

where $R(t)$ is the expression in brackets on the right side of (4.20). From (2.8), (2.10) and (4.19) we find,

(4.20)
$$R(t) = \sum_{r=2}^m p(r)e^{-2\gamma_r t} + p_1^{-1}(\sum_{r=2}^m p(r)e^{-\gamma_r t})^2$$
$$\leq \frac{(1-\pi(A))(1-\pi(A)-p((1))}{p(1)}\, e^{-2\gamma_2 t}.$$

Assume $\frac{\gamma_1}{\lambda_1} < 1$; from (2.9), (2.10), (4.19) and (4.20) we obtain,

$$(4.21)\qquad ||\pi_t - \alpha||^2_{\pi^{-1}} \leq (1 - \frac{\gamma_1}{\lambda_1})^{-3}(1 - \pi(A))(\frac{\gamma_1}{\lambda_1} - \pi(A))e^{-2(\gamma_2 - \gamma_1)t}.$$

Using (4.21) and an argument relating total variation distance to chi-square distance (Diaconis and Stroock (1991 p. 42)),

$$(4.22)\qquad \begin{aligned} \max_{B \subset A^c} |\pi_t(B) - \alpha(B)| &= \frac{1}{2}\sum_{j \in A^c} \frac{|\pi_t(j) - \alpha(j)|}{\pi(j)}\pi(j) \leq \frac{1}{2}||\pi_t - \alpha||_{\pi^{-1}} \\ &\leq \frac{1}{2}(1 - \frac{\gamma_1}{\lambda_1})^{-3/2}\,[(1 - \pi(A))(\frac{\gamma_1}{\lambda_1} - \pi(A))]^{1/2}\,e^{-(\gamma_2 - \gamma_1)t}. \end{aligned}$$

Next, it follows from Aldous and Brown ((1992, p.12)) and Brown ((1994, p.13)) that $\lambda_1 \leq \lambda_1^* < \gamma_2$ where λ_1^* is the second smallest eigenvalue of $-Q^*$. Thus,

$$(4.23)\qquad \gamma_2 - \gamma_1 > \lambda_1 - \gamma_1.$$

Finally, from (4.22) and (4.23), if $\gamma_1 < \lambda_1$ then

(4.24)

$$\max_{B \subset A^c} |P_\pi(X_t \epsilon B | T_A > t) - \alpha(B)| \leq \frac{1}{2}(1 - \frac{\gamma_1}{\lambda_1})^{-3/2}[(\frac{\gamma_1}{\lambda_1} - \pi(A)) \cdot (1 - \pi(A))]^{1/2}e^{-(\lambda_1 - \gamma_1)t}.$$

Inequality (1.5) of the introduction applies (4.24) to the reliability example, noting that,

$$\pi(A) = [\prod_1^4 \frac{\alpha_i}{\alpha_i + \beta_i}][1 + \sum_1^4 \frac{\beta_i}{\alpha_i}] \approx 4.089 \times 10^{-4}.$$

Example (vii) Our purpose in this example is to derive the inequality,

$$(4.25)\qquad \sup_s \frac{|P_w[T_A(t) > s) - P_\pi(T_A > s)|}{P_\pi(T_A > s)} \leq \frac{1}{2}\chi_1(w, \pi)(1 - \frac{\gamma_1}{\lambda_1})^{-1}e^{-\lambda_1 t}.$$

This inequality was applied to the reliability example resulting in (1.3).

In Section 3 we dealt with distributions on A^c or (for the star chain) on $\{1, \ldots, m\}$. Now it will be convenient to allow for mass at A or (for the star chain) at $\{0\}$. The modification is rather trivial. Represent a vector on $\{0,1,\ldots,m\}$ by $w = \binom{w(0)}{\tilde{w}}$ where $\tilde{w} = \begin{pmatrix} w(1) \\ \vdots \\ w(m) \end{pmatrix}$. The range of $\tilde{V}$ (the analog of V) is $R^{A^c \cup \{a\}}$, where a represents the collapsing of A to a single point. Define $\tilde{V}$ by,

$$\tilde{V}\begin{pmatrix} w(0) \\ \tilde{w} \end{pmatrix} = \begin{pmatrix} w(0) \\ V\tilde{w} \end{pmatrix}, \qquad \text{so that}$$

$(\tilde{V}w)(a) = w(0)$ and $(\tilde{V}w)(i) = (V\tilde{w})(i)$ for $i \epsilon A^c$. The inner products are then modified to,

$$(w,z)_{p-1} = \frac{w(0)z(0)}{\pi(A)} + \sum_1^m \frac{w(r)z(r)}{p(r)}, \quad \text{and}$$

$$(x,y)_{\pi^{-1}} = \frac{x(a)y(a)}{\pi(A)} + \sum_{A^c} \frac{x(i)y(i)}{\pi(i)}.$$

We see that,

$$\begin{aligned}(\tilde{V}w, \tilde{V}z)_{\pi^{-1}} &= \frac{w(0)z(0)}{\pi(A)} + (V\tilde{w}, V\tilde{z})_{\pi^{-1}} \\ &= \frac{w(0)z(0)}{\pi(A)} + (\tilde{w}, \tilde{z})_{p^{-1}} = (w,z)_{p^{-1}},\end{aligned}$$

so that the properties of V (given in Theorem (3.1)) extend to $\tilde{V}$.

Suppose now that $P_w(T_A > t) = \sum_{r=1}^m w^*(r)e^{-\gamma_r t}$. It follows that,

$$w^*(0) = 1 - \sum_1^m w^*(r) = P_w(T_A = 0) = w(A) = 1 - \sum_{A^c} w(i). \tag{4.26}$$

Moreover from Theorem (3.1)(v), and (4.25),

$$\begin{aligned}\chi_2^2(w^*,p) &\stackrel{\text{def}}{=} \sum_0^m \frac{(w^*(i)-p(i))^2}{p(i)} = \frac{(w^*(0)-\pi(A))^2}{\pi(A)} + \sum_1^m \frac{(w^*(r)-p(r))^2}{p(r)} \\ &\le \frac{(w^*(0)-\pi(A))^2}{\pi(A)} + \sum_{A^c} \frac{(w(i)-\pi(i))^2}{\pi(i)} \stackrel{\text{def}}{=} \chi_1^2(w,\pi).\end{aligned} \tag{4.27}$$

Next, define $TV(w^*,p)$ to be the total variation distance between w^* and p. Thus,

$$TV(w^*,p) = \frac{1}{2}\sum_0^m |w^*(r)-p(r)| = \sum_0^m (w^*(r)-p(r))^+ = \sum_0^m (w^*(r)-p(r))^-, \tag{4.28}$$

where $a^+ = \max(0, a), a^- = \max(0, -a)$.

Now,

$$P_w(T_A > s) - P_\pi(T_A > s) = \sum_1^m (w^*(r) - p(r))e^{-\gamma_r s} \le e^{-\gamma_1 s}\sum_1^m (w^*(r) - p(r))^+ \tag{4.29}$$
$$\le e^{-\gamma_1 s} TV(w^*, p), \quad \text{and}$$

$$P_\pi(T_A > s) - P_w(T_A > s) \le e^{-\gamma_1 s}\sum_1^m (w^*(r) - p(r))^- \le e^{-\gamma_1 s} TV(w^*, p). \tag{4.30}$$

Gathering together (4.27) - (4.30) and applying the Diaconis-Stroock argument as in (4.22) we obtain,

(4.31)

$$|P_w(T_A > s) - P_\pi(T_A > s)| \le TV(w^*, p)e^{-\gamma_1 s} \le \frac{1}{2}\chi_2(w^*, p)e^{-\gamma_1 s} \le \frac{1}{2}\chi_1(w, \pi)e^{-\gamma_1 s}.$$

Next, observe that,

$$P_w(T_A(t) > s) = \sum_{i,j} w(i)P_{ij}(t)P_j(T_A > s) = \sum_j w_t(j)P_j(T_A > s) \tag{4.32}$$
$$= P_{w_t}(T_A > s).$$

Applying (4.31) with w replaced by w_t (in view of (4.32)) we find,

$$|P_w(T_A(t) > s) - P_\pi(T_A > s)| \le \frac{1}{2}\chi_1(w_t, \pi)e^{-\gamma_1 s}. \tag{4.33}$$

Next, we recall a result of Fill ((1991) p.72),

$$\chi_1(w_t, \pi) \le \chi_1(w, \pi)e^{-\lambda_1 t}. \tag{4.34}$$

Finally (4.32) - (4.34) and (2.10) combine to give,

$$\frac{|P_w(T_A(t) > s) - P_\pi(T_A > s)|}{P_\pi(T_A > s)} \le \frac{1}{2}\chi_1(w, \pi)(1 - \frac{\gamma_1}{\lambda_1})^{-1}e^{-\lambda_1 t}. \tag{4.35}$$

Since the righthand side of (4.34) is independent of s, (4.25) now follows.

Example (viii) For our last example we will derive an upper bound for $P_w(T_A > t)$,

$$P_w(T_A > t) \le \frac{1}{2}[w(A^c) + (\pi(A^c)\sum_{A^c}\frac{w^2(i)}{\pi(i)})^{1/2}]e^{-\gamma_1 t}. \tag{4.36}$$

The coefficient of $e^{-\gamma_1 t}$ in (4.35) is the average of two quantities, the smaller of which is $w(A^c)$.

This inequality was applied to the reliability example, resulting in (1.2).

To prove (4.36) note that,

$$P_w(T_A > t) = \sum_1^m w^*(r)e^{-\gamma_r t} \le e^{-\gamma_1 t}\sum_1^m (w^*(r))^+ \tag{4.37}$$

Now,

$$\sum_1^m (w^*(r)^+ + \sum_1^m (w^*(r))^- = \sum_1^m |w^*(r)|, \quad \text{and}$$
$$\sum_1^m (w^*(r))^+ - \sum_1^m (w^*(r))^- = w(A^c), \quad \text{thus}$$

$$\sum_1^m (w^*(r))^+ = \frac{1}{2}[\sum_1^m |w^*(r)| + w(A^c)] \tag{4.38}$$

Next,

$$\begin{aligned}\sum_1^m |w^*(r)| &= \pi(A^c)[\frac{1}{\pi(A^c)}\sum_1^m(\frac{|w^*(r)|}{p(r)})p(r)] \\ &\le \pi(A^c)[\frac{1}{\pi(A^c)}\sum_1^m\frac{(w^*(r))^2}{p(r)}]^{1/2} = [\pi(A^c)\sum_1^m\frac{(w^*(r))^2}{p(r)}]^{1/2}.\end{aligned} \tag{4.39}$$

Finally from Theorem (3.1)(v),

$$\sum_1^m \frac{(w^*(r))^2}{p(r)} \le \sum_{A^c}\frac{w^2(i)}{\pi(i)}. \tag{4.40}$$

The result (4.36) now follows from (4.37) - (4.40).

To derive a related result, denote the right side of (4.36) as $\tilde{w}e^{-\gamma_1 t}$. It follows from (4.36) that,

$$w^*(1) = \lim_{t\to\infty}[e^{\gamma_1 t}P_w(T_A > t)] \le \lim_{t\to\infty}[e^{\gamma_1 t}(\tilde{w}e^{-\gamma_1 t})] = \tilde{w},$$

thus,

$$0 \le w^*(1) \le \frac{1}{2}[w(A^c) + (\pi(A^c)\sum_{A^c}\frac{w^2(i)}{\pi(i)})^{1/2}]. \tag{4.41}$$

Section 5. Comments and Additons.

Section 5.1. For the repairable system of n independent components with failure rates α_i and repair rates $\beta_i, i = 1, \ldots, n$, describe a state x by,

$$x_i = 1, i\epsilon B, x_i = 0, i\epsilon B^c \tag{5.1}$$

where B is a subset of $\{1, 2, \ldots, n\}$. (We allow B to be empty). The transition intensity rates are given by,

$$q(x,y) = \begin{cases} \alpha_i, y_i = 0, y_j = x_j, j \ne i, & \text{for some } i\epsilon B \\ \beta_i, y_i = 1, y_j = x_j, j \ne i, & \text{for some } i\epsilon B^c \\ -(\sum_B \alpha_i + \sum_{B^c}\beta_j), y = x \\ 0, elsewhere. \end{cases}$$

We now show that the eigenvalues of $-Q$ are,

$$\{\sum_{i\epsilon B}(\alpha_i + \beta_i), B \subset \{1, 2, \ldots, n\}\}. \tag{5.2}$$

The matrix, $-Q$, has a single eigenvalue equal to zero, and has as its second smallest eigenvalue the quantity,

$$\lambda_1 = \min_i(\alpha_i + \beta_i).$$

To prove (5.2) we appeal to the spectral representation,

$$P_t(x,x) = Pr(X(t) = x|X(0) = x) = \pi(x) + \sum \psi_k^2(x)e^{-v_k t} \tag{5.3}$$

where $\pi(x)$ is the stationary probability of x, $\{v_k\}$ are the non-zero eigenvalues of $-Q$, and $\{\psi_k\}$ are the eigenvectors corresponding to $\{v_k\}$.

Expression (5.3) follows from the spectral representation of the matrix Q, by an argument very similar to that given for Q_A in Section 2.1.

Furthermore,

$$\sum_{\{k:v_k=\lambda_r\}} \sum_x \psi_k^2(x) = \text{multiplicity of } \lambda_r$$

where $\{\lambda_r\}$, is the set of distinct values from $\{v_k\}$.

It follows that λ is an eigenvalue of $-Q$ if and only if for some $x \epsilon I, P_t(x,x)$ possesses a term of the form $ce^{-\lambda x}$, with $c > 0$.

For the repairable system model, consider x, defined by (5.1). Note that,

$$(5.5) \qquad P_t^{(i)}(0,0) \stackrel{\text{def}}{=} Pr(X_i(t) = 0 | X_i(0) = 0) = \frac{\alpha_i}{\alpha_i + \beta_i} + \frac{\beta_i}{\alpha_i + \beta_i} e^{-(\alpha_i + \beta_i)t},$$

and

$$(5.6) \qquad P_t^{(i)}(1,1) \stackrel{\text{def}}{=} Pr(X_i(t) = 1 | X_i(0) = 1) = \frac{\beta_i}{\alpha_i + \beta_i} + \frac{\alpha_i}{\alpha_i + \beta_i} e^{-(\alpha_i + \beta_i)t}.$$

Next,

$$(5.7) \qquad P_t(x,x) = (\prod_{i \epsilon B} P_t^{(i)}(1,1))(\prod_{i \epsilon B^c} P_t^{(i)}(0,0)).$$

Define $\mathcal{D}$ to be the collection of non-empty subsets of $\{1, 2, \ldots, n\}$. Then substituting (5.5), and (5.6) into (5.7) we find,

$$(5.8) \qquad P_t(x,x) = \Pi(x) + \sum_{D \epsilon \mathcal{D}} (\prod_{D_1} \frac{\beta_i}{\alpha_i + \beta_i})(\prod_{D_2} \frac{\alpha_i}{\alpha_i + \beta_i}) e^{-(\sum_D (\alpha_i + \beta_i))t}.$$

where $D_1 = (D \cap B^c) \cup (D^c \cap B)$ and $D_2 = (D \cap B) \cup (D^c \cap B^c)$.

As the coefficient of $e^{-(\sum_D (\alpha_i + \beta_i))t}$ is positive for all $D \epsilon \mathcal{D}$ we see that zero and every $\sum_D (\alpha_i + \beta_i)$ are eigenvalues of $-Q$ and that there are no other eigenvalues. The smallest positive eigenvalue of $-Q$ is thus, $\min(\alpha_i + \beta_i)$.

The general principle that follows from this argument is that if $\mathbf{X}(t) = (X_1(t), \ldots, X_n(t))$ is a Markov chain, obtained from the product of independent, finite state, time reversible chains, then the distinct eigenvalues of $-Q$ are the distinct values from among,

$$\{\sum_{i=1}^{n} \lambda_{r_i}^{(i)}, r_i \epsilon M_i, i = 1, \ldots, n\}$$

where M_i is the set of distinct eigenvalues of $-Q_i, i = 1, \ldots, n$. Moreover, the smallest positive eigenvalue of $-Q$ is given by $\min_i \min_{M_i \cap (0,\infty)} \lambda_j^{(i)}$. In the repairable system example, $M_i = \{0, \alpha_i + \beta_i\}$.

For most chains, $P_t(x, x)$ is difficult to explicitly compute. Thus, the above probabilistic approach to finding eigenvalues will not work in these cases.

Section 5.2. Some of our results require $\gamma_1 < \lambda_1$. This property holds if A is a singleton set. We can always "collapse" A to a single point without changing the distribution of $\mathcal{L}_w T_A$ (see Aldous and Brown, (1992) p.5). The collapsing leads to a chain with transition matrix Q'; γ_1 remains the same, and λ_1 is replaced by $\lambda_1' \geq \lambda_1$, with $\lambda_1' > \gamma_1$. For example consider the Markov chain with state space $\{0, 1, 2\}$ and transition intensity matrix,

$$Q = \begin{pmatrix} -1 & p & q \\ c & -c & 0 \\ c & 0 & -c \end{pmatrix}$$

with $0 < p < 1, q = 1 - p$, and $0 < c < 1$. The eigenvalues of $-Q$ are $0, c$ and $1 + c$, thus $\lambda_1 = c < 1$. Define $A = \{1, 2\}$, then $\gamma_1 = 1 > c = \lambda_1$. The collapsed chain has,

$$Q' = \begin{pmatrix} -1 & 1 \\ c & -c \end{pmatrix}.$$

The eigenvalues of $-Q'$ are 0 and $1 + c$ thus $\lambda_1' = (1 + c) > 1 = \gamma_1 > c = \lambda_1$.

The above example illustrates that for the study of questions related to T_A, λ_1' is more appropriate, and leads to sharper inequalities, than λ_1. However, in the reliability example, λ_1 was available without computation and lead to excellent bounds. Only a slight improvement would have been achieved by using $\lambda_1' (\lambda_1 = 18.9, \lambda_1' = 19.0312)$.

Another consideration is that λ_1' varies with A, while λ_1 does not. It is convenient to have a single relaxation time (λ_1^{-1}), independent of the choice of A. My recommendation, especially in the repairable system model is to use λ_1 in obtaining bounds and inequalities for T_A. If the bounds are not satisfactorily small, then attempt to compute λ_1'.

Section 6. References

Aldous, D.J. (1982). "Markov chains with almost exponential hitting times." *Stochastic Process. Appl.,* **13**, 305-310.

Aldous, D.J. and Brown, M. (1992), "Inequalities for rare events in time reversible Markov chains I." *Stochastic Inequalities*, Shaked, M. and Tong, Y.L., editors; Institute of Mathematical Statistics Lecture Notes – Monograph Series, Volume 22.

Aldous, D.J. and Brown, M. (1993), "Inequalities for rare events in time reversible Markov chains II". *Stochastic Processes and their Applications,* **44**, 15-25.

Aldous, D.J. and Fill, J.A. (1997). *Reversible Markov Chains and Random Walks on Graphs.* To appear.

Barlow, R.E., Fusell, J.B. and Singurwalla, N.D. (1975). *Reliability and Fault Tree Analysis.* SIAM, Philadelphia.

Brown, M. (1983). "Approximating IMRL distributions by exponential distributions with applications to first passage times." *Ann. Probab.*, 419-427.

Brown, M. (1999). "Interlacing eigenvalues in time reversible Markov chains." *Mathematics of Operations Research,* **24**, 847-864.

P. Diaconis and J.A. Fill (1990). "Strong stationary times via a new form of duality." *Ann. Probab.,* **18**, 1483-1522.

Diaconis, P. and Stroock, D. (1991). "Geometric bounds for eigenvalues of Markov chains." *Ann. Appl. Probab.,* **1**, 36-61.

Fill, J.A. (1991). "Eigenvalue bounds on convergence to stationarity for non-reversible Markov chains with an application to the exclusion process." *Ann. Appl. Prob.,* **1**, 62-87.

Gertsbakh, I.B. (1984). Asymptotic methods in reliability theory: A review. *Adv. in Appl. Probab.* **16**, 147-175.

Gertsbakh, I.B. (1989). *Statistical Reliability Theory.* Marcel Decker, N.Y.

Gnedeko, B.V., Belyaev, Yu, K., and Solovyev, A.D. (1969). *Mathematical Methods in Reliability Theory.* Academic Press, New York.

Keilson, J. (1975). "Systems of independent Markov components and their transient behavior." In, *Reliability and Fault Tree Analysis*, Barlow, R.E., Fussel, J.B. and Singpurwalla, N.D., ed. SIAM, Philadelphia.

Keilson, J. (1979). *Markov Chain Models, Rarity and Exponentiality*, Springer-Verlag, New York.

Solovyev, A.D. (1971). Asymptotic behaviour of the time of first occurrence of a rare event. *Engrg. Cybernetics* **9** 1038-1048.

Solvyev, A.D. (1972). "Asymptotic distribution of the moment of first crossing of a high level by a birth and death process." *Proc. 6th Berkeley Symp. Math. Statist. Prob.,* **3**, 71-86.

INFORMATION THEORETIC APPROACH TO STATISTICS

by

J.N.Kapur

Mathematical Sciences Trust Society, C-766, New Friends Colony, New Delhi-110065.

ABSTRACT

[In this paper, we study the applications of information-theoretic concepts to characterise probability distributions as maximum entropy or minimum cross-entropy probability distributions. We also develop an entropic measure of stochastic dependence and apply it to obtain the measure of dependence in some multivariate distributions and also to measure dependence in contingency tables. We also derive the principle of maximum likelihood from both maximum entropy and minimum cross-entropy principles. We also compare entropic method of estimating parameters with Fishers and Pearson's methods. We also find probability distribution of a family which is closest to a mixture of distributions of some members of the same family].

INDEX TERMS

CHARACTERIZATION OF PROBABILITY DISTRIBUTIONS/ ENTROPIC MEASURE OF STOCHASTIC DEPENDENCE/ CONTINGENCY TABLES/ PARAMTERIC ESTIMATION/ MIXTURE OF DISTRIBUTIONS/

1. MATHEMATICAL STATISTICS.

Mathematical statistics is concerned with using the theory of probability for drawing inferences about a population from a knowledge of a random sample drawn from the population. The only earlier knowledge about the population may be whether the variate is continuous or discrete and what the range of the variate is or which are the values taken by the random variate. We may also know the form of the density function containing one or more parameters and we may like to use the knowledge of a random sample from to population to estimate the value of the parameter or parameters. This estimation will be uncertain because we are dealing with random variates. In statistics, we also want to estimate the degree of uncertainty about the values of parameters. We also introduce axioms like the principle of maximum likelihood to go from inductive inference to deductive inference so that the powerful method of deductive logic used in mathematics can be applied. We can also use these methods in non-parametric estimation, testing of hypotheses, sequential analysis and Bayesian inference in order to enable us to take decisions under conditions of uncertainty.

2. INFORMATION THEORY.

Information theory also deals with drawing inferences under conditions of uncertainty. It starts by developing measures of uncertainty. The most important measure of uncertainty of a probability distribution $P=(p_1,p_2,\ldots\ldots\ldots,p_n)$ was developed by Shannon [8] in 1948 as $-\sum_{i=1}^{n} p_i \ln p_i$. Earlier Laplace had given his principle of insufficient reason which stated that if there is no information which makes one outcome more likely than another, then we should take $p_1=p_2=\ldots\ldots\ldots=p_n=1/n$, that is we should consider all outcomes as equally likely because there is no reason to choose any other probability distribution. The uniform distribution maximizes the measure of uncertainty given by Shannon. Later in 1957 Jaynes [1] modified Laplace's principle to the case when some information is available about the probability distribution in the form of knowledge of some moments or probabilities or some inequalities about moments or probailities. He stated his principle of maximum entropy that when we have some information about the probabilities, we should choose that probability distribution which satisfies all the available information, but which otherwise maximizes Shannon's measure of uncertainty or entropy. This principle could also be obtained by using Kullback-Leibler [7] measure of directed divergence $\sum_{i=1}^{n} p_i \ln(p_i/q_i)$ or discrepancy of the probability distribution P from the apriori probability distribution $Q=(q_1,q_2,\ldots\ldots,q_n)$. In the special case when Q is the uniform distribution $U=(1/n,\ 1/n,\ 1/n,\ldots\ldots\ldots\ldots\ldots,1/n)$, this measure of directed divergence (or cross-entropy) becomes

$$(D:U) = \ln n - \left(-\sum_{i=1}^{n} p_i \ln p_i\right) \qquad (1)$$

so that Shannon's measure of entropy is $\ln n - D(P:U)$, so that the nearner P is to the most uncertain distribution namely U, the greater is its uncertainty.

If we know the priori distribution Q which does not satisfy the constraints on probabilities, then Kulback's [6] principle of minimum discrimination information states, that we should choose a distribution which should be as near as possible to the known prior distribution, subject to the constraints on probabilities being satisfied. In the particular case when Q is the uniform distribution, this principle reduces to Jaynes principle of maximum entropy.

For a continuous random variate varying over the interval [a,b], Shannon's and Kullback-Leibler measures are

$$-\int_a^b f(x)\ln f(x)dx \text{ and } \int_a^b f(x)\ln f(x)/g(x)dx \qquad (2)$$

3. CHARACTERIZATION OF PROBABILITY DISTRIBUTIONS AS MAXIMUM ENTROPY OR MINIMUM CROSS-ENTROPY PROBABILITY DISTRIBUTIONS.

Here we first want to find the maximum entropy probability distribution when information is available in the form of some moments about the distribution. Thus suppose we know that a continuous random variate varies from $-\infty$ to $+\infty$ and its mean and variance are known as m and σ^2, then maximizing the entropy, $-\int_{-\infty}^{\infty} f(x) \ln f(x) dx$ subject to

$$\int_{-\infty}^{\infty} f(x)dx = 1, \int_{-\infty}^{\infty} xf(x)dx = m, \int_{-\infty}^{\infty} (x-m)^2 f(x)dx = \sigma^2 \tag{3}$$

by using calculus of variations, we get

$$f(x) = \frac{1}{\sqrt{2\pi}\sigma} e^{-\frac{1}{2}(x-m)^2/\sigma^2} \tag{4}$$

which shows that the maximum entropy distribution, when mean and variance are known for the random continuous variate varying from $-\infty$ to ∞ is given by the normal distribution with mean m and variance σ^2. In other words, the normal distribution is characterized by the two simplest moments viz. the mean and the variance.

In general, most of the useful probability distributions can be characterized as maximum entropy probability distributions, when one or two simple moments like

$$E(x), E(x^2), E(\ln x), E(\ln(1-x)),, E(\ln(1+x^2)) \text{ etc.} \tag{5}$$

are specified.

In this way go easily get the following maximum entropy distributions

Name	Distributions Density Function	Entropy (in nats)
Beta	$f(x) = x^{p-1}(1-x)^{q-1} / B(p,q); 0 \le x \le 1$ where $B(p,q) = \Gamma(p)p(q)/\Gamma(p+q); p,q > 0$	$\ln B(p,q) - (p-1)[\psi(p) - \psi(p+q)]$ $-(q-1)[\psi(q) - \psi(p+q)]$
Cauchy	$f(x) = (\lambda/\pi)(\lambda^2 + x^2)^{-1}, -\infty < x < \infty, \lambda > 0$	$\ln(4\pi\lambda)$
Chi-square	$f(x) = x^{(n/2)-1} \exp(-x(2\sigma^2)) / [2^{n/2}\sigma^n\Gamma(n/2)]; x > 0$	$\ln[2\sigma^2\Gamma(n/2)] + (1-n/2)\psi(n/2) + n/2$
Erlang	$f(x) = [\beta^n/(n-1)!]x^{n-1} \exp(-\beta x); x, \beta > 0$	$(1-n)\psi(n) + \ln[\Gamma(n)/\beta] + n$

Exponential	$f(x)=\sigma^{-1}exp(-x/\sigma); x,\sigma>0$	$1+ln\,\sigma$
Laplace	$f(x)=(1/2)\phi^{-1}exp(1-\|x-\theta\|/\phi);-\infty<x<\infty,\phi>0$	$1+ln(2\phi)$
Logistic	$f(x)=e^{-x}\left(1+e^{-x}\right)^{-2};-\infty<x<\infty$	2
Lognormal	$f(x)=\left[\sigma x\sqrt{(2\pi)}\right]^{-1}exp(-log\,x-m)^2/\left(2\sigma^2\right); x>0$	$m+(1/2)ln\left(2\pi e\sigma^2\right)$
Maxwell-Boltzmann	$f(x)=\left[4\pi^{-1/2}\beta^{3/2}\right]x^2 exp\left(-\beta x^2\right); x,\beta>0$	$(1/2)ln(\pi/\beta)+\gamma-1/2$
Normal	$f(x)=\left[\sigma\sqrt{2\pi}\right]^{-1}exp\left(-x^2\right)/\left(2\sigma^2\right),-\infty<x<\infty,\sigma>0$	$(1/2)ln\left(2\pi e\sigma^2\right)$
Generalized Normal	$f(x)=\left[2\beta^{\alpha/2}/\Gamma(\alpha/2)\right]x^{a-1}exp\left(-\beta x^2\right); x,\alpha,\beta>0$	$ln\left[\Gamma(\alpha/2)/\left(2\beta^{1/2}\right)\right]-[(\alpha-1)/2]\psi(\alpha/2)+\alpha/2$
Pareto	$f(x)=ak^a/x^{a+1}; x\geq k>0,a>0$	$ln(k/a)+1+1/\alpha$
Rayleigh	$f(x)=\left(x/b^2\right)e^{-x^2/\left(2b^2\right)}; x,b>0$	$1+ln\left(b/\sqrt{2}\right)+\gamma/2$
Uniform	$f(x)=1/(\beta-\alpha);\alpha<x<\beta$	$ln(\beta-\alpha)$
Weibull	$f(x)=(c/\alpha)x^{c-1}e^{-x^c}/\alpha; x,c,\alpha>0$	$(c-1)\gamma/c+log\left(\alpha^{1/c}/c\right)+1$

4. CHARACTERIZATION OF DISCRETE UNIVARIATE AND MULTIVARIATE PROBABILITY DISTRIBUTIONS.

In these cases, in addition to giving the ranges of the variate and some moments, we also require some apriori probability distributions. In this way the following distributions, among others have been characterized as minimum discrimination information distributions [3, 13].

Univariate distributions; binomial distribution; Poisson distribution; Riemann's zeta function distribution, generalized geometrical distribution, negative binomial distribution, generalized negative binomial distribution, binomial delta distribution, Poisson -delta distribution, generalized Poisson distribution, negative-binomial-negative binomial distribution, Bose - Einstein distribution, Fermi-Dirac distribution, multinomial distribution.

Discerete multivariate distributions: Multinomial distribution, multivariate generalized geometric distribution, multivariate negative binomial distribution, generalized multivariate negative binomial distribution, multivariate Poisson distribution, multivariate generalized negative binomial distribution, multivariate binomial delta distribution, multivariate binomial Poisson distribution, multivariate binomial negative binomial distribution, multivariate Poisson delta distribution, multivariate generalized Poisson distribution, multivariate Poisson binomial distribution, multivariate Poisson negative binomial distribution, multivariate negative binomial delta distribution, multivariate negative binomial Poisson distribution, multivariate negative binomial negative binomial distribution, multivariate Poisson rectangular distribution, multivariate Poisson distribution.

The following continuous multivariate probability distributions have been derived as maximum entropy probability distributions [13] multivariate normal distribution, multivariate lognormal distribution, multivariate polynomial distribution, multivariate exponential sums distribution, Dirichlet distribution, multivariate beta distribution of the second type, multivariate logistic distribution, multivariate generalized Cauchy distribution, multivariate Pareto distribution, multivariate gamma distribution and multivariate rectangular distribution.

The following additional multivariate distribution have been derived as maximum entropy distributions [15]; multivariate gamma distribution, multivariate beta distribution, multivariate exponential distribution, multivariate distributions for continuous ordered random variables, discrete analogue of multivariate gamma distribution, translated discrete multivariate gamma distribution, discerete version of multivariate exponential distributions, multirectangular multivariate distributions, multivariate Yule distributions, multivariate generalized Yule distribution.

5. MEASURE OF DEPENDENCE.

In statistics, the measure of dependence used is the correlation coefficient which gives the linear dependence between two random variates. We also use there partial and multiple correlation coeffetients which also give linear dependence in multivariate cases. However many times we need one measure of depedence between a large number of variates. If we have m variates, we can find $m(m-1/2)$correlation coefficients, some of which will be positive while others will be negative and they will all lie between -1 and +1, but these coefficients do not give us an idea about how dependent the m variates are among themselves.

For this purpose, I developed an entropic measure of dependence based on the fact that for independent variates, the entropy of the joint distribution is equal to the sum of the entropies of the marginal distributions. If these two entropies are different, that is if $D=S_1+S_2+-S>0$. the m variates are dependent and D gives us a measure of this dependence. This is a measure of dependence of all m variates among themselves, but it is always non-negative and in fact a measure of dependence should be non-negative.

I used this measure [8] in pattern recognition to find the $m\times n$ matrix A, so that the linear transformation $Y=AX$ transforms the normally distributed random variate $X=(x_1x_2,\dots\dots x_n)$ to the random variate $(y_1,y_2,\dots\dots,y_n)$, $m<<n$ and $y_1,y_2,\dots\dots,y_m$ are as independent as possible. I was able to show that the matrix A can be obtained by using the m eigen-vectors of the correlation matrix

of the random variables X corresponding to the m largest eigen values of this matrix. The earlier criteria of minimum loss of information or of minimum loss of power of discremination had led to the variance covariance matrix instead of the correlation matrix.

For the multivariate normal distribution the density function is given by

$$f(x_1,x_2,\ldots\ldots,x_m)=\frac{1}{(2\pi)^{\frac{m}{2}}|\Sigma|^{\frac{1}{2}}}exp\left[-\frac{1}{2}(X-\mu)^T\Sigma^{-1}(X-\mu)\right],$$

where

$$X=\begin{bmatrix}x_1\\x_2\\\vdots\\x_m\end{bmatrix}\quad \mu=\begin{bmatrix}m_1\\m_2\\\vdots\\m_m\end{bmatrix},\quad \Sigma=\begin{bmatrix}\sigma_1^2 & \rho_{12}\sigma_1\sigma_2 & \cdots & \rho_{1m}\sigma_1\sigma_m\\ \rho_2\sigma_2\sigma_1 & \sigma_2^2 & \cdots & \rho_{2m}\sigma_2\sigma_m\\ \vdots & \vdots & \vdots & \vdots\\ \rho_{m1}\sigma_m\sigma_1 & \rho_{m2}\sigma_m\sigma_2 & \cdots & \sigma_m^2\end{bmatrix}$$

so that $$S=-\int\ldots\int f(x_1,x_2,\ldots\ldots\ldots,x_m)\begin{bmatrix}-\frac{n}{2}ln(2\pi)-\frac{1}{2}ln|\Sigma|\\-\frac{1}{2}(X-\mu)^T\Sigma^{-1}(X-\mu)\end{bmatrix}dx_1dx_2\ldots\ldots dx_n$$

$$=\frac{m}{2}ln\,2\pi+\frac{1}{2}ln|\Sigma|+\frac{1}{2}m\;=\frac{m}{2}ln\,2\pi e+\frac{1}{2}ln|\Sigma|$$

Also $$S_i=\frac{1}{2}ln\,2\pi+\frac{1}{2}ln\,\sigma_i^2+\frac{1}{2},i=1,2,\ldots\ldots,m$$

so that $$D=\frac{1}{2}ln\,\sigma_1^2\sigma_2^2\ldots\ldots\sigma_m^2-\frac{1}{2}ln\,\sigma_1^2\sigma_2^2\ldots\ldots\sigma_m^2\begin{vmatrix}1 & \rho_{12} & \cdots & \rho_{1m}\\ \rho_{21} & 1 & \cdots & \rho_{2m}\\ \vdots & \vdots & \cdots & \vdots\\ \rho_{m1} & \rho_{m2} & \cdots & 1\end{vmatrix}$$

$$=-\frac{1}{2}ln\begin{vmatrix}1 & \rho_{12} & \cdots & \rho_{1m}\\ \rho_{21} & 1 & \cdots & \rho_{2m}\\ \vdots & \vdots & \cdots & \vdots\\ \rho_{m1} & \rho_{m2} & \cdots & 1\end{vmatrix}$$

Thus, this entropic measure of dependence depends on the $m(m-1)/2$ correlation coefficients between pairs of random variates. If all these correlation coefficients are zero, then $D=0$, and if $D=0$, then all the correlation coefficients are zero. This result has been proved for multivariate normal distributions and is not necessarily true for all multivariate distributions.

Thus for the case of Pareto multivariate distribution

$$f(x_1, x_2, \ldots\ldots\ldots, x_n) = -\frac{a(a+1)\ldots\ldots\ldots\ldots(a+m-1)}{\theta_1\theta_2\ldots\ldots\ldots\theta_m}$$

$$\left(\frac{x_1}{\theta_1}+\frac{x_2}{\theta_2}\ldots\ldots\ldots+\frac{x_m}{\theta_m}-(m-1)\right)^{-(\alpha+m)} \qquad x_i \geq \theta_i,$$

it can be shown that all the correlation coefficients are zero, but the measure of dependence is not zero, so that the variates are dependent. Thus the vanishing of the correlation coefficients does not imply that the variates are independent though when the variates are all independent the correlation coefficients are zero. This also shows a weakness of the correlation coefficients as a measure of dependence.

6. CONTINGENCY TABLES.

In an $m\times$n contingency table

$$\begin{matrix} a_{11} & a_{12} & \cdots\cdots & a_{1n} \\ a_{21} & a_{22} & \cdots\cdots & a_{2n} \\ \cdots\cdots & \cdots\cdots & \cdots\cdots & \cdots\cdots \\ a_{m1} & a_{m2} & \cdots\cdots & a_{mn} \end{matrix}\Bigg] \begin{matrix} a_1 \\ a_1 \\ \cdot\cdot \\ a_n \end{matrix}$$
$$\begin{matrix} b_1 & b_2 & & b_n & N \end{matrix}$$

the measure of dependence is given by

$$D=S_1+S_2-S$$

where S_1, S_2 are the entropies for the probability distribution $(a_1/N, a_2/N, \ldots\ldots\ldots\ldots, a_n/n)$ and $(b_1/N, b_2/N, \ldots\ldots, b_n/N)$ and S is the entropy of the joint distribution with probabilities as a_{ij}/N. If a_i's and b_j's are kept fixed, then minimizing D is equal and to maximizing S, so that to minimize D is is maximize S, subject to a_i's and b_j's remaining constant. Using Lagrange methods this gives

$$a_{ij}/N=(a_i/N)\,(b_j/N),$$

It shows the two atributes of classification are independent. In general because of random errors, this will be >0 and S will not be equal to S_1+S_2. However the value of D will give us an idea of the dependence in the table. It can be shown that D is a very good approximation for

$$x^2 = \frac{\left(a_{ij}-a_ib_j/N\right)^2}{a_ib_j/N}$$

so that this approach to dependence also leads to the chi-square distribution.

If there are k attributes, we shall get a k dimensional contingency table and if the marginal totals are kept fixed, then when S is maximum, D will be equal to zero. Again D can be approximated by a *chi-square* and *chi-square* distribution can be used to test the significance of the dependence in the table. Thus chi-square distribution, maximum entropy distribution and minimum dependence distribution, are closely related in the case of contingency tables.

7. FISHER'S METHOD OF MAXIMUM LIKELIHOOD.

We derive below this principle from Kullback's principle of minimum cross-entropy[ ].

Let the random sample be $x_1, x_2, \ldots\ldots, x_n$, where without loss of generality we assume that

$$x_1 < x_2 < x_3 < \ldots. < x_n.$$

Let $f(x)$ be the density for the observed distribution and let $F(x)$ be its cumulative density function, so that

$$F(x)=0, \qquad x<x_1$$

$$F(x)=\frac{1}{n}, \qquad x_1 \le x < x_2$$

$$F(x)=\frac{2}{n}, \qquad x_2 \le x < x_3$$

$$\ldots\ldots\ldots\ldots \qquad \ldots\ldots\ldots\ldots$$

$$F(x)=1, \qquad x \ge x_n.$$

Now we appeal to Kullback's principle of minimum cross-entropy and choose θ so that the distribution with density function $g(x,\theta)$ is as close as possible to the distribution with density function $f(x)$. We now seek to choose θ to minimize the cross-entropy.

$$\int f(x) \ln \frac{f(x)}{g(x,\theta)} dx = \int f(x) \ln f(x) dx - \int f(x) \ln g(x,\theta) dx$$

$$= \int f(x) \ln f(x) dx - \int \ln g(x,\theta) dF.$$

The first term on the R.H.S does not depend on θ, so that we have to minimize

$$-\ln g(x_1,\theta)\frac{1}{n} - \ln g(x_2,\theta)\frac{1}{n} - \ldots.. - \ln g(x_n,\theta)\frac{1}{n},$$

i.e. we have to maximize

$$\sum_{i=1}^{n} \ln g(x_i,.\theta) = \ln\left[g(x_1,\theta)g(x_2,\theta)\ldots\ldots g(x_n,\theta)\right]$$

$$= \ln L,$$

where L is Fisher's likelihood function. This gives us Fisher' principle of maximum likelihood, so that this principle can be regarded as a special application of Kullback's Maximum Cross Entropy principle. It may be noted that we are using this principle in a special sense. There are no linear constraints to be satisfied. Instead our choice is restricted to all probability distributions with density functions of the form $g(x,\theta)$. We shall have an infinite number of density functions to choose from, since θ can take an infinity of values.

In [14] we have given five other information - theoretic methods for estimating the parameters of a probability distributions in terms of a random sample from the population. These methods are as reasonable as Fisher's method, but these lead to more complicated calculations and more complicated estimators, so that the proofs of consistency, efficiency and other properties of these estimators will relatively be much more difficult to prove. However these proofs are open problems which those interested can try.

8. MAXIMUM ENTROPY PRINCIPLE AND FISHER'S AND PERSONS METHODS OF ESTIMATION.

(a) According to Max Ent, (Maximum Entropy Principle), we maximize the entropy subject to all the information given to us. Suppose the information is provided by the random sample $x_1, x_2, \ldots\ldots\ldots, x_n$. Now in this instance, the moment constraints are not specified as in the case of MaxEnt. As such, we choose the parameters $\theta_1, \theta_2, \ldots\ldots\ldots\ldots, \theta_m$ in the probability density function $f(x;\theta_1,\theta_2,\ldots\ldots\ldots\ldots,\theta_m)$ of a population in such a way that the entropy that remains after the sample values are known is as large as possible. In other words, the entropy of the sample itself has to be a minimum. This entropy is given by

$$-\int f(x,\Theta)\ln f(x,\Theta)dx = -\int \ln f(x,\Theta)dF,$$

where $\Theta = (\theta_1,\theta_2,\ldots\ldots\ldots\ldots,\theta_m)$.

According to the knowledge given by the sample,

$$F(x,\Theta) = 0 \quad \text{when} \quad x < x_1$$

$$F(x,\Theta) = 1/n \quad \text{when} \quad x_1 \le x < x_2$$

$$F(x,\Theta) = 2/n \quad \text{when} \quad x_2 \le x < x_3$$

$$\vdots \qquad \vdots \quad \vdots$$

$$F(x,\Theta) = r/n \quad \text{when} \quad x_r \leq x < x_{r+1}$$

$$F(x,\Theta) = 1 \qquad \text{when} \quad x_n \leq x.$$

This gives the entropy of the sample as

$$-\frac{1}{n}\left[\ln f(x_1,\Theta) + \ln f(x_2,\Theta) + \ldots\ldots + f(x_n,\Theta)\right]$$

$$= -\frac{1}{n}\left[L(x_1, x_2, \ldots\ldots\ldots\ldots, x_n;\ \theta_1, \theta_2, \ldots\ldots\ldots, \theta_m)\right]$$

This, to minimize the entropy of the sample, we have to maximize the likelihood function

$$L(x_1, x_2, \ldots\ldots\ldots\ldots, x_n;\ \theta_1, \theta_2, \ldots\ldots\ldots, \theta_m)$$

MaxEnt has thus led us to the principle of maximum likelihood although it predates the explicit statement of the former principle.

(b) The classical theory of inference due to Fisher has been in existence for a long time and it is therefore worthwhile to make comparisons between the method and the later theory of MaxEnt. Fisher's theory of estimation is implemented by the following steps:

1. Specify $f(x; \theta_1, \theta_2, \ldots\ldots\ldots, \theta_m)$ on the basis of experience, intuition, or theory. We specify the function, but do not specify the values of $\theta_1, \theta_2, \ldots\ldots\ldots, \theta_m$.

2. Write the likelihood function $L(x_1, x_2, \ldots\ldots\ldots\ldots, x_n;\ \theta_1, \theta_2, \ldots\ldots\ldots, \theta_m)$.

3. Find the values of $\hat{\theta}_1, \hat{\theta}_2, \ldots\ldots\ldots, \hat{\theta}_m$ for which the likelihood function is maximized. These values will be functions of the sample values.

4. The estimated density function is then

   $$f(x,\ \hat{\theta}_1, \hat{\theta}_2, \ldots\ldots\ldots, \hat{\theta}_m)$$

To implement the MaxEnt method of estimation, we take the following steps:

1. Specify m characterizing functions, $g_1(x), g_2(x), \ldots\ldots\ldots\ldots, g_m(x)$.

2. Use the MaxEnt to find $f(x)$,

3. Find estimates

   $$\hat{a}_r = \frac{1}{n}\left[g_r(x_1) + g_r(x_2) + \ldots\ldots + g_r(x_n)\right], \qquad r = 1, 2, \ldots\ldots\ldots, m$$

4. Use it to find $\hat{\lambda}_0, \hat{\lambda}_1, \ldots\ldots, \hat{\lambda}_m$

5. The estimated function is then

$$exp\left[-\hat{\lambda}_0 - \hat{\lambda}_1 g_1(x) - \hat{\lambda}_2 g_2(x) - \ldots - \hat{\lambda}_m g_m(x)\right]$$

(c) Pearson's method is yet another method of estimation. He suggested that in order to estimate , $\theta_1, \theta_2, \ldots\ldots\ldots\ldots, \theta_m$ we find the first *m* algebraic moment of the population which will be functions of $\theta_1, \theta_2, \ldots\ldots\ldots\ldots, \theta_m$ and then equate the first *m* algebraic moments of the sample with these functions and solve for $\hat{\theta}_1, \hat{\theta}_2, \ldots\ldots\ldots\ldots, \hat{\theta}_m$.

Fisher criticized this method because it sometimes gaves quite different results from those obtained by his theory of estimation, which was based on the use of the principle of maximum likelihood. He proved theoretically that his estimates had the nice properties of consistency, efficiency, and sufficiency.

Had Pearson used the *m* maximum-entropy characterizing moments, instead of the first *m* algebraic moments, his results would have coincided with those of Fisher's theory, and there would have been no room for controversy. Unfortunately, the MaxEnt method was not known at that time. From the vantage point of MaxEnt, one can conclude that Fisher had unintentionally invoked this principle in his use of the maximum likelihood principle, and thus his success can be attributed to this fact. This again demonstrates the great foresight of Fisher whose proverbial insight into statistical problems has been a source of perennial inspiration to the classical statisticians.

9. FISHER'S MEASURE OF INFORMATION.

Let $f(x,\theta)$ be the density function for a probability distribution. Here θ is a parameter. Let $f(x,\theta+\Delta\theta)$ be the density function for a neighbouring probability distribution of the same family with parameter $\theta+\Delta\theta$, then

$$D(f(x,\theta) : f(x,\theta+\Delta\theta)) = \int_a^b f(x,\theta) \ln \frac{f(x,\theta)}{f(x,\theta+\Delta\theta)} dx$$

$$= -\int_a^b f(x,\theta) \ln \left[1 + \Delta\theta \frac{1}{f}\frac{\partial f}{\partial\theta} + \frac{(\Delta\theta)^2}{f}\frac{1}{2}\frac{\partial^2 f}{\partial\theta^2} \ldots\ldots \right] dx$$

$$= -\int_a^b f(x,\theta) \left[\Delta\theta \frac{1}{f}\frac{\partial f}{\partial\theta} + \frac{(\Delta\theta)^2}{f^{//}}\frac{1}{2}\frac{\partial^2 f}{\partial^2\theta^2} - \frac{(\Delta\theta)^2}{2}\left(\frac{1}{f}\frac{\partial f}{\partial\theta}\right)^2 \ldots\ldots \right] dx$$

Since $\int_a^b f(x,\theta) = 1$

$$\int_a^b \frac{\partial f}{\partial \theta} = 0, \quad \int_a^b \frac{\partial^2 f}{\partial \theta^2} dx = 0$$

$$D(f(x,\theta)) : f(x, \theta + \Delta\theta)$$

$$= \frac{1}{2}(\Delta\theta)^2 \int_a^b \frac{1}{f}\left(\frac{\partial f}{\partial \theta}\right)^2 dx + \theta(\Delta\theta)^3$$

Thus the discrepancy or directed divergence of $f(x,\theta)$ from $f(x,\theta+\Delta\theta)$ is proportional to

$$\int_a^b \frac{1}{f}\left(\frac{\partial f}{\partial \theta}\right)^2 dx = \int_a^b f\left(\left(\frac{1}{f}\frac{\partial f}{\partial \theta}\right)^2\right) dx$$

This is known as Fisher's information measure

For most measures of directed divergence, the directed divergence of $f(x,\theta)$ from $f(x,\theta+\Delta\theta)$ will be found to be proportional to Fisher's information measure, so that Fisher's information is quite a robust measure of this discrepancy, independently of the measure of directed divergence used.

If the probability density function depends on a number of parameters $\theta_1, \theta_2, \ldots\ldots\ldots, \theta_m$, we get for the discrepancy as

$$D[f(x,\theta_1,\theta_2,\ldots\ldots\ldots,\theta_m) : f(x,\theta_1+\Delta\theta_1,\theta_2+\Delta\theta_2+\ldots\ldots\ldots+\theta_m+\Delta\theta_m)]$$

$$= \frac{1}{2}\int AFA^T dx$$

where $A = (\Delta\theta_1, \Delta\theta_2, \ldots\ldots\ldots, \Delta\theta_k)$

is a row matrix and

$$F = \begin{bmatrix} \frac{1}{f}\left(\frac{\partial f}{\partial \theta_1}\right)^2 & , \frac{1}{f}\frac{\partial f}{\partial \delta_1}\frac{\partial f}{\partial \theta_2} & , \ldots\ldots\ldots\ldots\ldots, & \frac{1}{f}\frac{\partial f}{\partial \theta_1}\frac{\partial f}{\partial \theta_m} \\ \frac{1}{f}\left(\frac{\partial f}{\partial \theta_2}\right)\left(\frac{\partial f}{\partial \theta_1}\right), & \frac{1}{f}\left(\frac{\partial f}{\partial \theta_2}\right)^2 & , \ldots\ldots\ldots\ldots\ldots, & \frac{1}{f}\frac{\partial f}{\partial \theta_2}\frac{\partial f}{\partial \theta_m} \\ \ldots\ldots\ldots\ldots & \ldots\ldots\ldots\ldots & \ldots\ldots\ldots\ldots & \ldots\ldots\ldots\ldots \\ \frac{1}{f}\left(\frac{\partial f}{\partial \theta_m}\right)\frac{\partial f}{\partial \theta_m}, & \frac{1}{f}\frac{\partial f}{\partial \theta_m}\frac{\partial f}{\partial \theta_2} & , \ldots\ldots\ldots\ldots\ldots, & \frac{1}{f}\left(\frac{\partial f}{\partial \theta_m}\right)^2 \end{bmatrix} \quad (28)$$

is known as Fisher's information matrix.

10. FINDING A DISTRIBUTION WHICH IS CLOSEST TO A MIXTURE OF DISTRIBUTIONS.

A distribution with density function $\sum_{j=1}^{m} \lambda_j g_j(x)$ where $\lambda_j \geq 0$ and $\sum_{j=1}^{m} \lambda_j = 1$ will be called mixture of distributions with density functions $g_1(x), g_2(x), \ldots\ldots\ldots\ldots, g_m(x)$.

A case of particular interest arises when $g_1(x), g_2(x), \ldots\ldots\ldots\ldots, g_m(x)$ belong to the same family and differ only in the values of the parameters used in different distributions.

Thus these may be normal distributions with parameters (μ_j, σ_j) or exponential distributions with parameters m_j or gamma distributions with parameters a_j, λ_j $(j=1,2,\ldots\ldots.,m)$ and so on .

There are two distributions which are closest to the mixture distributions and they have density functions $\sum_{j=1}^{m} \lambda_j g_j(x)$ and

or
$$\frac{\prod_{j=1}^{m} g_j^{\lambda_j}(x)}{\int_a^b \prod_{j=1}^{m} g_j^{\lambda_j}(x)dx}$$

but these distributions need not belong to the family to which $g_1(x), g_2(x), \ldots\ldots\ldots\ldots, g_m(x)$ belong. In general we are interested in getting the probability distribution closest to the mixture, but which also belong to the family. We discuss some special cases below:

Let
$$g_j(x) = \frac{1}{\sqrt{2\pi}\sigma_j} e^{-\frac{1}{2}(x-\mu_j)^2/\sigma_j^2}, \quad (j = 1,2\ldots\ldots,m)$$

Here we have to find μ and σ so that

$$f(x) = \frac{1}{\sqrt{2\pi}\sigma} e^{\frac{1}{2}(x-\mu)^2/\sigma^2}$$

gives a probability distribution closest to $\sum_{j=1}^{m} \lambda_j g_j(x)$, so that we have to choose μ and σ to minimize

$$\int_{-\infty}^{\infty} \left(\sum_{j=1}^{m} \lambda_j g_j(x) \right) ln \frac{\sum_{j=1}^{m} \lambda_j g_j(x)}{\frac{1}{\sqrt{2\pi}\sigma} e^{-\frac{1}{2}(x-\mu)^2/\sigma^2}} dx$$

or to minimize

$$\int_{-\infty}^{\infty} \left(\sum_{j=1}^{m} \lambda_j g_j(x) \right) \left(ln + \sigma \frac{1}{2} \frac{(x-\mu)^2}{\sigma^2} \right) dx$$

or
$$\ln\sigma + \frac{1}{2\sigma^2}\sum_{j=1}^{m} \lambda_j\left[(\sigma_j^2+\mu_j^2) - 2\mu\sum_{j=1}^{m} \lambda_j\mu_j + \mu^2\right]$$

Differentiating with respect to μ and σ, we get

$$\sum_{j=1}^{m} \lambda_j\mu_j = \mu \quad \frac{1}{\alpha}\frac{1}{\sigma^3}\left[\sum_{j=1}^{m} \lambda_j(\sigma_j^2+\mu_j^2) - \mu^2\right] = 0$$

so that the parameters of the closest normal distribution are given by

$$\mu = \sum_{j=1}^{m} \lambda_j\mu_j \quad \sigma^2+\mu^2 = \sum_{j=1}^{m} (\sigma_j^2+\mu_j^2)$$

The closest exponential, gamma, log normal and other continuous various distribution can be obtained in the same way.

11. CONCLUDING REMARKS .

We have discussed in this paper less than 50% of the applications of information theory to mathematical statistics. Applications to non-parametric estimation, queueing theory, mathematical programming, Bayesian inference, approximation of complicated distributions by simpler distributions, non-linear models, logistic models, analysis of variance, optimal information from design of experiments, pattern recognition etc. will be discussed in another paper, but what has been stated in the paper should convince everyone, that information theory has a significant role to play in mathematical statistics and all mathematical statisticians should be aware of this role.

The references (1-18) below dicssuss some of these applications.

REFERENCES

1 E.T.Jaynes (1957): "Information Theory and Statistical Mechanics, Physical Reviews", 106, 620-630, 108, 171-197.

2. J.N.Kapur (1982): "Maximum Entropy Probability Distributions of a Continuous Random Variate over a finite interval,Journ. Math. and Phy, Sci. 16,(1), 97-103.

3. J.N.Kapur (1982): "Maximum Entropy Formalism", for some univariate and multivariate Lagrangium distributions, Aligarh Journal of Statistcs, Vol.2, Pages 1-16.

4. J.N.Kapur (1983): "Maximum Entropy Probability Distributions for Continuous Random Variate,Journal. Ind. Soc. Agri. Stat. 35(3), 91-103.

5. J.N.Kapur (1984): "Maximum Entropy Distributions for Contingency Tables, Acta Ciencia, Indica, 10M(3) Pages 166-174.

6. J.N.Kapur (1984): "The Role of Maixmum Entropy and Minimum Discrimination Information Principles in Statistics", Jour. Ind. Soc. Agri. Stat. 36 (3), 12-55.

7. J.N.Kapur (1985): "Two Genralisations of Fisher's Information Matrix," National Academy of Sciences Letters, 8(8) 249-251.

8. J.N.Kapur (1986): " Applications of Entropic Measures of Stochastic Dependence to Pattern Recognition, Pattern Recognition Vol.1916, 473-476.

9. J.N.Kapur (1987): "Maximum Entropy Principle in Queuing Theory', Ind. Jour. Manag. and Syst.. 3(1), 24-41.

10. J.N.Kapur (1988): "Generalised Cauchy and Students Distributions as Maximum Entropy Distributions", Proc. Nat. Acad. Sci. India. 58(a), 235-246.

11. J.N.Kapur (1988): " $\int_{-\infty}^{\infty}\left(\sum_{j=1}^{m}\lambda_j\, g_j\,(x)\right)\ln\frac{\sum_{j=1}^{m}\lambda_j\, g_j(x)}{\frac{1}{\sqrt{2\pi}\sigma}e^{-\frac{1}{2}(x-\mu)^2/\sigma^2}}dx$ On Normalised Measures of Stochastic Dependence" Proc. Acad. Sci. 58(a) 1, 103-112.

12. J.N.Kapur & A.K.Seth (1990): "A Comparative Assessment of Entropic and Non-entropic Methods of Estimation in Maximum Entropy and Baysian Methods, Edited by P.F. Fougere, 451-462. Kluwer, Academic Publishers, Boston.

13. J.N.Kapur (1992): "Maximum Entropy Models in Science and Engineering" Wiley Eastern New Delhi and John Wiley. New York.

14. J.N.Kapur & H.K.Kesavan (1994): "Entropy Optimization Principle and Their Applications, Academic Press USA.

15. J.N.Kapur (1995): "Some Multivariate Statistical Distributions", MSTS.

16. C.E.Shannon (1948): " A Mathematical Theory of Communication", Bell. System. Tech. J 27, 379-423 , 623-659.

17. S.Kullback and R.A.Leibler (1951): "On Information and Sufficiency. Ann Math. Statist 22, 79-86.

Mean Square Error Estimation in Survey Sampling

Arijit Chaudhuri*

ABSTRACT

The classical problem of providing a 'point estimator' for a survey population total along with an interval around it needs an appropriate estimator for its mean square error.
A brief resume is provided for modern approaches to solutions for this by model-motivated-cum-design-based methods covering multi-stage unequal probability sampling, small domain estimation randomized responses for sensitive issues and employing in particular adaptive sampling and bootstrap techniques. Relevant current thoughts and practices are especially accommodated.

Keywords and phrases: Adaptive Sampling; Bootstrap; Empirical Bayes; Mean Square Error; Small domains.

AMS Subject Classification: 62 D05

1. INTRODUCTION

In this chapter we consider estimators for the total and mean of a single real variable defined on a survey population with a known number of identifiable units on surveying a suitably chosen sample from it and estimators for the mean square errors of the considered estimators.

We start with Rao's (1979) work covering a wide class of the relevant procedures available for single-stage unstratified sampling with unequal probabilities for selection of the units. Estimators for strata totals along with respective estimators of mean square errors (MSE), added across the strata may simply cover the case of stratified uni-stage sampling. We first note extensions beyond Rao's (1979) coverage. Discussing the details in Section 2 we review certain newly emerging procedures concerning multi-stage sampling in Section 3. Therein we also report model-

---

* Applied Statistics Unit, Indian Statistical Institute
203, Barrackpur Trunk Road, Kolkata - 700035, India
Email: achau@isical.ac.in

The work has done as a visitor to the University of Mannheim, Mannheim, Germany during 1 August – 30 September 2001.

"assisted, motivated and dominated" methods in addition to the classical design-based ones. In Section 4 we note how 'randomized responses' (RR) covering sensitive issues as opposed to 'direct responses' (DR) concerned with the innocuous ones may be dealt with in a manner paralleling that for multi-stage sampling. In Section 3 itself we narrate how the principle of 'small domain statistics' computation by 'borrowing strength' from 'outside' may be helpful in estimation. In order to produce enough survey data on a relatively seldom occurring phenomenon like 'maternal mortality' or 'earning by knitting woolen garments' in a community a possible technique is to adopt 'Adaptive' sampling through appropriate network formations. In Section 5 we discuss MSE-estimation in such a context. In Section 6 we present a specific 'bootstrap' sampling applicable to certain unequal probability sampling situations. We conclude with a few remarks in Section 7.

2. LINEAR ESTIMATION IN UNI-STAGE SAMPLING

Let U=(1,..,i,..,N) denote a survey population of N labels which identify a known number of N distinct individuals. Let y be a real variable of interest with y_i as its value for i in U. We shall write $\sum$ to denote sum over i in U, $\sum\sum, \sum_i\sum_{<j}, \sum_i\sum_{\neq j}$ that over i, j in U with no restriction and with the indicated restrictions respectively. By s we shall denote a sample with $\upsilon(s)$ distinct units in it as drawn from U according to a sampling design p with a selection probability $p(s)$.

Let $I_{si} = 1$ if $i \,\varepsilon\, s$ and 0 otherwise; $I_{sij} = I_{si} I_{sj}$; $\pi_i = \sum_s p(s) I_{si}$, $\pi_{ij} = \sum_s p(s) I_{sij}$ writing $\sum_s$ for sum over every s with p(s) >0.

To estimate $Y = \sum y_i$ Rao (1979) employed the 'homogeneous linear estimator' (HLE) $t_b = \sum y_i b_{si} I_{si}$

with b_{si} as constants free of $\mathbf{Y}$=(y_1,..y_i,..y_N) to be suitably chosen by an investigator. Writing E_1, V_1 as operators for expectation, variance according to p, the MSE of t_b about Y is $M(t_b) = E_1\,(t\text{-}Y)^2 = \Sigma\Sigma y_i\, y_j\, d_{ij}$ with $d_{ij} = E_1(b_{si} I_{si} - 1)\,(b_{sj} I_{sj} - 1)$, $i,j \,\varepsilon\, U$.

Rao (1979) imposed on t_b the restriction "C" which pre-supposes the following:

"There exist non-zero constants w_i such that in case $z_i = \frac{y_i}{w_i}$ equals a constant C for every i in U, then

$$M_1(t_b) = \sum\sum w_i w_j d_{ij} z_i z_j = c^2 \sum\sum w_i w_j d_{ij}$$

equals zero"

Then Rao (1979) has the

Theorem 1. $M_1(t_b) = -\sum_{i}\sum_{<j} d_{ij} w_i w_j (\frac{y_i}{w_i} - \frac{y_j}{w_j})^2$

if "C" holds.

This leads to a convenient form for an unbiased estimator for $M_1(t_b)$ as

$$m_1(t_b) = -\sum_{i}\sum_{<j} d_{sij} I_{sij} w_i w_j (\frac{y_i}{w_i} - \frac{y_j}{w_j})^2$$

for which it is easy to check the property of being 'uniformly non-negative' (UNN) provided one may find constants d_{sij} free of **Y** such that $E_1(d_{sij} I_{sij}) = d_{ij} \quad \forall \ i,j \ \varepsilon$ U.

Two easy examples are (1) $d_{sij} = \frac{d_{ij}}{\pi_{ij}}$, if $\pi_{ij} > 0$ and (2) $d_{sij} = \frac{d_{ij}}{\binom{N-2}{n-2} p(s)}$

if $\upsilon(s)=n \quad \forall \ s$ with $p(s)>0$.

Rao (1979), again, has the

Theorem 2. If "C" holds then in the class of homogeneous quadratic unbiased estimators for M_1 (t_b) any one with the "UNN" property is necessarily of the form m_1 (t_b) above.

The literature on Survey Sampling is full of examples of such p, b_{si} , d_{sij} 's as one may check, for example from Chaudhuri and Stenger's (1992) monograph, Särndal, Swensson, Wretman's (SSW,1992) text and other sources.

We give one example and discuss what one should do if "C" fails to hold.

Recalling that a "necessary condition is $\pi_i > 0 \quad \forall \ i \ \varepsilon$ U" for the existance of an unbiased estimator $t=t(s,\mathbf{Y})$ for Y such that t is (1) free of y_j for $j \notin s$ but (2) may involve y_i for $i \in s$, let us assume $\pi_i > 0 \ \ \forall \ i \ \varepsilon$ U and consider for Y the Horvitz and Thompson's (HT, 1952) estimator HTE which is $t_H = \sum \frac{y_i}{\pi_i} I_{si}$.

Its variance given by HT is $V_1(t_H)$ which is

$$V_1 = \sum y_i^2 \frac{1-\pi_i}{\pi_i} + \sum_i \sum_{\neq j} y_i y_j \left(\frac{\pi_{ij} - \pi_i \pi_j}{\pi_i \pi_j} \right).$$

If we choose $w_i = \pi_i$ then t_H satisfies "C" if "$\upsilon(s)$ is a constant $\forall\ s, p(s)>0$" (2.1)

If (2.1) holds an alternative form of $V_1(t_H)$ is $V_2 = \sum_i \sum_{<j} (\pi_i \pi_j - \pi_{ij}) \left(\frac{y_i}{\pi_i} - \frac{y_j}{\pi_j} \right)^2$

due to Yates and Grundy (1953) and Sen (1953). Unbiased estimators for V_1, V_2 respectively are

$$v_1 = \sum y_i^2 \left(\frac{1-\pi_i}{\pi_i} \right) \frac{I_{si}}{\pi_i} + \sum_i \sum_{\neq j} y_i y_j \left(\frac{\pi_{ij} - \pi_i \pi_j}{\pi_i \pi_j} \right) \frac{I_{sij}}{\pi_{ij}}$$

and $$v_2 = \sum_i \sum_{<j} y_i y_j \left(\frac{\pi_i \pi_j - \pi_{ij}}{\pi_i \pi_j} \right) \left(\frac{y_i}{\pi_i} - \frac{y_j}{\pi_j} \right)^2 \frac{I_{sij}}{\pi_{ij}}$$

provided in both cases "$\pi_{ij} > 0 \quad \forall\ i,j \varepsilon U$" (2.2)

If p is such that $\pi_i \pi_j \geq \pi_{ij} \quad \forall\ i,j$ (2.3)

then "v_2 is UNN". To test "UNN" property for v_1 one has to examine if v_1 is a 'Non-negative definite quadratic form in y_i's, $i \varepsilon s \ \forall\ s, p(s)>0$' – a hard task to accomplish. If (2.1) is violated, Chaudhuri (2000) has given to $V_1(t_H)$ the third form

$$V_3 = V_2 + \sum \frac{y_i^2}{\pi_i} \alpha_i, \alpha_i = 1 + \frac{1}{\pi_i} \sum_{j \neq i} \pi_{ij} - \sum \pi_i \qquad (2.4)$$

and an unbiased estimator for $V_1(t_H)$ as $v_3 = v_2 + \sum \frac{y_i^2}{\pi_i} \alpha_i \frac{I_{si}}{\pi_i}$,

provided (2.2) holds.

For the "UNN" property v_3 needs in addition to (2.3), also "$\alpha_i \geq 0 \quad \forall i \in U$" (2.5)

Chaudhuri and Pal (2001) have illustrated when (2.3) and (2.5) may hold together, with $v_3 \neq v_2$.

For t_b, in case "C" does not hold, choosing arbitrary w_i ($\neq 0$), they gave the formula $M_1(t_b) = -\sum_i \sum_{<j} d_{ij} w_i w_j \left(\frac{y_i}{w_i} - \frac{y_j}{w_j} \right)^2 + \sum \frac{y_i^2}{w_i} \beta_i$, with $\beta_i = \sum_{j=1}^{N} d_{ij} w_j$

and an unbiased estimator for it as

$$m_1'(t_b) = -\sum_i \sum_{<j} d_{sij} I_{sij} w_i w_j \left(\frac{y_i}{w_i} - \frac{y_j}{w_j} \right)^2 + \sum \frac{y_i^2}{w_i} \beta_i \frac{I_{si}}{\pi_i}.$$

Särndal (1996), Brewer (1999, 2000), Deville (1999) and others feel it impracticable to employ MSE- and variance- estimators that for large $\upsilon(s)$ contain too many cross-product terms and especially discourage computation of π_{ij} 's which for many sampling schemes are difficult to work out accurately and their magnitudes vary widely over $(0,\pi_i)$ rendering MSE-estimators containing π_{ij}'s unstable.

Confining to schemes with $\upsilon(s)$ fixed at an integer n for every s with $p(s)>0$, Brewer(2000) works out $V_1(t_H)$ as

$$V_{Br}(t_H) = \sum \pi_i (1-\pi_i) \left(\frac{y_i}{\pi_i} - \frac{Y}{n} \right)^2 + \sum_i \sum_{\neq j} (\pi_{ij} - \pi_i \pi_j) \left(\frac{y_i}{\pi_i} - \frac{Y}{n} \right) \left(\frac{y_j}{\pi_j} - \frac{Y}{n} \right).$$

Approximating π_{ij} by $\pi_{ij}^* = \pi_i \pi_j \left(\frac{C_i + C_j}{2} \right)$ with C_i as one of

(i) $C_i = \frac{n-1}{n-\pi_i}$, (ii) $C_i = \frac{n-1}{n - \frac{1}{n}\sum \pi_i^2}$, and (iii) $C_i = \frac{(n-1)}{n - 2\pi_i + \frac{1}{n}\sum \pi_i^2}$,

he approximates $V_{Br}(t_H)$ by

$$V_{Br}^*(t_H) = \sum \pi_i (1 - C_i \pi_i) \left(\frac{y_i}{\pi_i} - \frac{Y}{n} \right)^2 \quad (2.6)$$

and calls it the "*Natural variance*" of t_H free of π_{ij}'s. Brewer (2000) then recommends for $V_1(t_H)$ the estimator

$$v_4 = \sum \left(\frac{1}{C_i} - \pi_i \right) \left(\frac{y_i}{\pi_i} - \frac{t_H}{n} \right)^2 I_{si} \quad (2.7)$$

Deville (1999) recommends for $V_1(t_H)$ the estimator

$$v_5 = \frac{1}{(1 - \sum a_i^2 I_{si})} \sum (1-\pi_i) \left(\frac{y_i}{\pi_i} - A_s \right)^2 I_{si} \quad (2.8)$$

writing $a_i = \frac{(1-\pi_i)}{\sum (1-\pi_i) I_{si}}$, $A_s = \sum a_i \frac{y_i}{\pi_i} I_{si}$.

Though the properties of v_4, v_5 are discussed in the literature one still needs to examine which of v_j (j=1,..,5) renders t_H the most accurate point estimator for Y for a given set of data in a sample actually chosen following a specific sampling scheme. Also, one may be curious about which of the 95% Confidence Intervals (CI),

$(t_H - 1.96\sqrt{v_j}),(t_H + 1.96\sqrt{v_j})$, j=1,..,5 may have the narrowest width, treating $(t_H - Y)/\sqrt{v_j}$, (j=1,..,5) as a "*standard normal deviate*".

Poisson's scheme of sampling, discussed by *Ha'jek* (1981) and others yields

$V_1(t_H)$ as $\qquad V_{Po}(t_H) = \sum y_i^2\left(\frac{1-\pi_i}{\pi_i}\right)$

which is free of π_{ij}'s admitting an unbiased estimator $v_6 = \sum y_i^2\left(\frac{1-\pi_i}{\pi_i}\right)\frac{I_{si}}{\pi_i}$.

For this scheme $\upsilon(s)$ varies over the entire range (0, N).

Recalling the well-known fact that $\upsilon = E_1(\upsilon(s)) = \sum \pi_i$ for any sampling scheme, Brewer and Gregoire (2000) consider for Y an estimator, as an alternative to t_H, viz.

$$t_{RH} = \frac{\sum \pi_i}{\upsilon(s)} \sum \frac{y_i}{\pi_i} I_{si}.$$

If x be a variable, well-correlated with y with known values x_i (>0 $\forall$ $i \in U$) and $X = \sum x_i$, then *Ha*^{1}*jek*'s (1971) ratio estimator for Y is

$$t_{Ha} = \frac{X \sum \frac{y_i}{\pi_i} I_{si}}{\sum \frac{x_i}{\pi_i} I_{si}}.$$

Then, t_{RH} is immediately recognized as a ratio estimator for Y with $x_i = \pi_i$, $i \in U$.

Writing $R = \frac{Y}{X}$ and $\hat{R} = \frac{\hat{Y}}{\hat{X}}$, with $\hat{Y}, \hat{X}$ as unbiased estimators with an identical form each for Y, X respectively, it is known that for large $\upsilon(s)$ one may write the MSE of $\hat{R}$ about R *as* $M_1(\hat{R}) = \frac{1}{X^2} V(\hat{Y} - R\hat{X})$, using first order Taylor series expansion on neglecting higher order terms. If moreover, $\hat{Y}, \hat{X}$ are linear respectively in **Y** and **X** = $(x_1,..,x_i,..,x_N)$, then, $M_1(\hat{R}) = \frac{1}{X^2} V_1(\hat{Y})\Big|_{y_i = y_i - Rx_i}$.

In such a case a usual estimator for $M_1(\hat{R})$ is $m_1(\hat{R}) = \frac{1}{\hat{X}^2} \hat{V}_1(\hat{Y})\Big|_{y_i = y_i - \hat{R}x_i}$

if $\hat{V}_1(\hat{Y})$ is an unbiased estimator for $V_1(\hat{Y})$. Taking $\hat{Y}_R = X\hat{R}$ as an estimator for Y, one has $m_1(\hat{Y}_R) = \frac{X^2}{\hat{X}^2}\hat{V}_1(\hat{Y})\big|_{y_i = y_i - \hat{R}x_i}$

as an estimator for $M_1(\hat{Y}_R)$.

$$\text{So, } m_{1j}(t_{Ha}) = \frac{X^2}{\left(\sum \frac{x_i}{\pi_i} I_{si}\right)^2} v_j \Bigg|_{y_i = y_i - \left(\frac{\Sigma \frac{y_i}{\pi_i} I_{si}}{\Sigma \frac{x_i}{\pi_i} I_{si}}\right) x_i}, \quad j = 1 \,..\, 5 \qquad (2.9)$$

may be taken as estimators for $M_1(t_{Ha})$.

$$\text{Also, } m_{1j}(t_{RH}) = \frac{\left(\sum \pi_i\right)^2}{\upsilon^2(s)} v_j \Bigg|_{y_i = y_i - \left(\frac{\Sigma \frac{y_i}{\pi_i} I_{si}}{\upsilon(s)}\right) \pi_i}, \quad j = 1 \,..\, 5 \qquad (2.10)$$

may be used to estimate $M_1(t_{RH})$.

Incidentally, though a ratio estimator being the ratio and hence a non-linear function of two random variables is still considered by many survey samplers as a "*linear*" estimator for Y because it is linear in y_i for i in s and is of the form $t_b = \sum y_i b_{si} I_{si}$ with b_{si}, being a function of s is permitted to be a random variable as so is I_{si} with y_i as constant.

3. MULTI- STAGE SAMPLING

Suppose the units i of U are treated as 'clusters' or first stage units (fsu) composed of M_i second stage units (ssu) with

$$y_i = \sum_{j=1}^{M_i} y_{ij}$$

as the total of the M_i values y_{ij} - the value of y for the jth ssu in the ith fsu. If y_i is not ascertainable for i in s but may be estimated on drawing a sample of m_i ssu's out of the M_i ssu's in the ith fsu then we have 'Two-Stage' sampling. Similarly we have 'multi-stage' sampling on extending the number of stages.

Let every i in s be 'independently' sub-sampled in subsequent stages yielding 'independent' estimators $\hat{y}_i$ for y_i satisfying the following conditions:

(1) $E_L(\hat{y}_i) = y_i$, (2) $V_L(\hat{y}_i) = V_i$ or (2)' $V_L(\hat{y}_i) = V_{si}$ if $i \in s$ (3) $\exists\, \upsilon_i$

such that $E_L(\upsilon_i)=V_i$ or (3)' $\exists\ \upsilon_{si}$ such that $E_L(\upsilon_{si})=V_i$ writing E_L , V_L as expectation , variance operators for sampling at stages 'later ' than the 'first'.

Let us begin by stipulating that t_b is subject to

$$E_1(b_{si}I_{si})=1 \quad \forall i \tag{3.1}$$

Then, $E_1(t_b)=Y$ and $V_1(t_b)=\sum y_i^2 c_i+\sum_i\sum_{\neq j} y_i y_j c_{ij}$

where $c_i=E_1(b_{si}^2 I_{si})-1$ and $c_{ij}=E_1(b_{si}b_{sj}I_{sij})-1$.

Let there exist $c_{si},$, c_{sij} as constants free of Y such that

$$E_1(c_{si}I_{si})=c_i \text{ and } E_1(c_{sij}I_{sij})=c_{ij}.$$

From Chaudhuri and Stenger (1992) for instance one may find examples for such p, b_{si}, c_{si}, c_{sij}'s. Using the operators $E=E_1E_L$ and $V=V_1E_L+E_1V_L$ one may observe the following

$e_b=\sum \hat{y}_i b_{si}I_{si}=t_b\big|_{yi=\hat{y}_i}$ satisfies $E(e_b)$ under (3.1)

and $V(e_b)=\sum y_i^2 c_i+\sum_i\sum_{\neq j} y_i y_j c_{ij}+E_1(\sum V_i b_{si}^2 I_{si})$

under (2) and $=\sum y_i^2 e_i+\sum_i\sum_{\neq j} y_i y_j c_{ij}+E_1(\sum V_{si}b_{si}^2 I_{pi})$ under (2)'.

Writing $\upsilon(t_b)=\sum y_i^2 c_{si}I_{si}+\sum_i\sum_{\neq j} y_i y_j c_{sij}I_{sij}$

for which $E_1\upsilon(t_b)=v_1(t_b)$ and $\upsilon(e_b)=\upsilon(t_b)\big|_{y_i=\hat{y}_i}$, under (3), (3)',

from Raj (1968) and Rao (1975) respectively, one has

$\upsilon_1(e_b)=\upsilon(e_b)+\sum \upsilon_i b_{si}I_{si}$ such that $E\upsilon_1(e_b)=V(e_b)$ and

$\upsilon_2(e_b)=\upsilon(e_b)+\sum \upsilon_{si}(b_{si}^2-c_{si})I_{si}$ such that $E\upsilon_2(e_b)=V(e_b)$.

Also, $\upsilon_3(e_b)=\upsilon(e_b)+\sum \upsilon i(b_{si}^2-c_{si})I_{si}$ under (3) satisfies

$E\upsilon_3(e_b)=V(e_b)$ as well.

Suppose y_i is not ascertainable but estimable by $\hat{y}_i$ subject to (1), (2), (3) and in addition $E=E_1E_L=E_LE_1$ along with $V=V_1E_L+E_1V_L=V_LE_1+E_LV_1$ as justified and earlier utilized by Chaudhuri, Adhikary, Dihidar (2000).

Then for t_b subject to (3.1) and t_H with $e_b=t_b\big|_{y_i=\hat{y}_i}$ and $e_H=t_H\big|_{y_i=\hat{y}_i}$, we have from Chaudhuri and Pal (2001)

$$V(e_b) = -\sum_{i<}\sum_{j} d_{ij} w_i w_j \left(\frac{y_i}{w_i} - \frac{y_j}{w_j}\right)^2 + \sum \frac{y_i^2}{w_i}\beta_i + E_1\left[\sum V_i b_{si}^2 I_{si}\right]$$

$$= E_L\left[-\sum_{i}\sum_{<j} d_{ij} w_i w_j \left(\frac{\hat{y}_i}{w_i} - \frac{\hat{y}_j}{w_j}\right)^2 + \sum \frac{(\hat{y}_i)^2}{w_i}\beta_i\right] + V_L\left(\sum v_i\right)$$

admitting an unbiased estimator

$$v(e_b) = -\sum_{i<}\sum_{j} d_{sij} I_{sij} w_i w_j \left(\frac{\hat{y}_i}{w_i} - \frac{\hat{y}_j}{w_j}\right)^2 + \sum \frac{(\hat{y}_i)^2}{w_i}\beta_i \frac{I_{si}}{\pi_i} + \sum v_i b_{si} I_{si}$$

and

$$V(e_H) = \sum\sum (\pi_i \pi_j - \pi_{ij}) \left(\frac{y_i}{\pi_i} - \frac{y_j}{\pi_j}\right)^2 + \sum \frac{y_i^2}{\pi_i}\alpha_i + E_1\left(\sum V_i b_{si}^2 I_{si}\right)$$

$$= E_L\left[\sum\sum (\pi_i \pi_j - \pi_{ij}) \left(\frac{\hat{y}_i}{\pi_i} - \frac{\hat{y}_j}{\pi_j}\right)^2 + \sum \frac{(\hat{y}_i)^2}{\pi_i}\alpha_i\right] + E_1\left(\sum v_i\right)$$

admitting an unbiased estimator

$$v(e_H) = \sum\sum (\pi_i \pi_j - \pi_{ij}) \frac{I_{sij}}{\pi_{ij}} \left(\frac{\hat{y}_i}{\pi_i} - \frac{\hat{y}_j}{\pi_j}\right)^2 + \frac{(\hat{y}_i)^2}{\pi_i}\alpha_1 \frac{I_{si}}{\pi_i} + \sum v_i b_{si} I_{si}.$$

Using (3.2) it follows that corresponding to v_4, v_5 in (2.7), (2.8) as estimators for $V_I(t_H)$ given by Brewer (2000) and Deville (1999) respectively corresponding estimators for $V(e_H)$ one may respectively take as

$$v_4(e_H) = \sum \left(\frac{1}{c_i} - \pi_i\right) \left(\frac{\hat{y}_i}{\pi_i} - \frac{e_H}{n}\right)^2 I_{si} + \sum \frac{v_i}{\pi_i} I_{si}$$

and $$v_5(e_H) = \frac{1}{(1 - \sum a_i^2 I_{si})} \sum (1 - \pi_i) \left(\frac{\hat{y}_i}{\pi_i} - A_s\Big|_{y_i = \hat{y}_i}\right)^2 I_{si} + \sum v_i \frac{I_{si}}{\pi_i}.$$

Again, using (3.2), corresponding to m_{1j} (t_{Ha}) in (2.9) and m_{1j} (t_{RH}), $j=1,\ldots\ldots,5$ as *MSE*- estimators for t_{Ha} and t_{RH} respectively the same for

$$e_{Ha} = \frac{X \sum \frac{\hat{y}_i}{\pi_i} I_{si}}{\sum \frac{x_i}{\pi_i} I_{si}} \quad \text{and } e_{RH} = \frac{\sum \pi_i}{\upsilon(s)} \left(\sum \frac{y_i}{\pi_i} I_{si}\right)$$

may be taken as $$m_j(c_{HA}) = \frac{X^2}{(\sum \frac{x_i}{\pi_i} I_{si})^2} v_j \Bigg|_{y_i = \hat{y}_i - \left(\frac{\sum \frac{\hat{y}_i}{\pi_i} I_{si}}{\sum \frac{x_i}{\pi_i} I_{si}}\right) x_i} + \sum v_i \frac{I_{si}}{\pi_i}$$

and as $$m_j(e_{RH}) = \frac{(\sum \pi_i)^2}{\upsilon^2(s)} v_j \Bigg|_{y_i = (\hat{y}_i - \frac{\sum \frac{\hat{y}_i}{\pi_i} I_{si}}{v(s)} \pi_i)} + \sum v_i \frac{I_{si}}{\pi_i}$$

for $j=1, .., 5$.

So far we considered only the classical design-based approach of 'point' and 'interval' estimation of Y in terms of 'sampling design-based' expectations and MSE's of point estimators and lengths of CI's covering Y with desired coverage probabilities determined by sampling designs. The case of $\overline{Y} = \frac{Y}{N}$ in uni-stage sampling is covered because N is known and of $\overline{\overline{Y}} = \frac{Y}{\sum M_i}$, in case of multi-stage sampling is covered because M_i may be taken as x_i in 'ratio' estimation.

Now we turn to the 'model-assisted' approach for which a crucial reference is SSW(1992). Cassel, Särndal and Wretman (CSW, 1976) gave us the 'generalized regression' (greg) estimator for Y as

$$t_g = \sum \frac{y_i}{\pi_i} g_{si} I_{si}, \; g_{si} = 1 + (X - \sum \frac{x_i}{Q_i} I_{si}) \frac{x_i \pi_i Q_i}{\sum x_i^2 Q_i I_{si}}$$

with Q_i as 'arbitrarily' assignable 'positive' constants, usually taken as

$\frac{1}{x_i}, \frac{1}{x_i^2}, \frac{1}{\pi_i x_i}, \frac{1-\pi_i}{\pi_i x_i}, \frac{1}{x_i^g}$ with g in [0,2].

Writing $b_Q = \frac{\sum y_i x_i Q_i I_{si}}{\sum x_i^2 Q_i I_{si}}, B_Q = \frac{\sum y_i x_i Q_i \pi_i}{\sum x_i^2 Q_i \pi_i}, e_i = y_i - b_Q x_i, E_i = y_i - B_Q x_i$

it is well known from Särndal (1982) that MSE (t_g) has formulae

$$M_1(t_g) = \sum E_i^2 \frac{1-\pi_i}{\pi_i} + \sum\sum E_i E_j \frac{\pi_{ij} - \pi_i \pi_j}{\pi_i \pi_j}$$

for general designs and $M_2(t_g) = \sum\sum (\pi_i \pi_j - \pi_{ij})(\frac{E_i}{\pi_i} - \frac{E_j}{\pi_j})^2$

for designs with constant $\upsilon(s)$.

Chaudhuri and Pal (2001) give another as

$$M_3(t_g) = \sum\sum (\pi_i \pi_j - \pi_{ij})(\frac{E_i}{\pi_i} - \frac{E_j}{\pi_j})^2 + \sum \frac{E_i^2}{\pi_i} \alpha_i$$

for general designs.

Three pairs of estimators for MSE(t_g) follow from these as

$$m_{1k}(t_g)=\sum a_{ki}^2 e_i^2(\frac{1-\pi_i}{\pi_i})\frac{I_{si}}{\pi_i}+\sum\sum a_{ki}a_{kj}e_ie_j(\frac{\pi_{ij}-\pi_i\pi_j}{\pi_i\pi_j})\frac{I_{sij}}{\pi_{ij}}, k=1,2$$

$$m_{2k}(t_g)=\sum\sum(\pi_i\pi_j-\pi_{ij})(\frac{a_{ki}e_i}{\pi_i}-\frac{a_{kj}e_j}{\pi_j})^2\frac{I_{sij}}{\pi_{ij}}, k=1,2$$

$$m_{3k}(t_g)=m_{2k}(t_g)+\sum\frac{(a_{ki}e_i)^2}{\pi_i}\alpha_i\frac{I_{si}}{\pi_i}; k=1,2; a_{1i}=1, a_{2i}=g_{si}.$$

Sändral (1996) recommends approximating $MSE(t_g)$ by

$$M_s(t_g)=\sum E_i^2(\frac{1-\pi_i}{\pi_i})$$

and estimating it by $m_s(t_g)=\sum\left(a_{ki}e_i\right)^2\left(\frac{1-\pi_i}{\pi_i}\right)\frac{I_{si}}{\pi_i}$.

In case y_i is not ascertainable but estimated by $\hat{y}_i$ through multi-stage sampling, then

$$e_g=t_g\Big|_{y_i=\hat{y}_i}=\sum\frac{\hat{y}_i}{\pi_i}g_{si}I_{si}$$

may be the revised greg estimator for Y.

Letting $\hat{e}_i=e_i\Big|_{y_i=\hat{y}_i}$ and using (3.2), (1), (2) and (3), one may use for MSE(e_g) the estimators corresponding to $m_{jk}(t_g)$ as respectively

$$m_{jk}(e_g)=m_{j1}(t_g)\Big|_{e_i=\hat{e}_i}+\sum\frac{v_i}{\pi_i}g_{si}I_{si}, j=1,2,3; k=1,2.$$

Corresponding to M$_s$(t_g) we have no recommendations about estimating

$$M_s(e_g)=M_s(t_g)\Big|_{y_i=\hat{y}_i}, i\in U$$

Next follows our proposals concerning estimation of Y in multi-stage sampling with the following.

'model motivated' approach with details for 'two-stage' sampling.

Let us postulate the model to write

$y_{ij}=\beta x_{ij}+\in_{ij}$ with x_{ij} as the value of x for j^{th} *ssu* of ith *fsu*, β is an unknown constant and $\in_{ij}$'s are random variables. With $y_i=\sum_1^{M_i}y_{ij}, x_i=\sum_1^{M_i}x_{ij}$

let us further postulate the model for which $y_i=\theta x_i+\eta_i$ with θ as an unknown constant and η_i's are random variables, $i\in U$. These models will be seen to allow us to strengthen our estimators for y_i and Y by borrowing strength from outside the specific clusters chosen in the sample.

We consider the specific 'Two-stage' sampling in which n *fsu's* are chosen from U applying the Rao, Hartley and Cochran's (RHC, 1962) scheme supposing normed size-measures $p_i (o \langle p_i \langle 1, \sum p_i = 1)$ are available. For this, n groups of N_i clusters (i=1,....,n) are formed at random that are disjoint together exhaustive with U. Writing $\sum_n$ as sum over the groups, $\sum_n N_i = N$. By Q_i is meant the sum of the N_i numbers p_i*'s* in the ith group. From the respective groups so formed only 1 cluster is chosen with a probability proportional to its p_i*-value*. The selection process is independently repeated over the n groups. Writing (y_i, p_i) as the value of y and normed size measure for the unit chosen from the ith group, $t_{RHC} = \sum_n y_i \frac{Q_i}{p_i}$

is an unbiased estimator for Y with a variance

$$V(t_{RHC}) = A \sum_n \sum_n p_i p_i' \left(\frac{y_i}{p_i} - \frac{y_i'}{p_i'} \right)^2$$

writing $\sum_n \sum_n$ as sum over the disjoint groups with no duplication. An unbiased estimator for $V(t_{RHC})$ is

$$v(t_{RHC}) = B \sum_n \sum_n Q_i Q_i' \left(\frac{y_i}{p_i} - \frac{y_i'}{p_i'} \right)^2$$

Here

$$A = \frac{\sum_n N_i^2 - N}{N(N-1)} \text{ and } B = \frac{\sum_n N_i^2 - N}{N^2 - \sum_n N_i^2}$$

as are given by RHC themselves.

Now supposing y_i to be non-ascertainable a sample of m_i *ssu's* out of M_i ssu's in the ith*fsu* is again selected by the RHC scheme utilizing known size-measures

$$p_{ij} (0 < p_{ij} < 1, \sum_{j=1}^{M_i} p_{ij} = 1 \ \forall \ i \in U).$$

So, y_i may be unbiasedly estimated by $\hat{y}_i = \sum_{m_i} y_{ij} \frac{Q_{ij}}{p_{ij}}$, with obvious notations which, admits an unbiased variance estimator similar to $v(t_{RHC})$. But estimating

$$\beta \text{ by } \hat{b} = \frac{\sum_n \sum_{m_i} y_{ij} x_{ij} R_{ij}}{\sum_n \sum_{m_i} x_{ij}^2 R_{ij}},$$

$$\text{with } R_{ij} = \frac{\left(1 - \frac{p_{ij}}{Q_{ij}}\right)}{\left(\frac{p_{ij} x_{ij}}{Q_{ij}}\right)},$$

writing Q_{ij}'s as sums of p_{ij}'s within respective m_i groups while applying *RHC* scheme in the second stage of sampling one may employ for y_i the greg estimator

$$\hat{y}_{ig} = \sum_{m_i} y_{ij} \frac{Q_{ij}}{p_{ij}} + \hat{b}\left(x_i - \sum_{m_i} x_{ij} \frac{Q_{ij}}{p_{ij}}\right).$$

Next taking

$$R_i = \left(1 - \frac{p_i}{Q_i}\right) \bigg/ \left(\frac{p_i x_i}{Q_i}\right)$$

one may estimate

$$\theta \text{ by } \hat{\theta} = \frac{\sum_n \hat{y}_{ig} x_i R_i}{\sum_n x_i^2 R_i}$$

and finally for Y employ the 'two-stage' greg estimator

$$e_{(RHC)g} = \sum_n \frac{Q_i}{p_i} \hat{y}_{ig} + \hat{\theta}\left(X - \sum_n \frac{Q_i}{p_i} x_i\right)$$

Taking account of what preceded an estimator for the $MSE(e_{(RHC)g})$ and for the $MSE(\hat{y}_{ig})$ is easy to write down and work out. For simplicity we shall write y_i^* for $\hat{y}_{ig}$ and $m(y_i^*)$ for its estimated *MSE*.

At this stage we may apply a 'model-based' method by postulating that $y_{ij} = \beta x_{ij} + \in_{ij}$ with $\in_{ij}$ as $N(0,B)$, 'independently' leading to $y_i^* \setminus y_i \sim N(y_i, m(y_i^*))$.

Letting $A = V_m\left(\sum_i \sum_j \in_{ij}\right)$, with V_m as 'model-based' variance operator, we have A=BM, writing $M = \sum M_i$.

Since $y_i \sim N(\beta x_i, BM_i)$, the marginal distribution of y_i^* is

$$N(\alpha x_i, BM_i + m(y_i^*)) = N(\alpha x_i, \frac{AM_i}{M} + m(y_i^*)).$$

Letting $\psi_i = \dfrac{AM_i}{\sum_n \frac{Q_i}{p_i} M_i} + m(y_i^*)$,

$$\tilde{\beta} = \frac{\sum_n \frac{Q_i}{p_i} y_i^* x_i / \psi_i}{\sum_n \frac{Q_i}{p_i} x_i^2 / \psi_i}$$

and iteratively solving for A the equation

$$\sum_n \frac{Q_i}{p_i} (y_i^* - \tilde{\beta} x_i)^2 / \psi_i = \sum_n \frac{Q_i}{p_i} - 1,$$

let $\hat{\beta}, \hat{A}$, be the resulting estimators for β, A; also let

$$\hat{\psi}_i = \frac{\hat{A}M_i}{\sum_n \frac{Q_i}{p_i} M_i} + m(y_i^*) = \frac{\hat{A}M_i}{\hat{M}} + m(y_i^*).$$

Then,

$$\hat{y}_i(EB) = \left(\frac{\hat{A}M_i}{\hat{M}\hat{\psi}_i} \right) y_i^* + \left(\frac{m(y_i^*)}{\hat{\psi}_i} \right) \hat{\beta} x_i ,$$

following Fay and Herriot (1979) may be taken as an 'Empirical Bayes' estimator (EBE) for y_i. Then, for Y the final EBE may be proposed as $\hat{Y}(EB) = \sum_n \hat{y}_i(EB) \frac{Q_i}{p_i}$.

Writing

$$a_i = \frac{\hat{A}M_i}{\hat{M}\hat{\psi}_i}, d = \frac{2}{n^2} \sum_n \frac{Q_i}{p_i} \hat{\psi}_i ,$$

following Prasad and Rao (1990) for $\hat{y}_i(EB)$ we may employ the *MSE* estimator

$$m(\hat{y}_i(EB)) = a_i m(y_i^*) + \frac{(1 - a_i)^2 x_i^2}{\sum_n \frac{Q_i}{p_i} x_i^2 / \hat{\psi}_i} + 2d \frac{m^2(y_i^*)}{\hat{\psi}_i^3}.$$

Then following the approach discussed so far one may estimate $MSE(\hat{Y}(EB))$ by

$$m(\hat{Y}(EB)) = B\sum_{n}\sum_{n} Q_i Q_{i'}\left(\frac{\hat{y}_i(EB)}{p_i} - \frac{\hat{y}_{i'}(EB)}{p_{i'}}\right)^2 + \sum_{n} m(\hat{y}_i(EB))\frac{Q_i}{p_i}.$$

This may be regarded as our model-dominated approach of borrowing strength from outside clusters as is done in small domain computation.

4. SENSITIVE QUERIES: RANDOMIZED RESPONSE

Essentially the methods for Two-stage sampling may be employed to cover when y relates to a stigmatizing issue like drunken driving, tax evasion, practicing fraud etc. Here for a sampled person i, no matter how selected, it is often difficult to generate direct response (DR) about his/her y_i-*value*. Instead, a suitable 'randomized response' (RR) device may be employed to procure an RR, say, z_i , which may be suitably transformed to yield a random variable r_i for which the following may be held to be true. Writing E_R , V_R for expectation, variance operators for an RR scheme :

(1) $E_R(r_i) = y_i$,

(2) $V_L(r_i) = A_i y_i^2 + B_i y_i + C_i$ with A_i ,B_i ,C_i as known constants; then for

$$v_i = \frac{1}{(1+A_i)}(A_i r_i^2 + B_i r_i + C_i),$$

if $1 + A_i \neq 0, E_L(v_i) = V_L(r_i)$.

One example of an RR device is to ask a respondent i to randomly choose from a box with T cards bearing numbers $a_{i.},\ldots a_j, \ldots a_T$ one say marked a_j and 'independently' choose from a second box with L cards numbered $b_i,..b_k ,\ldots,b_L$ one card numbered b_k say and report the value

$$z_i = a_j y_i + b_k.$$

Then, $E_R(z_i) = y_i(\frac{1}{T}\sum_1^T a_j) + (\frac{1}{L}\sum_1^T b_k) = y_i\bar{a} + \bar{b}$,

say, and $r_i = \dfrac{z_i - \bar{b}}{\bar{a}}$, provided $\bar{a} \neq 0$, satisfies the above requirements.

If corresponding to a contemplated estimator $t_b = \sum y_i b_{si} I_{si}$ one then employs $r_b = \sum r_i b_{si} I_{si}$ then estimation of its *MSE* follows immediately with the approaches elaborated above. Chaudhuri and Mukerjee's (1988) text is a useful source.

5. ADAPTIVE SAMPLING

Sometimes y may relate to a seldom occurring phenomenon like incidence of maternal mortality in i^{th} household, number of earners through 'rope tricking' in i^{th} village and so on. Then, estimating the total count $Y = \sum y_i$ of units of U with such rare characteristics, with *'y_i equal to 0'* for many i in U, becomes really a problem of capturing enough relevant units in a chosen sample s from U. Chaudhuri (2000), following Thompson (1992) and Thompson and Seber (1996) has discussed how from an initial sample s chosen employing usual sampling schemes with varying or equal probabilities one may implement on 'Adaptive' sampling scheme with formation of suitable networks so as to effectively enhance the capture of relevant units with positive y_i -values. He has also shown that estimating Y and the related variance or MSE- estimation for an Adaptive sample are simple matters. Briefly we may describe as follows.

With every unit i of U is defined a neighbour of units; for example the villages with a common boundary with i^{th} village are together its neighbourhood. For a sample unit i with positive y_i , in 'Adaptive sampling' one is to extend observation to all units in its neighbourhood. If for any in the neighbourhood positive y is encountered the observation is to extend to those in the latter's neighbourhood and the process is to continue until a unit is reached with zero-valued y in every unit in a neighbourhood. The 'unique ' collection of units linked with a specific unit through the system of neighbourhoods each with a positive y is a 'Network' for the unit. The collection of all the units in the union of these neighbourhoods is a cluster containing this unit. The units in the cluster with *'zero-y's'* are the 'edge units', to be regarded as 'Singleton networks'. All the networks are non-overlapping and they together exhaust the population. Denoting by $A(i)$, the network containing i and by $N(i)$ its cardinality , let

$$t_i = \frac{1}{N(i)} \sum_{j \in A(i)} y_j$$

It easily may be checked that $T = \sum_{i=1}^{N} t_i$ equals $Y = \sum y_i$.

Corresponding to $\upsilon(s)$ for the original sample the effective size $\sum_{i \in s} N(i) = A(s)$ of the 'Adaptive' sample may be considerably larger. So, though t_i for $i \in s$ may be used for estimating *T=Y* in suitable ways along with easily derived *MSE*-estimators discretion must be properly exercised in keeping *A(s)* under control by appropriate definitions of 'networks' and 'neighbourhoods'.

6. BOOTSTRAP SAMPLING

When employing a complicated, non-linear estimator for *Y* a standard procedure to estimate its *MSE* is to apply linearization or delta-method based on first order Taylor series expansion. This was actually done above in the case of ratio estimator and the greg estimator. Another is replicated sampling producing independently distributed unbiased estimators for *Y* based on independently drawn samples and using their average to estimate *Y* and average of paired differences in estimating the variance. A variation of this is jackknifing employed originally by Quenouille (1949, 1956) as a bias- reduction technique later better utilized by Tukey (1958) in MSE estimation. We have no space here to elaborate on them. Another procedure is 'bootstrap' which we may briefly illustrate to show its alternative use in employing the greg estimator for *Y*. In the simplest case with a single auxiliary variable *x* on which *y* has a linear regression through the origin

$$t_g = \sum \frac{y_i}{\pi_i} I_{si} + \left(X - \sum \frac{x_i}{\pi_i} I_{si} \right) \frac{\sum (y_i x_i Q_i \pi_i) \frac{I_{si}}{\pi_i}}{\left(\sum x_i^2 Q_i \pi_i \frac{I_{si}}{\pi_i} \right)}, Q_i > 0 = f(t_y, t_x, t_{yxQ\pi}, t_{x^2Q\pi})$$

a non-linear function of 4 HT estimators of 4 population totals of 4 variables namely $y, x, yxQ\pi$ and $x^2Q\pi$ as we already discussed.

The principle of 'bootstrap' sampling demands that from the initial sample *s* for which the *HTE's* $t_y, t_x, t_{yxQ\pi}$ and $t_{x^2Q\pi}$ have already been calculated a large number *B=1000*, say, bootstrap samples $s_b^*, (b = 1,.., B)$ be 'drawn independently' in a suitable and identical way. Then, calculating $t_y(s_b^*), t_x(s_b^*), t_{yxQ\pi}(s_b^*)$ and $t_{x^2Q\pi}(s_b^*)$ and hence $t_g(s_b^*) = f(t_y(s_b^*), t_x(s_b^*), t_{yxQ\pi}(s_b^*), t_{x^2Q\pi}(s_b^*)$ is to be calculated.

Then, $\bar{t}_g = \frac{1}{B}\sum_{b=1}^{B} t_g(s_b^*)$ is taken as a 'Bootstrap' estimator for Y initiated on the 'greg' estimator. Then, $v_B = \frac{1}{B-1}\sum_{b=1}^{B}(t_g(s_b^*) - \bar{t}_g)^2$

is taken as the "Bootstrap" variance-estimator or *MSE*-estimator for $\bar{t}_g$.

To cover such a function f (.,.,.,.) of HTE's Rao and Wu (1988) have given a method of drawing 'bootstrap' samples when the original sampling design P to draw s is subject to 2 restrictions, namely

(i) Every sample s has a constant $\upsilon(s)$ and

(ii) $\pi_i \pi_j \geq \pi_{ij} \forall i.j(i \neq j)$.

We present here a modification of their 'bootstrap' sampling scheme covering HTE's only when (A) only (i) is relaxed but (ii) holds and when (B) both (i) and (ii) are violated.

Case (A). Out of $\upsilon(s)(\upsilon(s)-1)$ 'ordered' pairs of units $i, j(i \neq j)$ in s let us choose a 'Bootstrap' sample s_1^* of pairs (i^*, j^*) in *'m'* draws 'with replacement' with probabilities $\lambda_{i^*j^*}(i^* \neq j^*)$ such that $\lambda_{i^*j^*} = \lambda_{j^*i^*}$ with their values to be 'assigned' as below. Let us choose numbers $k_{i^*j^*} = k_{j^*i^*}$ in a manner to be described below. Constructing the bootstrap statistic

$t_1 = \frac{1}{m}\sum_{(i^*, j^*) \in s_1^*}\sum k_{i^*j^*}(\frac{y_{i^*}}{\pi_{i^*}} - \frac{y_{j^*}}{\pi_{j^*}})$ and writing E_*, V_* generically to denote expectation, variance operators for bootstrap sampling we have

$$E_*(t_1) = \sum_{i \neq j}\sum_{\in s} \lambda_{ij} k_{ij}(\frac{y_i}{\pi_i} - \frac{y_j}{\pi_j}) = 0 \text{ and}$$

$$V_*(t_1) = \frac{1}{m}\sum_{i \neq}\sum_{j \in s} \lambda_{ij} k_{ij}^2(\frac{y_i}{\pi_i} - \frac{y_j}{\pi_j})^2$$

Choosing $m = \upsilon(s)(\upsilon(s)-1), \lambda_{ij} = \frac{1}{m}$ and

$k_{ij} = m(\frac{\pi_i \pi_j - \pi_{ij}}{2\pi_{ij}})^{1|2} i, j(i \neq j \in s)$, we have

$$V_*(t_i) = v_2 = \sum_{i<}\sum_{j}(\pi_i\pi_j - \pi_{ij})\left(\frac{y_i}{\pi_i} - \frac{y_j}{\pi_j}\right)^2 \frac{I_{sij}}{\pi_{ij}}.$$

We need to have now a second bootstrap sample s_2^* out of the 'distinct' $\upsilon(s)$ units in s 'independently' of how s_1^* is chosen . Let this be drawn by Poisson's scheme with l_i as the probability of 'success' associated with i in s .

Writing i^* as elements of s_2^*, let

$$t_2 = \sum_{i^* \in s_2^*} \frac{y_i^*}{\pi_i^*} \frac{I_{s^* i^*}}{l_{i^*}},$$

then, $E_*(t_2) = \sum \frac{y_i}{\pi_i} I_{si} = t_H(y)$

and $V_*(t_2) = \sum \left(\frac{1}{l_i} - 1\right)\left(\frac{y_i}{\pi_i}\right)^2 I_{si}$

Letting $t=t_1+t_2$ we have $E_*(t)=t_H$, $v_*(t)=v_2+V_*(t_2)$

choosing $l_i = \dfrac{1}{1+\alpha_i}$ with α_i in (2.4),

since by (2.5), $\alpha_i \geq 0$, $0 \leq l_i \leq 1$ $\forall i$,

we have $V_*(t_2) = v_3$,

provided for the original sample drawn, $v_3 \geq 0$.

Thus, we modify Rao and Wu's (1988) 'bootstrap method' of equating the "Bootstrap" variance of a statistic, namely t in this case, to an estimate of $v_1(t_H)$, *namely* v_3 in this case (A).

Case(B). First we note that though it is impossible to have "$\pi_i\pi_j \leq \pi_{ij}$ $\forall i,j$ $(i \neq j)$" in case $\upsilon(s) = n$

$\forall$ s with $p(s) > 0$, it is quite possible to hold in case $\upsilon(s)$ varies with s especially

(a) if the largest number of draws n satisfies $n \geq 1 + E(\upsilon(s)) - \pi_i$ $\forall$ i, and / or

(b) if $Var(\upsilon(s)) \geq \sum \pi_i(1-\pi_i)$, for examples.

In this case let (1) from s a bootstrap sample s_1^* be drawn by Poisson scheme with k_{i*} as the 'probability for success' for a unit i^* in s.

Again, (2) 'independently' of the draw of s_1^*, let a 'bootstrap' sample s_2^* be drawn from the $\upsilon(s)(\upsilon(s)-1)$ 'ordered' pairs of distinct units of s again by Poisson scheme with $\lambda_{i^*j^*}$ as the 'probability of success' for the (i^*,j^*) - *paired unit*; $(i^*\neq j^*\in s)$.
Let us construct the bootstrap statistic

$$t=\left(\sum\frac{y_i^*}{\pi_i^*}\times\frac{1}{k_i^*}\times I_{s_1i^*}\right)+\left(\sum_{i^*}\sum_{\neq j^*}\frac{\sqrt{\frac{y_i^*}{\pi_i^*}\frac{y_j^*}{\pi_j^*}}}{\lambda_{i^*j^*}}I_{s_2i^*j^*}-\sum_{i}\sum_{\neq j}\sqrt{\frac{y_i}{\pi_i}\frac{y_j}{\pi_j}}I_{sij}\right)$$

Then, $E_*(t)=\sum\frac{y_i}{\pi_i}I_{si}$

and $V_*(t)=\sum\frac{y_i^2}{\pi_i^2}\left(\frac{1}{k_i}-1\right)I_{si}+\sum_{i}\sum_{\neq j}\left(\frac{1}{\lambda_{ij}}-1\right)\frac{y_i}{\pi_i}\frac{y_j}{\pi_j}$

choosing (1) $k_i=\frac{1}{2-\pi_i}$ and (2) $\lambda_{ij}=\frac{1}{2-\left(\frac{\pi_i\pi_j}{\pi_{ij}}\right)}$

this V* (t) is equated to v_1, provided for the original sample drawn v_1 happens to turn out non-negative.

A remark: If $v_3<0$ or $v_1<0$ our proposed schemes do not work. Similarly Rao and Wu's (1988) method does not work if at least one of (i) and (ii) is violated. So, further research seems necessary to cover all possible cases.

7. CONCLUDING REMARKS

Statistics Canada employs a 'Generalized Estimation System' as discussed among others by Särndal (1996) that predominantly involves the application of the greg estimation with one or more auxiliary correlated variables and its MSE-estimator with one or more auxiliary correlated variables and its MSE-estimation with simplifications avoiding π_{ij}'s and hence the cross-product terms. In Indian National Sample Survey Organization (NSSO), however the first stage units within strata are chosen as two equal-sized "half-samples" by probability proportional to size (PPS) circular systematic sampling (CSS) method and the second stage units by single CSS method with equal probabilities. So, variance estimation is accomplished by computing 'one-forth of the squared difference between the 2 half-sample estimators'.

In other organizations like US Bureau of Census, Canadian Labour Force Surveys, British Population Censuses and Surveys variance or MSE estimation receive attention in diverse appropriate ways. The above discussions in Sections 1-6 mainly serve theoretical purposes but may also be put to practical uses as some of the procedures have been applied in certain case studies undertaken in Indian Statistical Institute, Calcutta with active participation by the present author.

REFERENCES

Brewer, K.R.W. Cosmetic Calibration with Unequal Probability Sampling. Survey Meth. 25(2), 205-212,(1999).

___. Deriving and estimating an approximate variance for the Horvitz-Thompson estimator using only first order inclusion probabilities. Contributed to Second International Conference on Establishment Surveys, Buffalo, N.Y. June 17-21.(2000)

Brewer, K.R.W. and Gregoire, T.G. Estimators for use with Poisson Sampling and Related Selection Procedures. Invited paper in Second International Conference on Establishment Surveys, Buffalo, N.Y., June 17-21.(2000).

Cassel, C.M., Särndal, C.E. and Wretman, J.H. Some results on generalized difference estimation and generalized regression estimation for finite populations. Biometrika, 63, 615-620.(1976).

Chaudhuri, Arijit. Network and Adaptive Sampling with unequal probabilities. Cal. Stat. Assoc. Bull 50, 237-253.(2000).

Chaudhuri, Arijit, Adhikary, Arun Kumar and Dihidar, Shankar. Mean square Error estimation in multi-stage sampling. Metrika, 52,115-131.(2000).

Chaudhuri, Arijit and Mukherjee , Rahul. Randomized Response: Theory and Techniques. Marcel Dekker, N.Y.(1988).

Chaudhuri, Arijit and Pal, Sanghamitra, On certain alternative Mean Square Error estimators in Complex Survey Sampling. To appear in Jour. Stat. Plan. Inf. (2001).

Chaudhuri, Arijit and Stenger, Host. Survey Sampling: Theory and Methods. Marcel Dekker, N.Y. (1992).

Deville, Jean-Claude. Variance estimation for complex statistics and estimators: Linearization and residual techniques. Survey Meth. 25(2), 193-203 (1999).

Fay R.E. and Herriot, R.A. Estimates of income for small places: an application of James-Stein procedures to Census data. Jour. Amer. Stat. Assoc. 74, 269-277. (1979).

Ha'jek, J. Comment on a paper by Basu, D. In Foundations of Statistical Inference. Ed. Godambe, V.P. and Sprott, D.A. Holt, Rinchart, Winston, Toronto, 203-242. (1971).

_____, Sampling from a finite population . Marcel Dekker, N.Y. (1981).

Horvitz, D.G. and Thompson, D.J. A generalization of sampling without replacement from a finite universe. Jour. Amer. Stat. Assoc. 47, 663-685. (1952).

Prasad, N.G.N. and Rao, J.N.K. The estimation of mean squared errors of small-area estimators. Jour. Amer. Stat. Assoc. 85, 163-171. (1990).

Quenouille, M.H. Approximate tests of correlation in time-series, Jour. Roy. Stat. Soc. B, 11, 68-84. (1949).

_____. Notes on bias in estimation. Biometrika, 43, 353-360. (1956).

Raj, D. Sampling Theory. McGraw-Hill. N.Y. (1968).

Rao, J.N.K. Unbiased variance estimation for multi-stage designs. Sankhyā, C, 37, 133-139. (1975).

______. On deriving mean square errors and other non-negative unbiased estimators in finite population sampling. Jour. Ind. Stat. Assoc. 17, 125-136. (1979).

Rao, J.N.K., Hartley, H.O. and Cochran, W.G. On a simple procedure of unequal probability sampling without replacement. Jour. Roy. Stat. Soc. B, 24 482-491. (1962).

Rao, J.N.K. and Wu, C.F.J. Resampling inference with complex survey data. Jour. Amer. Stat. Assoc. 83, 231-241. (1988).

Särndal, C.E. Implications of survey design for generalized regression estimation of linear functions. Jour. Stat. Plan. Inf. 7, 155-170. (1982).

_____. Efficient estimators with simple variance in unequal probability sampling. Jour. Amer. Stat. Assoc. 91, 1289-1300. (1996).

Särndal, C.E., Swensson, B.E. and Wretman, J.H. Model Assisted Survey Sampling. Springer Verlag, N.Y. (1992).

Sen, A.R. On the estimator of the variance in sampling with varying probabilities. Jour. Ind. Soc. Agri. Stat. 5(2), 119-127. (1953).

Thompson, S.K. Sampling. John Wiley & Sons. Inc. N.Y. (1992).

Thompson, S.K. and Seber, G. A.F. Adaptive Sampling. John Wiley & Sons. Inc. N.Y. (1996).

Tukey, J.W. Bias and Confidence in not-quite large samples (Abstract). Ann. Math. Stat. 29, 614. (1958).

Yates, F. and Grundy, P.M. Selection without replacement from within strata with probability proportional to size. Jour. Roy. Stat. Soc. B, 15, 253-261. (1953).

Linear Programming: Recent Advances

S.K. Sen
Supercomputer Education and Research Centre, Indian Institute of Science, Bangalore 560 012, India
e-mail: sksen@serc.iisc.ernet.in

Abstract A linear program (LP) *Minimize* $\mathbf{c}^t\mathbf{x}$ *subject to* $A\mathbf{x}=\mathbf{b}$, $\mathbf{x}\geq\mathbf{0}$ (null column vector), where A is an m×n real matrix, **c** and **b** are n- and m-vectors, respectively .is a problem of great importance in numerous physical problems involving linear optimization such as diet problems, transport problems, industrial production problems. The algorithms such as simplex method, self-dual parametric algorithm, decomposition algorithm, primal-dual algorithm to solve an LP have been non-polynomial time. In order to appreciate the recent advances in this area the present chapter provides a background based on the simplex methods which completely ruled the scene during sixties, seventies and early eighties. Although the simplex methods are non-polynomial-time in the worst case, they did perform excellently in most real-world problems and behaved like a fast (polynomial-time) algorithm. The chapter then focuses on the development of several fast (polynomial-time) algorithms during the last two decades. It then briefly highlights heuristic and evolutionary approach to solve LPs including errorfree implementation.

Key Words Basic variables, heuristic algorithm, linear programming, polynomial-time algorithm, projective transformation, simplex method.

1. Introduction

The process of getting the best result, e.g., minimum or maximum values, under given conditions (constraints) is called optimization. The optimum seeking methods belong to the discipline of ***mathematical programming*** [1](MP) which is a branch of ***operations research*** (OR). OR is a branch of mathematics applied to decision-making problems and obtain the best solutions while linear programming is a branch of MP. The OR methods may be classified as follows.

OR Methods

Statistical Methods	Stochastic Process Methods	MP Methods
•Regression Analysis	•Statistical Decision Theory	•Methods of Calculus
•Cluster Analysis	•Markov Processes	•Calculus of Variations

[1] The term *programming* as used above referred originally to the scheduling of events or activities. There is no immediate connection with computer programming, mathematical programming problems are solved on a digital computer though.

- Design of Experiments
- Factor Analysis . .
- Renewal Theory
- Queuing Theory
- Reliability theory
- Simulation Methods . .
- Linear Programming
- Geometric Programming
- Dynamic Programming
- Nonlinear Programming
- CPM and PERT . .

We limit ourselves in the area of linear programming and recent advances. A linear program (LP) may be defined as *Minimize (Min)* $\mathbf{c}^t\mathbf{x}$ *subject to* A$\mathbf{x}$ = $\mathbf{b}$, $\mathbf{x}\geq\mathbf{0}$ (null column vector of appropriate order), where A =(a_{ij}) is a given m×(m+n) real matrix, $\mathbf{c}$=(c_j) and $\mathbf{b}$=(b_i) are specified n- and m-column vectors respectively, and x is an (m+n)-column vector to be computed. An equivalent LP is *Maximize(Max)* $-\mathbf{c}^t\mathbf{x}$ *subject to* A$\mathbf{x}$=$\mathbf{b}$, $\mathbf{x}\geq\mathbf{0}$. The cases and standard forms below may be considered from a practical point of view (e.g., to form an appropriate simplex tableau..

Case 1: $a_{1,n+1}=..=a_{m,n+1}=1$; ***Standard Form 1:*** $a_{1,n+1}=..=a_{m,n+1}=1$; $b_i\geq 0$ i=1(1)m.

This case arises when the constraints are originally $a_{i1}x_1+..+a_{in}\leq b_i$ i=1(1)m. and the slack variables $x_{n+1},..,x_{n+m}$ are introduced to transform the inequalities into equations.

Case 2: $a_{1,n+1}=..=a_{m,n+1}=-1$; ***Standard Form 2:*** $a_{1,n+1}=..=a_{m,n+1}=-1$; $b_i\geq 0$ i=1(1)m.

This case arises when the constraints are originally $a_{i1}x_1+..+a_{in}\geq b_i$ i=1(1)m. and the slack variables $x_{n+1},..,x_{n+m}$ are introduced to transform the inequalities into equations.

Case 3: $a_{1,n+1}=..=a_{m,n+1}=0$; ***Standard Form 3:*** $a_{1,n+1}=..=a_{m,n+1}=0$; $b_i\geq 0$ i=1(1)m.

This case arises when the constraints are originally in the form of equations in $x_1,...,x_n$.

Other form: If not all constraints belong to the same category as above then a combination of these cases arises.

We present in Sec. 2 the simplex algorithm [2, 3, 4, 5, 6, 17, 18, 19, 26, 27, 28] due to Dantzig (1963) which is an exterior-point method and is the first milestone in solving an LP. An ***exterior-point method*** is one in which the n-dimensional solution point $\mathbf{x}$ will always lie on the boundary or at a corner of the convex region (polytope) defined by A$\mathbf{x}$ = $\mathbf{b}$, $\mathbf{x}\geq\mathbf{0}$ and not inside the convex region. Although this algorithm is not polynomial time, it has been generally the only method for over two decades (1960s and 1970s) to solve LPs, often behaves like an O(n^3) polynomial algorithm, and has been remarkably successful in an intelligent computer implementation based on the nature of the LP. Even to-day it is possibly the most widely used algorithm to solve real-world linear optimization problems. A clear conceptual knowledge of the simplex algorithm helps us to appreciate the more recent development in the interior-point methods. An **interior-point *method*** is one in which the solution point $\mathbf{x}$ will move inside the polytope and continue to remain within it or at best touch the boundary or a corner point of the polytope. Some of these methods have been proved to be polynomial-time[9, 10, 20, 25]. The polynomial-time algorithms ,viz., the ellipsoid method [10] due to Kachiyan (1979) and the projective transformation method [9, 28] due to Karmatkar (1984) will be presented in Secs. 3 and 4. These algorithms are interior-point methods and did perform, in practice, much worse than the popular simplex method for most reasonably large linear programs. However, since these (specifically, Karmarkar algorithm) are polynomial-time, it may be shown that for some sufficiently large LP, these methods would perform better than the non-polynomial time algorithms. In Sec. 5, we present a variation on Karmarkar algorithm along with the detection of the basic variables [1, 22], which provides deeper insight into the geometry of an LP. We talk about other algorithms – heuristic, evolutionary (probabilistic), and deterministic including inequality sorting and error-free implementation [7, 12, 14, 15, 16, 21, 23] in Sec. 6.

2. The Simplex Algorithms

2.1 Basic solutions Consider the system of equalities $A\mathbf{x}=\mathbf{b}$, where A is an m×k matrix ($k \geq m$), **x** is a k-vector. Select m linearly independent columns of A (such m columns exist if the rank of A is m). Call the m×m matrix formed by these columns B. The matrix B is then nonsingular. Solve $B\mathbf{x}_B=\mathbf{b}$ for the m-vector $\mathbf{x}_B$. The vector $\mathbf{x}_B$ has the components x_i associated to the columns i, where $i \in [1, k]$. The matrix B having linearly independent columns (if they exist) is called a ***basis.*** The solution of $B\mathbf{x}_B=\mathbf{b}$ is called a ***basic solution*** of the system $A\mathbf{x}=\mathbf{b}$ with respect to the basis B. The components of **x** associated with the columns of B are called ***basic variables***. If one or more basic variables in a basic solution have value zero then that solution is called a ***degenerate basic solution***. A vector **x** that satisfies $A\mathbf{x}=\mathbf{b}$ is called a ***feasible solution***. The collection of all feasible solutions is called the ***feasible region*** (necessarily convex). If a feasible solution of $A\mathbf{x}=\mathbf{b}$ is also basic then it is a ***basic feasible solution***. If this basic feasible solution is degenerate then it is a ***degenerate basic feasible solution***

Let $A=(A', B)$ and $k=n+m$. The ***slack variables*** or, simply, slacks $x_{n+1}, \ldots, x_{n+i0}, \ldots, x_{n+m}$ which label the rows usually are basic variables and form $\mathbf{x}_B$. The columns associated with $x_{n+1}, \ldots, x_{n+m}$ usually form the basis B. The variables $x_1, \ldots, x_{j0}, \ldots, x_n$ which label the columns are nonbasic variables. The last column $[b_1, \ldots, b_{i0}, \ldots, b_m]^t$ contains the values of the basic variables $x_{n+1}, \ldots, x_{n+m}$, i.e., $x_{n+i}=b_i$ $i=1(1)m$.the values of the nonbasic variables being zero., i.e., $x_i=0$ $i=1(1)n$. A slack variable with a negative sign (usually introduced in a typical inequality $a_{i1}x_1+ \ldots + a_{in} \geq b_i$) is sometimes called the ***surplus variable***. The simplex algorithm, to start with, needs the knowledge of the basis B and that of the values of the basic variables, which are usually readily available.

Consider the LP *Min* $f=\mathbf{c}^t\mathbf{x}$ *subject to* $A\mathbf{x}=\mathbf{b}$, $\mathbf{x} \geq \mathbf{0}$. A feasible solution of the constraints $A\mathbf{x}=\mathbf{b}$, $\mathbf{x} \geq \mathbf{0}$ that achieves the minimum value of the objective function f is called an ***optimal feasible solution. An optimal feasible solution or, simply, an optimal solution (of the foregoing LP) obtained by the simplex algorithm lies at one of the corners of the feasible region.*** If this optimal feasible solution is basic then it is an ***optimal basic feasible solution***.

2.2 Fundamental theorem of linear programming Consider the LP *Min* $f=\mathbf{c}^t\mathbf{x}$ *subject to* $A\mathbf{x}=\mathbf{b}$, $\mathbf{x} \geq \mathbf{0}$, where A is an m×k matrix ($k \geq m$) of rank m. (a) If there is a feasible solution then there is a basic feasible solution and (b) if there is an optimal feasible solution then there is an optimal basic feasible solution.

For proof of this theorem, refer Luenberger (1984) [17]. The fundamental theorem considers ${}^kC_m$ possible combinations of the variables x_i, computes ${}^kC_m$ solutions, and chooses that solution that gives us the optimal solution of the LP. From the fundamental theorem of linear programming, there are ${}^kC_m = k!/(m!(k-m)!)$ ways of selecting m of k columns (of A and of $\boldsymbol{x}$), and hence ${}^kC_m$ solutions of the linear system $A\boldsymbol{x} = \boldsymbol{b}$. One of these finite number of solutions will be the required solution of the LP provided the nonnegativity condition $\boldsymbol{x} \geq \boldsymbol{0}$ is satisfied and there is a minimum value of the objective function. Let the LP be *Compute* $\boldsymbol{x}=[x_1 \ x_2 \ x_3 \ x_4]^t$ *that minimizes* $z = \boldsymbol{c}^t\boldsymbol{x} = [1 \ -2 \ 3 \ 1]\boldsymbol{x}$ *subject to* $A\boldsymbol{x} = \boldsymbol{b}, \boldsymbol{x} \geq \boldsymbol{0}$, where

$$A = \begin{bmatrix} 1 & 2 & 3 & 4 \\ -7 & 1 & -2 & 6 \end{bmatrix}, \ \boldsymbol{b} = \begin{bmatrix} 7 \\ 0 \end{bmatrix}.$$

Here m = 2, k = 4. Hence there are ${}^4C_2 = 4!/(2!(4-2)!) = 6$ ways of selecting 2 of 4 columns of A and of $\boldsymbol{x}$ and thus 6 solutions of the linear system $A\boldsymbol{x} = \boldsymbol{b}$. The six systems of linear equations are

$$\begin{bmatrix} 1 & 2 \\ -7 & 1 \end{bmatrix}\begin{bmatrix} x_1 \\ x_2 \end{bmatrix} = \begin{bmatrix} 7 \\ 0 \end{bmatrix}, \quad \begin{bmatrix} 1 & 3 \\ -7 & -2 \end{bmatrix}\begin{bmatrix} x_1 \\ x_3 \end{bmatrix} = \begin{bmatrix} 7 \\ 0 \end{bmatrix}, \quad \begin{bmatrix} 1 & 4 \\ -7 & 6 \end{bmatrix}\begin{bmatrix} x_1 \\ x_4 \end{bmatrix} = \begin{bmatrix} 7 \\ 0 \end{bmatrix},$$

$$\begin{bmatrix} 2 & 3 \\ 1 & -2 \end{bmatrix}\begin{bmatrix} x_2 \\ x_3 \end{bmatrix} = \begin{bmatrix} 7 \\ 0 \end{bmatrix}, \quad \begin{bmatrix} 2 & 4 \\ 1 & 6 \end{bmatrix}\begin{bmatrix} x_2 \\ x_4 \end{bmatrix} = \begin{bmatrix} 7 \\ 0 \end{bmatrix}, \quad \begin{bmatrix} 3 & 4 \\ 1 & -2 \end{bmatrix}\begin{bmatrix} x_3 \\ x_4 \end{bmatrix} = \begin{bmatrix} 7 \\ 0 \end{bmatrix}.$$

The six solution vectors are $[x_1 \ x_2]^t=[.4667 \ .2667]^t$, $[x_1 \ x_3]^t=[-.7368 \ 2.5789]^t$, $[x_1 \ x_4]^t=[1.2353 \ 1.4412]^t$, $[x_2 \ x_3]^t=[2 \ 1]^t$, $[x_2 \ x_4]^t=[5.2500 \ -.8750]^t$, $[x_3 \ x_4]^t=[1.4 \ .7]^t$.

In the first equation, x_1, x_2 are the basic variables while x_3, x_4 are the nonbasic variables whose values are taken as zero in the original equation $A\boldsymbol{x} = \boldsymbol{b}$. In the second equation, x_1 is negative; while x_2, x_4 are the nonbasic variables whose values are taken as zero in the original equation $A\boldsymbol{x} = \boldsymbol{b}$. Since this solution does not satisfy the nonnegativity condition, we reject this solution. In the third equation, x_1, x_4 are basic variables and x_2, x_3 are nonbasic variables whose values are taken as zero. Thus there are four solutions, viz., the first, the third, the fourth, and the sixth solutions, each of which satisfies the nonnegativity condition. If we compute the objective function value $z = \boldsymbol{c}^t\boldsymbol{x}$ for each of the four values of the solution vector $\boldsymbol{x}$ then we obtain the value of z as $-.0667$, 2.6765, -1, 4.9, respectively. The minimum value of the objective function is $z = -1$ which corresponds to the fourth equation. Therefore, $\boldsymbol{x} = [x_1 \ x_2 \ x_3 \ x_4]^t = [0 \ 2 \ 1 \ 0]^t$

is the required solution of the LPP. This algorithm with computational complexity $O({}^kC_m \times k^3)$ is combinatorial (not polynomial-time) and thus is slow. We did not have a polynomial-time algorithm for solving LPs till 1978. Since 1979, several polynomial-time algorithms for solving LPs have been developed. These polynomial algorithms are fast while some are faster than the others. ***For solving small LPs, a slow algorithm may be more economical than the fast ones***. Yet we would be interested in the fast ones and not in the slow ones. In fact, with the advent of high-performance computing devices (including the supercomputer ones), solving small problems is never a serious problem. The desired goal is to have a fast algorithm for truly large problems where slow algorithms will certainly be unacceptably expensive and thus useless. We discuss some of these algorithms later.

2.3 The simplex method in 'restricted tableau' Consider the standard form 1 of the LP *Max* $f=c_1x_1+\ldots+c_nx_n+0.x_{n+1}+\ldots+0.x_{n+m}$ *subject to* $bj\geq 0$ $j=1(1)m$, $x_i\geq 0$ $i=1(1)n+m$, and

$$
\begin{array}{l}
a_{11}x_1+a_{12}x_2+\ldots+a_{1n}x_n+x_{n+1}=b_1\\
a_{21}x_1+a_{22}x_2+\ldots+a_{2n}x_n+x_{n+2}=b_2\\
\bullet \qquad\qquad\qquad\qquad \bullet\\
a_{m1}x_1+a_{m2}x_2+\ldots+a_{mn}x_n+x_{n+m}=b_m
\end{array}
$$

The restricted simplex tableau for the foregoing LP can be written as

	x_1	x_{j0}	x_n	
x_{n+1}	a_{11} ..	a_{1j0} ..	a_{1n}	b_1
	•	•	•	•
x_{n+i0}	a_{i01} ..	a_{i0j0} ..	a_{i0n}	b_{i0}
	•	•	•	•
x_{n+m}	a_{m1} ..	a_{mj0} ..	a_{mn}	b_m
	$-c_1$ ..	$-c_{j0}$ ..	$-c_n$	0

S1 ***(Pivot selection)*** Let $-c_{j0}$ be negative. Consider then, for all positive a_{ij0}, the ratios b_i/a_{ij0} and take a smallest. If this is obtained for i0 then call $p=a_{i0j0}$ the pivot (marked with a plus in the example later).

S2 ***(Next-tableau Computation)*** Having interchanged x_{j0} and x_{n+i0} obtain the next tableau as follows.

	x_1		x_{n+i0}		x_n	
x_{n+1}			$-a_{1j0}/p$			
			•			
x_{j0}	a_{i01}/p	..	$1/p$	..	a_{i0n}/p	b_{i0}/p
			•			
x_{n+m}			$-a_{mj0}/p$			
			$+c_{j0}/p$			

The blank positions are filled in as follows. Replace *i-th row* (excluding the pivot row and the elements of the pivot column) of the tableau by *i-th row* $-$ $a_{ij0}\times$*pivot row*. The pivot row is the row containing the pivot while the pivot column is the column containing pivot).

S3 ***(Stopping Condition)*** If the bottom row (i.e., $-c_j$-row) excluding the last element is nonnegative, the solution is reached – terminate. Else go to the step **S1**.

Consider the problem *Max* $f=-2x_1-7x_2+2x_3$ *subject to* $x_1+2x_2+x_3\le 1$, $-4x_1-2x_2+3x_3\le 2$, $x_1, x_2, x_3\ge 0$. We write the solution in restricted tableau (Vajda 1975) as follows.

Restricted Tableau 0 (i0=2, j0=3)

	x_1	x_2	x_3	
x_4	1	2	1	1
x_5	−4	−2	3^+	2
	2	7	−2	0

Restricted Tableau 1 (i0=1, j0=1)

	x_1	x_2	x_5	
x_4	$7/3^+$	8/3	−1/3	1/3
x_3	−4/3	−2/3	1/3	2/3
	−2/3	17/3	2/3	4/3

Restricted Tableau 2

	x_4	x_2	x_5	
x_1	3/7	8/7	−1/7	1/7
x_3	4/7	6/7	1/7	6/7
	2/7	45/7	4/7	10/7

Since the last row is nonnegative (here positive), the solution is reached. The solution is: $x_1=6/7$, $x_2=0$, $x_3=6/7$, $f=10/7$.

2.4 Checking Rule for a Simplex Restricted Tableau Consider as an example the restricted tableau

	(c_1)	(c_6)	(c_2)	(c_4)	
	x_1	x_6	x_2	x_4	
(c_3) x_3	p_{11}	p_{12}	p_{13}	p_{14}	v_1
(c_5) x_5	p_{21}	p_{22}	p_{23}	p_{24}	v_2
	d_1	d_2	d_3	d_4	f

Then $c_3p_{11}+c_5p_{21}-c_1=d_1$; $c_3p_{12}+c_5p_{22}-c_6=d_2$; $c_3p_{13}+c_5p_{23}-c_2=d_3$; $c_3p_{14}+c_5p_{24}-c_4=d_4$; $c_3v_1+c_5v_2=f$.. Such a relationship holds in all tableaux. This relationship is referred to as the ***Checking Rule for a Tableau***. Satisfaction of this rule is necessary for a restricted tableau to be correct but it is not sufficient (i.e., the rule may be satisfied even if a computational mistake occurs).

2.5 Dual Simplex Method ***When to use*** Consider Case 2 (Sec. 1). Let all c_j $j=1(1)n$ be nonpositive so that, in the first tableau (Tableau 0), the first n elements in the bottom row are nonnegative (if we maximize). We call such a tableau ***dual feasible***. If, in addition, all b_i $i=1(1)m$ are nonnegative then the result is reached. Else, apply the dual simplex method as follows.

S1 ***(Pivot selection)*** Let b_{i0} be negative.Consider for all $a_{i0j}<0$, $|c_j/a_{i0j}|$ and take a smallest. If this is obtained for j0 then a_{i0j0} is the pivot.

S2 ***(Next tableau computation)*** Same as in the foregoing simplex algorithm.

S3 ***(Stopping condition)*** If the bottom row (i.e., c_j-row)excluding the last element is nonnegative then the solution is reached – terminate. Else, go to the step **S1**.

Consider the problem [27] *Max* $f=-x_1-2x_2$ *subject to* $x_1-4x_2\geq 2$, $2x_1-2x_2\geq 7$, $x_1+3x_2\geq -2$, $x_1, x_2\geq 0$. Introducing slacks with negative sign, we obtain *Max* $f=-x_1-2x_2$ *subject to* $x_1-4x_2-x_3=2$, $2x_1-2x_2-x_4=7$, $x_1+3x_2-x_5=-2$, $x_1, x_2, x_3, x_4, x_5\geq 0$. Multiplying each equation by -1, we get $-x_1+4x_2+x_3=-2$, $-2x_1+2x_2+x_4=-7$, $-x_1-3x_2+x_5=2$. Hence the tableaux are

Restricted Tableau 0

	x_1	x_2	
x_3	-1	4	-2
x_4	-2^+	2	-7
x_5	-1	-3	2
	1	2	0

Restricted Tableau 1

	x_4	x_2	
x_3	$-1/2$	3	3/2
x_1	1	-1	7/2
x_5	$-1/2$	-4	11/2
	½	3	$-7/2$

Since the last row is nonnegative (here positive), the solution is reached. The solution is: $x_1=7/2$, $x_2=0$, $f=-7/2$.

2.6 Artificial basis technique If the LP is in neither of the three standard forms (Sec. 2 and cannot be transformed into one of them then the method is as follows (note that the LP is neither primal feasible nor dual feasible):

S1 Multiply both sides of those constraints by -1, in which b_i is negative. The coefficients of x_{n+i} i=1(1)m are then $1, -1$, or 0.
S2 In the later two cases add a variable a_i called the ***artificial variable*** to the left-hand side.
S3 Subtract Ma_i from the objective function f to be maximized (Add Ma_i to the objective function f to be minimized). M is considered to be a value larger than any other with which it is compared during the computations.
S4 Set up the simplex tableau. Using the ***checking rule***, obtain the last row (its elements are to be multiplied by M). Note that the sum of the a_i-rows is the row of the objective function $f = \mathbf{c}^t\mathbf{x} + M\,\Sigma a_i$.
S5 Consider the last row to be $-c_j$-row for the next tableau computation.

Why artificial variables (i) *Contradictory constraints* The most ` mportant role of the artificial variables is to detect inconsistency (contradiction), if any, among the original constraints. If we do not use artificial variables then we might end up getting a solution where it does not exist in view of the contradictory (original) constraints. The original constraints are contradictory if it is impossible to make the artificial variables zero. (ii)

Redundant constraints If a constraint is redundant then an artificial variable may remain in the final basis with the value zero.

Consider the LP [27] *Max* $f=-x_1+x_2$ *subject to* $x_1+2x_2\le 4$, $3x_1-x_2\ge 1$, $x_1+3x_2=4$, x_1, $x_2\ge 0$. Rewrite the problem as *Min* $f=x_1-x_2$ *subject to* $x_1+2x_2+x_3=4$, $3x_1-x_2-x_4=1$, $x_1+3x_2=4$, x_1, x_2, x_3, $x_4\ge 0$. In the second and third equations the slacks x_4 and x_5 have coefficients -1 and 0, respectively. Adding artificial variables v_1 and v_2 to these equations and subtracting Mv_1 and Mv_2 from the objective function, we write the problem as *Min* $f=x_1-x_2+Mv_1+Mv_2$ *subject to* $x_1+2x_2+x_3=4$, $3x_1-x_2-x_4+v_1=1$, $x_1+3x_2+v_2=4$, x_1, x_2, x_3, x_4, v_1, $v_2\ge 0$. No slack variable is to be added to the third equality constraint. The restricted tableaux now can be written as (positive slacks and artificial variables only are written in the left-most column)

Restricted Tableau 0

		1	−1	0	
		x_1	x_2	x_4	
0	x_3	1	2	0	4
M	v_1	3^+	−1	−1	1
M	v_2	1	3	0	4
		−1	1	0	0
	(M)	4	2	−1	5

Restricted Tableau 1

	v_1	x_2	x_4	
x_3	−1/3	7/3	1/3	11/3
x_1	1/3	−1/3	−1/3	1/3
v_2	−1/3	$10/3^+$	1/3	11/3
	1/3	2/3	−1/3	1/3
	−4/3	10/3	1/3	11/3
	↑ Omit			

The v_1 column is not useful. So it can be omitted. In fact, computation of v_1 is not necessary. The next (final) tableau is

Restricted Tableau 2

	v_2	x_4	
x_3	−7/10	1/10	11/10
x_1	1/10	−3/10	7/10
x_2	3/10	1/10	11/10
	−1/5	−2/5	−2/5
	−1	0	0
	↑ Omit		

The v_2 column is also not of interest. So it can be omitted.. The solution is $x_1=7/10$, $x_2=11/10$, $f=-2/5$. A method alternative to the artificial basis technique is the ***self-dual parametric method*** [27]. We will not present this method here.

2.7 Revised Simplex method The tableau at any point in the simplex procedure can be determined solely by a knowledge of which variables are basic. Denote by B the submatrix of the original matrix A having m columns of the m×n matrix A corresponding to the basic variables. These columna are linearly independent and hence the columns of B form a basis. Call B basis or basis matrix. Let B consist of the first m columns of A. Then, by partitioning A, $\mathbf{x}$, and $\mathbf{c}^t$ as $A=[B, D]$, $\mathbf{x}=[\mathbf{x}_B, \mathbf{x}_D]$, $\mathbf{c}^t=[\mathbf{c}^t_B, \mathbf{c}^t_D]$, the standard form of the LP becomes

$$\textit{Max } f= \mathbf{c}^t_B\mathbf{x}_B+\mathbf{c}^t_D\mathbf{x}_D \textit{ subject to } B\mathbf{x}_B+D\mathbf{x}_D=\mathbf{b},\ \mathbf{x}_B\geq\mathbf{0},\ \mathbf{x}_D\geq\mathbf{0} \qquad (1)$$

The basic solution which, we assume, is also feasible corresponding to the basis B is $\mathbf{x}=(\mathbf{x}_B, \mathbf{0})$, where $\mathbf{x}_B=B^{-1}\mathbf{b}$. The basic solution results from setting $\mathbf{x}_D=\mathbf{0}$. However, for any value of $\mathbf{x}_D$ the necessary value of $\mathbf{x}_B$ can be computed from (1) as

$$\mathbf{x}_B= B^{-1}\mathbf{b}-B^{-1}D\mathbf{x}_D \qquad (2)$$

and this general expression when substituted in the cost function yields

$$f=\mathbf{c}^t_B\,(B^{-1}\mathbf{b}-B^{-1}D\mathbf{x}_D) + \mathbf{c}^t_D\mathbf{x}_D= \mathbf{c}^t_B B^{-1}\mathbf{b}+ (\mathbf{c}^t_D- \mathbf{c}^t_B\, B^{-1}D)\mathbf{x}_D \quad (3)$$

which expresses cost of any solution of (1) in terms of $\mathbf{x}_D$. Thus

$$\mathbf{r}^t_D=\mathbf{c}^t_D- \mathbf{c}^t_B\, B^{-1}D \qquad (4)$$

is the relative cost vector (for nonbasic variables). It is the components of this vector that are used to determine which vector to bring into the basis. Having derived the vector expression for the relative cost, we can now write the simplex tableau in matrix form. The initial tableau takes the form

$$\left[\begin{array}{c|c} A & \mathbf{b} \\ \hline \mathbf{c}^t & 0 \end{array}\right] = \left[\begin{array}{c|c|c} B & D & \mathbf{b} \\ \hline \mathbf{c}^t_B & \mathbf{c}^t_D & 0 \end{array}\right] \qquad (5)$$

which is not, in general, in canonical form and does not correspond to a point in simplex tableau. If the matrix B is used as a basis then the corresponding tableau becomes

$$T= \begin{bmatrix} I & B^{-1}D & B^{-1}\mathbf{b} \\ 0 & \mathbf{c}^t_D- \mathbf{c}^t_B\, B^{-1}D & -\mathbf{c}^t_B\, B^{-1}\mathbf{b} \end{bmatrix} \qquad (6)$$

which is the matrix form we desire. Note that the equation (6) is obtained by premultiplying the right-hand side of the equation (5) by

$$\begin{bmatrix} B^{-1} & 0 \\ -\mathbf{c}^t_B\, B^{-1} & 1 \end{bmatrix}.$$

The simplex method is expected to converge to an optimal solution in about m or perhaps 3m/2 pivot operations. If $m \ll n$, i.e., if the matrix A has far fewer rows than columns then pivots will occur in only a small fraction of the columns during the course of optimization. Since the other columns are not explicitly used, the work spent in computing the elements in these columns after each pivot is an wasted effort. The revised simplex method avoids the unnecessary computations by ordering the computations (needed of the simplex method). Given B^{-1} of a current basis and the current solution $\mathbf{x}_B=y_0=B^{-1}\mathbf{b}$, the steps of the revised simplex method are as follows.

S1 Compute the current relative cost coefficients $\mathbf{r}^t_D=\mathbf{c}^t_D-\mathbf{c}^t_B B^{-1}D$. If $\mathbf{r}^t_D\geq\mathbf{0}$ then stop; the current solution is optimal.
S2 Determine which vector $\mathbf{a}_q$ is to enter the basis by selecting a negative cost coefficient and computing $\mathbf{y}_q=B^{-1}\mathbf{a}_q$ which gives the vector $\mathbf{a}_q$ expressed in terms of the current basis.
S3 If no $y_{iq}>0$ then stop; the LP is ***unbounded***. Else, compute the ratios y_{i0}/y_{iq} for $y_{iq}>0$ to determine which vector is to leave the basis.
S4 Update $B^{-1}$and the current solution $B^{-1}\mathbf{b}$. Return to the step **S1**.

Updating B^{-1} is done by the usual pivot operations applied to an array consisting of $B^{-1}$and $\mathbf{y}_q$, where the pivot is the appropriate element in $\mathbf{y}_q$.Of course $B^{-1}\mathbf{b}$ may be updated at the same time by adjoining it as another column. Consider the LP *Max* $3x_1+x_2+3x_3$ *subject to* $2x_1+x_2+x_3\leq 2$, $x_1+2x_2+3x_3\leq 5$, $2x_1+2x_2+x_3\leq 6$, $x_1, x_2, x_3\geq 0$. Adding the slacks x_4, x_5, x_6 we convert the inequalities to equations and write the *extended tableau 0 for reference.*

Extended Tableau 0

Coef. Of x_1	*Coeff. Of* x_2	*Coeff. Of* x_3	*Coeff. Of* x_4	*Coeff. Of* x_5	*Coeff. Of* x_6	**b**
2	1	1	1	0	0	2
1	2	3	0	1	0	5
2	2	1	0	0	1	6
−3	−1	−3	0	0	0	0 ← $-\mathbf{c}_j$-row

$$\mathbf{a}_1=\begin{bmatrix}2\\1\\2\end{bmatrix}\quad \mathbf{a}_2=\begin{bmatrix}1\\2\\2\end{bmatrix}\quad \mathbf{a}_3=\begin{bmatrix}1\\3\\1\end{bmatrix}$$

We start with an initial basic feasible solution and corresponding B^{-1} (unit matrix here) as

	B^{-1}			$\mathbf{b}=\mathbf{x}_B$
x_4	1	0	0	2
x_5	0	1	0	5
x_6	0	0	1	6

Compute $\mathbf{c}^t_B B^{-1}=[0\ \ 0\ \ 0]B^{-1}=[0\ \ 0\ \ 0]$ and then (referring extended tableau 0)
$\mathbf{r}^t_D=\mathbf{c}^t_D-\mathbf{c}^t_B B^{-1}D=[-3\ \ -1\ \ -3]-[0\ \ 0\ \ 0]D=[-3\ \ -1\ \ -3]$, where

$$D=\begin{bmatrix}2 & 1 & 1\\1 & 2 & 3\\2 & 2 & 1\end{bmatrix}$$

We decide to bring x_2 into the basis. Its current representation is found by multiplying by B^{-1}; thus we have

	B^{-1}			$\mathbf{b}=x_B$	$y_2=B^{-1}\mathbf{a}_2$
x_4	1	0	0	2	1^+
x_5	0	1	0	5	2
x_6	0	0	1	6	2

After computing the ratios in the usual manner, we select the pivot indicated. The updated tableau becomes

	B^{-1}			$\mathbf{b}=x_B$
x_2	1	0	0	2
x_5	−2	1	0	1
x_6	−2	0	1	2

$\mathbf{c}^t_B B^{-1}=[-1\ \ 0\ \ 0]B^{-1}=[-1\ \ 0\ \ 0]$.

(Refer extended tableau 0)

$$\mathbf{r}^t_D=\mathbf{c}^t_D-\mathbf{c}^t_B B^{-1}D=[-3\ \ -3\ \ 0]-[-1\ \ 0\ \ 0]D=\overset{\quad x_1\quad x_3\quad x_4}{[-1\ \ -2\ \ 1]}, \text{ where } D=\begin{bmatrix}2&1&1\\1&3&0\\2&1&0\end{bmatrix}$$

We select x_3 to enter the basis. We have the tableau

	B^{-1}			$\mathbf{b}=x_B$	$y_2=B^{-1}\mathbf{a}_3$
2	1	0	0	2	1
5	−2	1	0	1	1^+
6	−2	0	1	2	−1

Using the pivot indicated, we get

	B^{-1}			$\mathbf{b}=x_B$
2	3	−1	0	1
3	−2	1	0	1
6	−4	1	1	3

Now $\mathbf{c}^t_B B^{-1}=[-1\ \ -3\ \ 0]B^{-1}=[3\ \ -2\ \ 0]$.

(Refer extended tableau 0.)

$$\mathbf{r}^t_D=\mathbf{c}^t_D-\mathbf{c}^t_B B^{-1}D=[-3\ \ 0\ \ 0]-[3\ \ -2\ \ 0]D=\overset{x_1\quad x_4\quad x_5}{[-7\ \ -3\ \ 2]}, \text{ where } D=\begin{bmatrix}2&1&0\\1&0&1\\2&0&0\end{bmatrix}$$

We select x_1 to enter the basis. We have the tableau

	B^{-1}			$\mathbf{b}=x_B$	$y_2=B^{-1}\mathbf{a}_1$
x_2	3	−1	0	1	5^+
x_3	−2	1	0	1	−3
x_6	−4	1	1	3	−5

Using the pivot indicated we obtain

	B^{-1}			$b=x_B$
x_1	3/5	−1/5	0	1/5
x_3	−1/5	2/5	0	8/5
x_6	−1	0	1	4

Now $\mathbf{c}^t_B B^{-1}=[-3\ \ -3\ \ 0]B^{-1}=[-6/5\ \ -3/5\ \ 0]$.
$\mathbf{r}^t_D=\mathbf{c}^t_D-\mathbf{c}^t_B B^{-1}D=[-1\ \ 0\ \ 0]-[-6/5\ \ -3/5\ \ 0]D=[7/5\ \ 6/5\ \ 3/5]$, where $D=\begin{bmatrix}1 & 1 & 0\\ 2 & 0 & 1\\ 2 & 0 & 0\end{bmatrix}$

Since all the elements of $\mathbf{r}^t_D$ are nonnegative, we conclude that the solution $\mathbf{x}=[1/5\ \ 0\ \ 8/5\ \ 0\ \ 0\ \ 4]^t$ is optimal.

3. The Ellipsoid Algorithm

L.G. Khachian published a polynomially bounded algorithm [10] to solve an LP. Let

$$\mathbf{a}_i\mathbf{x}<b_i \quad i=1(1)m,\ \mathbf{a}_i\in Z^n,\ b_i\in Z \qquad (7)$$

be a system of strict linear inequalities with integral coefficients. Define

$$L=[\Sigma\log_2(|a_{ij}|+1)+\Sigma\log_2(|b_i|+1)+\log_2 nm]+1 \qquad (8)$$

where L=the space needed to state the problem=the number of bits (binary digits) required to represent the input of the system of inequalities. The leftmost summation runs over i=1(1)m and j=1(1)n while the rightmost summation runs over i=1(1)m.

The algorithm Define a sequence $\mathbf{x}_0, \mathbf{x}_1, \ldots \in R^n$ and a sequence of symmetric positive definite matrices $A_0, A_1, \ldots$ recursively as follows. $\mathbf{x}_0=\mathbf{0}$, $A_0=2^L I$ where I is the unit matrix of order n. Assume that $(A_k, \mathbf{x}_k)$ is defined. Check if $\mathbf{x}_k$ is a solution of (7). If it is, stop. Else, pick any inequality in (7) which is violated.If $\mathbf{a}_i\mathbf{x}_k\geq b_i$ then set

$$\mathbf{x}_{k+1}=\mathbf{x}_k-A_k\mathbf{a}^t_i/((n+1)\surd(\mathbf{a}_iA_k\mathbf{a}^t_i)),$$
$$A_{k+1}=(n^2/(n^2-1))(A_k-(2/(n+1))(A_k\mathbf{a}^t_i)(A_k\mathbf{a}^t_i)^t/(\mathbf{a}_iA_k\mathbf{a}^t_i))$$

It can be shown that approximations within e^{-10nL} preserve the validity of the following theorem.

Theorem If the algorithm stops then $\mathbf{x}_k$ is a solution of (7). If it does not stop in $6n^2L$ steps then (7) is not solvable.

To decide the solvability of a system of the form

$$\mathbf{a}_i\mathbf{x}\leq b_i \quad i=1(1)m,\ \mathbf{a}_i\in Z^n,\ b_i\in Z \qquad (9)$$

consider instead the system

$$[2^L]\mathbf{a}_i\mathbf{x}<[2^L]\, b_i +1 \quad i=1(1)m,\ \mathbf{a}_i \in Z^n,\ b_i \in Z \qquad (10)$$

To solve an LP *max* $\mathbf{c}^t\mathbf{x}$ *subject to* $A\mathbf{x}\leq\mathbf{b}$, $\mathbf{x}\geq\mathbf{0}$, consider the system of linear inequalities

$$\mathbf{c}^t\mathbf{x}=\mathbf{b}^t\mathbf{y},\ A\mathbf{x}\leq\mathbf{b},\ \mathbf{x}\geq\mathbf{0},\ A^t\mathbf{y}\geq\mathbf{0},\ \mathbf{y}\geq 0. \qquad (11)$$

Consider the inequality $x_1+x_2-4x_3<-1$. $L=\log_2(1+1)+\log_2(1+1)+\log_2(4+1)+\log_2(1+1)+\log_2(1\times3)+1=7.9069$. $x_0=[0\ 0\ 0]^t$, $A_0= 2^L I$.. $\mathbf{x}_0$ does not satisfy the inequality. m=1, n=3. $x_1=[-.9129\ \ -.9129\ \ 3.6515]^t$.

$$A_1=\begin{bmatrix} 262.5 & -7.5 & 30 \\ -7.5 & 262.5 & 30 \\ 30 & 30 & 150 \end{bmatrix}$$

The vector $\mathbf{x}_1$ satisfies the inequality. So it is a solution and we stop. However, in the ellipsoid algorithm, the space L needed to represent the input of the system of inequalities is large for a reasonably large real-world LP. As a result , 2^L that occurs in A_0 ($=2^L I$) could be too large for the (floating point) precision of the computer. Also, the convergence is too slow, i.e., the number of iterations is too many for such LPs. Hence the method is impracticable and is not meant to be used for solving an LP although the method is of great academic interest and has stimulated the thought process of many operations researchers. This stimulation has resulted in the landmark polynomial-time algorithm [9] based on projective transformation due to Karmarkar in 1984 and subsequently several other improved algorithms. We discuss specifically the Karmarkar's algorithm which is an exterior-point method unlike the simplex algorithms. For the underlying geometrical concepts, Karmarkar's paper [9] should be referred.

4. Karmarkar Form of Linear Program and Algorithm

4.1 The philosophy A linear program (LP), can be defined as *Minimize (Min)* $z = \mathbf{c}^t\mathbf{x}$ *subject to* $A\mathbf{x} \leq \mathbf{b}$, $\mathbf{x} \geq \mathbf{0}$, where A is an $m \times n$ matrix and $\mathbf{0}$ is the n-dimensional column vector (n-vector) of 0s. A form of LP equivalent to the foregoing LP and an algorithm (for this form), both due to N. Karmarkar, are presented here precisely and concisely. This Karmarkar form of LP (KLP) is *Min* $z = \mathbf{c}^t\mathbf{x}$ *subject to* $A\mathbf{x} = \mathbf{0}$, $\mathbf{e}^t\mathbf{x} = 1$, $\mathbf{x} \geq \mathbf{0}$, $\mathbf{x} = \mathbf{e}/n$ *is feasible, minimal z-value* = 0, where e is the n-vector of 1s. Both the form and the operational aspects of the algorithm presented here are more easily followed. The algorithm is readily implementable/programmable on a computer. The Karmarkar algorithm (KA) uses a transformation from projective geometry to create a set of transformed variables $\mathbf{y}$. This transformation f always transforms the current point into the centre of the feasible region in the space defined by the transformed variables. If f takes the point $\mathbf{x}$ into the point $\mathbf{y}$ then we write $f(\mathbf{x}) = \mathbf{y}$. The KA begins in the transformed space in a direction that tends to improve z without violating feasibility. This yields a point $\mathbf{y}^1$, close to the boundary of the feasible region, in the transformed space. The new point is $\mathbf{x}^1$ that satisfies $f(\mathbf{x}^1) = \mathbf{y}^1$. The procedure is iterated replacing $\mathbf{x}^0$ by $\mathbf{x}^1$ until the z-value for $\mathbf{x}^k$ is sufficiently close to 0. An intelligent implementation of KA, however, does need a deeper insight (into the algorithm) that avoids redundant/partial duplication of computation/codes and that possibly reduces the number of iterations. This projective transformation based polynomial-time interior-point iterative algorithm is claimed to be more efficient than the widely used exponential-time exterior-point iterative method called the simplex algorithm for large LPs. The simplex algorithm and its variations have been the most widely used methods in linear optimization for over three decades (sixties—eighties) and is still being extensively used certainly for small and medium LPs. The KA is increasingly finding its place in literature/textbooks on linear programming/operations research. It is also stimulating in terms of visualizing every derived mathematical step geometrically (maximum three dimensions can be visualized, higher dimensions are just straight-forward mathematical extensions and cannot be visualized) or achieving the desired geometrical path/destination using the appropriate mathematics. Thus, we believe that there is a scope for such a presentation for the readers who desire to get a quick feel about this landmark algorithm. A MATLAB program for the KA is appended for ready check and for a quick feel about its convergence.

4.2 Notations

We use the following convention and notations. A bold lower case letter (such as **c**, **b**, **x**) always indicates a column vector. A bold zero, viz., **0**, denotes a null column vector (i.e., a column vector of 0s) of appropriate order (including the order 1). An upper case letter (such as A, P) denotes a matrix and t, when used as a superscript, indicates the transpose. The specific symbols used here have the following meaning.

Symbol	Meaning
A	an $m \times n$ matrix $[a_{ij}]$
c	an n-dimensional vector or simply n-vector $[c_i] = [c_1\ c_2 \,..\, c_n]^t$
b	an m-vector $[b_j] = [b_1\ b_2 \,..\, b_m]^t$
e	a vector $[1\ \ 1\ \,..\, 1]^t$ of appropriate order
s	an m-vector $[s_j] = [s_1\ s_2 \,..\, s_m]^t$ of slack variables
v	an n-vector $[v_i] = [v_1\ v_2 \,..\, v_n]^t$ of surplus variables
x	an n-vector $[x_i] = [x_1\ x_2 \,..\, x_n]^t$
$\mathbf{x}^k$ or $\mathbf{y}^k$	k-th iterate of the vector **x** or **y**
x_u^k or y_u^k	k-th iterate of the u-th element of **x** or **y**
$\mathrm{diag}(\mathbf{x}^k)$	$n \times n$ diagonal matrix whose (i,i)-th element is x_i^k
$j = 1(1)n$	$j = 1, 2, .., n$
$\\| \ \\|$	Euclidean norm
α	a real positive number < 1
$\mathbf{x}, \mathbf{y}, \mathbf{s} \geq \mathbf{0}$	$\mathbf{x} \geq \mathbf{0}, \mathbf{y} \geq \mathbf{0}, \mathbf{s} \geq \mathbf{0}$
Min (Max)	Minimize (Maximize)
X^+	minimum-norm least-squares inverse (p-inverse) of the matrix X

4.3 **The Scope** Consider the LP

$$Min\ \mathbf{c}^t\mathbf{x}\ \ subject\ to\ \ A\mathbf{x} = \mathbf{b},\ \ \mathbf{x} \geq \mathbf{0}. \quad (12)$$

The LP (12) is solved by the simplex method/revised simplex method/a variation of the simplex method (exterior-point method) designed and developed by G. Dantzig during early 1950s [2, 3, 4, 5, 6, 11, 17, 18, 26]. This method dominated the linear programming scene solving millions of optimization problems in almost all scientific and engineering areas. However, considerable amount of research went into this area and many special-purpose algorithms were designed and used with a significant success. All these algorithms are exponential-time (nonpolynomial-time). The simplex method is *exponential-time* in the worst case. This implies that if an LP of size[2] n is solved by the simplex method, there exists a positive number p such that for any n, an LP of size n can be solved in at most $p2^n$ operations. The simplex mehtod may even enter into a cycling (infinite loop) though very rarely [2]. Efforts to develop a polynomial-time algorithm for LPs did not meet with any success till almost the end of 1970s. In 1979, L.G. Khachiyan reported the first known interior-point iterative algorithm called the Ellipsoid method [10] – not of great practical importance but of great academic interest – to solve LPs discussed in the previous section. Then, in 1984, N. Karmarkar proposed the second polynomial-time $O(n^{3.5})$ interior-point iterative method [8, 9, 19, 28] based on a projective transformation, which is of academic and of practical interest.

We provide here the conversion of any LP to the Karmarkar form of LP (KLP). We also present Karmarkar Algorithm (KA) precisely and concisely so that one could simply solve an LP just by mechanically following the steps. We omit the proof as well as much explanation which are available in Karmarkar's paper [9]. A MATLAB Version 5.1 program for the KA is included for ready verification and feel about the algorithm.

4.4. Conversion of an LP to KLP A standard LP (constraints in an equality form) or any LP whose constraints are in an inequality form can be converted to a KLP as follows. Consider the LP

$$Max\ z = \mathbf{c}^t\mathbf{x}\ \ subject\ to\ \ A\mathbf{x} \leq \mathbf{b},\ \mathbf{x} \geq \mathbf{0}. \quad (13)$$

The dual of LP (13) is

$$Min\ \ w = \mathbf{b}^t\mathbf{y}\ \ subject\ to\ A^t\mathbf{y} \geq \mathbf{c},\ \mathbf{y} \geq \mathbf{0}. \quad (14)$$

From the duality theorem, we know that if the n-vector $\mathbf{x}$ is feasible in (13), the m-vector $\mathbf{y}$ is feasible in (14), and the z-value in (13) equals the w-value in (14) then $\mathbf{x}$ is maximal for (13). This implies that any feasible solution of the following set of constraints will produce the maximal solution of (13).

---

[2] The size of an LP could be defined as the number of symbols needed to represent the LP in binary notation.

$$\mathbf{c}^t\mathbf{x} - \mathbf{b}^t\mathbf{y} = 0,\ \mathbf{A}\mathbf{x} \le \mathbf{b},\ \mathbf{A}^t\mathbf{y} \ge \mathbf{c},\ \ \mathbf{x}, \mathbf{y} \ge \mathbf{0} \qquad (15)$$

Inserting slack and surplus variables into (15) we get

$$\mathbf{c}^t\mathbf{x} - \mathbf{b}^t\mathbf{y} = 0,\ \mathbf{A}\mathbf{x} + \mathbf{I}_m\mathbf{s} = \mathbf{b},\ \mathbf{A}^t\mathbf{y} - \mathbf{I}_n\mathbf{v} = \mathbf{c},\ \ \mathbf{x}, \mathbf{y}, \mathbf{s}, \mathbf{v} \ge \mathbf{0}, \quad (16)$$

where $\mathbf{s} = [\,s_1\ \ s_2\ \ldots\ s_m]^t$ is the m-vector of slack variables, $\mathbf{v} = [\,v_1\ \ v_2\ \ldots\ v_n]^t$ is the n-vector of surplus variables, I_m is the unit matrix of order m, and I_n is the unit matrix of order n. We now append to (16) yet another constraint such that the feasible solution of (16) satisfies the equation

$$\mathbf{e}^t\mathbf{x} + \mathbf{e}^t\mathbf{y} + \mathbf{e}^t\mathbf{s} + \mathbf{e}^t\mathbf{v} + d_1 = k, \qquad (17)$$

where k is to be found/supplied such that the sum of the values of all the variables $\le$ k. The variable $d_1 \ge 0$ is a dummy (slack) variable. This yields

$$\mathbf{c}^t\mathbf{x}-\mathbf{b}^t\mathbf{y}=0,\ \ \mathbf{A}\mathbf{x}+\mathbf{I}_m\mathbf{s}=\mathbf{b},\ \ \mathbf{A}^t\mathbf{y}-\mathbf{I}_n\mathbf{v}=\mathbf{c},\ \ \mathbf{e}^t\mathbf{x}+\mathbf{e}^t\mathbf{y}+\mathbf{e}^t\mathbf{s}+\mathbf{e}^t\mathbf{v}+d_1=k,\ \ \mathbf{x}, \mathbf{y}, \mathbf{s}, \mathbf{v}, d_1 \ge \mathbf{0},\ (18)$$

To make nonzero right-hand sides of (18) zero, we introduce yet another dummy variable d_2, where $d_2 = 1$. Thus, we obtain

$$\mathbf{c}^t\mathbf{x}-\mathbf{b}^t\mathbf{y}=0,\ \mathbf{A}\mathbf{x}+\mathbf{I}_m\mathbf{s}-\mathbf{I}_m\mathbf{b}d_2=\mathbf{0},\ \mathbf{A}^t\mathbf{y}-\mathbf{I}_n\mathbf{v}-\mathbf{I}_n\mathbf{c}d_2=\mathbf{0},\ \mathbf{e}^t\mathbf{x}+\mathbf{e}^t\mathbf{y}+\mathbf{e}^t\mathbf{s}+\mathbf{e}^t\mathbf{v}+d_1-kd_2=0,$$
$$\mathbf{e}^t\mathbf{x}+\mathbf{e}^t\mathbf{y}+\mathbf{e}^t\mathbf{s}+\mathbf{e}^t\mathbf{v}+d_1+d_2 = k + 1,,\ \ \mathbf{x}, \mathbf{y}, \mathbf{s}, \mathbf{v}, d_1, d_2 \ge \mathbf{0} \qquad .. \ (19)$$

Allowing the following change of variables $\mathbf{x}=(k+1)\mathbf{x}'$, $\mathbf{y}=(k+1)\mathbf{y}'$, $\mathbf{s}=(k+1)\mathbf{s}'$, $\mathbf{v}=(k+1)\mathbf{v}'$, $d_1=(k+1)d_1'$, $d_2=(k+1)d_2'$ we obtain

$$[\mathbf{x}\ \ \mathbf{y}\ \ \mathbf{s}\ \ \mathbf{v}\ \ d_1\ \ d_2] = (k+1)[\,\mathbf{x}'\ \ \mathbf{y}'\ \ \mathbf{s}'\ \ \mathbf{v}'\ \ d_1'\ \ d_2']\text{: } \mathbf{c}^t\mathbf{x}'-\mathbf{b}^t\mathbf{y}'=\mathbf{0},\ \mathbf{A}\mathbf{x}'+\mathbf{I}_m\mathbf{s}'-\mathbf{I}_m\mathbf{b}d_2'=\mathbf{0},$$
$$\mathbf{A}^t\mathbf{y}'-\mathbf{I}_n\mathbf{v}'-\mathbf{I}_n\mathbf{c}d_2'=\mathbf{0},\mathbf{e}^t\mathbf{x}'+\mathbf{e}^t\mathbf{y}'+\mathbf{e}^t\mathbf{s}'+\mathbf{e}^t\mathbf{v}'+d_1'-kd_2'=0,\mathbf{e}^t\mathbf{x}'+\mathbf{e}^t\mathbf{y}'+\mathbf{e}^t\mathbf{s}'+\mathbf{e}^t\mathbf{v}'+d_1'+d_2'=1,$$
$$\mathbf{x}',\mathbf{y}',\mathbf{s}',\mathbf{v}',d_1',d_2'\ge\mathbf{0}. \qquad .. \ (20)$$

We now enforce that a solution (geometrically, a point in [2n + 2m + 2] dimensional polytope [16] defined by (20)) that sets all variables equal is feasible in (20). This is achieved by adding the third dummy variable d_3' to the last but one constraint in (20) and then adding a multiple of d_3' to each of its preceding constraints. This multiple is chosen so that the sum of the coefficients of all variables in each constraint (except the last two) equals zero. This yields KLP (21).

Min d_3' subject to

$$\mathbf{c}^t\mathbf{x}'-\mathbf{b}^t\mathbf{y}' - (\mathbf{e}^t\mathbf{c}-\mathbf{e}^t\mathbf{b})d_3'=0,\ \ \mathbf{A}\mathbf{x}'+\mathbf{I}_m\mathbf{s}'-\mathbf{I}_m\mathbf{b}d_2' - [\mathbf{A}\mathbf{e} + \mathbf{I}_m(1-d_2')\mathbf{e}]d_3'=\mathbf{0},$$
$$\mathbf{A}^t\mathbf{y}'-\mathbf{I}_n\mathbf{v}'-\mathbf{I}_n\mathbf{c}d_2'-[\mathbf{A}^t\mathbf{e}-\mathbf{I}_n(1-d_2')\mathbf{e}]d_3'=\mathbf{0},\ \ \mathbf{e}^t\mathbf{x}'+\mathbf{e}^t\mathbf{y}'+\mathbf{e}^t\mathbf{s}'+\mathbf{e}^t\mathbf{v}'+d_1'-kd_2'-(2n+2m+1-k)d_3'=0,$$
$$\mathbf{e}^t\mathbf{x}'+\mathbf{e}^t\mathbf{y}'+\mathbf{e}^t\mathbf{s}'+\mathbf{e}^t\mathbf{v}'+d_1'+d_2' + d_3'= 1,\ \ \ \mathbf{x}', \mathbf{y}', \mathbf{s}', \mathbf{v}', d_1', d_2', d_3' \ge \mathbf{0} \qquad .. \ (21)$$

Observe that we cannot write the expression $\mathbf{e}^t\mathbf{x}'+\mathbf{e}^t\mathbf{y}'+\mathbf{e}^t\mathbf{s}'+\mathbf{e}^t\mathbf{v}'$ as $\mathbf{e}^t(\mathbf{x}'+\mathbf{y}'+\mathbf{s}'+\mathbf{v}')$ since the order of $\mathbf{e}^t$ differs from $\mathbf{x}'$ to $\mathbf{y}'$, in general. In the KLP (21) the solution (point) $[x_1'$

$x_2' \,..\, x_n' \;\; y_1' \;\; y_2' \,..\, y_m' \;\; s_1' \;\; s_2' \,..\, s_n' \;\; v_1' \;\; v_2' \,..\, v_m' \;\; d_1' \;\; d_2' \;\; d_3']^t = (1/(2n+2m+3))e^t$ is feasible. Since d_3' should be zero in a feasible solution of (20), we need to minimize d_3' in (21). If (20) is feasible then the minimum value of d_3' in KLP (21) will be zero and the remaining 2n+2m+2 variables in a minimal solution of (21) will give a feasible solution to (20). The values of $x_1, x_2, .., x_n$ in the minimal solution of (21) will produce an optimal solution of the original LP (13). The KLP (21) is now ready for solution by the KA. Consider the LP Max $\mathbf{c}^t\mathbf{x}$ subject to $\mathbf{Ax \le b, x \ge 0}$, where

$$A = \begin{bmatrix} 1 & 2 & 1 \\ -4 & -2 & 3 \end{bmatrix}, \mathbf{b} = \begin{bmatrix} 1 \\ 2 \end{bmatrix}, \ \mathbf{c} = \begin{bmatrix} -2 \\ -7 \\ 2 \end{bmatrix}, \ \mathbf{x} = [x_1 \ x_2 \ x_3]^t, \ m = 2, \ n = 3.$$

From KLP (21), we have, choosing $k=20$ *(conservatively) and setting* $x_j=21x_j' \; j = 1(1)n$, $y_i=21y_i' \; i = 1(1)m$, $s_i=21s_i' \; i = 1(1)m$, $v_j=21v_j' \; j = 1(1)n$, $d_1=21d_1'$, $d_2=21d_2'$,

Min d_3' *subject to*

$$\begin{bmatrix} -2 & -7 & 2 & -1 & -2 & 0 & 0 & 0 & 0 & 0 & 0 & 0 & 10 \\ 1 & 2 & 1 & 0 & 0 & 1 & 0 & 0 & 0 & 0 & 0 & -1 & -4 \\ -4 & -2 & 3 & 0 & 0 & 0 & 1 & 0 & 0 & 0 & 0 & -2 & 4 \\ 0 & 0 & 0 & 1 & -4 & 0 & 0 & -1 & 0 & 0 & 0 & 2 & 2 \\ 0 & 0 & 0 & 2 & -2 & 0 & 0 & 0 & -1 & 0 & 0 & 7 & -6 \\ 0 & 0 & 0 & 1 & 3 & 0 & 0 & 0 & 0 & -1 & 0 & -2 & -1 \\ 1 & 1 & 1 & 1 & 1 & 1 & 1 & 1 & 1 & 1 & 1 & -20 & 9 \\ 1 & 1 & 1 & 1 & 1 & 1 & 1 & 1 & 1 & 1 & 1 & 1 & 1 \end{bmatrix} \begin{bmatrix} x_1' \\ x_2' \\ x_3' \\ y_1' \\ y_2' \\ s_1' \\ s_2' \\ v_1' \\ v_2' \\ v_3' \\ d_1' \\ d_2' \\ d_3' \end{bmatrix} = \begin{bmatrix} 0 \\ 0 \\ 0 \\ 0 \\ 0 \\ 0 \\ 0 \\ 1 \end{bmatrix}$$

All variables ≥ 0.

The foregoing LP is the required KLP for the KA. Thus, without any confusion or loss of generality, the general form of KLP can be written as

Min $z=\mathbf{c}^t\mathbf{x}$ *subject to* $A\mathbf{x}=\mathbf{0}$, $\mathbf{e}^t\mathbf{x} = 1$, $\mathbf{x} \ge \mathbf{0}$, $\mathbf{x} = \mathbf{e}/n$ is feasible, *minimal z-value = 0, (22)*

where the matrix A is $m \times n$. We will be using this general form for the KA.

4.5. The Karmarkar Algorithm (KA) Consider the KLP (22). Assume that a feasible solution having a minimal z-value $< \varepsilon$ (ε is a small positive value compared to the average element of A, **b**, **c**) is acceptable. The KA is then as follows.

Step 1 Input A, **b**, **c**, m, n. Set n-vector $\mathbf{e} = [1 \;\; 1 \,..\, 1]^t$.

Step 2 Set the feasible point (solution) $\mathbf{x}^0 = \mathbf{e}/n$, the iterate $k = 0$.
Step 3 If $\mathbf{c}^t\mathbf{x}^k < \varepsilon$ then stop else go to Step 4.
Step 4 Compute the new point (an n-vector) $\mathbf{y}^{k+1}$ in the transformed n-dimensional unit simplex S (S is the set of points $\mathbf{y}$ satisfying $\mathbf{e}^t\mathbf{y} = 1, \mathbf{x} \geq \mathbf{0}$) given by

$$\mathbf{y}^{k+1} = \mathbf{x}^0 - \alpha\mathbf{c}_p/[\sqrt{(n(n-1))}\|\mathbf{c}_p\|],$$

where

$$\mathbf{c}_p = (I_n - P^t(PP^t)^+P)[\mathrm{diag}(\mathbf{x}^k)]\mathbf{c}, \quad P = \begin{bmatrix} A[\mathrm{diag}(\mathbf{x}^k)] \\ \mathbf{e}^t \end{bmatrix}, \quad 0<\alpha<1.$$

$\alpha = 0.25$ is known to ensure convergence. P is the $(m + 1) \times n$ matrix whose last row $\mathbf{e}^t$ is a vector of 1s. $(PP^t)^+$ is the p-inverse [13] of the matrix PP^t.
Step 5 Compute now a new point $\mathbf{x}^{k+1}$ in the original space using the Karmarkar Centring transformation to determine the point corresponding to the point $\mathbf{y}^{k+1}$:

$$\mathbf{x}^{k+1} = \mathbf{q}/(\mathbf{e}^t\mathbf{q}),$$

where

$$\mathbf{q} = [\mathrm{diag}(\mathbf{x}^k)]\mathbf{y}^{k+1}.$$

Increase k by 1 and return to Step 3.

Remark The computation of $\mathbf{x}^{k+1}$ in Step 5 may equivalently be written as $x_j^{k+1} = x_j^k y_j^{k+1}/\Sigma(x_t^k y_t^{k+1})$ $j = 1(1)n$, where the summation runs from $t = 1$ to n.

Example *Consider the example of Sec. 4.4. If we now call the 8×13 matrix A, the left-hand side 13- vector* $\mathbf{x}$, *and the right-hand side 8-vector* $\mathbf{b}$ *then the KA gives us, in the first iteration,*

$$\mathbf{y}^1 = [.0672\ .0683\ .0701\ .0753\ .0709\ .0706\ .0770\ .0692\ .0824\ .0733\ .0769\ .0750\ .0781]^t,$$

$$\mathbf{x}^1 = [\ .0068\ 0\ .0408\ .0136\ .0272\ 0\ 0\ 0\ .3061\ 0\ .5578\ .0476\ 0]^t.$$

To obtain 4 decimal places accuracy in the elements of $\mathbf{x}$, we need to go up to 1247 iterations. Thus, retaining the elements of $\mathbf{x}$ correct up 4 places, we have

$$\mathbf{x}^{1247} = [\ .0068\ 0\ .0407\ .0139\ .0272\ .0003\ .0001\ .0006\ .3066\ .0001\ .5560\ .0476\ 0]^t.$$

Observe that here $\mathbf{x} = [\ x_1'\ x_2'\ x_3'\ y_1'\ y_2'\ s_1'\ s_2'\ v_1'\ v_2'\ v_3'\ d_1'\ d_2'\ d_3']^t$. Hence $x_1' = .0068$, $x_2' = 0, \ldots, d_3' = 0$. Thus the required (true) solution correct up to 3 places is, noting that $k = 20$, $x_1 = 21x_1'$, $x_2 = 21x_2', \ldots, d_2 = 21d_2'$,

$$[x_1\ x_2\ x_3\ y_1\ y_2\ s_1\ s_2\ v_1\ v_2\ v_3\ d_1\ d_2\ d_3']^t$$
$$= [\ .143\ 0\ .855\ .292\ .571\ .001\ .003\ .012\ 6.439\ .002\ 11.675\ 1\ .001]^t.$$

4.6 Conclusions ***Need for*** $\mathbf{d_3'}$ It is not readily seen *a priori* that the original LP is feasible. If it is known that the LP is feasible then we need not bring d_3' in the KA at all. If the LP is not feasible due to inconsistency in the constraints and we do not use d_3' then we will end up getting incorrect solution. While the simplex algorithm needs artificial variables to tackle/detect inconsistency in the constraints, the KA needs d_3'.

Enhanced dimension of KLP If the original LP is in an inequality form ($A\mathbf{x} \le \mathbf{b}$) then the corresponding KLP will have 2(n+m)+3 variables where A is $m \times n$. Clearly there has been an increase of n+2m+3 variables (and hence the increase in the dimension of the polytope defined by $A\mathbf{x} \le \mathbf{b}$, $\mathbf{x} \ge \mathbf{0}$) over the original LP. If, on the other hand, the original LP is in an equality form ($A\mathbf{x} = \mathbf{b}$) then the corresponding KLP will have relatively small dimension.

Non-feasibility of error-free computation The KA needs the computation of $\sqrt{(n(n-1))}$ which cannot be computed exactly, in general. Hence, unlike simplex and other methods [12, 14, 15], the KA is not amenable to error-free computation.

Polynomial-time noniterative algorithm – an open problem The KA is polynomial-time iterative needing clearly too many iterations compared to the simplex algorithm. A mathematically noniterative (direct) polynomial algorithm for an LP is still an open problem. However, a heuristic direct polynomial algorithm which is significantly useful in solving many real world LPs does exist [Sen and Ramful 2000]. It may be seen that the nonnegativity condition ($\mathbf{x} \ge \mathbf{0}$) is the real difficulty in the way of developing direct algorithm.

Parallel implementation The KA is relatively easy to be implemented/programmed on a parallel machine unlike the simplex method.

General Observe that *max* $\mathbf{c}^t\mathbf{x}$ is the same as *min* $-\mathbf{c}^t\mathbf{x}$. There has been a surge of interest among scientists/operations researchers to relook into the LP after the publication of the KA in 1984 [1, 8, 15, 16, 19, 20, 21, 22, 23. Consequently, there have been several inerior-point polynomial-time iterative algorithms (which are indeed excellent) reported in the literature. We do feel that a through conceptual knowledge of the KA, specifically from the geometrical point of view, is not only refreshing and enjoyable but also an important basis for further research in linear optimization.

4.7 MATLAB Program for Karmarkar Algorithm (KA) A MATLAB 5.1 version program for the KA is presented below for the reader to readily check the algorithm for different kind of LPs including extreme ones (not large) and get a feel of it. No effort has been made to make the program more efficient so as to differ from the KA presented here. Observe that a MATLAB program is not meant to solve really large LPs. The inputs to this program are A, b, c, k (a parameter that differs from problem to problem), m, n.

```
function[ ] = karmarkar(A,b,c,k,m,n);
%A is an mxn matrix; e=n-vector of 1s needed later.
%This is KA for the LP min z= c^tx s.t. Ax=0, x>=0, e^tx=1, x=e/n is feasible,
```

```
%minimal z-value=0.  k = 20  here. k differs from problem to problem.
e=ones(n,1); x0=e/n; x=x0; alp=0.25; I=eye(n); eps=0.00005; n2=sqrt(n(n-1));
%eps=0.00005 should be replaced by eps=0.00005*(average of the elements of A,b, & c)
%for 4 significant digit accuracy in the solution (not the true solution) by KA.

for j=1:3000
    if c'*x<eps
        string  'eps, iteration no., x '
        eps, j, x'
    break
    end
P=[A*diag(x);e'];
cp=(I-P'*pinv(P*P')*P)*diag(x)*c;
y=x0-alp*cp/(n2*norm(cp));
q=diag(x)*y;
x=q/(e'*q);
string 'The iteration no. and solution are'
j, x'
end;
xt=(k+1)*x;
string 'The iteration no. and true solution are'
j, xt'
```

5. Variation on Karmarkar Algorithm: Detection of Basic Variables

5.1 Introduction The projective transformation algorithm due to Karmarkar has brought about a resurgence of interest in linear programs (LPs). More recently Barnes [1] has developed a concise algorithm that can be applied to the standard form of an LP for which the minimum value of the objective function need not be known in advance. The Barnes algorithm is reviewed and some alternative proofs are provided. Also included is a sufficient condition for the algorithm to produce a bounded solution for an LP. The monotonic convergence of the solution vector and hence the detection of the basic variables during the execution of the algorithm are discussed [22].

5.2 The Algorithm Let the LP be

$$\mathit{Min}\ \mathbf{c}^t\mathbf{x}\ \mathit{subject\ to}\ A\mathbf{x}=\mathbf{b},\ \mathbf{x}\geq\mathbf{0}. \quad (23)$$

where $A=[a_{ij}]$ is an m×n matrix of rank m, **c**, **x**, and **b** are as defined in the previous section. Denote the j-th column of A by $\mathbf{a}_j$. In order to solve the LP (23), we solve the problem

$$\mathit{Min}\ \mathbf{c}^t\mathbf{x}\ \mathit{subject\ to}\ A\mathbf{x}=\mathbf{b},\ \Sigma(x_i-y_i)^2/y_i^2=R^2 \qquad (24)$$

Iteratively, where $\mathbf{y}=[y_1\ \ y_2\ ..\ \ y_n]^t>\mathbf{0}$ is a feasible solution, R is a positive constant defined later, and the summation is over I from 1 to n. Let $\mathbf{x}^0>\mathbf{0}$ be given. In general, if $\mathbf{x}^k$ is known then define

$$D_k=\mathrm{diag}(x_1^k\ x_2^k\ ..\ x_n^k) \qquad (25)$$

Compute $\mathbf{x}^{k+1}>\mathbf{0}$ by the formula

$$\mathbf{x}^{k+1}=\mathbf{x}^k-R_kD_k^2(\mathbf{c}-A^t\lambda_k)/\|D_k(\mathbf{c}-A^t\lambda_k)\| \qquad (26)$$

where

$$\lambda_k=(AD_k^2A^t)^{-1}AD_k^2\mathbf{c} \qquad (27)$$

and

$$R_k=\min_{(c_i-\mathbf{a}_i^t\lambda_k>0)} \|D_k(\mathbf{c}-A^t\lambda_k)\|/(x_i^k(c_i-\mathbf{a}_i^t\lambda_k))-\alpha_k \qquad (28)$$

for some $\alpha_k>0$ so that $R_k>0$.

The fact that the algorithm converges to a solution is based on the following theorems and lemma (presented without proof). For proof, refer [22].

Theorem 1 Let the LP (23) have a bounded solution. Then the algorithm defined by the equations (25)-(28) converges to a solution of (23).

For an $m\times(m+1)$ matrix A of (23), the following theorem states that at least two of the variables including the nonbasic one converges monotonically.

Theorem 2 Consider the LP (23) with $n=m+1$. Then the sequence $\{x_i^k\}$, where i is such that the detection of the i-th column from A results in a nonsingular matrix, converges monotonically.

The following lemma specifies a sufficient condition for a problem to have bounded solutions.

Lemma 1 Let $fd_k=\mathbf{c}^t\mathbf{x}^k-\mathbf{c}^t\mathbf{x}^{k+1}$. The LP (23) has a bounded solution if the sequences $\{\mathbf{c}^t\mathbf{x}^k\}$ and $\{fd_k\}$ are decreasing.

Consider the LP *Min* $f=x_1-x_2$ *subject to* $x_1+2x_2\le4$, $3x_1-x_2\ge1$, $x_1+3x_2=4$, $x_1, x_2\ge0$. Adding the slack x_3 to the first equation, subtracting the slack x_4 (called a surplus variable) from the second equation, and adding the artificial variables x_5 and x_6 to the second and the third equations respectively, we get the LP *Min* $f=x_1-x_2+0x_3+0x_4+Mx_5+Mx_6$ *subject to* $x_1+2x_2+x_3=4$, $3x_1-x_2-x_4+x_5=1$, $x_1+3x_2+x_6=4$, all variables≥0. The value M is considered to be larger than any other with which it is compared during the computation. *We take here arbitrarily M=50.* Observe that artificial variables have to be added to '≥' as well as '=' constraints but not to '≤' constraints (assuming $b_i\ge0$ for all i) for consistency (of constraints) check. The artificial variable values must occur in the optimal solution with values zero if the LP is contradiction-free (consistent). ***If we do not use artificial variables for '≥' and '=' constraints assuming b≥0 then we could get a solution for the inconsistent LP where no solution exists. For '≤' constraints no artificial variables (except slacks) are needed.***

Let the initial feasible solution be

$$\mathbf{x}^0 = [.01\ \ .01\ \ (4-.01-2\times.01)\ \ .01\ \ (1-3\times.01+.01+.01)\ \ (4-.01-3\times.01)]^t$$
$$= [.01\ \ .01\ \ 3.97\ \ .01\ \ .99\ \ 3.96]^t \geq \mathbf{0} \text{ (null column vector).}$$

$$A = \begin{bmatrix} 1 & 2 & 1 & 0 & 0 & 0 \\ 1 & -1 & 0 & -1 & 1 & 0 \\ 1 & 3 & 0 & 0 & 0 & 1 \end{bmatrix},\quad \mathbf{b} = \begin{bmatrix} 4 \\ 1 \\ 4 \end{bmatrix},\quad \mathbf{c} = [1\ \ -1\ \ 0\ \ 0\ \ 50\ \ 50]^t.$$

Iteration 0 (k=0)

D_0=diag($\mathbf{x}^0$)=the 6×6 diagonal matrix whose diagonal elements are those of $\mathbf{x}^0$.
$\lambda_0=[-.0025\ \ 49.9444\ \ 49.9968]^t$. *We have chosen arbitrarily* α_k=.01>0 *for all k.* In fact, to get a positive R_k, we should choose an α_k appropriately. D_0=diag($\mathbf{x}^0$)=the 6×6 diagonal matrix whose diagonal elements are those of $\mathbf{x}^0$. Substituting the values of x_1, c_1, and the vector $\mathbf{a}_1$=the first column of the matrix A, we obtain one value R_0^1 and the corresponding one value $v_1=c_1-\mathbf{a}_1{}^t\lambda_0$. Similarly, we obtain R_0^2 and $v_2= c_2-\mathbf{a}_2{}^t\lambda_0$. Thus we have

$$[R_0^1\ R_0^2\ R_0^3\ R_0^4\ R_0^5\ R_0^6]=[-1.1599\ \ -2.2727\ \ 226.3848\ \ 4.5675\ \ 41.4905\ \ 189.3568],$$
$$[v_1\ \ v_2\ \ v_3\ \ v_4\ \ v_5\ \ v_6]=[-198.8273\ \ -101.0410\ \ .0025\ \ 49.9444\ \ .0556\ \ .0032].$$

The minimum $R_0=R_0^4$=4.5675 for which v_4=49.9444>0. Hence we take R_0=4.5675. We are now having all the required values to compute $\mathbf{x}^1$ which is

$$\mathbf{x}^1=[.0497\ \ .0302\ \ 3.8899\ \ 0\ \ .8810\ \ 3.8597]^t,\ f= 237.0545.$$

Iteration 1 (k=1)

D_1=diag($\mathbf{x}^1$)=the 6×6 diagonal matrix whose diagonal elements are those of $\mathbf{x}^1$.
$\lambda_1=[-.0440\ \ 48.2694\ \ 49.9490]^t$.
$[R_1^1\ R_1^2\ R_1^3\ R_1^4\ R_1^5\ R_1^6]=[-1.0725\ \ -3.3180\ \ 59.7917\ \ 9631.2\ \ 6.7020\ \ 52.0274]$,
$[v_1\ \ v_2\ \ v_3\ \ v_4\ \ v_5\ \ v_6]=[-193.732\ \ -102.4898\ \ .0440\ \ 48.2694\ \ 1.7306\ \ .0510]$.

The minimum $R_1=R_1^5$=6.7020 for which v_5=1.7306>0. Hence we take R_1=6.7020. We are now having all the required values to compute $\mathbf{x}^2$ which is

$$\mathbf{x}^2=[.3633\ \ .0913\ \ 3.4540\ \ 0\ \ .0013\ \ 3.3626]^t,\ f= 168.4688.$$

Iteration 2 (k=2)

D_2=diag($\mathbf{x}^2$)=the 6×6 diagonal matrix whose diagonal elements are those of $\mathbf{x}^2$
λ_2=[−.2680 −14.9562 49.5961]^t.
[R_2^1 R_2^2 R_2^3 R_2^4 R_2^5 R_2^6]=[−12.0557 −1.0195 16.3490 −46022 177.0070 11.1385],
[v_1 v_2 v_3 v_4 v_5 v_6]=[−3.4596 −164.2085 .2680 −14.9562 64.9562 .4039].

The minimum $R_2=R_2^6$=11.1385 for which v_6=.4039>0. Hence we take R_2=11.1385. We are now having all the required values to compute $\mathbf{x}^3$ which is

$\mathbf{x}^3$=[.6993 1.0992 1.1022 0 .0012 .0030]^t, f= −.1874.

We continue the iteration till we get a desired accuracy, say, 4 significant digit accuracy. The exact solution of the LP is **x**=[.7 1.1 1.1 0 0 0]^t, f=−.4.

5.3 Detection of Basic Variables The most difficult part in solving an LP is the lack of knowledge of the basic variables in the LP. If we know them a priori then the LP can be solved (substituting zero for the nonbasic variables) just like the way a linear system is solved in $O(n^3)$ operations noniteratively (or iteratively). In fact, if there is an $O(n^3)$ noniterative algorithm to detect the basic variables then it is clearly an achievement (a milestone) in the area of LPs.

However, we discuss here how the foregoing iterative algorithm could be used to detect basic variables. The detection is made possible on the basis of the monotonic convergence of the variables including the nonbasic ones. We discuss the following cases for the detection of basic variables, the consequent optimal solution, and the problems involved therein.

Case 1 If both the primal and its dual are nondegenerate then the LP has a unique solution having exactly m variables basic nonzero (see corollary of Property 1 later). The detection of basic variables and the consequent optimal solution are as follows. After a sufficient number of iterations, the m columns of the coefficient matrix associated with those m (basic) variables which have the larger values are taken and the resulting linear equations with the m×m nonsingular coefficient matrix are solved. The remaining n−m variables (nonbasic) are set to zero.

Case 2 If the primal is degenerate and its dual is nondegenerate then the problem has a unique solution (see Theorem 4 later) having k<m variables nonzero. The basis here is not unique and may not be detectable. However, the k variables which are basic and can be detected give the optimal solution as the remaining n−k variables are zero. The optimal solution in this case is computed as follows. After a sufficient number of iterations, m columns including the k columns are chosen and the resulting linear equations with the m×m coefficient matrix are solved by choosing the values of the arbitrary variables, if any, and those of the remaining n−m variables as zero.

Case 3 If the primal is nondegenerate and its dual is degenerate then the problem has multiple solutions (see Property 1 later) having $k \geq m$ variables nonzero. If the algorithm converges to a solution with exactly m variables nonzero (which, in general, does not happen) then these are basic. Otherwise, the detection is difficult as is highlighted in Lemma 2 and Theorem3 presented later.

Case 4 If both the primal and its dual are degenerate then also the problem has multiple solutions (see Property 1 later) and hence the detection of basic bvariables is hard here too.

The detection of basic variables in Cases 3 and 4 is discussed later in this section. The following lemma illustrates that the solution of the LP (23) obtained by the algorithm need not be an extreme point.

Lemma 2 If the LP (23) has a bounded solution then the solution obtained by the algorithm need not be an extreme point.

The proof of Lemma 2 follows from considering a counter example – the LP *Min* $-2x_1-x_2$ *subject to* $2x_1-2x_2+x_3=1$, $2x_1-3x_2+x_4=1$, $2x_1+x_2+x_5=2$, all variables≥ 0. An artificial variable x_6 is introduced to construct an initial feasible solution vector.with all elements 1.The following theorem indicates a class of problems for which the algorithm may fail to give an extreme point.

Theorem 3 If an LP has multiple solutions then the algorithm need not give an extreme point of the constraint set $\{\mathbf{x}: A\mathbf{x}=\mathbf{b}, \mathbf{x}>\mathbf{0}\}$.

The proof follows from Lemma 2 and the foregoing discussion. The following property expresses the relationship between multiple solutions and degeneracy.

Property 1 A primal has multiple solutions if and only if its dual is degenerate.

Corollary 1 If the primal and its dual are nondegenerate then the problem has a unique solution which is an extreme point.

The motive of the following theorem is to highlight the fact that the degeneracy of the primal does not pose any difficulty in computing the optimal solution.

Theorem 4 If the dual of a given LP is nondegenerate then the algorithm converges to an extreme point.

We have seen that if the dual is degenerate then the primal has multiple solutions (from Property 1) and hence the algorithm applied to the primal will not, in general, give an extreme point and, therefore detection of basic variables is not possible. As a remedy, the algorithm may be applied (i) to the dual if the primal is nondegenerate or (ii) to a problem which differs from the original one in the cost vector **c** so that the dual is nondegenerate and the solution of the perturbed LP is also a solution of the original LP. In the perturbed technique, the perturbed problem is solved in the same way as in Case 1 if the original LP belongs to Case 3 and as in Case 2 if the original LP belongs to Case 4.

It is not necessary to know beforehand whether the primal is degenerate or not. We halt at some iteration of the algorithm and seeve out the basic variables keeping in view that the nonbasic variables tend to zero. To halt, we choose an exit parameter which is any numerical zero whose choice depends on the measure of degeneracy of the dual; in our numerical experiments, it is chosen as $10^{-7} \times \|\mathbf{x}^k\|$. The iteration may be continued till anyone of the variables including the artificial ones has a value less than the exit parameter or till a specified number of iterations chosen here as 2n have gone through, whichever is satisfied earlier. If the dual is near-degenerate, the choice of numerical zero may not be effective, i.e., it may not allow us to recognize the correct basis. A smaller exit parameter (subject, however, to the precision of the computer used) as well as alarger number of iterations are then called for.

6. About Other Algorithms –Deterministic, Heuristic, and Probabilistic

6.1 ***Shrinking Polytope Algorithm: Deterministic*** Let the LP be *Max* $\mathbf{c}^t\mathbf{x}$ *subject to* $A\mathbf{x} = \mathbf{b}$, $\mathbf{x} \geq \mathbf{0}$, where $A = [a_{ij}] = [\mathbf{a}_1^t\ \mathbf{a}_2^t\ \ldots\ \mathbf{a}_m^t]^t$ is an $m \times n$ known matrix of rank r with i-th row $\mathbf{a}_i^t = [\ a_{i1}\ a_{i2}\ \ldots\ a_{in}]$, Geometrically, $\mathbf{a}_i^t\mathbf{x} = b_I$ is the i-th hyperplane of dimension n (i = 1, 2, . ., m). That portion of the intersection of these m hyperplanes, that lies in the nonnegartive quadrant (i.e., the first quadrant defined by $\mathbf{x} \geq \mathbf{0}$ called the nonnegativity condition) constitutes convex region called here a polytope – the region which is of interest and is searched/spanned by the variables x_i of the vector $\mathbf{x}$ [16]. If one of the corners of the polytope happens to be the required point (solution) $\mathbf{x}$, obviously that produces the maximum value of the objective function (OBJ) $\mathbf{c}^t\mathbf{x}$, will have usually (nodegenerate case) positive values of some of the variables of x_i while other values of the variables will be zero. Those variables x_i which have positive values are called basic variables while the remaining variables of $\mathbf{x}$ are called nonbasic variables which will always have zero values. There is absolutely no way to know a priori the basic variables. For, if the basic variables are known a priori then the LP can be readily solved noniteratively just like solving linear equations in $O(n^3)$ operations. Which variables will become basic depend on the constraints $A\mathbf{x} = \mathbf{b}$, the OBJ $\mathbf{c}^t\mathbf{x}$, as well as on the nonnegativity condition $\mathbf{x} \geq \mathbf{0}$. Once the basic variables constituting the vector $\mathbf{x}_B$ are known then the required solution is $\mathbf{x}_B = A_B^{-1}\mathbf{b}$, where A_B is the matrix A without those columns corresponding to the nonbasic variables, and there is no need to know $\mathbf{c}^t\mathbf{x}$.

While a corner point of the polytope, that gives the required maximum and that makes n – r variables x_i nonbasic, is most desired, a noncorner point could also maximize the objective function, i.e., it could give the same maximum value of the objective function as the former one. In the later case the number of positive variables x_i will be more than the number of basic variables. Although the LP is certainly solved (in the later case) with more positive variables, such a solution is not often desired in practice.

The first hurdle in solving the LP is computing/obtaining a point in the polytope, i.e., obtaining a nonnegative solution of the constraints $\mathbf{Ax} = \boldsymbol{b}$ *while the second hurdle is to obtain that* $\boldsymbol{x}$ *which maximizes* $\mathbf{c}^t\mathbf{x}$ *and which is preferably a corner (point) of the polytope defined by* $\mathbf{Ax} = \boldsymbol{b}$, $\boldsymbol{x}\geq\boldsymbol{0}$. *To cross the first hurdle noniteratively in a polynomial time without increasing the dimension (columns) of A (e.g., without inserting artificial variables in* $\mathbf{Ax} = \boldsymbol{b}$*) is an open problem. Equally open problem is to cross the second hurdle noniteratively in a polynomial time. However, these problems have been solved iteratively in a polynomial time by, say, Karmarkar* $O(n^{3.5})$ *projective transformation algorithm [9] or noniteratively in exponential (combinatorial) time using, for example, the fundamental theorem of linear programming, i.e., by searching over* ${}^{n}C_{m}$ *basic solutions. The simplex method is an improvement over the method of proof of the theorem.*

Once we have found a point (solution) inside the polytope we have crossed the first hurdle, i.e., we have obtained a nonnegative solution of $\mathbf{Ax} = \mathbf{b}$. We know the exact direction of search, viz., the direction of the **c** vector (**c**-direction), for a maximum but we do not know the point from which we proceed in that direction. If we start moving in the **c**-direction from a point which is different from the foregoing required point and which does not lie on the **c**-direction that passes through this required point then we will hit a side (bounding hyperplane) or a corner (an intersection of two or more hyperplanes) of the polytope corresponding to the OBJ value less than the maximum OBJ value.

In the absence of the knowledge of the required point from which we start our search in the **c**-direction, a sensible/logical way is to start from a centre of the polytope or from somewhere in the middle region of the polytope. Unlike a multidimensional sphere, a polytope does not have a unique centre. Hence we consider the point of intersection of the minimum number of normals (directed inward the polytope) as a centre. A rigid weighted centre as found by P.M. Vaidya [25] is computationally more complex and strictly not necessary. From our centre we proceed in the **c**-direction and hit a hyperplane. A hyperplane from this point of hit and perpendicular to the **c**-direction encloses a much smaller space called here a shrunk polytope (which is within the current polytope). We again go to a centre of this shrunk polytope and proceed once again in the **c**-direction. This results in still smaller polytope. We continue this process till we get the required solution. A situation in this process that may crop up is a full rank linear system with k equations in k variables. This will, however, produce the required solution through solving these equations noniteratively – no further iteration (successive approximation) is needed at this stage.

6.2 $O(n^3)$ Noniterative Algorithm: Heuristic There exists no mathematically direct (noniterative) algorithm to solve LPs like the ones (e.g., Gauss reduction with partial pivoting) to solve linear systems. Sen and Ramful (2000) [23] developed an $O(n^3)$ mathematically noniterative heuristic procedure that needs no artificial variables and that includes an optimality test for solving LPs. Numerical experiments depicts that this algorithm is of considerable practical utility. An errorfree implementation of this algorithm is also developed [15].

6.3 Probabilistic Algorithm: Evolutionary An evolution-inspired linear program (LP) solver [7, 24] is presented. The solver has been called "evolutionary" or "genetic" although the actual resemblance to natural genetics was *minimal*. The evolutionary algorithm (EA) computes a solution of the LP *Maximize* $\mathbf{c}'^t\mathbf{x}'$ *subject to* $A'\mathbf{x}' \le \mathbf{b}'$, $\mathbf{k}' \le \mathbf{x}' \le \mathbf{k}$, where $\mathbf{k}' = [k'_1\ k'_2\ ..\ k'_n]^t$ and $\mathbf{k} = [k_1\ \ k_2\ \ ..\ k_n]^t$ ($\ge \mathbf{k}'$) are n-vectors of real numbers and $A' = [a'_{ij}]$ is an m×n real matrix. The EA is inherently highly parallel and is readily implementable on a parallel machine and needs no slack/surplus (for conversion of inequalities to equations) and artificial variables (for consistency check). A sequential implementation of the algorithm is easy but cannot compete with the popular deterministic exterior/interior-point methods in terms of computing resource requirements and accuracy. A sequential MATLAB version of the solver is included for a quick feel about this evolutionary algorithm. The result here is evidently not claimed to produce basic variables and to be enough accurate, though, for many practical problems, such a result is useful. A parallel version, however, can possibly be competitive and is relatively easy to comprehend. The implementation of this algorithm, much unlike that of deterministic procedures, to solve nonlinear programs (NLPs) as well as integer NLPs and integer LPs is straightforward.

6.4 MATLAB EA to Solve LP A MATLAB 5.1 version of the evolutionary algorithm is given below to obtain an approximate (not very accurate) solution of the given LP. The solution vector may have more than the number of basic variables nonzero.

```
function[]=opt34a(m,n,A,b,c,size,d,s,ka,kb);
string 'ORIGINAL m,n,A,b,c,size,d,s,ka,kb'
m,n,A,b,c,size,d,s,ka,kb
%
%Compute solution vector x that maximizes c'x subject
%to Ax<=b,0<=ka<=x<=kb,Matrix A is mxn.
%size=population size (100, say); 1/d=fraction for
%perturbation (d=50, say); s=max # of seeds (10, say).
%n-D vector ka is the lower bound of the vector x.
%n-D vector kb is the upper bound of the vector x.
%
bd=b;cd=c;Ad=A;maxgen=70;
b=b-A*ka;
for j=1:n
   c(j)=c(j)*(kb(j)-ka(j)); A(:,j)=A(:,j)*(kb(j)-ka(j));
end;
string 'CONVERTED A,b,c'
A,b,c
%
maxfit=0;
for seed=1:s
   %max s different populations allowed;
   %each derived from a different but related seed.
   %From each initial population, called Generation 0,
   %the next generation (of the same size) is
   %derived from the immediate preceding generation
   %based on perturbations of each member of the population/generation.
%
   [X,z]=population(m,n,A,b,c,size);
```

```
    for g=1:maxgen %g denotes generation no.(Max maxgen (70) generations
       %for each seed (initial population, i.e., Generation 0)
       %has been allowed here. maxgen may be changed to any
       %other value, say 50.
       %Each successive superior generation is obtained
       %by perturbations of each member of the preceding generation.

       maxfit1=maxfit;

       for k=1:size %scanning over each member of one generation.
          xx=X;
          for j=1:n
             alp=X(j,k)/d;X(j,k)=X(j,k)+alp;x=X(:,k);
             fitnessp(j)=fitnessf(A,b,c,x,m);
             X(j,k)=X(j,k)-2*alp;x=X(:,k);
             fitnessn(j)=fitnessf(A,b,c,x,m);
             X(j,k)=X(j,k)+alp;x=X(:,k);
             fitness(j)=fitnessf(A,b,c,x,m);
          end

          [fitpmax,jp]=max(fitnessp);
          [fitnmax,jn]=max(fitnessn);
          [fitmax,j0]=max(fitness);

          if (fitpmax>=fitnmax) & (fitpmax>=fitmax)
             jj=jp;X(jj,k)=xx(jj,k)+xx(jj,k)/d;
             if X(jj,k)>1,X(jj,k)=1;end;
          elseif (fitnmax>=fitpmax)&(fitnmax>=fitmax)
             jj=jn;X(jj,k)=xx(jj,k)-xx(jj,k)/d;
             if X(jj,k)<0,X(jj,k)=0; end;
          elseif (fitmax>=fitpmax)&(fitmax>=fitnmax)
             jj=j0;X(jj,k)=xx(jj,k);
          end
       end
       [X,z]=generation(m,n,A,b,c,X,size);
       [maxfit,kk]=max(z);
       string 'Generation/population #,Its best member#'
       g,kk
       string 'Fitness Value of this member, Member'
       maxfit, X(:,kk)

       if (abs(maxfit-maxfit1)/maxfit)<.0000001|g>=maxgen
          maxfitg(g)=maxfit;Xg(:,g)=X(:,kk);
          break
       end
    end

    [maxfitglobal,gmax]=max(maxfitg);
    string 'Seed #, Best fit for the seed, best member'
    seed, maxfitglobal,Xg(:,gmax)
    fitnessv(seed)=maxfitglobal;XXg(:,seed)=Xg(:,gmax);
end
[bestfit,bestk]=max(fitnessv);
string 'best fitness value, best member'
bestfit,XXg(:,bestk)
%
```

```
for j=1:n
   xvalue(j)=XXg(j,bestk)*(kb(j)-ka(j));
end;
xvalue=xvalue'+ka;
objfnvalue=cd'*xvalue;
string 'objective function value'
objfnvalue
string 'solution vector'
xvalue
%
count=0;
A=Ad;b=bd;
%
for i=1:m
   if A(i,:)*xvalue<=b(i),count=count+1;end;
end;
if count<m
   string 'Not all constraints are strictly satisfied.'
   string 'Number of constraints satisfied are'
   count
else string 'All the constraints have been satisfied'
   break
end;
count1=0;
for i=1:m
   lhs=A(i,:)*xvalue;
   if lhs<=b(i), count1=count1+1;
   else diff=abs((b(i)-lhs)/b(i))*100;constraintno=count1+1;
      string 'Constraint no., b(constraintno)'
      constraintno, b(constraintno)
      string 'exceeds b(constraintno) by (percent of b(constraintno)'
      diff
   end;
end

function[points]=reward(A,b,x,m);
rpv=0.5*ones(m,1);points=0;
for i=1:m
   if A(i,:)*x<=b(i)
      points=points+rpv(i);
   end
end;

function[fitness]=fitnessf(A,b,c,x,m);
points=reward(A,b,x,m);
if points>=m*.5-.00001
   fitness=c'*x+points;
else fitness=points;
end;

function[X,z]=population(m,n,A,b,c,size);
%Generation 0,i.e., initial population of size, say, 100
%n is the dimension of each member of the population
X=rand(n,size);
for k=1:size
   x=X(:,k);
   z(k)=fitnessf(A,b,c,x,m);
```

```
end;

function[X,z]=generation(m,n,A,b,c,X,size);
%generation produces fitness of each member of successive generations,
%i.e., Generation 1 onwards
%n=dimension of each member of the generation
%size=number (constant, say, 100) of members in a generation,
%i.e., no population increase or decrease in a generation

for k=1:size, x=X(:,k); z(k)=fitnessf(A,b,c,x,m); end
%Vector z gives fitness of each of the members in a generation
```

References

1. Barnes, E.R., A variation of Karmarkar's algorithm for solving linear programming problems, *Math. Program.*, **36**, 1986, 174-182.
2. Beale, E.M.L., Cycling in the dual simplex algorithm, *Naval Research Logistics Quartly,* **2**, 1955, 269-275
3. Beale, E.M.L., Mathematical Programming in Practice, Pitman, London, 1968
4. Dantzig, G., Linear Programming and Extensions, Princeton University Press, Princeton, New Jersey, 1963.
5. Gass, S.I., Linear Programming: Methods and Applications, McGraw-Hill, New York, 1969.
6. Gass, S.I., Illustrated Guide to Linear Programming, McGraw-Hill, New York, 1970.
7. Goldberg, D.E., Genetic Algorithms in Search, Optimization, and Machine Learning, Addison-Wesley, Reading, Massachusetts, 2000.
8. Hooker, J.N., Karmarkar's linear programming algorithm, *Interfaces*, **16**(4), 1986, 75-90.
9. Karmarkar, N., A new polynomial-time algorithm in linear programming, Technical Report, AT & T Bell Labs., New Jersey, 1984, also *Combinatorica,* **4**, 1984, 373-395.
10. Khachiyan, L.G., A polynomial algorithm in linear programming, *Doklady Akademia Nauk SSSR*, 244, 1979, 1093-1096.
11. Krishnamurthy, E.V.; Sen, S.K., Numerical Algorithms: Computations in Science and Engineering, Affiliated East-West Press, New Delhi, 2000.
12. Lakshmikantham, V.; Sen, S.K.; Sivasundaram, S., An inequality sorting algorithm for a class of linear programming problems, *J. Math. Anal. Appl.*, **174**(2), 1993, 450-460.
13. Lakshmikantham, V.; Sen, S.K.; Howell, G., Vectors versus matrices: p-inversion, cryptographic application, and vector implementation, *Neural, Parallel, and Scientific Computations*, **4**, 1996, 129-140.
14. Lakshmikantham, V.; Maullo, A.K.; Sen, S.K.; Sivasundaram, S., Solving linear programming problems exactly, *Appl. Maths. Comput.*, **81**, 1997, 69-80.
15. Lakshmikantham, V.; Sen, S.K.; Jain, M.K.; Ramful, A., O(n^3) noniterative heuristic algorithm for linear programs with error-free implementation, *Applied Maths. Comput.*, **110**, 2000, 53-81.
16. Lord, E.A.; Venkaiah, V. Ch.; Sen, S.K., A shrinking polytope method for linear programming, *Neural, Parallel and Scientific Computations*, **4**, 1996, 325-340.
17. Luenberger, D.G., Introduction to Linear and Nonlinear Programming, 2nd Edition,

Addison-Wesley, 1984.
18. Murty, K.G., Linear and Combinatorial Programming, Wiley, New York, 1976.
19. Murty, K.G., Linear Complementarity, Linear and Nonlinear Programming, Verlag, Berlin, 1989.
20. Renegar, J., A polynomial-time algorithm, based on Newton's method for linear programming, *Math. Prog.*, **40**, 1988, 59-93.
21. Sen, S.K.; Du, H.; Fausett, D.W., A center of a polytope: an expository review and a parallel implementation, *Internat. J. Math. & Math. Sci.,* **16**(2), 209-224, 1993.
22. Sen, S.K.; Sivasundaram, S.; Venkaiah, V.Ch., Barnes' algorithm for linear programming: on detection of basic variables, 1993 (Unpublished)
23. Sen, S.K.; Ramful, A., A direct heuristic algorithm for linear programming, *Proc. Indian Acad. Sci. (Math. Sci.)***110** (1), 2000, 79-101.
24. Sen, S.K.; Sen, S., Linear program solver: evolutionary approach, 46th Congress of ISTAM (an International Meet), R.E.C., Hamirpur, Himachal Pradesh, 2001 (*to appear*).
25. Vaidya, P.M., An algorithm for linear programming which requires $O(((m+n)n^2+(m+n)^{1.5}n)L)$ arithmetic operations, *Proc. ACM Annual Symposium on Theory of Computing*, 29-38, 1987; also *Mathematical Programming*, 1990, **47**, 175-201.
26. Vajda, S., Theory of Linear and Nonlinear Programming, Longman, London, 1974.
27. Vajda, S., Problems in Linear and Nonlinear Programming, Charles Griffin, London, 1975.
28. Winston, W.L., Operations Research: Applications and Algorithms, Duxbury Press, California, 1994.

OPERATIONS RESEARCH IN THE DESIGN OF CELL FORMATION IN CELLULAR MANUFACTURING SYSTEMS

E. Stanley Lee
Dept. of Industrial and Manufacturing Systems Engineering
Kansas State University, Manhattan, Kansas 66506 USA

Ping-Feng Pai
Department of Industrial Engineering, Da-Yeh University
Da-Tsuen, Changhua 515 Taiwan

ABSTRACT

Cell formation problems are practically important and are NP hard, which is very difficult to solve. Various operations research techniques, from the early use of the various mathematical programming techniques to the more recent neural fuzzy approaches, have been proposed to use to solve this problem. This chapter presents these operations research approaches. To save space and also to introduce the approaches in reasonably detail, at least one numerical example is used for each type of the technique discussed. A detailed list of references is also given.

1. Introduction

Various systems, such as just-in-time, flexible manufacturing, cellular manufacturing system, and etc., have been proposed to increase the efficiency in manufacturing. These systems yield many advantages in different ways. For example, just-in-time manufacturing, also known as the pull system, has been implemented in industry to improve the productivity by reducing in-processing inventory. Flexible manufacturing system, which is a compromise between the flexibility of cellular manufacturing and the higher production rate of the specially designed manufacturing system, is designed for medium volume manufacturing where computer control is used. In cellular manufacturing, increased productivity is achieved by forming cells or groups with similar properties or similar processing requirements.

A major task in the design of cellular manufacturing system is cell formation, which includes the identification of part families and the formation of associated machine cells. The problem is how to design part families and associated machine cells such that all parts and machines in a cell have high similarity. This similarity can be based on various factors such as geometry, functioning aspects, material, processing, tools needed, and even the operator required. Thus, many factors can be or need to be considered in forming the cellular manufacturing system. Two basic approaches, namely, part coding analysis and production flow analysis, have been proposed. The former uses the information in the parts' attributes based on parts' coding and the latter uses information of the relationships between parts and machines.

Please address all correspondences to E. S. Lee, email: eslee@ksu.edu
Supported by the Taiwan NSC Project #89-2213-E-212-057

In part coding approach, the code, which characterizing the parts, can be represented by real (crisp) data, fuzzy data, or interval data. Crisp data are data that can be measured and defined precisely. Length of the part and type of material are examples of this type. Fuzzy data are data that cannot be defined precisely and is usually expressed linguistically. Some examples of interval data are tolerance level and the degree of surface finish, which are frequently represented approximately by intervals.

In the production flow approach, part-machine matrices are used to represent the relationships between the parts and the machines. There are three types of part-machine matrices, binary part-machine matrix, weighted part-machine-part, and non-binary part-machine matrix. A binary part-machine matrix only shows machines needed to process a certain part. It does not present information concerning processing sequence of a given part. A weighted part-machine matrix presents not only information of machines needed to processing a certain part but also the level of this processing, which represents the level of machine loading, production volume, or machining hours. Information of processing sequence of a certain part can be obtained from a non-binary part-machine matrix.

From the operations research or mathematical algorithm standpoint, the above two basic approaches of cell formation can be considered approximately as the clustering approach and the classification approach. With clustering approach, the similarities of parts or machines are usually used as the indices of performance. The objective is to minimize the differences inside a cluster and to maximize the differences among the clusters. With the classification approach, the basic idea is to establish the relationship between the parts and the machines. As discussed above, a part-machine matrix can be used to represent the relation. A typical part-machine matrix is illustrated in Figure 1. A binary part-machine matrix is usually a one-to-one relation. Most traditional approaches assume the existence of a one-to-one mapping between machines and part types. However, in actual practice, a certain part may be most suitable to be processed by, say, the first type of machines; but, it can also be processed, although less desirable, by a second type of machines. Suppose the first type of machine is very busy while the second type of machine is idle; from the standpoint of better machine utilization, the second type of machine should be used to process this part. In other words, from the machine utilization standpoint, the one-to-one mapping should not be absolute. The recent proposed fuzzy approaches can be used to achieve this purpose.

MACINES PARTS	M_1	M_2	$\cdots$	M_j
P_1	X_{11}	X_{12}	$\cdots$	X_{1j}
P_2	X_{21}	X_{22}	$\cdots$	X_{2j}
$\vdots$	$\vdots$	$\vdots$	$\vdots$	$\vdots$
P_i	X_{i1}	X_{i2}	$\cdots$	X_{ij}

Figure 1. Part-machine matrix

Since cell formation is essentially an optimization or decision-making problem; various operations research techniques, which are summarized in the top row of Table 1, have been used to solve this problem. As has been discussed before, in forming this decision-making problem, many different aspects of the process can be considered as the most important factor. Some of these factors typically considered are listed in the left most column of the table. Only some the recent research papers are listed to illustrate the approach. These approaches will be discussed in the various sections. Figure 2 gives an overview of the approaches and the various connections of the approaches.

From a more basic operations research standpoint, the various techniques used can be classified as deterministic, stochastic and fuzzy approaches. Deterministic approaches assume that the information concerning the process is known without uncertainty and fuzzy approaches assume a more practical situation where the information is only fuzzily represented. Historically, mathematical programming was first proposed and fuzzy approach was a fairly recent proposition. Because of the vagueness or fuzziness, neural network learning was proposed to up-date or to model the system more accurately.

In this chapter, the various operations research techniques used to solve the cell formation problem will be summarized with emphasis on the recent developments. Mathematical programming approach is first discussed in the next section with two numerical examples, one of which investigates the difficult problem of dealing with exceptional elements. Because cell formation is basically an integer problem, integer programming is most suited. Although a finite integer programming problem has a finite number of solutions, the number of this finite number of solution can still be very large even for a reasonably small problem. In fact, it has been proved that most cell formation problem is a non-polynomial problem [Kamrani et al 1995] and is NP-complete [Garey and Johnson 1979]. To overcome this difficulty, various heuristic approaches have been proposed. These heuristic approaches, with emphasis on recent developments, are discussed in Section 3. The remaining sections in this chapter discuss the use of the recently developed neural network and fuzzy decision making approaches for cell formation.

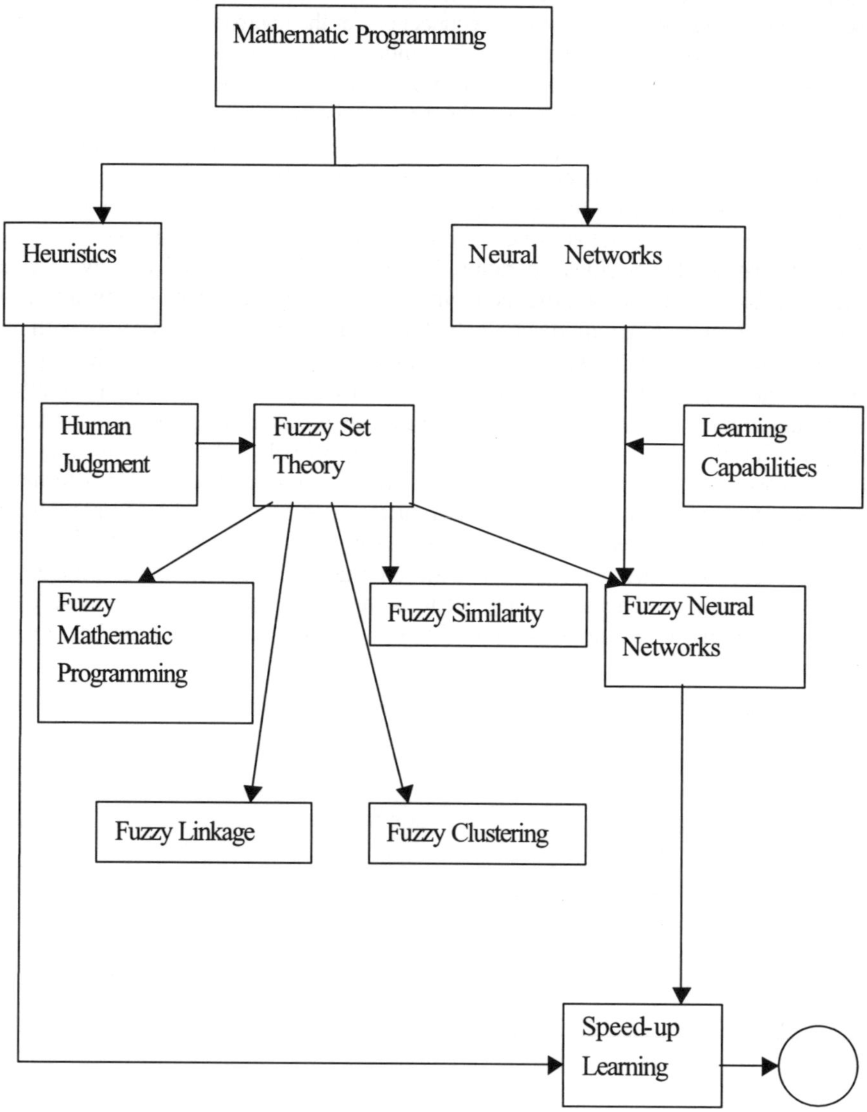

Figure 2. Operations research approaches in cell formation

Table 1. Operations research techniques and factors considered in cell formation

Approaches / Factors	Mathematical Programming	Heuristics	Neural Networks	Fuzzy Set Theory	Fuzzy Neural Networks
Routing	Cao and McKnew (1998), Selim et al (1998).	Hwang and Sun (1996), Joines et al.(1996), Ang and Hegi(1997), Chan et al. (1998), Spiliopoulos, and Sofianopoulou (1998), Chang and Lee(2000), Onwubolu and Mutingi (2001),	Kao and Moon(1991), , Kaparthi and Suresh(1992), Rao and Gu(1994), Kulkarni and Kiang (1995), Chen and Cheng(1995), Enkes et al. (1998) .	Chu and Hayya(1991), Zhang and Wang (1992), Tsai et al.(1994), Su(1995), Leem and Chen(1996), Wen et al. (1996), Szwarc et al. (1997), Sen and Dave(1999)	Suresh and Kaparthi (1994), Burke and Kamal (1995), Kamal and Burke (1996), Suresh et al. (1999)
Throughput		Hwang and Sun (1996)			Kamal and Burke (1996)
Productivity	Abdelmola et al. (1998)	Hwang and Sun (1996)			Kamal and Burke (1996)
Capacity	Akturk and Wilson (1998)	Suresh et al. (1995), Taboun et al.(1998),		Tsai et al. (1994), Szwarc et al. (1997)	
Part Attributes		Taboun et al. (1998)	Kao and Moon (1991), Kaparthi and Suresh(1991), Moon and Roy(1992), Chakraborty and Roy (1993), Liao and Chen(1993), Chung and Kusiak (1994), Wu and Jen (1996), Pilot and	Mital et al. (1988), Xu and Wang(1989), Ben-Arieh and Triantaphyllou (1992), Zhang and Wang (1992),Su(1995) , Ben-Arieh et al. (1996), Narayanaswamy et al.(1996), , Masnata and Settineri (1997), Gengor and Arikan(2000), Liao(2001)	Lee and Fischer (1999), Pai and Lee (2001 a,b), Kuo et al. (2001)

			Knosala (1998)		
Costs	Berardi et al. (1999)	Taboun et al. (1998), Zhou and Askin (1998), Abdelmosa and Taboun(1999) , Moon and Gen (1999),	Rao and Gu (1994)	Tsai et al. (1994), Tsai et al. (1997), Szwarc et al. (1997)	
Number of Cells		Hwang and Sun (1996), Chan et al.(1998)			Pai and Lee (2001 a)

2. Mathematical Programming in Cell Formation

Various approaches have been proposed to form machine cells based on mathematical programming. Almost all mathematical programming techniques, such as linear programming, quadratic programming, integer programming, dynamic programming, mixed integer programming, goal programming and etc., have been used. Depending on the different emphasis of the various factors, different optimization problem with different objectives and constraints can be formed. For example, problems may be formulated to minimize the following various factors or costs: intercellular travel, setup time, exceptional element cost, total production cost, machine idle time, the number of inter-cell transfer, the number or cost of machine duplication, the number of exceptional elements, inventory cost, machine relocation cost, equipment and tooling investment, floor space, intra and inter movements of the operator, and etc.. Example problems to maximizing objective functions are: maximizing machine utilization, maximizing similarity or compatibility measure, maximizing the number of parts completed in a cell, maximizing capacity utilization, and etc. Some of the constraints proposed are: the number of parts in a cell, the number of machines in a cell, the number of operators per cell, the number of parts per operator, time availability, number of tool type available, annual operating budget, tool life, and etc. For more details, the reader can refer the various review papers such as the review by Selim et al [1998].

Several recent developments using mathematical programming are summarized briefly in the following. One of these developments is the application of the Lagrangian relaxation algorithm, which has been shown to be effective for solving large combinatorial problems. Cao and McKnew [1998] used this relaxation algorithm with a partial early termination technique to terminate some sub-models in order to reduce the computational effort. Deutsch et al. [1998] applied an improved p-median approach to maximize the similarities between the different parts. Abdelmola et al. [1998] used a two-stage model to handle the cellular manufacturing productivity problem. Binary integer programming was used in the first stage and integer programming was used to optimize the total productivity in the second stage. Berardi et al. [1999] employed a mixed integer programming approach to evaluate the influences of exceptional parts based on alternative machine clusters.

To illustrate the approach, two numerical examples will be formulated and solved in the following. The first example illustrates the general approach and the second deals with the problem of exceptional elements. Because of the integer nature of the cell formation problem, integer or mixed integer programming is usually the most appropriate. Although integer programming problem with bounded feasible region is guaranteed to have a finite number of solutions, the number of the finite number of solutions can be very large even for relatively small problems. Thus, it cannot be solved easily. To overcome this problem, heuristic approaches, which will be discussed in the next section, are frequently used.

Example 1 [Parapat Gultom, 1996]

To illustrate the approach, machine cells will be formed using integer programming. The problem considered has eight parts and five machines. The objective is to minimizing the total cost, which consists of processing cost and the cost of machines. The number of cells and the maximum number of parts in a cell are both assumed as three. Table 2 summarizes, for each part, the operating sequence and the required processing time for each operation. The bottom row of Table 2 shows the yearly demand of each part. Table 3 shows, for each machine, the processing cost of each operation, the availability in hours per year, and the cost.

Table 2

Operation No.	Part No.							
	1	2	3	4	5	6	7	8
1	1.0		3.0	2.0				5.0
2		2.0		4.0	2.0		2.0	2.0
3	2.0	2.0	1.0	3.0	5.0	1.0	3.0	
4	4.0				3.0	2.0		4.0
5	4.0	4.0	4.0	3.0		4.0	4.0	2.0
Demand	400	300	400	400	300	200	400	200

Table 3

Machine	Operation					Machining Hour Available	Maintenance Cost
	1	2	3	4	5		
1	20	15	25	40	20	5000	6000
2	30	30	40	20	25	6000	8000
3	25	10	30	40	20	8000	7000
4	40	25	10	20	40	6000	8000
5	50	25	30	25	40	8000	6000

The objective and constraints for this cell formation problem are:

$$Minimize \quad \sum_{i=1}^{8}\sum_{l=1}^{5}\sum_{j=1}^{5}\sum_{k=1}^{3} P_{il} C_{jl} q_i x_{ijk} + \sum_{j=1}^{5}\sum_{k=1}^{3} f_j y_{jk} \tag{1}$$

Subject to

Processing time constraint:

$$\sum_{i=1}^{8}\sum_{k=1}^{3}\sum_{l=1}^{5} P_{il} q_i x_{ijk} \le T_j \qquad \forall (j) \tag{2}$$

Each part is allocated to one cell only:

$$\sum_{k=1}^{3} x_{ik} = 1, \qquad \forall i \tag{3}$$

Maximum number of parts allowed in a cell:

$$\sum_{i=1}^{8} x_{ik} \le 3, \qquad \forall k \tag{4}$$

Assignment of part i to cell k:

$$\sum_{j=1}^{5} x_{ijk} - x_{ik} \ge 0, \qquad \forall j \in S_i, \forall (i,k) \tag{5}$$

Each machine can only assigned to one cell:

$$\sum_{k=1}^{3} Y_{jk} = 1, \qquad \forall j \tag{6}$$

Maximum number of machines in a cell:

$$\sum_{j=1}^{5} Y_{jk} \le 3, \qquad \forall k \tag{7}$$

Assignment of part:

$$\sum_{i=1}^{8} x_{ijk} \le (8-1) Y_{jk}, \qquad \forall j \in S_i, \forall k \tag{8}$$

Decision variables must be integers:

$$x_{ijk}, x_{ik}, Y_{jk} \in (0,1) \quad \forall (i,j,k) \tag{9}$$

where:

$i - index \quad of \quad part, i = 1, \ldots, n$

$j - index \quad of \quad machine, j = 1, \ldots, m$

$k - index \quad of \quad group, k = 1, \ldots, g$

$l - index \quad of \quad operation, l = 1, \ldots, r$

C_{lj} : processing cost of operating ℓ on machine j

f_j : annual fixed cost rate of machine j

q_i : annual production requirement of part i

P_{il} : process time for operation of l of part

S_i : *Set of machines needed to process part i*

B_k : *maximum number of parts in cell k*

U_k : *maximum number of machines in cell k*

$$x_{ijk} = \begin{cases} 1 & \text{if part } i \text{ on machine } j \text{ belongs to group } k \\ 0 & \text{otherwise} \end{cases}$$

$$xik = \begin{cases} 1 & \text{if part } i \text{ belongs to group } k \\ 0 & \text{otherwise} \end{cases}$$

$$y_{ik} = \begin{cases} 1 & \text{if machine } j \text{ belongs to group } k \\ 0 & \text{otherwise} \end{cases}$$

This model was solved using the LINDO software. The optimal solution is summarized in Table 4.

Table 4

Cell No.	Part No.	Machine No.	Total Cost
1	P3,P7	M3	$646,500
2	P1,P6,P8	M2,M5	
3	P2,P4,P5	M1,M4	

Example 2. Exceptional Elements Problem in Cell Formation [Berardi et al. 1999]

A mix integer programming model was proposed by Berardi et al. [1999] to investigate the problem of exceptional elements. In cellular manufacturing, the ideal situation is that all operations of parts in a family should be carried out within a single machine cell. In other words, cells in the part-machine matrix should be totally independent of each other. But, in actual practice, this ideal situation can seldom be achieved. Parts that are processed by more than one machine cell and machines that are required by two or more part families are known as exceptional parts and exceptional or bottleneck machines, respectively. These exceptional parts and bottleneck machines are known collectively as exceptional elements. The problem of exceptional elements is very difficult to solve. In fact, this 0-1 binary clustering problem is a traveling salesman problem, which is NP-complete.

Exceptional elements, which cause additional operations and additional cost, are undesirable and should be reduced, or, should be handled in such a way that the cost due to these elements can be reduced. In order to achieve these purposes, various approaches have been proposed to handle the problem of exceptional elements. The most frequently used ones are the following three approaches: the use of machine duplication, the use of intercellular movement, and the use of part subcontracting. If only one of these approaches is used, Berardi et al called the approach pure strategy. Mixed strategy implies the use of more than one approach. In order to compare the optimal costs, Berardi et al [1999] formed the following mixed integer problem with the costs of these three approaches as the objective function

$$\text{Min}\sum_{f}\left\{\sum_{i\in G_f} X_i S_i + \sum_{k\in H_f} Y_{kf} A_k + \sum_{i\in G_f} Z_{ik} I_i\right\} \tag{10}$$

Subject to:

$$Z_{ik} = D_i - X_i - (C_k M_{ik} / P_{ik}) \quad \forall EEs, \tag{11}$$

$$\sum_{i\in G_f} M_{ik} \le Y_{kf} \quad \forall k, f, \tag{12}$$

where the variables X_i, Y_{kf}, Z_{ik} are all integer and

X_i = units of part i to be subcontracted,

Y_{kf} = number of machines of type k to be purchased for cell f,

Z_{ik} = number of intercellular transfers required by part i because of no machine type k available within the part cell,

M_{ik} = number of machines of type k dedicated to the production of part i (utilization of machine type k to produce part i)

A_k = annual total cost of a machine of type k,

S_i = incremental cost of subcontracting a unit of part i,

I_i = incremental cost for moving part i outside of a cell,

C_k = annual capacity of machine type k,

D_i = annual demand for part i,

P_{ik} = processing time of part i on a machine type k,

G_f = set of exceptional parts in cell f,

H_f = set of bottleneck machines required by parts in cell f.

The aim is to minimize the summation of the subcontracting cost, the machine duplication cost, and the intercellular transfer cost. All of which are due to the present of exceptional elements. Equation (11) represents a logical balance on the number of intercellular transfers for exceptional elements. The optimization model represented by Equations (10)-(12) assumes that the part-machine cells, which are represented by the machine-cell grouping matrix, are already in existence. The purpose is to obtain the optimal costs using the above model and to study the influences of the various approaches to handle the exceptional elements. If there is no exceptional elements, the cells in the desired part-machine grouping matrix is arranged in mutually exclusive groups along the diagonal of the matrix. With the presence of the exceptional elements, mutually exclusive cells cannot be obtained. With a given problem, in order to take care of the exceptional elements, many different machine-cell grouping matrices can be obtained. Thus, the above optimization model can also used to study the different machine-cell matrices formed due to the presence of the exceptional elements. There are many algorithms to obtain the desired machine-cell grouping matrices. Berardi et al, based on the numerical values listed in Figure 3, obtained six alternative part-machine grouping matrices by using the two following different clustering algorithms. The single linkage clustering analysis developed by Sneath [Sokal and Sneath 1968] for use in the field of numerical taxonomy and the rank order clustering developed by King [1980] for the purpose of part-machine grouping. For problems without exceptional elements, these clustering algorithms essentially consist of exchanging rows and columns in the part-machine matrix so that an entry in the matrix is contained in mutually exclusive groups arranged along the diagonal of the matrix.

The data used is summarized in Figure 3. The main part of Figure 3 lists the processing sequence and processing time of a given part. For example, part 2 is processed in machines 2, 3 and 8 with corresponding processing times 5.18, 4.29, and 5.32 minutes, respectively. The remaining rows and columns list the various costs and capacities. The six alternative part-machine clusters obtained by using the numerical values listed in Figure 3 is summarized in Figure 4. The first four clusters (alternatives 1-4) were obtained by using the single linkage clustering approach and the last two were obtained by the order rank clustering algorithm.

Using the model represented by Equations (10)-(12), the optimal solutions of these six clusters listed in Figure 4 were obtained [Berardi et at 1999] and the results are summarized in Table 5, where the number of cells formed and the number of exceptional elements obtained from Figure 4 are also listed. The numerical results were obtained by Berardi et al [1999] using IBM's optimization subroutine library on an RS/6000 model 530 workstation.

The optimal results for the mixed strategy, which was obtained by using Equation (10) as the objective function, are listed under the columns labeled MP. The optimal results using the pure strategy, or, only use one of the three costs as the objective function, are also listed in this table. As can be seen from the table, the mixed strategy for any of the alternative cost less than any one of the pure approaches. For example, for Alternative 2, the total cost for the mixed strategy is only 483313 while the least cost for the three pure strategies is 590016.

M \ P	1	2	3	4	5	6	7	8	9	10	A(k)	C(k)
1	2.95		2.20							4.61	50,784	2,000
2	2.76	5.18	1.89	3.89		5.14					67,053	2,000
3	5.54	4.29									43,944	2,000
4	2.91			1.97	2.59	4.01		2.70			67,345	2,000
5				4.28		4.51					42,414	2,000
6	1.92						2.23		5.52		75,225	2,000
7					3.40		1.16	4.72		2.49	52,741	2,000
8		5.32						3.75	3.85		63,523	2,000
9							4.04			1.83	50,632	2,000
S(i)	4.2	4.3	3.5	4.4	5	3.9	4.4	4.6	5	5		
D(i)	32,128	27,598	20,651	11,340	18,707	17,040	46,196	45,384	16,409	22,000		
I(i)	3.7	2.8	2.8	3.3	2.8	3.5	2.8	2.6	3.4	3.2		

Figure 3. Numerical data used, example 2

ALTERNATIVE 1

MACHINES \ PARTS	1	2	3	4	5	6	7	8	9	10
1	1		1							1
2	1	1	1	1		1				
3	1	1								
4	1			1	1	1		1		
5				1		1				
6	1						1		1	
7					1		1	1		1
8		1						1	1	
9							1			1

(a)

ALTERNATIVE 2

MACHINES \ PARTS	7	10	5	8	1	4	6	2	3	9
9	1	1								
7	1	1	1	1						
4			1	1	1	1	1			
2					1	1	1	1	1	
5						1	1			
3					1			1		
1		1			1				1	
6	1				1					1
8				1				1		1

(b)

ALTERNATIVE 3

MACHINES \ PARTS	1	2	3	4	6	7	9	10	5	8
1	1		1					1		
2	1	1	1	1	1					
3	1	1								
4	1			1	1				1	1
5				1	1					
6	1					1	1			
8		1					1			
9						1		1		
7						1		1	1	1

(c)

ALTERNATIVE 4

MACHINES \ PARTS	10	7	5	8	4	6	3	1	2	9
9	1	1								
7	1	1	1	1						
4			1	1	1	1		1		
2					1	1	1	1	1	
5					1	1				
3								1	1	
1	1						1	1		
8				1					1	1
6		1						1		1

(d)

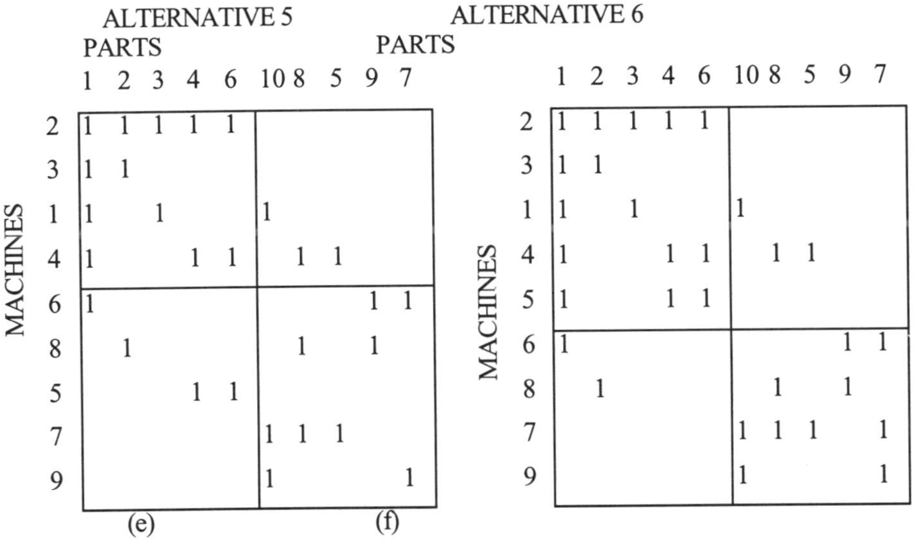

Figure 4. Alternative clusters

Table 5. Numerical results, example 2

Alt.	No. of cells	No. of EEs	MP cost	total MP cost components			Pure strategy cost		
				Machine Duplication	Part subcontracting	Intercellular moves	Machine duplication	Part subcontracting	Intercellular moves
1	3	8	$460183.60	$185182.00	$134937.60	$140064.00	$641937.00	$782262.00	$652861.60
2	3	7	$483313.60	$332102.00	$61566.40	$89645.20	$590016.00	$869172.80	$684273.20
3	3	8	$462761.40	$309618.00	$61566.40	$91577.00	$567531.00	$869172.80	$754673.20
4	3	8	$524710.60	$332102.00	$180237.00	$12370.80	$641020.00	$869172.80	$761547.60
5	2	7	$364364.20	$235768.00	-	$128596.20	$472573.00	$782262.40	$533988.00
6	2	5	$317797.60	$193354.00	-	$124443.60	$387745.00	$665910.40	$436926.00

3. Heuristic Approaches in Cell Formation

Cell formation problems are essentially discrete search problems or integer programming problems, which are difficult to solve even for a medium size problem. Thus, various heuristic approaches have been proposed to solve these problems. The heuristics proposed can be roughly divided into the various ad hoc approaches and the evolutionary approaches. Some of the important algorithms for the latter approaches are genetic algorithm, simulated annealing, and neural network. Neural network, combined with fuzzy logic, will be discussed in latter sections.

Some of the typical ad hoc approaches are summarized briefly in the following. Suresh et al [1995] developed a capacitated hierarchical heuristics to deal with the cell formation problem. The proposed approach is capable of solving problems with large amount of part-machine data and with multi-objective functions. Ang and Hegi [1997] presented an algorithm to deal with the improper part-components assignment problem. Spiliopoulos and Sofianopoulou [1998] presented a tree search heuristic for dealing with the cell formation problem. A procedure was proposed to reduce the size of the tree. Results showed that the proposed algorithm is very efficient. Taboun et al. [1998] proposed a two-stage model to deal with part family and machine formation. A heuristic was proposed for the first stage to determine the number of cell and thus reduces the number of constraints in the second stage. Chang and Lee [2000] using the idea of nearest neighborhood and presented a heuristic with emphasis on the use of the decision maker's judgment.

Various investigators have proposed algorithms to use the simulated annealing in cell formation. Simulated annealing originated from the field of metallurgy. Kirkpatrick et al. [1983] proposed the algorithm based on the analogy between the annealing of solids and the problem of solving combinatorial optimization problems. Some of the researches using simulated annealing are briefly summarized in the following. Zolfaghari and Liang [1998] proposed a simulated annealing approach by considering processing time, machine capacity and machine duplication. To promote faster convergence, these authors used an improved Hopfield network to generate reasonably good starting solutions. Zhou et al. [1998] employed simulated annealing heuristics to improve the greedy heuristics and to minimize the increment heuristics in cell formation problems. The results showed that when the size of the problem increases, the proposed heuristics outperforms the integer programming model significantly. Abdelmola and Taboun [1999] proposed a simulated annealing approach, which outperforms the nonlinear 0-1 integer-programming model. Caux et al. [2000] combined the simulated annealing algorithm with branch and bound. The former focussed on the generation of partitions and the latter solved the routing assignment problem.

Inspired by the natural evolution process, Holland [1975] proposed the genetic algorithm, which is a somewhat organized random search technique and which imitates the biological evolution process. Onwubolu and Mutingi [2001] used the genetic algorithm to solve the cell formation problem with the upper and lower bounds of the cell size determined by the designer. The results compared favorably to the results of using the traveling salesman heuristics. Hwang and Sun [1996] combined genetic algorithms with the greedy heuristic. The approach consists of two phases. The first phase identifies machine cells and the second phase identifies the associated part families. Moon and Gen [1999] presented a genetic algorithm based heuristic with the simultaneous consideration of processing time, production volume, the number of cells, cell size, and machine capacity. The problem was first formulated as binary integer programming and solved by the proposed heuristics.

Another heuristic approach is the tabu search algorithm [Glover 1986], which is very useful for solving combinatorial optimization problems. Sun et al [1995] applied tabu search heuristic to handle the cell formation problem. A binary-tree data structure and a look-ahead scheme were employed to improve the search efficiency. The results showed that the proposed algorithm is able to generate good cell configuration within an acceptable computation time.

Heuristic approaches are generally problem dependent. A given heuristic may be very effective for certain problems but is very inefficient for others. In fact, even for the same problem, different parameter settings result in different efficiencies. Thus, it is difficult to design a heuristic approach, which is effective for all the problems. However, for the evolutionary approaches, some general approximate conclusions can be obtained from the standpoint of effectiveness for cell formation. The parameters that influence this effectiveness are the mutation rate of genetic algorithm, the forbidden rules of tabu search, and the rate of temperature decrease in simulated annealing. Another way to increase the effectiveness is to use a combination of different heuristics.

To illustrate the approach, a numerical example is solved in the following using simulated annealing.

Example 3. Simulated Annealing

Simulated annealing is a random search technique based on the annealing of metallurgical solids, where the metal solid is first heated to its melting point and then slowly cool down to room temperature. It is hoped that, during this cooling process, the energy of the metal will eventually reach an absolute minimum. If the cooling down is too fast, the energy of the metal may reach a local minimum, which has a much higher energy than the absolute minimum. However, due to thermal agitation, there exists a chance that the metal will eventually jump out this local minimum. Thus, as time goes on, the system may eventually reach the absolute minimum. Notice that the process always changes in the direction of decreasing energy and the probability of jumping out the local minimum depends on the temperature level. Thus, by varying the temperature parameter, the probability of jumping out thc local minimum can be changed. The probability, p, of changing the energy state from E_t to E_{t+1}, where t represents the iteration number, obeys the equation:

$$p = \frac{1}{1+\exp(-\Delta E / T)} \tag{13}$$

where $\Delta E = E_t - E_{t+1}$ corresponds the energy change and T is the temperature or a parameter. Suppose we wish to maximize the total productivity (TP), then the TP can be expressed as a function of the energy state, $TP(E_t)$. Thus, the general procedure of this approach can be summarized in Figure 5 [Pham and Karaboga, 2000] and approximately classified into the following steps:

Initiation. Set the various parameters such as the initial and final temperatures, the incremental change of temperature, the cooling rate, etc.

Generating neighboring or new solution. This corresponds to jumping out the local minimum.

Evaluation. This can be accomplished based on the machine cells and exceptional elements costs, etc.

Change temperature level or not? If not, go back to "generating neighboring solution" step. Otherwise, go to next step.

Incremental change of temperature. Change the temperature and go to next step.

Stop or not? This step is based on the final temperature. If the current temperature is equal to or less than the final temperature, go to next step. Otherwise, go to "generating neighboring solution" step.

Stop and calculate the final solution

where the steps needed to keep the iteration counter have been omitted.

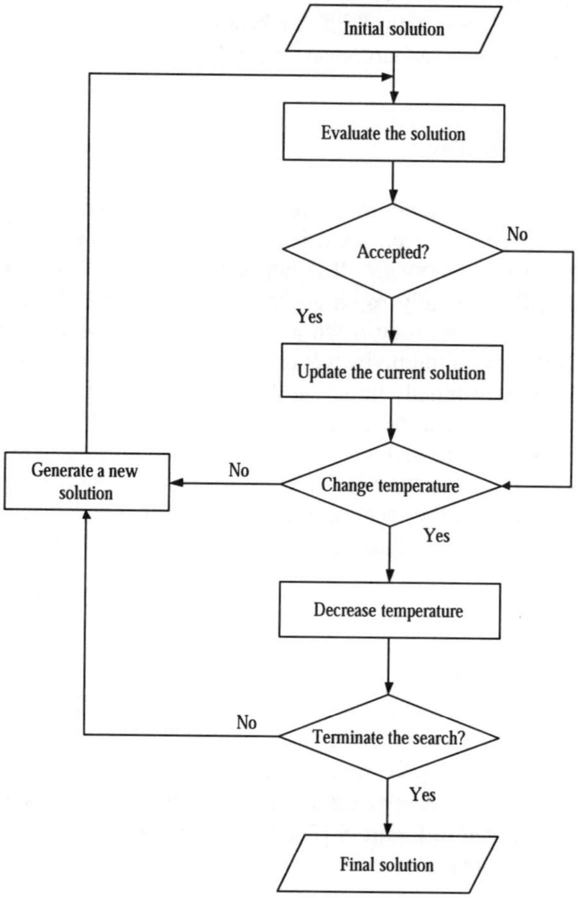

Figure 5. Simulated annealing algorithm [Pham and Karaboga, 2000]

Abdelmola and Taboum [1999] solved a cell formation problem with 10 machines and 10 parts by the simulated annealing approach. The problem is to maximize the total productivity (TP) and is represented by the following nonlinear 0-1 integer-programming model:

$$TP = \frac{\sum_i \sum_k D_i S_i Y_{ik}}{\sum_i \sum_k \sum_j NM_i \times IMC \times D_i \times Y_{ik} + \sum_i \sum_k \sum_j (1 - X_{jk}) b_{ij} Y_{ik} \times EMC \times D_i} \tag{14}$$

Subject to

$$\sum_{j} X_{jk} < NM \qquad \forall k \tag{15}$$

$$\sum_{k} X_{jk} = 1 \qquad \forall j \tag{16}$$

$$\sum_{K} Y_{ik} = 1 \qquad \forall i \tag{17}$$

$$X_{jk} = 0 \quad or \quad 1 \quad \forall (j,k) \tag{18}$$

$$Y_{ik} = 0 \quad or \quad 1 \quad \forall (i,k) \tag{19}$$

where:
i=1,2,…,pparts index
j=1,2,…,mmachine index;
k=1,2,…,ccells index.
b_{ij} =1, if part type i require machine type j,
=0, otherwise ;
D_i = annual demand of part i;
EMC= inter-cell material handing cost;
IMC= intra-cell material handing cost;
NM_i = number of machines required by part type i ;
S_i =sales price of product i
X_{jk} =1, if machine type j is used in cell k,
= 0,otherwise;
Y_{ik} =1, if part i belongs to cell k,
=0,otherwise

The numerator in the objective function, Equation (14), represents the total sale price of the parts produced and the first and the second terms in the denominator represent the intra cell and inter cell material handling cost, respectively. Equations (15) and (16) state that each machine is assigned to only one cell and the maximum number of machines in each cell cannot over a given number, respectively.

Using Equations (14)-(19), Abdelmola and Taboum [1999] solved a cell formation problem with numerical values listed in Table 6, where the part sequence and other values used are listed. For comparison purposes, these authors also solve this problem using the mathematical programming approach with the LINGO software. The results by mathematical programming are listed in Table 7.

Since the simulated annealing approach is essentially a heuristic algorithm, several parameters such as the initial and final temperatures, the amount of perturbation or the incremental change of temperature, and the cooling rate, influence the convergence rate.
The cooling rate controls the number of iterations in each temperature level or determines when to change temperature. This rate factor is connected with the actual problem being solved. For the current approach, the number of iterations in each temperature level is assumed to be proportional to the number of machines in the system, or

(Maximum iterations at a temperature level) = k (number of machines) where the proportional factor or the cooling rate factor, k, determines the cooling rate or the number of iterations in each temperature level.

After some experimental analysis, the authors used the values of 0.99, 50, 0.1, and 32 for the incremental change of temperature, the initial temperature, the final temperature, and the cooling rate factor, respectively. The incremental change of temperature, 0.99, is a multiplication factor. In other words, the temperature decreases by one percent each time. The results obtained by simulated annealing are listed in Table 8. The computational results show the simulated annealing algorithm outperforms the mathematical programming method in terms of obtaining a better objective function value and computation time.

Table 6. The numerical values used, example 3

Part Type	Sales Price($)	Demand D_i	Machine Number									
			M1	M2	M3	M4	M5	M6	M7	M8	M9	M10
P1	14	299	1	0	0	0	0	0	0	0	1	0
P2	14	291	0	0	1	1	0	0	0	1	0	0
P3	11	239	0	0	0	0	1	1	0	0	0	0
P4	10	210	1	0	0	0	0	0	0	0	0	0
P5	10	203	0	0	0	0	0	0	1	0	1	0
P6	14	281	1	0	0	0	0	0	1	0	1	0
P7	12	248	0	0	1	0	0	0	1	1	1	0
P8	13	260	0	0	0	0	0	1	0	0	0	1
P9	13	237	0	1	1	1	0	0	0	1	0	0
P10	11	255	0	1	1	1	0	0	0	0	0	0

Table 7. Solution by mathematical programming

TP=1.8698		CPU time=73 sec
Cell	Parts	Machines
1	1,7,9	1,4,5,6
2	2,3,4,8	2,7,9,10
3	5	3
4	6,10	8

Table 8. Solution by simulated annealing

TP=1.9927		CPU time=55.42 sec
Cell	Parts	Machines
1	2,7,9,10	2,3,4,8
2	8	10
3	3	5,6
4	1,4,5,6	1,7,9

4. Fuzzy Set Theory in Cell Formation

The basic concept of the traditional cell formation approach is that each part belongs to exactly one family. Even assume that there exists no exceptional part in the cell formation problem, this basic concept is not ideal because difficulties encountered in practice are ignored. The first difficulty is the fact that in cell formation the information used for assignment of cells is frequently vague or linguistic, which is fuzzy, not well defined, and cannot be expressed exactly in numerical terms. For example, one important parameter in part coding is the length parameter, which is frequently described in linguistic terms such as *very long, long, average length, short, very short* and etc. A second problem is that some characteristics of the part itself cannot be described exactly. For example, the primary shape of a part is frequently not cylindrical or prismatic but is somewhere in between, which is difficult to represent exactly..

A third problem is machine utilization. For example, a given part may be most suitable to be processed in the first type of machines, but also can be processed, even though not as efficient, in a second type of machines. If the machines in the first type are very busy but the machines in the second type are idle; then, for the purpose of better machine utilization, this given part should be processed in the second type of machines. Thus, for machine utilization purpose, this given part should belong to both types of machines to a certain degree.

All the above problems can be handled and have been handled by the use of the fuzzy set theory, which was developed by Zadeh to overcome the limitations encountered in two-value logic. Many investigators have proposed the use of the fuzzy concept to solve the cell formation problem. Mital et al. [1988] proposed the use of fuzzy numbers to represent part features and used membership grade to classify the parts. Based on fuzzy similarity, Xu and Wang [1989] developed a computer program for classifying part families. Several rotational parts from the industry have been classified and the results were proved satisfactory. The fuzzy clustering algorithm, fuzzy c-mean, was first adopted by Chu and Hayya [1991] to deal with cell formation problem. Ben-Arieh and Triantaphyllou [1992] presented a methodology for handling crisp and fuzzy part features in a unified manner. The proposed methodology is based on a modification of the revised analytical hierarchy process. Zhang and Wang [1992] proposed two fuzzy methods: fuzzy set based single linkage 'cluster analysis and fuzzy rank order clustering. With the inclusion of fuzziness in the production flow analysis, both methods were applicable to the machine-component grouping problem. A fuzzy integer programming approach was proposed by Tsai et al [1994] to deal with cell formation problems. Different membership functions are examined to analyze the impacts on computational performance. Fuzzy clustering approach was employed by Gindy et al. [1995] to obtain the optimal number of groups. An industrial case was used to demonstrate the performance of the proposed algorithm and the results showed that the presented algorithm outperformed existing algorithms in the literature. Su [1995] proposed a multi-criteria fuzzy approach, which includes both the geometric features and the production routing information. Ben-Arieh et al. [1996] used fuzzy numbers to represent the coding information. Fuzzy relation and average linkage methods were used to form part families. In this paper, the authors classified the part attributes into three types for coding: continuous and crisp attributes, fuzzy attributes, and interval attributes. Leem and Chen [1996] presented a fuzzy clustering algorithm for machine-cell formation. A similarity coefficient was used for machine grouping. The objective of the algorithm was to minimize the intercellular movement. Szwarc et al. [1997] used fuzzy nonlinear mathematical models to solve the cell formation problem, which considers both the fuzzy demand and the machine capacity. The objective function was to minimize material handing and processing cost. To reduce computation time, alternative crisp and fuzzy nonlinear mathematical models were used. Several examples were solved. Since the solution strategy was heuristic, optimality cannot be guaranteed. However, the solutions obtained were found to be near the optimum. Sen and Dave [1999] applied the noise clustering technique [Dave 1991] to solve the cell formation problem. The identification of bottleneck was considered as the isolation of noise and outliers. Gungor and Arikan [2000] used a fuzzy decision model to solve the cell formation problem, which considers the design, manufacturing attributes, and operation sequences as factors. The approach emphasizes human judgment than pure mathematical aspects. Based on similarity measurement, Liao [2001] proposed an approach to deal with part family formation problem in a fuzzy environment. An example was used to demonstrate the feasibility of the proposed approach.

Example 4. Fuzzy Linear Programming [Tsai et al 1994]

To illustrate the fuzzy approach, the problem solved by Tsai et al [1994] will be summarized. For comparison purpose, both the crisp (non-fuzzy) and fuzzy versions were solved. The equations for the non-fuzzy or traditional approach are listed in the following:

Minimize:

$$\sum_{k}\sum_{i} A_i R_{ik} + \sum_{k}\sum_{(i,j)\in sp} I_j Z_{ijk} + \sum_{j}\sum_{(i,j)\in sp} O_{ijk} S_j \tag{20}$$

Subject to:

$$\sum_{k=1}^{c} X_{ik} = 1, \quad \forall i \tag{21}$$

$$\sum_{k=1}^{c} Y_{jk} = 1, \quad \forall j \tag{22}$$

$$\sum_{i=1}^{m} X_{ik} \le NM, \quad \forall k \tag{23}$$

$$\sum_{j=1}^{n} Y_{jk} \le NP, \quad \forall k \tag{24}$$

$$U_{ijk}+V_{ijk}\le 1,\ \ \forall (i,j)\in sp,\ \ \forall k \tag{26}$$

$$O_{ijk}+Z_{ijk}+\frac{C_i}{P_{ij}}M_{ijk}=D_jU_{ijk},\ \ \forall (i,j)\in p,\ \ \forall k \tag{27}$$

$$\sum_{(i,j)\in sp} M_{ijk}\le R_{ik},\ \ \forall i,\ \ \forall k \tag{28}$$

$$\sum_k \sum_{(i,j)\in sp} \frac{P_{ij}}{C_i}Z_{ijk}\le Q_i-\sum_{(i,j)\in sp}\frac{P_{ij}D_j}{C_i}(1-\sum_k V_{ijk}),\ \ \forall k \tag{29}$$

where X_{ik}, Y_{jk}, U_{ijk}, V_{ijk} are 0, 1 integers; and Q_I and R_{ik} are integers, and the notations used are:

i machine index; i=1,…..,m

j part index; j=1,…….,n

k cell index; k=1……c

D_j Annual demand for part j.

P_{ij} Processing time of machine type i needed to produce part j.

I_j Incremental cost for moving a unit of part j within two cells.

S_j Incremental cost of subcontracting a unit of part j for an operation.

A_i Annual cost of acquiring a machine type i.

C_i Annual capacity of machine type i.

NM The maximum number of machine allowed each cell.

N P The maximum number of part allowed in each cell.

S P Set of pairs (i,j) such that a_{ij}=1.

X_{ik} 1,if machine i is assigned to cell k; 0, otherwise.

Y_{jk} 1,if part j is assigned to cell k; 0 ,otherwise.

U_{ijk} 1,if a_{ij}=1, Y_{jk}=1, and X_{ik}=0; 0, otherwise.

V_{ijk} 1,if a_{ij}=1, Y_{jk}=0, and X_{ik}=1; 0, otherwise.

R_{ik} Number of machine type i to be dedicated in cell k.

Q_i Numbers of machine type i needed to process the corresponding parts.

O_{ijk} Units of part j to be subcontracted as a result of machine type i not being available within cell k.

M_{ijk} Numbers of machine i in cell k for producing part j.

Z_{ijk} Numbers of intercellular transfers required by part j as a result of machine type i not being available within cell k.

The objective function, Equation (20), minimizes the costs of duplicating a machine, intercellular transfer, and the cost of subcontracting. Equation (23) is to prevent the assignment of more than *NM* machines and Equation (24) serves the same purpose for the number of parts. Equations (21) and (22) prevent the duplication of parts and machines.

For many complex practical problems, the objective function near the optimum is fairly flat. As a result, the plant supervisor frequently does not require the optimum but only requires the achievement of a certain goal, which is near the optimum from past operating experiences. Thus instead of the objective function, Equation (20), we have the following inequality:

$$\sum_k \sum_i A_i R_{ik} + \sum_k \sum_{(i,j)\in sp} I_j Z_{ijk} + \sum_j \sum_{(i,j)\in sp} O_{ijk} S_j = B_0 x \le Z \tag{30}$$

where Z is the minimum goal, x represents the decision variable vector and B_0 the corresponding coefficient vector.

The problem now becomes the system of inequalities, Equations (21)-(30). However, a fixed goal is not very reasonable. We would like to make the goal as near the minimum as possible. One way to achieve this is to use the following membership function:

$$\mu_0(x) = \begin{cases} 1, & B_0 x \le Z \\ 1 - \dfrac{B_0 x - Z}{T_0}, & Z \le B_0 x \le Z + T_0 \\ 0, & B_0 x \ge Z + T_0 \end{cases} \tag{31}$$

where T_0 is the tolerance allowed for the minimum goal.

Furthermore, suppose we wish to allow some tolerances concerning the maximum numbers of machines and parts in each cell and let the following membership functions to represent these tolerances:

$$\mu_i(x) = \begin{cases} 1, & B_i x \le d_i \\ 1 - \dfrac{B_i x - d_i}{T_i}, & d_i \le B_i x \le d_i + T_i \\ 0, & B_i x \ge d_i + T_i \end{cases} \tag{32}$$

where $i=1,2$ correspond Equations (23) and (24), respectively; x and B are the decision variables and the coefficients of the decision variables, respectively; d represents the right hand side of the equation, and T_1 and T_2 represent the tolerances on the maximum number of machines and maximum number of parts, respectively.

In order to satisfy all the inequalities, we must take the minimum, or the intesection, among all the membership functions. Thus

$$\lambda = \min_i \mu_i(x) \tag{33}$$

However, ë is a membership function. We would like to obtain the maximum of the membership function and satisfies the tolerances, $T_i, I=0,1,2$. Thus, the problem becomes the maximization of ë,

$$Max \;\; \lambda = \max \{ \min_i \mu_i(x) \} \tag{34}$$

with the original constraints, Equations (21), (22), (25)-(29), and the following new constraints:

$$\sum_{k}\sum_{i} A_i R_{ik} + \sum_{k}\sum_{(i,j)\in sp} I_j Z_{ijk} + \sum_{j}\sum_{(i,j)\in sp} O_{ijk} S_j + \lambda T_0 \le Z + T_0 \tag{35}$$

$$\sum_{i=1}^{m} X_{ik} + \lambda T_1 \le NM + T_1, \quad \forall k \tag{36}$$

$$\sum_{j=1}^{n} Y_{jk} + \lambda T_2 \le NP + T_2, \quad \forall k \tag{37}$$

and with the original integer restrictions on the variables. The symbols, T_i, $I= 0, 1, 2,$ represent the tolerances allowed. The above three equations are obtained by considering the maximum satisfaction of the membership functions. The approach follows that of Zimmerman [1987].

Both the original crisp problem and the fuzzy problem are linear programming problems. Tsai et al [1994] solved both problems. The numerical values used by these authors are listed in Table 9, which shows the processing time and machine sequence of each part, the costs involved, the part demanded and machined capacity. This example contains nine machines and nine parts. To solve the original crisp or non-fuzzy problem, which is represented by Equations (20)-(29), the desired number of cells is set at three and the maximum number of machines as well as parts allowed in each cell are set no more than four.

The numerical results obtained by Tsai et al are summarized in Table 10, where the symbol "TS" represents "total similarity", n(i) represents the number of machines needed for machine type i, and [i] denotes the duplication of machine type i. The results listed for Model I are for the crisp or non-fuzzy case and the results for Model II are for the fuzzy case. From Table 10, it can be seen that the cost for the fuzzy case is much less than those obtained for the non-fuzzy case.

Table 9. Numerical values used, Example 4

		1	2	3	4	5	6	7	8	9	A(I)	C(I)
M	1	4.82	3.07	0	0	2.18	0	0	0	0	$17709	2000
A	2	3.96	2.02	0	0	0	2.84	0	0	3.78	$15224	2000
C	3	0	0	2.56	0	0	0	3.7	2.78	0	$38616	2000
H	4	0	3.05	4.74	3.97	0	0	0	3.80	0	$20472	2000
I	5	3.4	0	0	4.6	4.4	0	0	2.49	0	$44903	2000
N	6	0	3.92	0	0	0	2.13	0	0	2.51	$39557	2000
E	7	0	0	2.26	0	0	0	3.02	3.23	0	$17558	2000
S	8	0	0	4.2	2.2	2.3	0	2.91	3.56	0	$23555	2000
	9	0	3.22	0	0	0	1.74	0	0	2.77	$43621	2000
S(J)		$4.73	$4.25	$3.57	$4.18	$4.32	$4.15	$4.41	$4.02	$3.65		
D(J)		68172	43657	58449	54073	45955	70309	77248	75183	73901		
I (J)		$3.86	$3.14	$2.8	$3.15	$2.11	$2.73	$2.69	$3.47	$3.65		

Table 10 Numerical results, Example 4

Model	EE	TS+	TOTAL COST	CELL #	Clustered Results {machines/parts}	# of Machines needed *
Model I	6	9.65	$282878	1 2 3	{2,6,9/2,6,9} {3,4,7,8/3,4,7,8} {1,5/1,5}	3(2);3(6);2(9) 2(3);4(4);2(7);3(8);[5] 2(1);3(5);[2];[8]
Model II	3	9.65	$161930	1 2	{1,2,6,9/1,2,6,9} {3,4,5,7,8/3,4,5,7,8 }	2(1);4(2);3(6);2(9);[5] 2(3);4(4);3(5);2(7);4(8);[1]
Model III	6	9.65	$282878	1 2 3	{2,6,9/2,6,9} {3,4,7,8/3,4,7,8} {1.5/1.5}	3(2);3(6);2(9) 2(3);4(4);2(7);3(8);[5] 2(1);3(5);[2];[8]
Model IV	6	9.05	$305544	1 2 3	{1,2,6,9/1,2,6,9} {3,4,7,8/3,4,7,8} {5/5}	2(1);4(2);3(6);2(9);[5] 2(3);4(4);2(7);3(8);[5] 2(5);[1];[8]

5. Fuzzy Neural Network in Cell Formation

The powerful clustering ability of neural network forms an ideal approach for dealing with cell formation problems. However, in order to handle the vague and linguistic problems encountered in cell formation, the recently developed fuzzy neural network is even more appropriate. Neural network can be combined with fuzzy set in many different ways. Depending on the degree of fuzzification, Buckley and Hayashi [1994] classified fuzzy neural networks into the following three types:

1. Neural networks with real number as input signals but fuzzy weights

2. Neural networks with fuzzy set as input signals and real number weights
3. Neural networks with both fuzzy set as input signals and fuzzy weights.

Gupta and Ding [1994] classified the current fuzzy-neural computations into two major categories. One is the fuzzy logic-based neural network, where fuzzy logic is combined with the parallel neural network concept. With this approach, the membership functions in the fuzzy logic system can be up-dated or adopted. An example of applying this adoptive fuzzy system to cell formation will be introduced to illustrate the approach. The second category is the employment of the neural network to realize the fuzzy structures such as the membership function and the fuzzy logic operators, *and* and *or*. In this approach, several forms of fuzzy neurons have been proposed to approximate the problem with fuzzy uncertainties. The approach is known as the neural network-based fuzzy system.

Fuzzy neural network has been applied to many practical areas and many investigators have applied this approach to solve the cell formation problems. For example, self-organizing feature map or self-organizing map (SOM) due to Kohonen is able to perform the mapping of an external signal into different representational spaces without any human intervention. Kuo et al. [2001] proposed a fuzzy self-organizing map neural network to deal with part clustering problem. The results obtained are more accurate than those obtained with fuzzy c-means algorithm. Pai and Lee [2001a] modified the classic SOM with fuzzy weights so that the network is able to deal with fuzzy part attributes. A new training algorithm was also proposed to train the fuzzy weights. By the use of fuzzy weights between the input and the output layers, more meaningful linguistic information for the final trained weights could be obtained. The fuzzy adaptive system was employed by Pai and Lee [2001b] to deal with cell formation problems. Fuzzy rules were used in part-machine mapping. Influences of different parameters for training were illustrated.

Example 5. Fuzzy Self Organizing Map Network [Pai and Lee, 2001]

To illustrate the approach, a cell formation problem solved by Pai and Lee [2001b] by the use of the fuzzy SOM is summarized in the following. The fuzzy SOM network is an SOM network with fuzzy weights between the input and the output layers. The network is able to deal with crisp, fuzzy, and interval input data. The fuzzy SOM network is shown in Figure 6. Due to the use of fuzzy weight and fuzzy input data, a fuzzy learning algorithm that is different from the conventional SOM learning procedure is proposed in this example. The fuzzy SOM algorithm can be summarized approximately as:

Step 0. Extract part attributes descriptions from experienced operators and represent them as fuzzy numbers.
Step 1. Feed the input data into the input layer, where the input data can be crisp, interval or fuzzy numbers.
Sep 2: Compute the "distance" or the weight vector Wij between each Kohonen node i and the input vector Xj
Step 3. Determine the winning Kohonen node.
Step 4: Adjust the weight vector of the winning node to be closer to the input vector

In this example, a SOM network (see Figure 6) with fuzzy weights was used to cluster 100 parts with two fuzzy attributes, namely part shape and tolerance. The L-R fuzzy numbers are used to represent the fuzzy attributes. Figure 7 shows the scatter plot of the data, which were generated by the use of "Excel" software with uniformly distributed random numbers. The clustering results to be discussed later are also shown on the figure by the use of dotted lines. In this example, the number of output nodes is set at 30 and the winner-take-all algorithm is adopted.

The initial learning rate, α, was set equal to 0.001. Two thousand epochs were carried out. The convergent behavior is shown in Figure 8. As can be seen from this figure, convergence or training is essentially completed after approximately 1200 epochs. After convergence, the clustered ten groups obtained are shown in Figure 7 by the use of dotted lines. The fuzzy weight associated with each winning node represents the particular characteristics of the group. After convergence, the final weights of the classified ten groups in the form of triangular membership functions are listed in Tables 11 and 12 by the use of the triangular nomenclature in the form of (C, L, R), where C represents the most desirable value and its membership value is equal to one. The letters L and R represent the left and right spread of the triangle from the most desirable value, C, respectively. At the positions L and R, the membership values are zero.

In the results, several groups are fairly near each other. When two groups are very near or sufficiently similar, they can be combined into a single group. For example, Groups 6 and 9 for primary shape in Table 11 can be combined into one group with new membership functions of (0.192,0.494,0.619). Similarly, Group 5 and 6 for tolerance can probably be combined with new membership functions of (0.879,0.941,1). The criterion of "sufficiently similar" is fairly arbitrary and depends on the particular problem under consideration.

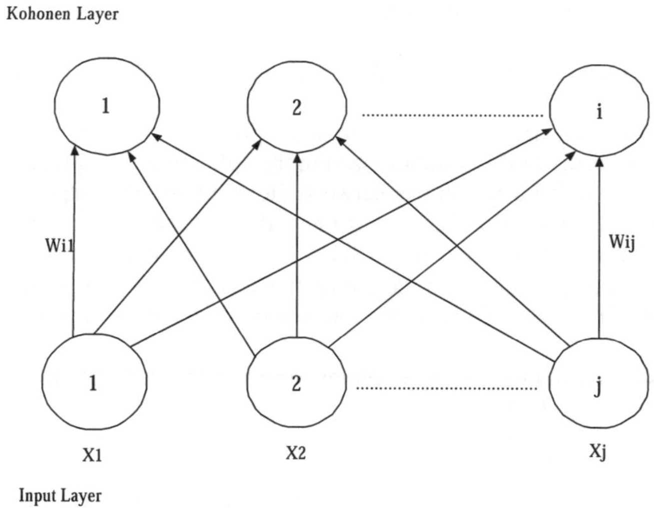

W_{ij} : Fuzzy Weights
X_j : Input (fuzzy, interval, or crisp data)

Figure 6. Self-organizing map network

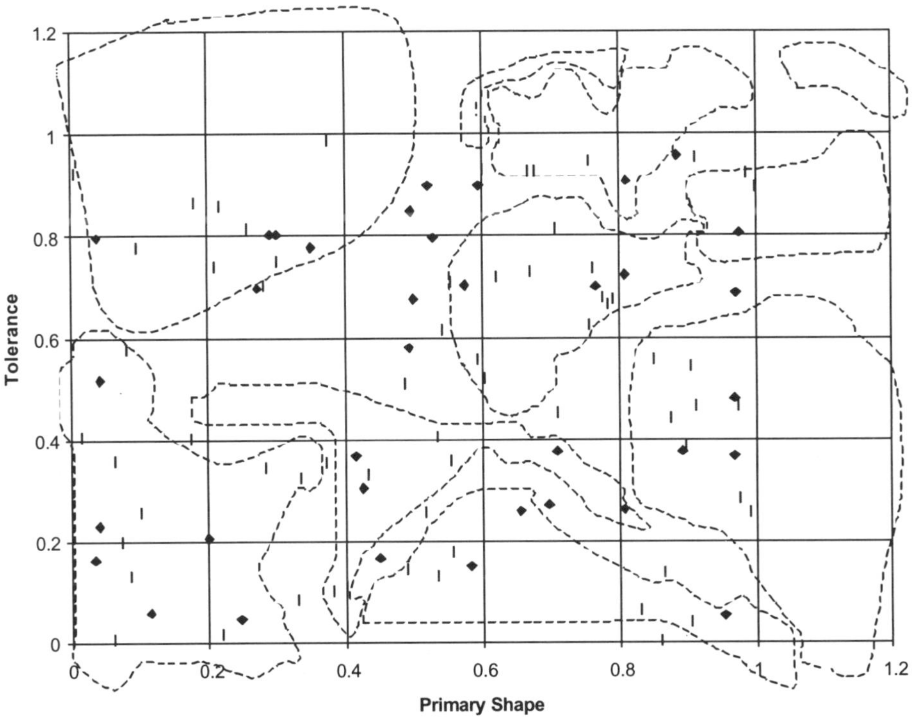

Figure 7. Scatter plot of data

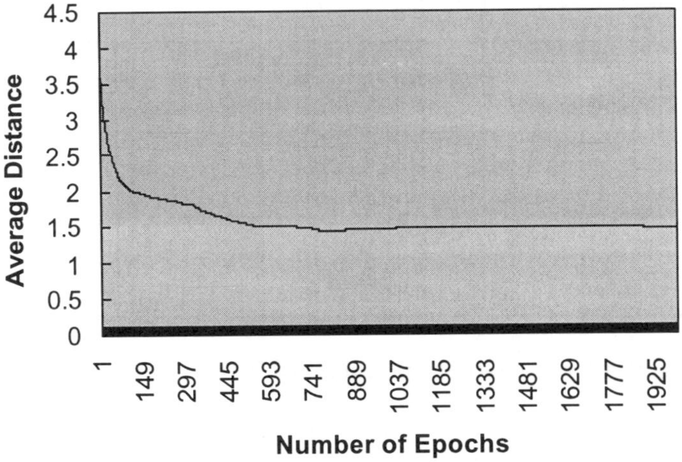

Figure 8. Convergent behavior

Table 11. Final obtained weights in triangular membership functions

Primary Shape / Tolerance	|C-L| = 0 C=0.123 C+R=0.243	|C-L| =0.192 C = 0.489* C+R = 0.591	|C-L| =0.215 C= 0.499* C+R =0.619	|C-L| = 0.218 C= 0.513 C+R = 0.678
|C-L| = 0 C = 0.155 C+R= 0.168			Group #9	
|C-L| = 0 C = 0.128 C+R = 0.54	Group #10			
|C-L| = 0.21 C = 0.34 C+R = 0.7613				Group #3
|C-L| = 0.6436 C= 0.788 C+R= 0.985	Group #1			
|C-L| = 0.879 C= 0.947* C+R = 1		Group #6		

Table 12. Final obtained weights in triangular membership functions

Primary Shape / *Tolerance*	|C-L| = 0.313 C=0.585 C+R=0.701	|C-L| =0.561 C = 0.846 C+R = 0.935	|C-L| =0.683 C= 0.887 C+R =1	|C-L| = 0.218 C= 0.513 C+R = 0.678
|C-L| = 0.141 C = 0.298 C+R 0.632			Group #8	
|C-L| = 0.431 C = 0.618 C+R = 0.8563	Group #2			
|C-L| = 0.532 C = 0.679 C+R= 0.896		Group #7		

Example 6. Fuzzy Neural Adaptive Network [Pai and Lee 2001]

Fuzzy neural adaptive system is based on fuzzy rules and fuzzy logic with training or learning ability. To establish the fuzzy rules for this example and also to simply the presentation, only two important attributes, namely primary shape and tolerance will be considered. In actual practice, primary shape of a part is described linguistically and tolerance is expressed in intervals. In order to establish the fuzzy rules, linguistic description will be adopted for both attributes. The linguistic variable, primary shape, can be represented by the following linguistic terms: cube (C), like cube (LC), like cylinder (LCY), cylinder (CY). These linguistic terms are represented by Gaussian fuzzy membership functions, which are differentiable and are illustrated in Figure 9. Similarly, the linguistic variable, tolerance, is represented by the four linguistic terms, very precise (VP), precise (P), rough (R), very rough (VR). Using Gaussian membership functions, a similar graph as that show in Figure 9 can be obtained. Obviously, if more accurate representation is needed, more linguistic terms can be added. For example, we could use, very precise, precise, more or less precise, not precise, average, not rough, rough, and etc.

Based on these fuzzy terms, 16 fuzzy IF-Then rules (or 16 machine groups) can be formed as follows:

If Primary Shape is C and Tolerance is VP, then use MG 1 (rule 1)
If Primary Shape is LC and Tolerance is VP, then use MG 2 (rule 2)
If Primary Shape is LCY and Tolerance is VR, then use MG 15 (rule 15)
If Primary Shape is CY and Tolerance is VR, then use MG 16 (rule 16)

where MG is the acronym for Machine Group. These fuzzy rules form a rule matrix, which is listed in Table 13.

Table 13. Fuzzy rule matrix

	VP	P	R	VR
C	rule 1(MG 1)	rule 5 (MG 5)	rule 9(MG 9)	rule 13(MG 13)
LC	rule 2(MG 2)	rule 6 (MG 6)	rule 10(MG 10)	rule 14(MG 14)
LCY	rule 3(MG 3)	rule 7 (MG 7)	rule 11(MG 11)	rule 15(MG 15)
CY	rule 4(MG 4)	rule 8 (MG 8)	rule 12(MG 12)	rule 16(MG 16)

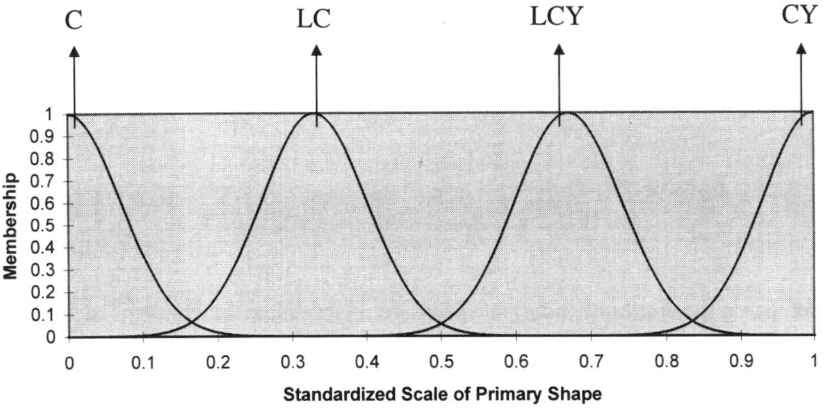

Figure 9 Membership functions of the primary shape

The linguistic variables, primary shape and tolerance, are the input variables for the fuzzy system. The output variable is the machine group (MG), which, again, is a linguistic variable and with the linguistic terms, MG 1, MG 2, MG 3, … etc. These linguistic terms also assume the shapes of the Gaussian membership functions.

Depending on how the fuzzy number is handled, various fuzzy logic systems have been proposed. If we assume the input is not fuzzy, then the crisp number must be first fuzzified. After manipulating through the fuzzy rules and fuzzy inference engine, the resulting fuzzy number must be defuzzified. Thus, to form the fuzzy logic system, we must consider the following four components or operations:

1. Method for fuzzification
2. Method for fuzzy inference, or, method used first to combine the antecedent and the consequent of each rule and, then, to combine the various fuzzy rules
3. Membership function used
4. Method for defuzzification

There are various ways to carry out these operations. In this example, singleton fuzzifier, product inference rule, Gaussian membership function, and center average defuzzification was employed. Using the center average defuzzification and with M rules, we obtain the output of the fuzzy logic system as:

$$\frac{\sum_{r=1}^{M} Y^r\_bar[\mu c'(Y^r\_bar)]}{\sum_{r=1}^{M} \mu c'(Y^r\_bar)} \tag{38}$$

where Y^r_bar is the center of the output fuzzy set for the fuzzy rule r. The expression μ c'(Y^r_bar) is the aggregation of the output membership functions. Using product inference, this aggregation can be expressed as:

$$\prod_{i=1}^{n} \mu_{F^r(x_i)} \tag{39}$$

where $\mu_{F^r(xi)}$ is the membership function of the premise section for rule r and for the ith attribute or the ith linguistic variable, i = 1,2, . . ., n. Using Gaussian function, the numerator of Equation (38) becomes:

$$\sum_{r=1}^{M} Y^r\_bar\{\prod_{i=1}^{n} \exp[-(\frac{(x_i - x_i^r\_bar)^2}{\sigma_i^{r^2}})]\} \tag{40}$$

where x_i^r_bar is the center of the input Gaussian membership function for the i-th attribute and rth rule, and σ_i^r is the standard deviation of this input membership function.

Using the product inference rule and Gaussian membership function, the denominator of Equation (38) becomes:

$$\sum_{r=1}^{M}\{\Pi_{i=1}^{n}\exp[-(\frac{(x_i - x_i^r\_bar)^2}{\sigma_i^{r^2}})]\} \tag{41}$$

Substituting the above equations into Equation (38), finally the fuzzy logic system equation was obtained as following:

$$F(x)=\frac{\sum_{r=1}^{M} Y^r\_bar\{\Pi_{i=1}^{n}\exp[-(\frac{(x_i - x_i^r\_bar)^2}{\sigma_i^{r^2}})]\}}{\sum_{r=1}^{M}\{\Pi_{i=1}^{n}\exp[-(\frac{(x_i - x_i^r\_bar)^2}{\sigma_i^{r^2}})]\}} \tag{42}$$

The three parameters in Equation (42) are Y^r_bar, $x_i^r\_bar$, and , $\sigma_i^{r^2}$, which correspond to the center of the output membership function, the center of the input membership function, and the variance of the input membership function, respectively. These parameters are adjustable. The problem is to adjust these parameters so that certain given input-output pair can be represented. In the following, we shall first establish an adaptive fuzzy network, which represents Equation (42) and then use back propagation to obtain these parameters.

Following Wang [1944], Equation (42) can be represented by the fuzzy adaptive network as that show in Figure 10. There are three layers in Figure 10. Equation (42) is functional equivalent to Figure 10. Using Figure 10, back propagation algorithm can be derived and the three parameters can be trained based on given data pairs. The three parameters are:

1. Y^r_bar represents the center or the maximum value of the output membership function for fuzzy rule r.
2. $x_i^r\_bar$ represents the center or the maximum value of input fuzzy membership function for ith linguistic variable and rth rule.
3. $\sigma_i^{r^2}$ represents the standard deviation of input fuzzy membership function for ith attribute and for the rth rule.

Figure 10. Network representation of fuzzy logic system [Wang 1994]

Suppose we have a given data pair (X^d, D^d), we wish to adjust these three parameters so that the following square of the error is minimized:

$$\text{error}=0.5(F(X^d\text{-}D^d)) \tag{43}$$

Differentiation by the chain rule, the learning rules for the three parameters can be obtained as:

$$Y^r\_bar(t+1) = Y^r\_bar(t) - \alpha\frac{F(x)-D}{B}z^r \tag{44}$$

$$x_i^r\_bar(t+1) = x_i^r\_bar(t) - \alpha \frac{F(x)-D}{B}(y^r\_bar - F(x))z^r \frac{2(x_i^p - x_i^r\_bar(t))}{\sigma(t)_i^{r^2}} \quad (45)$$

$$\sigma_i^r(t+1) = \sigma_i^r(t) - \alpha \frac{F(x)-D}{B}(y^r\_bar - F(x))z^r \frac{2(x_i^p - x_i^r\_bar(t))^2}{\sigma_i^{r^3}(t)} \quad (46)$$

The influences of the learning rate and the number of rules on convergence rate were investigated. The number of rules used are 9, 16 and 25. The 16 rules fuzzy logic was discussed earlier and these rules are listed in Table 13. The cases of the 9 and 25 rules can be obtained in a similar manner. For each different number of rules, three hundreds data point were generated. Each data point has three numbers, which represent the three parameters. Backpropagation was carried out by using these generated data points. To measure the performance of the approach, the following three indices were used:

Training Error (TRE)

$$TRE = \sum_{i=1}^{150}\left(\frac{|Y_i - Dtr_i|}{Dtr_i}\right) \quad (47)$$

where Yi is the actual output of the training data pair i and Dtri is the desired output of the training data pair i.

Testing Error (TEE)

$$TEE = \sum_{i=1}^{150}\left(\frac{|Y_i - Dtr_i|}{Dtr_i}\right) \quad (48)$$

Total Error (TTE)

$$TTE = TRE + TEE \quad (49)$$

With five different learning rates, the results are summarized in Table 14, where L represents the learning rate and NR represents the number of rules. The best performance results are indicated by the use a "*" in the table. For the cases of 9 and 16 fuzzy rules, learning rate of 0.00005 give the best performance and for learning rate for 25 rules, the best is 0.00003. The convergence rates for the three best performance sets are shown in Figure 11. From Table 14, we can see that the more rules we have, the less the training errors, but the larger the testing error

Table 14. Parameter influence

	NR=9	NR=16	NR=25
L=0.00001	TRE:0.16580	TRE:0.10280	TRE:0.03514
	TEE:0.13964	TEE:0.10864	TEE:0.67560
	TTE:0.30544	TTE:0.21144	TTE:0.71074
L=0.00003	TRE:0.06100	TRE:0.14000	TRE:0.00565
	TEE:0.00129	TEE:0.17000	TEE:0.10750
	TTE:0.06229	TTE:0.31000	TTE:0.11315*
L=0.00005	TRE:0.00768	TRE:0.00583	TRE:0.11997
	TEE:0.02868	TEE:0.07390	TEE:0.11651
	TTE:0.03636*	TTE:0.07973*	TTE:0.23648
L=0.00007	TRE:0.00757	TRE:0.03580	TRE:0.05360
	TEE:0.03100	TEE:0.07760	TEE:0.11240
	TTE:0.03857	TTE:0.11340	TTE:0.16600
L=0.0001	TRE:0.02510	TRE:0.03780	TRE:0.07600
	TEE:0.10470	TEE:0.08166	TEE:0.11727
	TTE:0.12980	TTE:0.11946	TTE:0.19327

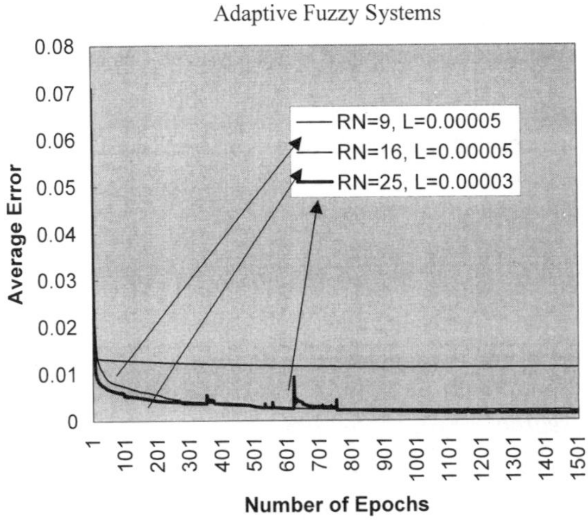

Figure 11. Convergent behavior

6.Conclusions

This chapter presents the application of operations research techniques to the cell formation problem. Emphasis is placed on the frontier area where recently developed operations research techniques are used. Mathematical programming was first applied to solve the cell formation problem. However, due to difficulties to solve this NP complete problem, heuristics the recently developed evolutionary approaches were proposed. From the practical standpoint, vague linguistics and human judgments are usually involved in cell formation problems. Fuzzy set theory possesses powerful ability for linguistic representation and thus the fuzzy approach was adopted to solve the cell formation problem. The recently developed fuzzy neural network has been shown to be a useful approach to solve practically encountered cell formation problems. To save space, all the approaches are illustrated by actual examples. The approaches presented in this chapter should provide reasonably good foundations for future research in the application of operations research techniques to solve cell formation problems.

References

[1] Abdelmola, A. I. and S. M. Taboun, 1999, "Productivity model for the cell formation problem: a simulated annealing algorithm," *Computers ind. Engng.*, 37, 327-330.

[2] Abdelmola, A. I., S. M. Taboun and S. Merchawi, 1998, "Productivity optimization of cellular manufacturing systems," *Computers ind. Engng.*, 35(3-4), 403-406.

[3] Amirahmadi, F. and F. Choobineh, 1996, "Identifying the composition of a cellular manufacturing system," *International Journal of Production Research*, 34(9), 2471-2488.

[4] Ang, D.S. and Charles E.Hegji, 1997, "An algorithm for part families identification in cellular manufacturing," *International Journal of Materials and Production Technology*, 12(4-6), 320-328.

[5] Akturk, M. S. and G. R. Wilson, 1998, "A hierarchical model for the cell loading problem of cellular manufacturing systems," *International Journal of Production Research*, 36(7), 2005-2023.

[6] Ben-Arieh, David and Evangelos Triantaphyllou , 1992, "Quantifying Data for Group Technology with Weighted Fuzzy Features," *International Journal of Production Research*, 30(6), 1285-1299.

[7] Ben-Arieh, D., S. E. Lee and P. T. Chang, 1996, "Theory and Methodology: Fuzzy part coding for group technology," *European Journal of Operational Research*, 92, 637-648.

[8] Berardi, V. L., G. Zhang and O. F. Offodile, 1999, "A mathematical programming approach to evaluating alternative machine clusters in cellular manufacturing," *International Journal of Production Economics*, 58, 253-264.

[9] Billo, R. E., 1998, "A design methodology for configuration of manufacturing cells," *Computers ind. Engng.*, 34(1), 63-75.

[10] Boctor, F.F., 1990, "A linear formation of the machine-part cell formation problem", *International Journal of Production Research*, 29, 343-356.

[11] Buckley, James and Yoichi Hayashi, 1994, "Fuzzy Neural Networks: A Survey", *Fuzzy Sets and Systems*, 66, 1-13

[12] Burke, L. I. And S. Kamal, 1995, "Neural networks and the part family/machine group formation problem in cellular manufacturing: A framework using fuzzy ART," *Journal of Manufacturing Systems*, 14(3), 148-159.

[13] Cao, Q. and Mark A. Mcknew,1998, "Partial termination rule of lagrangian relaxation for manufacturing cell formation problems," *Computers Ops. Res.*, 25(2), 159-168.

[14] Caux, C. , R. Bruniaux and H. Pierreval, 2000, "Cell formation with alternative process plans and machine capacity constraints: A new combined approach", *International Journal of Production Economic*, 64, 279-284.

[15] Carpenter, G.A. and S.Grossbergt,1987, "A massively parallel architecture for a sefl-organizing neural pattern recognition machine"., 37, 54-115

[16] Chakraborty, K. and U. Roy, 1993, "Connectionist models for part-family classifications," *Computers ind. Engng.*, 24(2), 189-198.

[17] Chan, F. T. S., K. L. Mak, L. H. S. Luong and X. G. Ming, 1998, "Machine-component grouping using genetic algorithms," *Robotics and Computer-Integrated Manufacturing*, 14, 339-346.

[18] Chang P.T. and E.S. Lee, 2000, "A multi-solution method for cell formation – exploring practical alternatives in group technology manufacturing," *Computers and mathematics with applications.*, 40, 1285-1296.

[19] Chen, S.J. and C.S. Cheng,1995, "A neural network-based cell formation algorithm in cellular manufacturing" *International Journal of Production Research,*33(2),293-318.

[20] Chester, Michael, 1993, "Neural networks–a tutorial", *Prentice-Hall Inc.*

[21] Chu, C. H., 1997, "An improved neural network for manufacturing cell formation," *Decision Support Systems*, 20, 279-295.

[22] Chu, C.H. and J.C. Hayya,1991,"A Fuzzy Clustering Approach to Manufacturing Cell Formation," *Internationan Journal of Production Research*,29(7),1475-1487

[23] Chung, Y. and A. Kusiak, 1994, "Grouping parts with a neural network," *Journal of Manufacturing Systems*, 13(4), 262-275.

[24] Crama, Y. and M. Oosten, 1996, "Models for machine-part grouping in cellular manufacturing," *International Journal of Production Research*, 34(6), 1693-1713.

[25] Dave , R. N.,1991 ,"Characterization and detection of noise in clustering", *Pattern Recognition Letters*, 12, 657-664.

[26] Deutsch, S.J., S.F. Freeman and M.Helander,1998, "Manufacturing cell formation using an improved p-median model" *Computers ind. Engng.*, 34(1), 135-146.

[27] Enke, D., K. Ratanapan and C. Dagli., 1998, "Machine-part family formation utilizing an ART1 neural network implemented on A parallel neuro-computer," *Computers ind. Engng.*, 34(1), 189-205.

[28] Fausett, Laurene, 1994, "Fundamentals of Neural Networks", Prentice Hall Inc.

[29] Garey, M.R. and D.S. Johnson,1979,"Computers and Intractability : A Guide to the Theory of NP-Completeness," Freeman, New York.

[30] Gindy, N.N.Z,. T.M. Ratchev and K.Case,1995, "Component grouping for GT applications-a fuzzy clustering approach with validity measue" *International Journal of Production Research*, 33(9), 2493-2509.

[31] Glover, F., 1986, "Future paths for integer programming and links to artificial intelligence", *Computers and Operation Research*, 13 , 533-549.

[32] Gultom Parapat, 1996, "Fuzzy Set Theory Applied To the Design of Cell Formation in Cellular Manufacturing Systems", Ph. D. Dissertation, Kansas State University, USA.

[33] Gungor, Z and Fryzen Arikan,2000, "Application of fuzzy decision making in part-machine grouping" *International Journal of Production Economics*, 63, 181-193.

[34] Gupta, M.M. and H. Ding, 1994,"Foundations of Fuzzy Neural Computations,"in F. Aminzadeh and M. Jamshidi, eds.," Soft Computing : Fuzzy Logic, Neural Networks, and Distributed Artificial Intelligence,"165-195.

[35] Gwiazda, A. and R. Knosala, 1997, "Group technology using neural nets," *Journal of Materials Processing Technology*, 64, 181-188.

[36] Hansen, P.,1986,"The steepest ascent mildest decent heuristic for combinatorial programming, *Conf. On Numerical Methods in Combinatorial optimization, Capri, Italy.*

[37] Harhalakis, G., R. Nagi and J. M. Proth, 1990, "An efficient heuristic in manufacturing cell formation for group technology applications," *International Journal of Production Research*, 28, 185-198.

[38] Heragu, S. S. and S. R. Kakuturi, 1997, "Grouping and placement of machine cells," *IIE Transactions*, 29, 561-571.

[39] Ho, Y. C. and C. L. Moodie, 1996, "Solving cell formation problems in a manufacturing environment with flexible processing and routing capabilities," *International Journal of Production Research*, 34(10), 2901-2923.

[40] Holland, J.H,1975, "Adaptation in Natural and Artificial Systems", *University of Michigan Press, Ann Arbor, MI.*

[41] Hwang, H. and J. U. Sun, 1996, "A genetic-algorithm-based heuristic for the GT cell formation problem," *Computers ind. Engng.*, 30(4), 941-955.

[42] Joines, J. A., C. T. Culbreth and R. E. King, 1996, "Manufacturing cell design: an integer programming model employing genetic algorithms," *IIE Transactions*, 28, 69-85.

[43] Kamal, S. and L. I. Burke, 1996, "FACT: A new neural network-based clustering algorithm for group technology," *International Journal of Production Research*, 34(4), 919-946.

[44] Kamrani, AK, Parsaei HR, and Liles DH, 1995, "Planning, design, and analysis of cellular manufacturing systems", *New York: Elsevier.*

[45] Kao, Y. and Y. B. Moon, 1991, "A unified group technology implementation using the backpropagation learning rule of neural networks " *Computers ind.Engng*, 20(4), 425-437.

[46] Kao, Y. and Y. B. Moon, 1998, "Feature-based memory association for group technology," *International Journal of Production Research*, 36(6), 1653-1677.

[47] Kaparthi, S. and Suresh, N.C,1991, "A neural network system for shape-based classification and coding of rotational parts," *International Journal of Production Research,*29(9),1771-1784

[48] Kaparthi, S. and Suresh, N.C,1992, "Machine-component cell formation in group technology :a neural network approach ," *International Journal of Production Research,*30(6),1353-1367

[49] Kaparthi, S. Nallan C.Suresh and Robert P.Cerveny,1993, "An improved neural network leader algorithm for part-machine grouping in group technology ," *Eroupean Journal of Operational Research,*69,342-356.

[50] Kiang, M.Y., Uday R. Kulkarni,and Kar Yan Tam,1995,"Self-organizing Map Network as Interactive Clustering Tool-An Application to Group Technology," *Decision Support Systems,*15,351-374.

[51] Kirkpatrick, S., Gelatt, C.D. Jr and Vecchi, M.P.,1983, "Optimization by simulated annealing," *Science*, 220(4598) ,671-680.

[52] Kosko, B., 1992, "Fuzzy systems as universal approximators", IEEE International conference on Fuzzy Systems, 1153-1162.

[53] Kulkarni, U. R. and M. Y. Kiang, 1995, "Dynamic grouping of parts in flexible manufacturing systems—A self-organizing neural networks approach," *European Journal of Operational Research*, 84, 192-212.

[54] Kuo, R. J. , S. C. Chi and P.W. Teng,2001"Generalized part family formation through fuzzy self-organizing feature map neural network", Computer &industrial Engineering,40,79-100.

[55] Lee, M. K., H. S. Luong and K. Abhary, 1997, "A genetic algorithm based cell design considering alternative routing," *Computer Integrated Manufacturing Systems*, 10(2), 93-107.

[56] Lee, C.C, 1990, "Fuzzy logic in control systems: Fuzzy logic control-Parts I and II", *IEEE Transactions on Systems, Man and Cybernetics* 20, 404-435.

[57] Lee, S. Y. and G. W. Fischer, 1999, "Grouping parts based on geometrical shapes and manufacturing attributes using a neural network," *Journal of Intelligent Manufacturing*, 10, 199-209.

[58] Leem, C. W. and J. J. Chen, 1996, "Fuzzy-set-based machine-cell formation in cellular manufacturing," *Journal of Intelligent Manufacturing*, 7, 355-364.

[59] Liao, T.W. and L.J. Chen,1993, "An evaluation of ART1 neural models for GT part family and machine cell forming," *Journal of Manufacturing Systems*, 12(4),282-290

[60] Liao, T.W.,2001, "Classification and coding approaches to part family formation under a fuzzy environment" *Fuzzy sets and Systems*, 122, 425-441.

[61] Mamdani, E.H. and S. Assilian, 1975, "An experiment in linguistic synthesis with a fuzzy logic controller", *International Journal of Man-machine Studies* 7, 1-13.

[62] Michie,D. and D.J. Spiegelhalter, and C.C Talor,1994,"Machine arning, Neural and statistical classification," Ellis Horwood Limited.

[63] Mital, A., S.Kromodihardjo and C.Channaveeraiah,1988, "Increasing the sensitivity of parts classification system" *Fuzzy Sets and Systems*, 28, 1-13.

[64] Mital, A. and L. Settineri, 1997, "Increasing the sensitivity of parts classification system" *International Journal of Production Research*, 35(4), 1077-1094.

[65] Moon, Y. B. and S. C. Chi, 1992, "Generalized part family formation using neural network techniques," *Journal of Manufacturing Systems*, 11(3), 149-159.

[66] Moon, C. and M. Gen, 1999, "A genetic algorithm-based approach for design of independent manufacturing cells," *Int. J. Production Economics*, 60-61, 421-426.

[67] Moon, Y.B. and U.Roy , 1992, "Learning Group-technology part families from solid models by parallel distributed processing," *International Journal of Advanced Manufacturing Technology*,7,109-118.

[68] Narayanaswamy, P., C. R. Bector and D. Rajamani, 1996, "Theory and Methodology: Fuzzy logic concepts applied to machine-component matrix formation in cellular manufacturing," *European Journal of Operational Research*, 93, 88-97.

[69] Needy, K. L., R. E. Billo and R. C. Warner, 1998, "A cost model for the evaluation of alternative cellular manufacturing configurations," *Computers ind. Engng.*, 34(1), 119-134.

[70] Onwubolu, G.C. and M. Mutingi,2001, "A genetic algorithm approach to cellular manufacturing systems", *Computers and industrial engineering*,39,125-144.

[71] Pham, D.T. and Karaboga, 2000, "Intelligent optimization techniques", *Springer-Verlag Inc.*

[72] Pai, Ping-Feng and E.S. Lee, 2001, "Adaptive fuzzy systems application in group technology", *Computers and Mathematics with Applications*, volume 42, Issue 10/11,Pages 1393-1400.

[73] Pai, Ping-Feng and E.S. Lee, 2001" Parts clustering by self organizing map neural network in a fuzzy environment", *Computers and Mathematics with Applications*, Volume 42, Issue 1/2, Pages 179-188.

[74] Pham, D.T. and D. Karaboga, 2000, "Intelligent optimization techniques", Springer-Verlag London Limited.

[75] Pilot, T. and R. Knosala, 1998, "The application of neural networks in group technology," *Journal of Materials Processing Technology*, 78, 150-155.

[76] Prasad, R. and V.N.Rajan,1994. "Group technology cell formation using the ART neural network paradigm" *Intelligent Engineering Systems*,4,1085-1089

[77] Rardin R.L., 1998 "Optimization in operations research", *Prentice Hall Inc.*

[78] Rajamani, D., N. Singh and Y. P. Aneja, 1996, "Design of cellular manufacturing systems," *International Journal of Production Research*, 34(7), 1917-1928.

[79] Rao, Harish and P. Gu, 1994, "Expert self-organizing neural network for the design of cellular manufacturing systems", Journal of Manufacturing Systems, 13(5) 346-358.

[80] Sarker, B. R. and C. V. Balan, 1996, "Cell formation with operation times of jobs for even distribution of workloads," *International Journal of Production Research*, 34(5), 1447-1468.

[81] Selim, H.M., R.G. Askin and A.J. Vakharia,1998, "Cell formation in group technology: review, evaluation and directions for feature research" *Computers ind. Engng*, 34(1), 3-20.

[82] Sen, S. and Rajesh N.D.,1999, "Application of noise clustering in group technology" *Annual Conference of the North American Fuzzy Information Processing Society - NAFIPS 1999*, p 366-370

[83] Snead, Charles S.,1989, "Group Technology Foundation for Competitive Manufacturing", *Van Nostrand Reihold.*

[84] Spiliopoulos, K. and S. Sofianpoulou,1998,"An optimal tree search method for the manufacturing system cell formation problem," *European Journal of Operational Research,105,537-551.*

[85] Spiliopoulos, K. and S. Sofianopoulou,1998, "An optimal tree search met" *Computers ind. Engng,* 34(1), 3-20.

[86] Su, C. T. and C. M. Hsu, 1996, "A two phased genetic algorithm for the cell formation problem," *International Journal of Industrial Engineering*, 3(2), 114-125.

[87] Su, Chwne-Tzeng, 1995, "A fuzzy approach for part family formation, "*Intermational IEEE/IAS Conference*, 289-292.

[88] Sun, D L.Lin and R.Batta,1995, "Cell formation using tabu search," *Computers ind. Engng.*, 28(3), 485-494.

[89] Sundaram, R. M. and K. Doshi, 1993, "Cellular manufacturing system design with alternative routing consideration," *Computers ind. Engng.*, 25(1-4), 477-480.

[90] Suresh, N. C. and S. Kaparthi, 1994, "Performance of fuzzy ART neural network for group technology cell formation," *International Journal of Production Research*, 32(7), 1693-1713.

[91] Suresh, N. C. J.Slomp and S. Kaparthi, 1995, "The capacitated cell formation problem:a new hierarchical methodology" *International Journal of Production Research*, 33(6), 1761-1784.

[92] Suresh, N. C., J. Slomp and S. Kaparthi, 1999, "Sequence-dependent clustering of parts and machines: a Fuzzy ART neural network approach," *International Journal of Production Research*, 37(12), 2793-2816.

[93] Szwarc.D,D. Rajamani and C.R. Bector,1997,"Cell Formation Considering Demand and Machine Capacity, "*International Journal of Production Research*,13,134-147.

[94] Takagi, T. and M. Sugeno, 1985, "Fuzzy identification of systems and its applications to modeling and controlling" *IEEE Trans. On Systems, Man, and Cybern,* SMC-15(1), 116-132.

[95] Taboun S.M. N.S. Merchawi and T. Ulger,1998, "An two-stage model for cost effective part family and machine cell formation," *Computers ind.Engng*,34(4),759-776.

[96] Tsai, C. C., C. H. Chu and A. T. Barta, 1994,"Fuzzy Linear Programming Approach to Manufacturing Cell Formation," *IEEE World Congress on Computational Intelligence*,2,1406-1411.

[97] Tsai, C. C., C. H. Chu and A. T. Barta, 1997, "Modelling and analysis of manufacturing cell formation problem with fuzzy mixed-integer programming," *IIE Transactions*, 29(7), 533-547.

[98] Venugopal, V. and T. T. Narendran, 1992, "A genetic algorithm approach to the machine-component grouping problem with multiple objectives," *Computers ind. Engng.*, 22(4), 469-480.

[99] Venugopal, V. and T. T. Narendran, 1992, "A neural network approach for designing cellular manufacturing systems," *Advances in Modelling and Analysis*, 32(2), 13-26.

[100] Wang, Li-Xin, 1994, "Adaptive fuzzy systems and control", *Prentice Hall Inc.* New Jersey.

[101] Wang, Li-Xin and J.M. Mendel, 1992, "Fuzzy systems are universal approxiamtors", *IEEE International Conference on Fuzzy systems*, 1163-1170.

[102] Wen H.J., C.H.Smith and E.D. Minor,1996, "Formation and dynamic routing of part families among fixable manufacturing cells" *International Journal of Production Research*, 34(8), 2229-2245.

[103] Wu, M.C. and S.R. Jen,1996,"A Neural Network Approach to The Classification of 3D Prismatic Parts," *Interational Journal of Advanced Manufacturing Technology*,(11),325-335.

[104] Xu.H. and H.P. Wang,1989,"Part Family Formation for GT Applications Based on Fuzzy Mathematics, "*International Journal of Production Research*,27(9),1637-1651.

[105] Zadeh, L.A., 1965, "Fuzzy sets", *Information and Control*, 338-353.

[106] Zhang, C. and H. Wang, 1992, "Concurrent formation of part families and machine cells based on the fuzzy set theory," *Journal of Manufacturing Systems*, 11(1), 61-67.

[107] Zhou, M. and R. G. Askin, 1998, "Formation of general GT cells: An operation-based approach," *Computers ind. Engng.*, 34(1), 147-157.

[108] Zimmerman, H-J, 1984, *Fuzzy Set Theory and its Applications*, Kluwer Publishing, Boston.

[109] Zolfaghari,S. and M. Liang,1998,"Machine cell/part family formation considering processing times and machine capacities: A simulated annealing approach " *Computers ind. Engng.*, 34(4), 813-823.

SCHEDULING PROBLEMS IN LARGE ROAD TRANSPORT CORPORATIONS: SOME EXPERIENCES IN STRUCTURING AND MODELING

SURESH ANKOLEKAR
Indian Institute of Management, Ahmedabad

V. L. MOTE
Arvind Mills, Ahmedabad

N. R. PATEL
Massachusetts Institute of Technology

JAHAR SAHA
Indian Institute of Management, Ahmedabad

Scheduling problems form the core of the operational planning problem in typically large State Road Transport Corporations in India. The problems include, scheduling of trips to satisfy the traffic demand, allocation of trips to depots, and scheduling of buses and crews to operate the trips while satisfying various operational constraints and efficiency considerations. The size and the structural complexity of these hard problems involve solution approaches that emphasize interplay between modelling, algorithms, and their efficient computer implementation, requiring blending of ideas from Transportation Science, Operations Research and Computer Science. The solutions need to be complete and closer to the real-life practice for effective implementation. This paper presents our experiences in addressing these issues and highlights the insights gained from our efforts to implement the solutions in real-life.

Keywords: Bus and Crew Scheduling, Fleet-Size Optimization, Heuristics.

Introduction

The paper describes some experiences in structuring and modeling the scheduling problems that arise in large Road Transport Corporations. We realized the importance of these problems in course of a consulting assignment with an Indian State Road Transport corporation on determining optimal size of its divisions and depots. The corporation was divided into divisions that were further sub-divided into depots. The size of a division or a depot is the number of buses attached to the division or the depot. It became soon clear to us that determining the optimal size of a division or a depot would involve resolving some scheduling problems, which were complex. We decided to probe these issues further in a research programme over a period of more than a decade. In this paper we are presenting our total experience in an integrated form for the first time, although some of the works were earlier published independently.

In India, the state-owned Road Transport Corporations (SRTCs) meet the major demand for movement of people between cities. The first step in this process is to define 'trips', each of which indicates that a bus should be provided for moving passengers at a given time from a place to reach another place at a specified time. Given a set of trips, the problem of concern is to devise a vehicle- schedule, which will use minimum number of vehicles to operate these trips. In many situations there is flexibility in choosing trip timings. There is also a flexibility to reassign a trip to an alternate depot if it results in reduction in number of vehicles required. Buses, during its operations, require daily routine maintenance at its parent depots. This makes scheduling problems more complex. In addition the complex rules governing the services of crews make scheduling in SRTCs more difficult. In our view, it is essential that we look at all these problems in an integrated way and develop an operational model so that SRTC achieves greater operational efficiency.

In this paper we present our experiences in structuring and modeling these problems that resulted in an operational planning model for SRTCs. We believe that the use of this integrated model will have significant impact on the performance of SRTCS.

We have structured our presentation in 9 sections. In Section 2 we elaborate the structure of transport scheduling in SRTCs in terms of five sub-problems. The Sections 3 through 7 elaborate on each of the sub-problems in terms of model formulation, analysis, algorithms and data structures, and implementation and computational experiences. The sub-problems are integrated to form an integrated model of transport scheduling problems in Section 8. Finally, the Section 9 contains our concluding remarks.

Structure of the Transport Scheduling Problem

The SRTCs provide the transport services as a set of trips operated by the depots with a set of vehicles and crew allocated to them. A trip is specified in terms of the origin, destination, departure and arrival times, and the depot responsible for operating the trip. The transport scheduling seeks to minimize the number of vehicles and crew required to operate a given set of trips while satisfying various constraints implied by the specifications of the trip and the specific operational considerations of vehicle and crew.

The specification of the trips implies certain space, time, and depot compatibility constraints on the successive trips to be operated by a vehicle/crew. The space constraint states that the origin of the next trip must match the destination of the previous trip to be operated by a vehicle/crew. Unlike the case of urban transport, it is not possible to operate empty 'dead heading' trips between geographically widely scattered terminals in the SRTC context. The time constraint states that the departure time of the next trip must be later than the arrival time of the previous trip to be operated by a vehicle/crew. The depot compatibility constraint requires that the successive trips operated by a vehicle/crew must belong to the same depot. For a given set of trips, while the space constraint is fully specified in terms of the origin and destination of every trip, there is some flexibility to finalize the departure timings within an interval around the provisional timings indicated by the traffic needs. It is desirable, however, to minimize such perturbations around the provisional timings.

The SRTCs operate in a decentralized manner, and are organized in terms of divisions with operationally autonomous depots. Accordingly, the vehicles and crew can operate trips belonging to only their own home depot. Transport scheduling is carried out at a division level. It is customary to make the provisional trip or route assignment to the depots on the basis of geographical considerations, though there is considerable overlap among the terminals reached as origins and destinations of the set of trips operated by the individual depots. This, however, may not be operationally efficient, in the sense that the sum of the minimum fleet-sizes required by the individual depots may exceed the minimum fleet-size required to operate the pooled set of trips of all the depots put together. It is possible to reassign some of the trips to other depots to ensure operational efficiency. It is, however, desirable to minimize the number of reassignments.

The vehicles used for operating the trips are required to undergo a routine maintenance of about 90 minutes at its home depot, preferably daily, or at least once in two days. It covers the cleaning and some routine check-ups to ensure road-worthiness of the vehicle. The maintenance may be carried out at the beginning or at the end of the vehicle schedule or accommodated within the idle time between the two successive trips operated by the vehicle. Accordingly, the routine maintenance constraint on a vehicle schedule requires that either the schedule begins or terminates at the home depot, or an adequate maintenance gap is provided during one of its idle periods at the depot, at least once in two days. The scheduling of routine maintenance is also constrained by the capacity of the maintenance bay, limited to about 3-6 vehicles per hour depending upon the category of the depot.

The crew duty schedules are constrained by the duty conditions as per the agreement with the Trade Union. The signing-in of a crew takes place at the departure terminal of the first trip and the signing-off at the arrival terminal of the last trip of the schedule. The crew schedule should not lead to two consecutive signing-offs, called night-outs, at a terminal other than the home depot of the crew. The total steering duty that includes travel times of the trips in the schedule, signing-in and signing-off times, and the terminal idle times is required to be less than a pre-specified limit. There is also a constraint on the 'spread over', the elapsed time between the signing-in and signing-off of a crew performing a crew schedule. In addition, the crew duty schedules should contain the suitable meal-breaks and rest-periods before and after the lengthy trips. It is desirable to maximize the overlap between vehicle and crew schedules so that the crew is not required to change the vehicles too frequently to operate the trips.

Thus, the transport scheduling is a complex combinatorial optimization problem with multiple objectives. To make it somewhat tractable, we have decomposed the problem into a set of interrelated sub-problems as follows:

- Fixed-Schedule Fleet-Size Problem: This assumes that the trip timings are fixed, considers only the space and time constraints, and ignores all other operational considerations to minimize the fleet-size.
- Variable-Schedule Fleet-Size Problem: It extends the fixed-schedule problem by allowing the trip timings to vary within a specified interval around the most desirable timings. It also seeks to minimize the number of such perturbations, in addition to the primary objective of minimizing the fleet-size
- Depot Allocation Problem: This is an extension of the single-depot fixed-schedule fleet-size problem. It seeks to minimize the total fleet-size collectively required by all the depots by reassigning the provisional depot allocations of some of the trips while satisfying the depot compatibility constraint and minimizing the number of reassignments.
- Vehicle Scheduling Problem: This extends the fixed-schedule fleet-size problem by considering the operational constraint on the routine maintenance of vehicles while minimizing the fleet-size and maximizing the number of schedules with maintenance.
- Crew Scheduling Problem: It extends the fixed-schedule fleet-size problem to minimize the crew-size required for operating the given set of trips within the operational constraints due to the crew duty conditions.

Fixed-Schedule Fleet-Size Problem (P1)

Several researchers have explored the Fixed-Schedule Fleet-Size Problem. Bartlett[4] developed an algorithm for computing the minimum fleet-size required by analyzing chronologically ordered sequence of arrivals and departures, called the A-D sequence, occurring at each terminal. Saha[13] treated the problem as minimum chain decomposition problem in an acyclic graph. Following the results on minimal decomposition by Raghavachari and Mote[12], he formulated the problem as a Linear Programming problem. To overcome the size complexity of large-scale real-life problems, he formulated the problem as a bipartite network flow problem. His computational experience, however, with the labeling algorithm of Ford and Fulkerson to solve the network flow problem was not encouraging. He developed an algorithm for speeding the solution, which essentially involves assigning arrivals to departures in a first-in-first-out (FIFO) basis in the A-D sequence. Gertsbach and Gurevich[8] developed a simple analytical framework based on 'deficit function' defined over the A-D sequence as cumulative excess of departures over arrivals at a terminal. Ankolekar[2] extended the framework further to a more convenient 'surplus function' defined over the A-D sequence as minimum fleet available at the terminal after every arrival and departure event.

Formulation

The *linkability*, defined for a pair of trips that can be consecutively operated by a vehicle, is the key concept in the formulation of the Fixed-Schedule Fleet-Size Problem (P1) given in Table 1. Specifically, the *linkability sets* L_i and B_j are defined as sets of potentially succeeding and preceding trips that satisfy the space and time constraints associated with arrival and departure of the trip respectively. Accordingly, they are intimately related to the A-D sequence at a terminal in a sense that L_i corresponds to the departures succeeding the arrival of trip i and B_j corresponds to the arrivals preceding the departure of trip j in the A-D sequence.

The linking constraints assert that an arriving trip i can either be linked to a departing trip j from set L_i (i.e. $X_{ij}=1$) or the trip i remain unlinked (i.e. $\lambda_{i0}=1$). Similarly, an arriving trip i from set B_j can either be linked to the departing trip j (i.e. $X_{ij}=1$) or let the trip j remain unlinked (i.e. $\lambda_{0j}=1$). The fleet-size or the number of vehicle schedules is given by the sum of unlinked departure trips ($\sum_j \lambda_{0j}$) at the beginning of a vehicle schedule, or the sum of unlinked arrival trips ($\sum_i \lambda_{i0}$) at the end of a schedule.

Analysis

The objective function maximizing the number of *linkings* ($\sum_i \sum_j X_{ij}$) can be shown to be equivalent to minimizing the fleet-size ($\sum_i \lambda_{i0}$ or $\sum_j \lambda_{0j}$) as follows:
Summing up the linking constraints over all the trips,

$$\sum_j \sum_i X_{ij} + \sum_j \lambda_{0j} = \sum_j 1 \quad j \in N,\ i \in B_j$$

$$= |N| = \sum_i \sum_j X_{ij} + \sum_i \lambda_{i0} \quad i \in N,\ j \in L_i$$

Thus,

Linkings + Fleetsize = Number of Trips

Table 1. Fixed-Schedule Fleet-Size Problem (P1)

Formulation	Notations
Maximize: $\sum_i \sum_j X_{ij}$ $i \in N,\ j \in L_i$ Subject to: Linking Constraints: $\sum_j X_{ij}+\lambda_{i0} = 1$ $i \in N,\ j \in L_i$ $\sum_i X_{ij}+\lambda_{0j} = 1$ $j \in N,\ i \in B_j$ $X_{ij},\ \lambda_{i0},\ \lambda_{0j} = 0,\ 1$	N: Set of trips T: Set of terminals $o_i \in T$: Origin of trip i $d_i \in T$: Destination of trip i p_i: Departure time of trip i q_i: Arrival time of trip i L_i: Set of departure trips linkable to the arrival trip i $= \{j: o_j = d_i \text{ AND } p_j \geq q_i,\ j \in N\}$ B_j: Set of arrival trips to which the departure trip j is linkable to $= \{i: o_j = d_i \text{ AND } p_j \geq q_i,\ i \in N\}$ X_{ij} = 1 if the arrival trip i is linked to departure trip j = 0 otherwise λ_{i0} = 1 if the arrival trip i is not linked to any departure trip = 0 otherwise λ_{0j} = 1 if no arrival trip is linked to the departure trip j = 0 otherwise

Since the *linkability sets* L_i and B_j primarily correspond to the terminal-specific A-D sequences, we can express the above result at the level of a terminal. So,

$$\sum_j\sum_i X_{ij} + \sum_j \lambda_{0j} = \sum_j 1 \qquad j \in N_{b1}=\{k: o_k=b, k \in N\},\ i \in B_j$$

$$= |N_{b1}| \qquad N_{b1}\text{: Trips departing from terminal b}$$

Similarly,

$$\sum_i\sum_j X_{ij} + \sum_i \lambda_{i0} = \sum_i 1 \qquad i \in N_{b2}=\{k: d_k=b, k \in N\},\ j \in L_i$$

$$= |N_{b2}| \qquad N_{b2}\text{: Trips arriving at terminal b}$$

Obviously, $|N_{b1}| = |N_{b2}|$ due to the law of conservation and empty 'dead heading' trips are not allowed between any terminals.

To minimize the fleet-size, it is sufficient to maximize the *linkings* or minimize the unlinked trips departing or arriving within the A-D sequence at every terminal. If we perform linking of every departing trip in the A-D sequence of terminal from among preceding arriving trips, then the departures with cumulative count exceeding that of arrivals must remain *unlinked*. Accordingly, let

A_{bt} = Cumulative number of arrivals up to time t
$= |\{i: d_i=b, q_i \le t, i \in N\}|$

D_{bt} = Cumulative number of departures up to time t
$= |\{i: O_i=b, P_i \le t, i \in N\}|$

$$F_{bt} = D_{bt} - A_{bt}$$

$$F_b = \max_t \{D_{bt} - A_{bt}\}$$

The F_{bt}, known as the 'deficit function', indicates the shortfall of vehicles of at terminal b at time t. The F_b indicates the number of departures that must remain unlinked at the early part of the A-D sequence at the terminal b. A balanced set of trips, $|N_{b1}| = |N_{b2}|$, will ensure that the terminal b will also end up with exactly F_b unlinked arrivals at the end part of the A-D sequence. The time intervals between the peak 'deficit function' values, $F_{bt}=F_b$, partition the A-D sequence in terms of 'hollow zones' defined by Gertsbach and Gurevich[8]. The linking restricted within the 'hollow zones' would be optimal, leaving not more than F_b unlinked departures and arrivals at the first and last 'hollow zones' respectively at each of the terminals. The total minimum fleet-size is given by

$$F = \sum_b F_b \quad b \in T$$

The concept of 'deficit function' could be extended to a more convenient 'surplus function' defined over the A-D sequence as,

$$S_{bt} = F_b + A_{bt} - D_{bt}$$

S_{bt} indicates a minimum surplus of fleet-size required at any time t at a terminal b to take care of all the trips departing at or after time t from the terminal. As the 'surplus function' changes only on occurrence of arrival/departure event, it is sufficient to define it as a discrete function over the events rather than as a step-wise continuous function over time. Accordingly,

s_{j1} = Value of the 'surplus function' upon departure of trip j
$= F_b + A_{bt} - D_{bt} \quad j \in N_{b1},\ t=p_j$

s_{i2} = Value of the 'surplus function' upon arrival of trip i
$= F_b + A_{bt} - D_{bt} \quad i \in N_{b2},\ t=q_i$

The 'critical departure' trips, $N_{b*}=\{k: s_{k1}=0, k \in N_{b1}\}$, conveniently partition the A-D sequence in terms of 'hollow zones' for every terminal. Unlike the terminal-specific peak 'deficit function' value, F_b, the 'surplus function' has the same value of 0 for every 'critical departure' at any terminal. By definition, on linking of an arrival trip to a departure trip succeeding it in the A-D sequence, the 'surplus function' value of each of the intermediate arrival/departure events would have to reduced by 1. This indicates that the arrival trip could be optimally linked to a departure trip only within the 'hollow zone'. Else, the 'surplus function' value of 'critical departure' would become negative, requiring additional fleet to restore the value to 0.

The 'surplus function' value of an arrival trip, being the minimum surplus of fleet-size required at the time, also indicates the optimal number of linking choices for the arrival trip in terms of available succeeding departure trips after optimally linking all the succeeding arrival trips. Accordingly, the total number of optimal solutions to the Fixed-Schedule Fleet-Size Problem is given by,

$$S = (\prod_{i \in N} s_{i2})/(\prod_{b \in T} F_b!)$$

The denominator in the above expression accounts for the arrival trips that remain unlinked within the last 'hollow zone'. The number of optimal solutions, S, is usually a very large number even for a moderately sized problem. One of our real-life problems with 220 trips, had $6.74\text{x}10^{89}$ **optimal** solutions to the Fixed-Schedule Fleet-Size Problem!

Table 2. Some Algorithms and Data Structures to Solve P1

Let, $LIST_b(AD)$: A-D sequence at terminal b $LIST_b(\lambda_{0j})$: List of λ_{0j} as defined in P1 $LIST_b(\lambda_{i0})$: List of λ_{i0} as defined in P1 $QUEUE_b(A)$: Queue of *Arrivals* for linking $STACK_b(A)$: Stack of *Arrivals* for linking $LIST_b(A)$: List of *Arrivals* for linking $LIST_b(X_{ij})$: List of X_{ij} as defined in P1 ***FIFO Algorithm*:** Initialize: $LIST_b(AD)$, $LIST_b(\lambda_{0j})$ $LIST_b(\lambda_{i0})$, $QUEUE_b(A)$ For every terminal b ∈ T Scan $LIST_b(AD)$ until end if A then insert A in $QUEUE_b(A)$ else if $QUEUE_b(A)$ is not empty then **remove A, link to D, insert *linking* in $LIST_b(X_{ij})$** else insert D in $LIST_b(\lambda_{0j})$ *At the end of scan, empty the unlinked As in* $QUEUE_b(A)$ *into* $LIST_b(\lambda_{i0})$	***LIFO Algorithm*:** Initialize: $LIST_b(AD)$, $LIST_b(\lambda_{0j})$ $LIST_b(\lambda_{i0})$, $STACK_b(A)$ For every terminal b ∈ T Scan $LIST_b(AD)$ until end if A then push A in $STACK_b(A)$ else if $STACK_b(A)$ is not empty then **pop A, link to D, insert *linking* in $LIST_b(X_{ij})$** else insert D in $LIST_b(\lambda_{0j})$ *At the end of scan, empty the unlinked As in* $STACK_b(A)$ *into* $LIST_b(\lambda_{i0})$ ***General Algorithm*:** Initialize: $LIST_b(AD)$, $LIST_b(\lambda_{0j})$ $LIST_b(\lambda_{i0})$, $LIST_b(A)$ For every terminal b ∈ T Scan $LIST_b(AD)$ until end if A then insert A in $LIST_b(A)$ else if $LIST_b(A)$ is not empty then **pick *any* A, link to D, insert *linking* in $LIST_b(X_{ij})$** else insert D in $LIST_b(\lambda_{0j})$ *At the end of scan, empty the unlinked As in* $LIST_b(A)$ *into* $LIST_b(\lambda_{i0})$

The specific algorithms and associated data structures to optimally solve the Fixed-Schedule Fleet-Size Problem are presented in Table 2. We use the standard 'ordered list' data structure to accommodate the A-D sequence, *linkings* (X_{ij}s), and *unlinked* trips (λ_{0j}s and λ_{i0}s). To accommodate the candidate arriving trips for linking, we logically use queue, stack, and 'ordered list' data structures, to perform the linking using first-in-first-out (FIFO), last-in-first-out (LIFO), and 'general' linking algorithms respectively. Physically, all the data structures are implemented as a single 'ordered list'.

It is to be noted that the algorithms do not explicitly require us to use surplus functions or identify 'hollow zones'. As we chronologically select departure trips for linking, the candidate arrival trips simply get accumulated in the selected data structure. If we arbitrarily select the departure trips for linking, however, we must identify the 'hollow zones', to be able to restrict the optimal linking within.

The algorithms are identical except for the data structures. Of particular interest is the General Algorithm, specifically the key statement 'pick *any* A, link to D, insert linking in $LIST_b(X_{ij})$'. The statement holds the promise that if even the arbitrary choice of arrival trip would do for the fleet-size optimality, then we could potentially make a systematic choice for linking to satisfy any operational constraints without adversely affecting the fleet-size optimality. Consequently, the solution to the formidable vehicle and Crew Scheduling Problems would be a matter of elaborating the key statement in the specific contexts.

Implementation and Computational Experience

Our initial implementation of the General Algorithm to solve the Fixed-Schedule Fleet-Size Problem was on a PDP-11/70 minicomputer using FORTRAN-IV Plus language. The modest computing resource, with only 64KB of addressable memory available for the program of which only 32KB could be used for the data, forced us to develop efficient implementation of the algorithm and the data structures to be able to handle real-life size problems of over 1000 trips. We used a set of two-column arrays to accommodate trip data, surplus scores, and two-way linked 'ordered lists' for A-D sequence and the linking solution.

Table 3. Computational Experience of Fixed-Schedule Fleet-Size Problem

Trip	Terminal	Optimal Fleet-Size Solutions	Optimal Fleet-Size	M-Gaps	Schedules with maint.	Maint. feasible (YES/NO)	CPU (sec.)
105	23	$1.20\text{x}10^{20}$	21	19	14	NO	0.64
220	72	$6.74\text{x}10^{89}$	50	32	26	NO	2.10
246	43	$1.83\text{x}10^{58}$	31	22	18	NO	1.74
120	30	$7.25\text{x}10^{32}$	35	32	25	NO	0.80
92	35	$8.50\text{x}10^{23}$	22	7	6	NO	0.58
72	25	$1.30\text{x}10^{15}$	18	12	11	NO	0.40
66	27	$2.01\text{x}10^{14}$	23	19	16	NO	0.44
84	28	$5.02\text{x}10^{12}$	20	16	14	NO	0.48

The computational experience of our implementation with real-life data is given in Table 3. Interestingly, all of the solutions generated by the General Algorithm to optimally solve the Fixed-Schedule Fleet-Size Problem were maintenance-infeasible. In fact, our early attempt to solve the Vehicle Scheduling Problem by generating numerous solutions to the Fixed-Schedule Problem failed to yield any maintenance-feasible optimal fleet-size solution even after considering over 50000 solutions.

Variable-Schedule Fleet-Size Problem (P2)

The Variable-Schedule Fleet-Size Problem has been tackled using mathematical programming and heuristic approaches in the literature. Martin-Lof[10] has described a branch-and-bound approach to solve the problem for two terminals. Levin[9] has used branch-and-bound method of Land-and-Doig variety to solve his integer programming formulation. Bokinge and Hasselstrom[7] have given a heuristic approach that seeks to minimize active number of vehicles on the road at any given moment. We found that the Bokinge and Hasselstrom algorithm consistently performed even worse than the Fixed-Schedule Fleet-Size Problem, which is not surprising since the active number of vehicles forms only a lower bound on the fleet-size, whereas the fleet-size is essentially determined by the *linkability* among the trips. Ankolekar, Patel, and Saha[1] have used a heuristic method to identify and perturb trips to approach the lower bound on the variable-schedule fleet-size.

Formulation

In the formulation of the Variable-Schedule Fleet-Size Problem (P2) in Table 4, we extend the concept of *linkability* to include the trips that are 'potentially linkable'. The corresponding extended linkability sets, L_{iEL} and B_{jEL}, are defined over a modified A-D sequence where every trip is considered to be departing at its latest and arriving at its earliest timing. Linking of some of those trips eventually might turn out to be infeasible if the arrival timing is indeed later than the departure timing of the trip to which it is being considered for linking. The *linkability* set, L_{iLE}, identifies such potentially unlinkable trips among L_{iEL}.

The linking constraints on the extended *linkability sets*, L_{iEL} and B_{jEL}, would generally enable enhanced number of linkings in P2 compared to its Fixed-Schedule counterpart (P1), subject to the additional perturbation constraints. One set of perturbation constraints do not allow linking of a trip i to a trip j if the actual arrival time of trip i (p_i+t_i) is later than the actual departure time of trip j (p_j). It is sufficient to define such constraints over the set of potentially unlinkable trips (L_{iLE}). The second set of perturbation constraints allow the timings to be fixed within earliest and latest limits for each of the trips.

In addition to the primary objective function ($\sum_i\sum_j X_{ij}$) related to the fleet-size, the P2 also has a secondary objective function of maximizing the number of trips ($\sum_i Y_i$) with the most desirable timings. A set of constraints defined for each trip, called 'no perturbation is good', count such good trips.

Analysis

The mathematical programming approach to solve the Variable-Schedule Fleet-Size Problem is computationally not attractive due to the multiple objectives and large number of constraints involving integer variables. We extend the analytical framework developed earlier for the fixed-schedule counterpart of the P2 to solve the problem heuristically. Accordingly, we define following related parameters with respect to earliest, latest, and best timings of the trips.

$$A_{btE} = \left| \{i{:}d_i=b, q_{iE}\leq t, i\in N\} \right|$$

$D_{btL} = | \{i:O_i=b,P_{iL}\leq t,i\in N\} |$

$F_{btL} = D_{btL} - A_{btE}$

$F_{bL} = \max\{D_{btL}-A_{btE}\}$

$F_L = \sum_{b\in T} F_{bL}$

$A_{btB} = | \{i:d_i=b,q_{iB}\leq t,i\in N\} |$

$D_{btB} = | \{i:O_i=b,P_{iB}\leq t,i\in N\} |$

$s_{j1L} = F_{bL} + A_{btB} - D_{btB} \quad j\in N_{b1},\ t=p_{jB}$

$s_{i2L} = F_{bL} + A_{btB} - D_{btB} \quad i\in N_{b2},\ t=q_{iB}$

Table 4. Variable-Schedule Fleet-Size Problem (P2)

Formulation	Notations
Maximize:	$N, X_{ij}, \lambda_{i0}, \lambda_{0j}, o_i, d_i, p_i, q_i$ as in P1
$\sum_i \sum_j X_{ij}$ $\quad i\in N,\ j\in L_{iEL}$	
$\sum_i y_i$ $\quad i\in N$	p_{iE}, p_{iB}, p_{iL}: Earliest, Best, and Latest departure times for trip i
Subject to:	
Linking Constraints:	q_{iE}, q_{iB}, q_{iL}: Earliest, Best, and Latest arrival times for trip i
$\sum_j X_{ij}+\lambda_{i0} = 1 \quad i\in N, j\in L_{iEL}$	
$\sum_i X_{ij}+\lambda_{0j} = 1 \quad j\in N, i\in B_{jEL}$	t_i : Duration of trip i
Perturbation Constraints:	$=q_{iB}-p_{iB} = q_{iE}-p_{iE} = q_{iL}-p_{iL}$
$p_j-p_i-t_i\geq(X_{ij}-1)M \quad i\in N, j\in L_{iLE}$	$L_{iEL}=\{j:o_j=d_i,\ p_{jL}\geq q_{iE},\ j\in N\}$
$p_{iE} \leq p_i \leq p_{iL} \quad i\in N$	$B_{jEL}=\{i:o_j=d_i,\ p_{jL}\geq q_{iE},\ i\in N\}$
No Perturbation is Good:	$L_{iLE}=\{j:o_j=d_i,\ p_{jE}\leq q_{iL},\ j\in L_{iEL}\}$
$p_i - p_{iB} \leq(1 - Y_i)M \quad i\in N$	$Y_i=1$ if $p_i = p_{iB}$ $i\in N$
$p_{iB} - p_i \leq(1 - Y_i)M \quad i\in N$	$=0$ otherwise
$X_{ij},\ Yi, \lambda i0,\ \lambda_{0j} = 0,\ 1$	M : A large constant
$p_i, q_i \geq 0$	$p_{iE}, p_{iB}, p_{iL}, q_{iE}, q_{iB}, q_{iL}, t_i$: Trip related constants

Since the modified 'surplus function' is defined over the A-D sequence using lower bound on the peak 'deficit function' value, the critical departures and the trips around them would have infeasible (-ve) values for the terminals with the potential for fleet saving. The Perturbation Algorithm attempts to eliminate the infeasibility by advancing the arrivals and postponing the departures to cross over the infeasible part of the A-D sequence without creating additional infeasibility in the A-D sequences at the complementary ends of the trips being perturbed.

Implementation and Computational Experience

The implementation builds on the data structures and analytical modules used for the Fixed-Schedule Fleet-Size Problem. The Perturbation Algorithm attempts to extract maximum possible perturbation of a trip. If the perturbation is obstructed by a critical departure at the complementary end, the obstructing departures are recursively perturbed until no further obstruction is encountered, and unobstructed perturbation is carried out during backtracking phase on the obstructed trips. We used a 'stack' data structure for handling the recursive perturbation. We simply stack the obstructed trips facing temporary suspension of the perturbation, and reactivate the perturbation at the top of the stack during backtracking, removing the trip from the stack on perturbation.

The computational experience of our implementation with real-life data for a perturbation tolerance limit of ± 10 minutes is given in Table 5.

Table 5. Computational Experience of Variable-Schedule Fleet-Size Problem

Trip	Terminal	Fixed-Schedule Fleet-Size	Lower Bound (±10min.)	Fleet-Size Achieved	# Perturb.	CPU (Sec.)
105	23	21	20	20	1	1.30
220	72	50	47	47	8	4.48
66	27	23	22	22	1	0.64
120	30	35	34	35	0	1.80
246	43	31	26	26	10	3.80
92	35	22	22	22	0	1.14
84	28	20	20	20	0	0.90
72	25	18	18	18	0	0.60

Depot Allocation Problem (P3)

The Depot Allocation Problem has not been widely addressed in the literature. Ankolekar and Patel[3] have given a heuristic approach to identify and reassign the provisional depot allocations using a tree-search algorithm.

Formulation

In the Depot Allocation Problem (P3), the fleet-size problem is aggregated over multiple depots and then subjected to the depot compatibility and assignment constraints as shown in Table 6.

The depot compatibility constraints assert that the trip `i` can be *linked to* trip `j` if and only if both of them are assigned to the same depot `k` ($Y_{ik}=Y_{jk}=1$). The depot assignment constraints assert that a trip is assigned to only one of the depots. The constraints also ensure that for every terminal, the arrivals of trips assigned to a depot are equal to the departures of trips assigned to the same depot.

Table 6. Depot Allocation Problem (P3)

Formulation	Notations
Maximize: $\sum_i \sum_j X_{ij}$ $i \in N,\ j \in L_i$ $\sum_i Y_{ik}$ $i \in N_k$ Subject to: Linking Constraints: $\sum_j X_{ij}+\lambda_{i0} = 1\ i \in N, j \in L_i$ $\sum_i X_{ij}+\lambda_{0j} = 1\ j \in N, i \in B_j$ Depot Compatibility Constraints: $y_{ik}-y_{jk}+X_{ij} \leq 1\ i \in N, j \in L_i$ $y_{jk}-y_{ik}+X_{ij} \leq 1\ \ k \in K$ Depot Assignment Constraints: $\sum_k Y_{ik} = 1\ i \in N, k \in K$ $\sum_i Y_{ik} = \sum_j Y_{jk}\ i,j \in N, k \in K,$ $o_i = d_j = a \in T_a$ $X_{ij},\ \lambda_{i0},\ \lambda_{0j}, Y_{ik} = 0,\ 1$	$N,T,X_{ij},\lambda_{i0},\lambda_{0j},o_i,d_i,p_i,q_i,L_i,B_j$ as defined earlier in P1 a_i : depot to which trip i is assigned originally K : set of depots N_k : $\{i: a_i=k, i \in N\}$ Y_{ik} = 1 if trip i is assigned to $k \in K$ = 0 otherwise $T_a = \{b: F_b>0, b \in T\}$

Analysis

The Depot Allocation Problem has the fleet-size related primary objective function ($\sum_i \sum_j X_{ij}$) and a secondary objective that maximizes ($\sum_i Y_{ik}$) for the most desirable depot assignments. As the aggregation mutually enhances the potential *linkability sets* of each of the depots, the primary objective function for the aggregated set of trips ought to be more than the sum of linkings with the set of trips of individual depots. To prove this intuitive argument we extend the analytical framework developed earlier for the P1 to the multiple depots. Accordingly,

$$A_{bkt} = \left| \{i: d_i=b, q_i \leq t, i \in N, Y_{ik}=1\} \right|$$

$$D_{bkt} = \left| \{i: O_i=b, P_i \leq t, i \in N, Y_{ik}=1\} \right|$$

$$F_{bkt} = D_{bkt} - A_{bkt}$$

$$F_{bk} = \max_t \{D_{bkt}-A_{bkt}\}$$

For the aggregated set of trips, we have

$$F_b = \max_t \{\textstyle\sum_k (D_{bkt}-A_{bkt})\} \text{ and } F = \textstyle\sum_b F_b \quad b \in T,\ k \in K$$

It follows that $\sum_b \sum_k F_{bk} \geq F$ since,

$$\textstyle\sum_k F_{bk} = \sum_k \{\max_t (D_{bkt}-A_{bkt})\} \geq \max_t \{\sum_k (D_{bkt}-A_{bkt})\} = F_b$$

The mathematical programming approach to solve the Depot Allocation Problem too is computationally unattractive due to the multiple objectives and large number of constraints involving integer variables. As before, we extend the analytical framework developed earlier for the fixed-schedule counterpart of the P3 to solve the problem heuristically. The solution involves construction of A-D sequences and identification of the 'global hollow zones' for the aggregated set of trips. To achieve the optimal fleet-size for the aggregated set of trips while satisfying the depot-compatibility constraint, we must ensure that the 'local hollow zones' corresponding to the subset of trips of the specific depots are fully contained within the 'global hollow zone' so that the global optimality conditions are not violated. The 'local hollow zones' that span across the 'global hollow zones' lead to 'local unbalance' in terms of excess/deficit of arrivals and departures within the affected 'hollow zones'. The Depot Reassignment Algorithm uses a tree search technique to identify chains of such unbalanced trips and reassigns them to appropriate depot as described in Ankolekar and Patel[3].

Implementation and Computational Experience

The implementation builds on the data structures and analytical modules used for the Fixed-Schedule Fleet-Size Problem, and adds additional features to implement concepts specific to the Depot Allocation Problem, such as, 'local hollow zones', 'local unbalance', search-tree, and so on. The search-tree forest of potentially re-assignable chains was implemented as a 'queue' data structure, where the initial roots and subsequently branched leaves are placed at the tail of the queue, and the node for branching is picked up from the head of the queue. To eliminate potential cycling of the algorithm, repeat reassignment of a trip to a depot to which it was earlier assigned, was prohibited.The computational experience of our implementation with real-life data is given in Table 7.

Table 7. Computational Experience of Depot Allocation Problem

Depots	Trips	Initial Fleet-Size	Lower Bound (F)	Fleet-Size Achieved	Trips Re-assigned	CPU (sec.)
8	1005	220	217	217	10	29
8	1026	213	212	212	6	27
6	1226	273	265	265	77	88
5	1162	257	250	250	62	62
5	881	224	220	220	46	37
5	867	200	195	195	40	29
5	1044	226	219	219	56	46
5	1072	239	233	233	47	51
4	980	210	205	205	41	48
4	890	192	189	189	24	32
4	922	172	168	168	17	29
4	1008	223	217	217	34	35
4	950	185	181	181	15	28
4	1040	203	198	198	35	33
3	826	176	173	173	31	31
3	768	138	135	135	15	29
2	704	122	120	120	16	24
2	499	83	81	81	9	16

Vehicle Scheduling Problem (P4)

The Vehicle Scheduling Problem with routine maintenance constraint has not been widely addressed in the literature. Ankolekar, Patel, and Saha[1] have used a heuristic approach to solve large-scale real-life problems.

Formulation

Conceptually, the Vehicle Scheduling Problem (P4) is an extension of the Fixed-Schedule Fleet-Size Problem (P1) with the maintenance constraints on the vehicle schedules. This seemingly simple operational requirement adds numerous quadratic constraints to the P4 as formulated in Table 8.

Table 8. Vehicle Scheduling Problem (P4)

Formulation	Notations
Maximize: $\sum_i \sum_j X_{ij}$ $i \in N, j \in L_i$ $\sum_i V_i$ $i \in N$ Subject to: Linking Constraints: $\sum_j X_{ij}+\lambda_{i0} = 1$ $i \in N, j \in L_i$ $\sum_i X_{ij}+\lambda_{0j} = 1$ $j \in N, i \in B_j$ Identify (Vehicle Schedule) Chains: $R_{ii} = \lambda_{0i}$ $i \in N$ $R_{ij} = \sum_k R_{ik}.X_{kj}$ $i,j \in N, k \in B_j$ $U_{jj} = \lambda_{j0}$ $j \in N$ $U_{ij} = \sum_k X_{ik}.U_{kj}$ $i,j \in N, k \in L_i$ Identify Chains with Maintenance: $V_i = \lambda_{0i}$ $i \in N_{d1}$ $\sum_p \sum_q R_{iq}.X_{qp} + \sum_r R_{ir}.\lambda_{r0} \geq V_i$ $i \in N_{a1}$, $p \in N_{d1}$, $q \in B_{pM}, r \in N_{d2}$ $W_j = \lambda_{j0}$ $j \in N_{d2}$ $\sum_p \sum_q X_{qp}.U_{pj} + \sum_r \lambda_{0r}.U_{rj} \geq W_j$ $j \in N_{a2}$, $p \in N_{d1}$, $q \in B_{pM}, r \in N_{d1}$ Maintenance Constraint: $\sum_i V_i + \sum_j W_j \geq F_a$ $i \in N_{a1}, j \in N_{a2}, a \in T-d$ $X_{ij}, \lambda_{i0}, \lambda_{0j}, R_{ij}, U_{ij}, V_i, W_j = 0,1$	$N, T, X_{ij}, \lambda_{i0}, \lambda_{0j}, o_i, d_i, p_i, q_i, L_i$, B_j, F_b, H_{brs} as defined earlier in P3 $d \in T$: depot for maintenance g : maintenance gap $N_{a1} = \{j : o_j = a \in T-d, \ j \in N\}$ $N_{a2} = \{i : d_i = a \in T-d, \ i \in N\}$ $N_{d1} = \{j : o_j = d, \ j \in N\}$ $N_{d2} = \{i : d_i = d, \ i \in N\}$ R_{ij} = 1 if the trip j is contained in the chain *beginning* with the trip i = 0 otherwise U_{ij} = 1 if the trip i is contained in the chain *ending* with the trip j = 0 otherwise B_{jM} : Linking set with maintenance $= \{i : o_j = d_i = d \text{ AND } p_j - q_i \geq g, \ i \in N\}$ V_i = 1 if the chain *beginning* with the trip i has maintenance gap(s) = 0 otherwise W_j = 1 if the chain *ending* with the trip j has maintenance gap(s) = 0 otherwise

Unlike the P1, the P4 requires explicit identification of vehicle schedules to be able to express the maintenance constraints on them. A schedule can be identified by its starting trip i ($\lambda_{0i}=1$) and a set of non-starting trips associated with it ($\{j : R_{ij}=1\}$) or by its ending trip j ($\lambda_{j0}=1$) and a set of non-ending trips associated with it ($\{i : U_{ij}=1\}$). A schedule with maintenance can be identified as the one that starts at the depot, or the one starting at a terminal other than depot gets large enough *linking gap* at the depot (V_i). Similarly, a schedule with maintenance can also be identified as the one that ends at the depot, or the one ending at a terminal other than depot gets large enough *linking gap* at the depot (W_j). To ensure that a vehicle gets its maintenance at least once in two days, the number of schedules with maintenance ending at a terminal other than depot ($\sum_j W_j$), should be greater than or equal to the number of schedules starting from that terminal going without maintenance ($F_a - \sum_i V_i$).

The Vehicle Scheduling Problem has a fleet-size related primary objective function ($\sum_i \sum_j X_{ij}$) and a secondary objective that maximizes ($\sum_i V_i$) for the schedules with daily maintenance.

Analysis

Like its variable-schedule (P2) and the depot allocation (P3) counterparts, the Vehicle Scheduling Problem has multiple objectives. It also has a large number of quadratic constraints involving numerous integer variables. Consequently, it is very hard to solve the problem using the mathematical programming techniques. As before, we extend the analytical framework developed for the Fixed-Schedule Fleet-Size Problem to solve the Vehicle Scheduling Problem using heuristic techniques. The problem is solved in two phases. In phase one, a minimum fleet-size problem is solved with maximum number of maintenance gaps embedded among the vehicle schedules. The maintenance gaps are then appropriately redistributed among the chains in phase two, using a restructuring process.

The problem of embedding maximum number of maintenance gaps among the vehicle schedules is formulated as an assignment problem for linking arrivals with departures within 'hollow zones' as shown in Table 9.

The maintenance-gaps sub-problem is optimally solved using a Greedy Algorithm in which a departure selected in a chronological order simply grabs a maintenance opportunity within the 'hollow zone', if available. Else, it benevolently settles for linking with least idle gap, thereby increasing the chances of succeeding departures of grabbing the maintenance opportunity. The optimality also holds for the greedy procedure based on arrivals selected in reverse chronological order. Ankolekar, Patel, and Saha[1] have proved the optimality of the greedy procedure.

Table 9. The Maintenance-Gaps Sub-problem

Formulation	Notations
Minimize:	$X_{ij}, B_j, B_{jM}, d, H_{brs}, N_{d1}, N_{d2}$
$\sum_i \sum_j f_{ij}.X_{ij} \quad i \in H_{dmn2}, j \in H_{dmn1}$	as defined in Vehicle Scheduling Problem (P4)
Subject to:	$f_{ij} = 1$ if $i \in B_j - B_{jM}$
Precedence Constraints:	$= 0$ if $i \in B_{jM}$
$\sum_i X_{ij} = 1 \quad i \in H_{dmn2}, j \in H_{dmn1}$	$= \infty$ otherwise
$\sum_j X_{ij} = 1 \quad i \in H_{dmn2}, j \in H_{dmn1}$	H_{dmn} = Hollow zone at depot starting with arrival
	trip m and ending with departure trip n
$X_{ij} = 0, 1$	$= \{i, j : i \in N_{d2}, j \in N_{d1}, \; q_m \le q_i \le p_j \le p_n\}$
	$m \in N_{d2}, n \in N_{d1}$
	$H_{dmn1} = \{j : j \in N_{d1}, j \in H_{dmn}\}$
	$H_{dmn2} = \{i : j \in N_{d2}, i \in H_{dmn}\}$

Greedy linking at the depot results in a set of partial schedules, some with embedded maintenance gaps or starting/ending at the depot. The linking at the terminals other than depot follows a similar greedy procedure, where the partial chains without any maintenance so far, link to ones with maintenance to widely distribute the maximum maintenance opportunities during the first phase. The phase two reinforces the distribution further by breaking the 'rich schedules' with more maintenance opportunities and the 'poor schedules' with no opportunities, into partial schedules and swapping them to redistribute the maintenance opportunities more widely among the schedules.

Implementation and Computational Experience

The implementation builds on the data structures and analytical modules used for the Fixed-Schedule Fleet-Size Problem, and adds additional features to implement concepts specific to the Vehicle Scheduling Problem, namely, the greedy linking and the swapping of the partial schedules. The computational experience of our implementation with real-life data is given in Table 10.

Table 10. Computational Experience of Vehicle Scheduling Problem

Trips	Terminals	Optimal Fleet-Size	M-Gaps	Schedules with maintenance	Maintenance feasibility (YES/NO)	CPU (sec.)
105	23	20	27	20	YES	0.66
220	72	47	69	43	YES	25.4
246	43	26	40	26	YES	2.10
120	30	35	50	35	YES	1.06
92	35	22	24	18	YES	1.70
72	25	18	17	14	YES	0.86
66	27	22	24	18	YES	0.88
84	28	20	22	20	YES	0.58

Crew Scheduling Problem (P5)

Booler[6] formulated the crew scheduling as a linear programming problem and solved it using Dantzig and Wolfe decomposition principle, with suggestion of branch-and-bound approach to take care of integrality constraints. The formulation does not address crew duty conditions associated with steering duty and night-outs at terminals other than depots. Blais et. al.[5] have suggested a heuristic approach for crew scheduling in an urban transit system. The solution consists of a 'macro' stage of solving a simplified version as a linear programming problem, and a 'micro' stage to perform necessary 'fine tuning'.

Formulation

Conceptually, the Crew Scheduling Problem (P5) is similar to its Vehicle Scheduling counterpart (P4), except for different kinds of operational constraints, such as, limit on total steering and spread-over (elapsed time between sign-in and sign-off of the crew) duty, limit on consecutive night-outs at terminals other than depot, meal-breaks, and the rest pauses before and after long trips. For our formulation in Table 11, we ignore the meal-breaks and rest pauses constraints, their treatment being analogous to the *maintenance gaps* in the Vehicle Scheduling.

The night-out constraints simply assert that a crew must end up at the depot on starting after a night-out or must start at the depot if ending as a night-out. At the terminals with positive deficit function (F_a>0), the night-outs are inevitable. The night-outs might also result in terminals with no deficit (F_b=0), because the number of linkings would be substantial smaller in Crew Scheduling Problem (P5) compared to the Vehicle Scheduling (P4) due to more stringent operational constraints. There may be a set of terminals (T_b), however, where the night-out is not allowed at all. For such terminals, the linking constraint does not allow any starting or ending trips in the crew schedules.

Table 11. Crew Scheduling Problem (P5)

Formulation	Notations
Maximize: $\sum_i \sum_j {}^cX_{ij}$ $\quad i \in N,\ j \in L_i$ Subject to: Linking Constraints: $\sum_j {}^cX_{ij} + {}^c\lambda_{i0} = 1 \quad i \in N_{a2}, a \in T-T_b, j \in L_i$ $\sum_i {}^cX_{ij} + {}^c\lambda_{0j} = 1 \quad j \in N_{a1}, a \in T-T_b, i \in B_j$ $\sum_j {}^cX_{ij} = 1 \quad i \in N_{a2}, a \in T_b, j \in L_i$ $\sum_i {}^cX_{ij} = 1 \quad j \in N_{a1}, a \in T_b, i \in B_j$ Identify (Crew Schedule) Chains: ${}^cR_{ii} = {}^c\lambda_{0i} \quad i \in N_{a1}, a \in T-T_b$ ${}^cR_{ij} = \sum_k {}^cR_{ik}.{}^cX_{kj} \quad j \in N,\ k \in B_j$ ${}^cU_{jj} = {}^c\lambda_{j0} \quad j \in N_{a2}, a \in T-T_b$ ${}^cU_{ij} = \sum_k {}^cX_{ik}.{}^cU_{kj} \quad i \in N,\ k \in L_i$ Night-Out Constraints: $\sum_j {}^cU_{ij} \ge {}^c\lambda_{0i} \quad i \in N_{a1}, a \in T_a, j \in N_{d2}$ $\sum_i {}^cR_{ij} \ge {}^c\lambda_{j0} \quad j \in N_{a2}, a \in T_a, i \in N_{d1}$ Steering Duty Constraints: $\sum_j t_i.{}^cR_{ij} \le L1 \quad i \in N_{a1}, a \in T-T_b, j \in N$ Spread-over Duty Constraints: ${}^cR_{ij}.{}^cU_{ij}.(q_j-p_i) \le L2 \quad i \in N_{a1}, j \in N_{a2}$ $a \in T-T_b$ ${}^cX_{ij}, {}^c\lambda_{i0}, {}^c\lambda_{0j}, {}^cR_{ij}, {}^cU_{ij} = 0,1$	$N, T, X_{ij}, \lambda_{i0}, \lambda_{0j}, Rij, U_{ij}, N_{a1},$ $N_{a2},$ $N_{d1}, N_{d2}, o_i, d_i, p_i, q_i, L_i, B_j$ as defined earlier in P4 T_a : Set of terminals other than depot where crew night-out is permitted T_b : Set of terminals where crew night-out is NOT permitted $= T - T_a - d$ L1 : Steering duty limit L2 : Spreadover duty limit t_i : Steering time of the trip i ${}^cX_{ij}, {}^c\lambda_{i0}, {}^c\lambda_{0j}, {}^cRij, {}^cU_{ij}$ are crew counterparts of corresponding notations $X_{ij}, \lambda_{i0}, \lambda_{0j}, Rij, U_{ij}$ in problems P1-P4

The steering duty constraints impose the limit on the accumulated steering duties of trips associated with a crew schedule. Similarly, the spread-over duty constraints impose the limit on elapsed time between the sign-in of the first trip and sign-off after the last trip of a crew schedule.

While the formulation specifies only the primary objective function, $\sum_i\sum_j {}^cX_{ij}$, the crew scheduling involves a secondary objective of maximizing the overlap between vehicle and crew schedules so that the crew is not required to change the vehicles too frequently to operate the trips. That makes the interrelationship between the two problems even more intimate, necessitating an integrated view of the sub-problems. Accordingly, we tackle the crew scheduling using the basic vehicle scheduling framework to develop multi-crew schedules with *unlinked* trips in the first and last 'hollow zones', and subsequently cut them to pieces to yield single-crew schedules.

Implementation and Computational Experience

The implementation builds on the data structures and analytical modules used for the Vehicle Scheduling Problem, and adds additional features to implement concepts specific to the Crew Scheduling Problem. The computational experience of our implementation with real-life data is given in Table 12.

Table 12. Computational Experience of the Crew Scheduling Problem

Trips	Fleet-Size	Crew-Size Lower Bound	Crew-Size Achieved	Night-out infeasible Crew Schedules	Real-Life Crew-Size	Real-Life Night-out infeasible	CPU (sec.)
105	20	35	39	1	39	1	7.78
220	45	86	94	2	92	2	33.02
246	26	44	51	0	51	1	29.28
120	34	62	66	2	65	1	17.08
72	22	43	46	6	46	3	12.98
92	24	47	54	2	53	0	10.70
66	19	36	38	6	39	0	8.58
84	21	34	41	0	40	1	9.54

Integrated Transport Scheduling Problem (P6)

The problems (P2-P5) can be merged to formulate the Integrated Transport Scheduling Problem (P6) as given in Table 13, where the objective functions and the constraints of the sub-problems are simply merged together. The linking constraints and primary objective functions ($\sum_i\sum_j X_{ij}$ and $\sum_i\sum_j {}^cX_{ij}$) provide the necessary glue that binds the sub-problems together.

As mentioned earlier, the integrated problem has an additional secondary objective function that seeks to maximize the desirable overlap ($\sum_i \sum_j O_{ij}$) between the vehicle schedules X_{ij} and the crew schedule ${}^cX_{ij}$. A set of constraints defined for each potential linking, called 'Vehicle/Crew Overlap is Good', count such overlaps.

The Integrated Transport Scheduling Problem turns out to be a massive combinatorial optimization problem with six objective functions and large number of quadratic constraints. For example, a moderate size problem of about 1000 trips would involve over 5 million quadratic constraints, primarily accounted for the schedule identification and spread-over duty constraints of the Vehicle and Crew Scheduling sub-problems. Our common analytical framework for the sub-problems and the common set of data structures enables us to integrate the heuristic solutions to the sub-problems together to solve the integrated problem.

Table 13. Formulation of the Integrated Transport Scheduling Problem (P6)

Maximize: $\sum_i\sum_j X_{ij}$ $i\in N,\ j\in L_{iEL}$ $\sum_i\sum_j {}^cX_{ij}$ $i\in N,\ j\in L_{iEL}$ $\sum_i\sum_j O_{ij}$ $i\in N,\ j\in L_{iEL}$ $\sum_i y_i$ $i\in N$ $\sum_i Y_{ik}$ $i\in N_k$ $\sum_i V_i$ $i\in N$ Subject to: Linking Constraints: $\sum_j X_{ij}+\lambda_{io} = 1$ $i\in N,\ j\in L_{iEL}$ $\sum_i X_{ij}+\lambda_{oj} = 1$ $j\in N,\ i\in B_{jEL}$ $\sum_j {}^cX_{ij}+{}^c\lambda_{io} = 1$ $i\in N_{a2},$ $a\in T-T_b, j\in L_{iEL}$ $\sum_i {}^cX_{ij}+{}^c\lambda_{oj} = 1$ $j\in N_{a1},$ $a\in T-T_b, i\in B_{jEL}$ $\sum_j {}^cX_{ij} = 1$ $i\in N_{a2}, a\in T_b, j\in L_{iEL}$ $\sum_i {}^cX_{ij} = 1$ $j\in N_{a1}, a\in T_b, i\in B_{jEL}$ Perturbation Constraints: $q_i = p_i + t_i$ $p_j-q_i \geq (X_{ij}-1)M$ $i\in N, j\in L_{iLE}$ $p_j-q_i \geq ({}^cX_{ij}-1)M$ $i\in N, j\in L_{iLE}$ $p_{iE} \leq p_i \leq p_{iL}$ $i\in N$ No Perturbation is Good: $p_i - p_{iB} \leq (1 - Y_i)M$ $i\in N$ $p_{iB} - p_i \leq (1 - Y_i)M$ $i\in N$ Depot Compatibility Constraints: $y_{ik}-y_{jk}+X_{ij} \leq 1$ $i\in N, j\in L_{iEL}$ $y_{jk}-y_{ik}+X_{ij} \leq 1$ $k\in K$ $y_{ik}-y_{jk}+{}^cX_{ij} \leq 1$ $i\in N, j\in L_{iEL}$ $y_{jk}-y_{ik}+{}^cX_{ij} \leq 1$ $k\in K$ Depot Assignment Constraints: $\sum_k Y_{ik} = 1$ $i\in N, k\in K$ $\sum_i Y_{ik} = \sum_j Y_{jk}$ $i,j\in N, k\in K,$ $o_i=d_j=a\in T_a$	Identify (Vehicle Schedule) Chains: $R_{ii} = \lambda_{oi}$ $i\in N$ $R_{ij} = \sum_k R_{ik}.X_{kj}$ $i,j\in N, k\in B_{jEL}$ $U_{jj} = \lambda_{jo}$ $j\in N$ $U_{ij} = \sum_k X_{ik}.U_{kj}$ $i,j\in N,\ k\in L_{iEL}$ Identify Chains with Maintenance: $V_i = \lambda_{oi}$ $i\in N_{d1}$ $\sum_p \sum_q R_{iq}.X_{qp}+\sum_r R_{ir}.\lambda_{ro} \geq V_i$ $i\in N_{a1},$ $p\in N_{d1}, q\in B_{pM}, r\in N_{d2}$ $W_j = \lambda_{jo}$ $j\in N_{d2}$ $\sum_p \sum_q X_{qp}.U_{pj}+\sum_r \lambda_{or}.U_{rj} \geq W_j$ $j\in N_{a2},$ $p\in N_{d1}, q\in B_{pM}, r\in N_{d1}$ Maintenance Constraint: $\sum_i V_i+\sum_j W_j \geq F_a$ $i\in N_{a1}, j\in N_{a2}, a\in T-d$ Identify (Crew Schedule) Chains: ${}^cR_{ii} = {}^c\lambda_{oi}$ $i\in N_{a1}, a\in T-T_b$ ${}^cR_{ij} = \sum_k {}^cR_{ik}.{}^cX_{kj}$ $j\in N,\ k\in B_j$ ${}^cU_{jj} = {}^c\lambda_{jo}$ $j\in N_{a2}, a\in T-T_b$ ${}^cU_{ij} = \sum_k {}^cX_{ik}.{}^cU_{kj}$ $i\in N,\ k\in L_{iEL}$ Night-Out Constraints: $\sum_j {}^cU_{ij} \geq {}^c\lambda_{oi}$ $i\in N_{a1}, a\in T_a, j\in N_{d2}$ $\sum_i {}^cR_{ij} \geq {}^c\lambda_{jo}$ $j\in N_{a2}, a\in T_a, i\in N_{d1}$ Steering Duty Constraints: $\sum_j t_i.{}^cR_{ij} \leq L1$ $i\in N_{a1}, a\in T-T_b, j\in N$ Spreadover Duty Constraints: ${}^cR_{ij}.{}^cU_{ij}.(q_j-p_i) \leq L2$ $i\in N_{a1}, j\in Na2,$ $a\in T-T_b$ Vehicle/Crew Overlap is Good: $X_{ij} - {}^cX_{ij} + O_{ij} \leq 1$ $i\in N, j\in L_{iEL}$ ${}^cX_{ij} - X_{ij} + O_{ij} \leq 1$ $i\in N, j\in L_{iEL}$ $X_{ij}, \lambda_{io}, \lambda_{oj}, Rij, U_{ij}, V_i, W_j, Y_i, Y_{ik},$ ${}^cX_{ij}, {}^c\lambda_{io}, {}^c\lambda_{oj}, {}^cR_{ij}, {}^cU_{ij}, O_{ij}=0,1$ $p_i, q_i \geq 0$

Implementation and Computational Experience

We were able to integrate our algorithms and data structures together to solve the Integrated Transport Scheduling Problem. At a practical level, however, given the nature of the sub-problems, it is sufficient to consider the integration at the level of two subsets of sub-problems, namely, variable-schedule (P2) and depot allocation (P3) problems, and vehicle scheduling (P4) and crew scheduling (P5) problems. The former subset essentially deals with the pre-processing of the set of trips of multiple depots in terms of perturbation of timings and re-assignment of the depots to minimize the fleet-size. It is then sufficient to carry out the integrated vehicle and crew scheduling individually for each of the depots.

The integrated variable-schedule and depot allocation seeks to achieve global optimization of the fleet-size by increasing the potential for perturbation, while making the depot allocation somewhat harder due to increased unbalance among the 'local hollow zones', as is apparent from the number of perturbations and reassignments in the computational experience of the problem given in Table 14.

The integrated vehicle and crew scheduling seeks to achieve greater overlap between vehicle and crew schedules in three phases. In phase one, preliminary linking is performed using the Greedy Algorithm. In phase two, the Swapping Algorithm, enriched to handle crew duty constraints, performs restructuring of both the schedules considering both crew and vehicle duty constraints, but giving priority to the latter. In other words, restructuring based on crew scheduling considerations is performed only if it does not result in deterioration of vehicle scheduling considerations. In phase three, the Swapping Algorithm operates exclusively on crew scheduling considerations to give a crew feasible solution, while heuristically optimizing the crew-size requirement. We found that the computational experience of the integrated vehicle and crew scheduling with prior pre-processing by the integrated variable-schedule and depot allocation algorithm was similar to that of the specific sub-problems except for a minor reduction in maintenance gaps and schedules with maintenance as expected.

Table 14. Computational Experience of the Integrated Scheduling Problem

Depot	Trips	Perturb limit (min.)	Sum of Fixed-sch. Fleet-size	Fleet-size Lower bound	Achieved Fleet-Size	Trips perturb-ed	Trips Reassi-gned	CPU (sec.)
8	1005	0	220	217	217	0	10	29
		±5	220	208	210	20	32	90
		±10	220	206	207	25	66	74
		±15	220	204	206	26	56	42
8	1026	0	213	212	212	0	6	27
		±5	213	211	211	1	43	53
		±10	213	207	208	10	130	114
		±15	213	205	205	37	166	184

Concluding Remarks

In this paper we have presented our experiences of structuring and modeling of scheduling problems that arise in operations of large state road transport corporations. We started with a simplified version of the real problem. We looked at the complexities sequentially and looked at each problem independently and finally developed an integrated model of the real problem in large Road Transport Corporations.

Our thrust has been to develop methods, which can solve large problems efficiently. The problems described in the paper are complex and combinatorial in nature with multiple objectives. We could not always find methods to solve the problems optimally. Under those circumstances, we have devised heuristic solutions, which are most often very close to the bound for the optimal solution.

Secondly, as practitioners faced the problem, our efforts have been to devise solutions methods, which are efficient. From the evidences that we have provided in the paper, we are of the view that the methods suggested by us are efficient and could be used in practice.

Thirdly the solutions we arrive using our methods can be implemented in practice. We have seen this in the context of a few divisions and depots in two State Road Transport Corporations.

We believe that formal methods, besides having many advantages, also give economic solutions. In dealing with the scheduling problems described in the paper and as faced in a few State Road Transport Corporations, we have found that savings in bus and crew requirements were in the region of 5% to 15%. We are convinced that the methodology developed has potential to make the operations in large transport corporations more efficient.

Acknowledgements

The Gujarat State Road Transport Corporation (GSRTC) was the initial sponsor of this research. The research was carried out at the real-life context of the GSRTC. The research also benefited from computational experience with the data from the Karnataka State Road Transport Corporation (KSRTC). The Indian Institute of Management, Ahmedabad (IIMA), where all the authors have been on the faculty, generously supported this work as doctoral and faculty research. The refinement and implementation of the Depot Allocation algorithm was carried out at the Transportation Research Center of the University of Montreal during Ankolekar's UNDP sponsored fellowship study. The Maastricht School of Management, Netherlands, and the IIMA sponsored the pilot implementation study in the selected SRTCs. The Maharashtra State Road Transport Corporation (MSRTC), Andhra Pradesh State Road Transport Corporation (APSRTC), and the GSRTC participated in the pilot implementation. The authors wish to thank the SRTCs and the Institutions for their generous support for this research work.

References

1. Ankolekar S. R., Patel Nitin R., and Saha J. L. (1981) Optimization of Vehicle Schedules for a Road Transport Corporation, in N. K. Jaiswal (ed.) *Scientific Management of Transportation Systems.*, North-Holland Publishing Company.
2. Ankolekar Suresh, (1982) *Operational Planning for Large State Road Transport Corporations*, Doctoral dissertation, Indian Institute of Management, Ahmedabad.
3. Ankolekar Suresh, and Patel Nitin R. (1989) Optimal Trip Assignment to Depots in a Large Inter-City Bus System, *INFOR*, 27, No. 1, pp. 114-121.
4. Bartlett T. E. (1957) An Algorithm for Minimal Number of Transport Units to Maintain a Fixed Schedule, *Naval Research Quarterly*, 4, No. 2, pp. 139-149.
5. Blais J. Y., et. al. (1976) *The Problems of Assigning Drivers to Bus Routes in an Urban Transit System*, University of Montreal.
6. Booler J. M. P. (1975) Method for Solving Crew Scheduling Problem *Operational Research Quarterly*, 26, No. 1, pp. 55-62.

7. Bokinge Ulf, and Hasselstrom Dick (1980) Improved Vehicle Scheduling in Public Transport Through Systematic Changes in the Time-Table, *European Journal of Operational Research*, 5, No. 6, pp. 388-395.
8. Gertsbach I., and Gurevich Yu (1977) Constructing an Optimal Fleet for a Transport Schedule, *Transportation Science*, 11, No. 1, pp. 20-36
9. Levin Amos (1971) Scheduling and Fleet routing Models for Transportation Systems, *Transportation Science*, 5, No. 3, pp. 232-255.
10. Martin Lof Anders (1969) *An LP Algorithm for Scheduling Vehicles in a Transportation Network*, Research Report No. 5, Operations Research Center, Massachusetts Institute of Technology.
11. Patel Nitin R. (1979) Load-Factor Measurement for Road Transport Corporation, *OPSEARCH*, 16, No. 1, pp. 34-44
12. Raghavachari M., and Mote V. L. (1970) Generalization of Dilworth's Theorem on Minimal Decomposition, *Management Science*, 16, No. 7, pp. 508-511.
13. Saha J. L. (1970) Algorithm for Bus Scheduling Problem, *Operational Research Quarterly*, 21, No. 4, pp. 463-474.

Optimal Shutdown Policies for a Computer System based on the Power-Effective Design

Hiroyuki OKAMURA†, Tadashi DOHI† and Shunji OSAKI‡

† Department of Information Engineering, Hiroshima University, 1–4–1 Kagamiyama, Higashi-Hiroshima, 739–8527, JAPAN

‡ Department of Information and Communication Engineering, Nanzan University, 27 Seirei-cho, Seto, 489–0863, JAPAN

Abstract: The dynamic power management is one of the most effective technologies for reducing the power consumption in computer systems. Especially, the sleep function based on the shutdown policy is usually installed in the almost operating systems. In this paper, we propose a stochastic model based on the dynamic power management concept to determine the optimal shutdown policy. More precisely, introducing the so-called power effectiveness criterion by taking account of the processing efficiency, the optimal shutdown policies maximizing the power effectiveness can be derived in two cases; single-user operating system and multi-tasking operating system. In a numerical example, we calculate the optimal shutdown policies numerically and perform the sensitivity analysis of model parameters.

Keywords: dynamic power management, shutdown policy, power effectiveness, stochastic model, Markovian arrival process.

1 Introduction

Since ENERGY STAR was introduced by the US Environmental Protection Agency in 1992, the management of the electrical power consumed by computer systems has received considerable attention all over the world. As a computer system consists of a number of electric components and devices, the power management to reduce energy consumption has been discussed at each component level such as IC chip [22], microprocessor [25],

CPU, disk drive, display and so on. Recently, several measurement techniques for electrical power have been also developed with object to reduce energy consumption in real computer operation [16, 26]. In general, the power management should be carried out at each level of hierarchical computer design processes; circuit level, layout level, logic level, behavioral level, architectural level, *etc.* In particular, the system level power management techniques have emerged as one of the most useful design methodologies in practice, because they do not assume the development of new low-power devices. For the detail on the system level power management techniques, see [1, 3].

The dynamic power management, as it is generically known, can provide a control scheme that dynamically reconfigures an electric system to provide the requested services and to guarantee the desired performance level with minimum number of active components or minimum amount of workload on such components [1, 3]. The design method will be useful for operating systems and control systems of peripheral devices. Especially, since operating systems can monitor and control the application software programs which are executed on them, the dynamic power management plays an important role to achieve energy efficiency. However, it is known that typical operating systems like UNIX, WindowsOS and MacOS were not designed originally with energy efficiency in mind.

The dynamic power management basically supports three energy states, *busy*, *idle* and *inactive*, defined in the following:

- The busy state is defined as an active state. In the busy state, the system can process the requested tasks and therefore consumes higher electrical power.

- In the idle state, the system waits for an arrival request. Although the system does not process any task in this state, the amount of electrical power consumption is

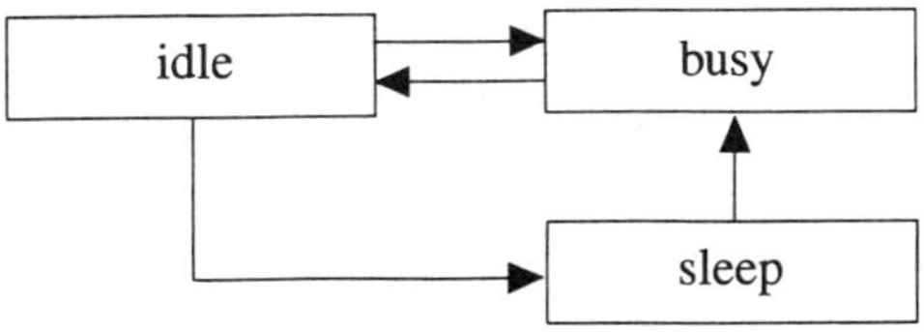

Figure 1: Configuration of the dynamic power management.

Table 1: An example of delay time in a CPU device

from		to	delay time
busy	$\rightarrow$	idle	~ 10 (μs)
idle	$\rightarrow$	busy	~ 10 (μs)
idle	$\rightarrow$	sleep	~ 90 (μs)
sleep	$\rightarrow$	busy	~ 160 (ms)

assumed to be same as that in the busy state.

- The inactive state, which is usually called the *sleep state*, provides the least amount of electrical power consumption. In general, the functions such as a sleep and a hibernation lead the computer system into the sleep state.

Figure 1 illustrates the state transition in the dynamic power management. Under the dynamic power management, it is reported that delay time can occur at transition among the energy states. Table 1 presents an example of the delay time in a CPU device [2]. In addition to the delay time, the large amount of energy consumption will be observed instantaneously at transition from the sleep state to the busy state. This instantaneous power is generally called *wake up power*. The simplest way to establish the power reduction in the current operating system is to add the ability to selectively shutdown useless peripheral devices. That is, if the system has waited for an arrival request in the idle state during a constant time period, the system goes to the sleep state automatically. The constant time period is called the *shutdown policy*. The system wakes up and goes to the busy state if an additional request occurs in the sleep state. In fact, this method, called a shutdown approach, is applied to the hard disk unit [23] and the VLSI circuits system [1, 6, 24] as an energy saving function. Typical examples for the shutdown approach can be found in mobile computers [8, 9, 10]. Since the capacity of a battery for the mobile computer is limited, the available electrical power should be carefully assigned among all of the components and the peripheral devices. For such systems, the shutdown approach is greatly effective and, in fact, is installed as a standard function for mobile computers.

However, the system designing based on the shutdown approach to reduce energy consumption is difficult due to the existence of both delay time and wake up power. For instance, if the system is designed such that it goes to the sleep state whenever it is in the idle state, the total energy consumption will become larger due to the excessive wake up power consumption. This implies that the optimal shutdown policy depends on the usage environment of the system. Thus, the problem to design the suitable shutdown function can be motivated. Okamura *et al.* [21] revisited the design of shutdown function in terms of the stochastic behavior. They assumed that the arrival request follows the general renewal process and derived the approximate expected electrical power consumption in the steady state and the approximate optimal shutdown policy minimizing it. Furthermore, they derived the exact optimal shutdown policy by applying the phase-type renewal process to the inter-arrival process [19, 20].

In this paper, we consider again the optimal design problem for the shutdown function. Our approach is based on a stochastic modeling technique under the criterion of the energy saving effectiveness. The criterion of the energy reduction is usually the electrical power consumption in the steady state. Although this criterion may be intuitive and reasonable, it does not focus on the performance of processing tasks. Since the processing efficiency may degrade due to the extreme energy reduction, we have to consider the trade-off. That is, one should be careful to the performance restrictions in terms of system usability. Thus, this paper introduces a criterion of optimality called the *power effectiveness* [17, 18] taking account of the processing ability. The power effectiveness indicates the possible mean time length of keeping the busy state per unit amount of electrical power consumption.

2 Single-User Operating System

2.1 Model Description

In this section, we discuss a stochastic shutdown model for a single-user operating system. As shown before, the dynamic power management may control three energy states: busy, idle and sleep states. The shutdown function for the single-user operating system is modeled under the following assumption:

Assumption A: The electrical power consumption per unit time in the idle state is same as that in the busy state.

Assumption B: The system needs delay time to go to the busy state from the idle state.

Assumption C: The wake up power occurs uniformly during the delay time period when the system goes to the busy state from the sleep state.

Assumption D: When the system is in the busy state, an arrival request is refused.

One of the most important factors in the design of shutdown policy is trade-off between the amount of electrical power reduced by shutdown and the amount of the wake up power consumption. In other words, the optimal design essentially depends on the difference between the normal electrical power consumption and the wake up power consumption. It does not depend on the difference between the electrical power consumption in the idle state and the busy state. Thus, the electrical power consumption in the idle state is well assumed to be same as the busy state. Note that it does not affect the optimal design of shutdown policy. The similar dependence can be found on delay time among the three states. That is, the optimal policy strongly depends on the delay time from the sleep state to the busy state. Also, comparing the delay time from the sleep state to the busy state with those in the other cases, it is short enough to be ignored. Therefore, we make Assumptions A and B. Assumption C concerns the wake up power. The behavior of the wake up power is naturally sharp rather than flat. When we focus on the wake up power consumed per unit time, it has a property of inversely proportional to the delay time, *i.e.*,

$$\text{(the wake up power consumed per unit time)} \times \text{(the delay time)}$$

is a constant approximately. This fact leads to Assumption C. Since the underlying system is assumed to be a single-user operating system, we make the assumption that an arrival request is refused in the busy state, namely, it is equivalent to Assumption D.

Furthermore, the arrival request process is assumed to follow the phase-type renewal process [13]. The phase-type renewal process is one of the most general class of stochastic process which can be tractable mathematically. It is governed by an irreducible Markov process.

In our modelling, the following notation is used:

$\{N_t;\ t \geq 0\}$: the cumulative number of arrival requests at time t,

$\{J_t;\ t \geq 0\}$: the phase of users' circumstances at time t,

S_k: processing time for the k-th task,

$\tau\ (> 0)$: delay time at the transition from the sleep state to the busy state,

t_0: the shutdown policy ($0 \leq t_0 < \infty$),

P_1 (> 0): the amount of electrical power consumption per unit time in the idle state and the busy state,

P_2 ($> P_1$): the amount of wake up power consumption per unit time during the delay time period when the system goes to the busy state from the sleep state.

Suppose that the phase process $\{J_t;\ t > 0\}$ is an irreducible Markov process. It has an infinitesimal generator $\boldsymbol{M}$ ($m \times m$ matrix) and initial probability vector $\boldsymbol{a}$ ($1 \times m$ row vector). $\boldsymbol{\lambda}$ ($m \times 1$) column vector denotes the arrival rate of requests. Let $\{N_t;\ t \geq 0\}$ denote the number of arrival requests at time t. The inter-arrival time is mutually and independently distributed with an identical probability distribution function, where $F(t)$ is the probability distribution function of the inter-arrival time. Then the probability distribution $F(t)$ is given by

$$F(t) = 1 - \boldsymbol{a}\exp(\boldsymbol{T}t)\boldsymbol{e}, \tag{1}$$

where $\boldsymbol{T} = \boldsymbol{M} - \text{diag}(\boldsymbol{\lambda})$ and $\boldsymbol{e}$ is a column vector whose all the elements are 1. The distribution represented by Equation (1) is called the *phase-type distribution.*

The dynamics of the underlying system can be summarized as follows. If the system is in the busy state, it processes the task requested by the last arrival. The system takes processing time S_k to process the task requested by the k-th arrival. The processing time S_k has the absolutely continuous probability distribution function $G(t)$ with finite mean $1/\mu$ (> 0) and variance σ_s^2 (> 0). When an arrival request occurs in the busy state, the request is refused. After the system completes processing the task, it goes to the idle state. In the idle state, if the amount of sojourn time in the idle state reaches a

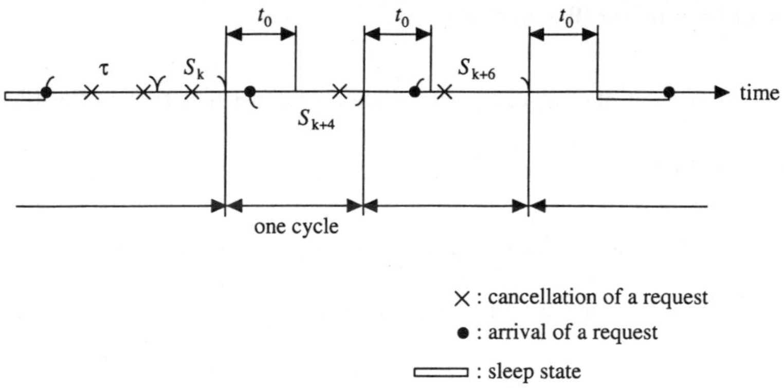

Figure 2: Possible realization of the single-user operating system with shutdown.

threshold t_0 before an arrival request occurs, the system goes to the sleep state at that time. Otherwise, the system begins processing the task requested by the arrival. In the sleep state, if an arrival request occurs, the system wakes up and goes to the busy state. It takes delay time τ for the system to go to the busy state. Of course, the requests arriving during the delay time period are refused. Figure 2 illustrates possible realization of the single-user operating system. From Assumptions A and C, the amount of electrical power consumption per unit time in both the busy and the idle states is denoted by P_1. The amount of the wake up power consumption per unit time is P_2, which is wasted during the delay time period when the system goes to the busy state from the sleep state. To simplify the mathematical analysis, the amount of electrical power consumption in the sleep state is assumed to be zero.

2.2 Formulation of Power Effectiveness

The power effectiveness criterion is defined as the mean length of available time per unit amount of electrical power consumption. The formal definition of the power effectiveness is given by

$$W(t_0) = \lim_{t\to\infty} \frac{\text{E[the total length of available time in } [0,t) \text{]}}{\text{E[the total amount of electrical power consumed in } [0,t) \text{]}}, \tag{2}$$

where the available state corresponds to the busy state and thus the length of available time means the length of sojourn time in the busy state.

Define the time interval from a task completion point until the next point as one cycle and the following expected values;

$\gamma_i(t_0)$: the mean length of available time during one cycle, provided that the phase is i at the beginning of the cycle.

$\alpha_i(t_0)$: the expected amount of electrical power consumption during one cycle, provided that the phase is i at the beginning of the cycle.

$\pi_i(t_0)$: the probability that the phase is i at the beginning of one cycle in the steady state.

Let $\boldsymbol{\gamma}(t_0)$, $\boldsymbol{\alpha}(t_0)$ and $\boldsymbol{\pi}(t_0)$ denote the vectors whose i-th element is $\gamma_i(t_0)$, $\alpha_i(t_0)$ and $\pi_i(t_0)$, respectively.

Proposition 2.1. The power effectiveness is given by

$$W(t_0) = \frac{\boldsymbol{\pi}(t_0)\boldsymbol{\gamma}(t_0)}{\boldsymbol{\pi}(t_0)\boldsymbol{\alpha}(t_0)}. \tag{3}$$

The proof of Proposition 2.1 is given in Appendix. Proposition 2.1 is an extension of Renewal Reward Theorem [4] on the Markov renewal reward process. Proposition 2.1 is quite similar to the familiar renewal reward theorem. The essential difference between them is whether the phase at the beginning of one cycle is considered or not.

From Proposition 2.1, we may focus only on the expected amount of electrical power consumption during one cycle and mean length of available time during one cycle.

Let $\boldsymbol{T}(t_0)$ denote a transition probability matrix for the phase at the beginning of one cycle. From the formulation of the phase-type distribution function, $\boldsymbol{T}(t_0)$ can be written in the form;

$$\begin{aligned}\boldsymbol{T}(t_0) =& \int_0^{t_0} \exp(\boldsymbol{T}t)\boldsymbol{\lambda}\mathrm{E}[\boldsymbol{a}\exp\{\boldsymbol{Q}S_k\}]dt \\ &+ \int_{t_0}^{\infty} \exp(\boldsymbol{T}t)\boldsymbol{\lambda}\mathrm{E}[\boldsymbol{a}\exp\{\boldsymbol{Q}(\tau + S_k)\}]dt \\ =&\left(\int_0^{t_0} \exp(\boldsymbol{T}t)\boldsymbol{\lambda}dt\boldsymbol{a} + \int_{t_0}^{\infty} \exp(\boldsymbol{T}t)\boldsymbol{\lambda}dt\boldsymbol{a}\exp\{\boldsymbol{Q}\tau\}\right) \\ &\times \int_0^{\infty} \exp\{\boldsymbol{Q}t\}dG(t), \end{aligned} \tag{4}$$

where $\boldsymbol{\lambda} = -\boldsymbol{T}\boldsymbol{e}$, $\boldsymbol{Q} = \boldsymbol{T} + \boldsymbol{\lambda}\boldsymbol{a}$ and $\boldsymbol{I}$ is the identity matrix. From the well-known argument on the Markov chain, the stationary probability vector $\boldsymbol{\pi}(t_0)$ can be derived by

$$\boldsymbol{\pi}(t_0)\boldsymbol{T}(t_0) = \boldsymbol{\pi}(t_0), \quad \boldsymbol{\pi}(t_0)\boldsymbol{e} = 1. \tag{5}$$

Equation (5) is equivalent to

$$\begin{aligned}\boldsymbol{\pi}(t_0) =& \boldsymbol{a}\int_0^{\infty} \exp\{\boldsymbol{Q}t\}dG(t) \\ &\times \left(\boldsymbol{I} + \exp\{\boldsymbol{T}t_0\}\boldsymbol{e}\boldsymbol{a}(\boldsymbol{I} - \exp\{\boldsymbol{Q}\tau\})\int_0^{\infty} \exp\{\boldsymbol{Q}t\}dG(t)\right)^{-1}. \end{aligned} \tag{6}$$

Also, the expected amount of electrical power consumption during one cycle is given by

$$\begin{aligned}\boldsymbol{\alpha}(t_0) =& \int_0^{t_0} \{P_1(t + 1/\mu)\}\exp(\boldsymbol{T}t)\boldsymbol{\lambda}dt \\ &+ \int_{t_0}^{\infty} \{P_1t_0 + P_2\tau + P_1/\mu\}\exp(\boldsymbol{T}t)\boldsymbol{\lambda}dt \\ =& (P_2\tau + P_1/\mu)\boldsymbol{e} - \{P_2\tau\boldsymbol{I} + P_1\boldsymbol{T}^{-1}\}\int_0^{t_0} \exp(\boldsymbol{T}t)\boldsymbol{\lambda}dt. \end{aligned} \tag{7}$$

On the other hand, since the system processes just one task in one cycle, the mean length of available time during one cycle is easily given by

$$\gamma(t_0) = (1/\mu)\boldsymbol{e}. \tag{8}$$

From Equations (6), (7) and (8) and Proposition 2.1, we can formulate the power effectiveness. Then the problem is to find the optimal shutdown policy t_0^* which maximizes the power effectiveness.

2.3 Optimal shutdown policies

Consider the case where requests arrive at the system according to the homogeneous Poisson process with rate λ (> 0). The phase-type distribution on the inter-arrival time is reduced to the exponential distribution, *i.e.*,

$$\boldsymbol{T} = -\lambda \qquad \text{and} \qquad \boldsymbol{a} = 1. \tag{9}$$

Then, the power effectiveness is explicitly given by

$$W(t_0) = \frac{\gamma(t_0)}{\alpha(t_0)}, \tag{10}$$

where

$$\gamma(t_0) = 1/\mu \tag{11}$$

and

$$\alpha(t_0) = P_2\tau + P_1/\mu - (P_2\tau - P_1/\lambda)\int_0^{t_0} \lambda \exp\{-\lambda t\}dt. \tag{12}$$

Theorem 2.1. Case (i): If $P_2/P_1 < 1/(\lambda\tau)$, the optimal shutdown policy is $t_0^* = 0$ and

$$W(0) = \frac{1}{P_2\mu\tau + P_1}. \tag{13}$$

Case (ii): If $P_2/P_1 \geq 1/(\lambda\tau)$, then $t_0^* \to \infty$ and

$$W(\infty) = \frac{\rho}{P_1(1+\rho)}, \tag{14}$$

where $\rho = \lambda/\mu$.

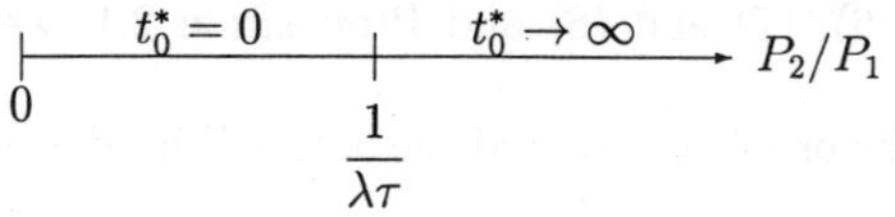

Figure 3: The optimal shutdown policy for the single-user operating system in the Poisson arrival case.

Figure 3 summarizes Theorem 2.1. The optimal shutdown policy in the case of the Poisson arrival is zero or infinity. That is, the simple on-off switching policy is optimal. In particular, it is remarkable that $1/(\lambda\tau)$ directly affects the optimal shutdown policy. Further, $P_2\tau$ and P_1/λ indicate the respective total amounts of the electrical power consumption in the delay time and the sojourn time of the idle state. Theorem 2.1 means that the optimal policy is to compare the overhead power consumption for the sleep, $P_2\tau$, with the useless electrical power consumption for the idle, P_1/λ. This may be an intuitive result and be valid in the physical meaning.

Next, we investigate the condition on which the optimal shutdown policy exists in the general arrival case. Since the power effectiveness converges to a constant as $t_0 \to \infty$, the result on the existence of the optimal shutdown policy can be obtained as follows.

Theorem 2.2. There exists a finite optimal shutdown policy which maximizes $W(t_0)$, if and only if there exists a finite $t \in [0, \infty)$ satisfying

$$\boldsymbol{\pi}(t)\{P_2\tau\boldsymbol{I} + P_1\boldsymbol{T}^{-1}\}\exp(\boldsymbol{T}t)\boldsymbol{e} < P_1\{\boldsymbol{\pi}(\infty) - \boldsymbol{\pi}(t)\}(-\boldsymbol{T})^{-1}\boldsymbol{e}. \tag{15}$$

Corollary 2.1. If

$$P_2\tau < P_1\boldsymbol{\pi}(\infty)(-\boldsymbol{T})^{-1}\boldsymbol{e}, \tag{16}$$

there exists a finite optimal shutdown policy $t_0^* \in [0, \infty)$ maximizing $W(t_0)$.

Proof of Theorem 2.2 and Corollary 2.1:

There exists a finite optimal shutdown policy, if and only if there exists a finite t satisfying

$$W(t) > W(\infty). \tag{17}$$

Inequality (17) is reduced to

$$\boldsymbol{\pi}(t)\{P_2\tau\boldsymbol{I} + P_1\boldsymbol{T}^{-1}\}\exp(\boldsymbol{T}t)\boldsymbol{e} < P_1\{\boldsymbol{\pi}(\infty) - \boldsymbol{\pi}(t)\}(-\boldsymbol{T})^{-1}\boldsymbol{e}. \tag{18}$$

Corollary 2.1 can be given by putting $t = 0$ in Equation (18). The proof is completed. □

Corollary 2.1 is the sufficient condition on which there exists a finite optimal shutdown policy. Since $\boldsymbol{\pi}(\infty)(-\boldsymbol{T})^{-1}\boldsymbol{e}$ means the expected inter-arrival time, it is obvious that Corollary 2.1 is related to Theorem 2.1.

3 Multi-tasking Operating System

In this section, we discuss a stochastic shutdown model for a multi-tasking operating system based on the dynamic power management. Since the present operating systems such as WindowsOS, UNIX, *etc.* provide the multi-tasking processing circumstance, the model proposed here will be useful in many practical applications.

3.1 Model Description

Consider a multi-tasking operating system. Unlike the single-user operating system, the multi-tasking operating system can receive and process all the arrival requests in the busy state. Thus, Assumption D in the single-user operating system has to be modified in the following;

Assumption D': When the system is in the busy state, the tasks requested by arrivals are stored in the buffer and then the system processes them under the first-come first-serve discipline.

The same notation as the single-user operating system is used. The dynamics of the multi-tasking operating system is summarized as follows. The inter-arrival time distribution is the phase-type distribution with parameters $\boldsymbol{\alpha}$ and $\boldsymbol{T}$. All the arrival requests are stored in the buffer. The processing time S_k will be needed to process the k-th task. The processing time has an absolutely continuous probability distribution function $G(t)$ with finite mean $1/\mu$ (> 0) and variance σ_s^2 (> 0). The arrival request during the processing time period is also stored in the buffer. If there is no task in the buffer, that is, the system completes processing all the tasks stored in the buffer, the system goes to the idle state. In the idle state, the behavior of the system is same as that in the single-user operating system. If the amount of sojourn time in the idle state reaches a threshold level t_0 before a request arrives, the system goes to the sleep state at that time. Otherwise, the system begins processing the task requested at the occurrence of the arrival request. In the sleep state, if an arrival request occurs, the system wakes up and goes to the busy state. The delay time τ is needed to go to the busy state. The arrival requests during the delay time period are stored in the buffer. Figure 4 depicts the possible behavior of the multi-tasking operating system. The amount of electrical power consumption per unit time in both the busy and the idle states is P_1. The amount of the wake up power consumption per unit time during the delay time period is P_2. The amount of electrical power consumption in the sleep state is assumed to be zero for convenience.

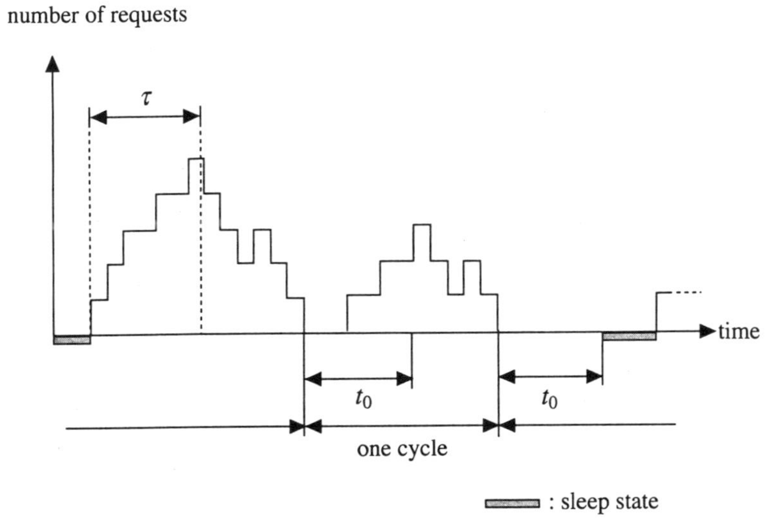

Figure 4: Possible realization of the multi-tasking operating system with shutdown.

3.2 Matrix-geometric analysis

The matrix-geometric analysis [14, 15] is a powerful tool to analyze the *M/G/1* type queueing system. Before formulating of the power effectiveness criterion, we describe the matrix-geometric analysis and some results being needed to formulate the power effectiveness.

Let us define a transition probability:

$$P_{ij}(n,t) = \Pr\{N_t = n, J_t = j | N_0 = 0, J_0 = i\} \tag{19}$$

and matrix $\boldsymbol{P}(n,t)$ whose (i,j)-element is $P_{ij}(n,t)$. The following Chapman-Kolmogorov forward equation holds

$$\begin{aligned} &\frac{d}{dt}\boldsymbol{P}(n,t) = \boldsymbol{P}(n,t)\boldsymbol{T} + \boldsymbol{P}(n-1,t)\boldsymbol{\lambda a}, \\ &\boldsymbol{P}(0,0) = \boldsymbol{I}, \\ &\boldsymbol{P}(-1,t) = \boldsymbol{O}, \end{aligned} \tag{20}$$

where $\boldsymbol{O}$ is a zero matrix. Thus, the matrix generating function $\boldsymbol{P}^*(z,t)$ is expressed in the following form:

$$\boldsymbol{P}^*(z,t) = \sum_{n=0}^{\infty} \boldsymbol{P}(n,t) z^n = \exp\{(\boldsymbol{T} + z\boldsymbol{\lambda a})t\}. \tag{21}$$

Since the behavior of both the number of arrival requests N_t and the phase J_t forms an embedded Markov chain at the completion point of a task, its transition probability matrix can be given by

$$\boldsymbol{P} = \begin{bmatrix} \boldsymbol{B}_0 & \boldsymbol{B}_1 & \boldsymbol{B}_2 & \cdots \\ \boldsymbol{A}_0 & \boldsymbol{A}_1 & \boldsymbol{A}_2 & \cdots \\ \boldsymbol{O} & \boldsymbol{A}_0 & \boldsymbol{A}_1 & \cdots \\ \boldsymbol{O} & \boldsymbol{O} & \boldsymbol{A}_0 & \cdots \\ \cdot & \cdot & \cdot & \\ \cdot & \cdot & \cdot & \end{bmatrix}, \tag{22}$$

where $\boldsymbol{A}_n$ and $\boldsymbol{B}_n$ are $m \times m$ matrices with respective elements $[A_n]_{ij}$ and $[B_n]_{ij}$. The element $[A_n]_{ij}$ is the conditional probability that the phase changes from i to j, provided that n arrival requests occur during processing time period for a task. On the other hand, the element $[B_n]_{ij}$ is the conditional probability that the phase changes from i to j, provided that n arrival requests occur for the period from the beginning of the idle period to the next completion point of a task. Thus, it is easy to obtain

$$\boldsymbol{A}_n = \int_0^{\infty} \boldsymbol{P}(n,t) dG(t) \tag{23}$$

and

$$\begin{aligned} \boldsymbol{B}_n = & \int_0^{t_0} \exp(\boldsymbol{T}t)\boldsymbol{\lambda} dt \boldsymbol{a} \boldsymbol{A}_n \\ & + \int_{t_0}^{\infty} \exp(\boldsymbol{T}t)\boldsymbol{\lambda} dt \boldsymbol{a} \sum_{k=0}^{n} \boldsymbol{P}(k,\tau)\boldsymbol{A}_{n-k}. \end{aligned} \tag{24}$$

Taking z-transform of Equations (23) and (24), we have

$$\boldsymbol{A}^*(z) = \sum_{n=0}^{\infty} \boldsymbol{A}_n z^n = \int_0^{\infty} \exp\{(\boldsymbol{T} + z\boldsymbol{\lambda a})t\} dG(t) \tag{25}$$

and

$$\begin{aligned} B^*(z;t_0) &= \sum_{n=0}^{\infty} B_n z^n \\ &= \left(\int_0^{t_0} \exp(Tt)\lambda dt a + \int_{t_0}^{\infty} \exp(Tt)\lambda dt a \exp\{(T + z\lambda a)\tau\} \right) A^*(z). \end{aligned} \quad (26)$$

Consider the fundamental probability matrix G. The element indicates the probability that the phase changes from i to j while the number of requests decreases to n from $n+1$. This time period is called the fundamental period. It is found that the fundamental period corresponds to the busy period in the ordinary *M/G/1* type queueing system without vacation. Further, we define an $m \times m$ matrix $K(t_0)$ which is a transition probability from i to j at time point when the number of requests is 0. For the fundamental matrix, the following equation holds (see Lucantoni, Meier-Hellstern and Neuts [11]):

$$G = \sum_{n=0}^{\infty} A_n G^n. \quad (27)$$

Equation (27) can be reduced to

$$G = \int_0^{\infty} \exp\{(T + \lambda a G)t\} dG(t), \quad (28)$$

and therefore we obtain

$$\begin{aligned} K(t_0) &= \sum_{n=0}^{\infty} B_n G^n \\ &= \left(\int_0^{t_0} \exp(Tt)\lambda dt a + \int_{t_0}^{\infty} \exp(Tt)\lambda dt a \exp\{(T + \lambda a G)\tau\} \right) G. \end{aligned} \quad (29)$$

The computation algorithm for the matrix G was proposed by Lucantoni and Ramaswami [12]. They introduced the computation algorithm for the fundamental matrix G as follows.

Computation of the fundamental matrix [12]:

The matrix G is efficiently computed by the following recursive scheme. First, start with $G_0 = O$. Next, for $k = 0, 1, 2, \cdots$, compute

$$H_{n+1,k} = [I + \theta^{-1}(T + \lambda a G_k)] H_{n,k}, \qquad n = 0, 1, 2, \cdots, \quad (30)$$

$$\boldsymbol{G}_{k+1} = \sum_{n=0}^{\infty} \gamma_n \boldsymbol{H}_{n,k}, \tag{31}$$

where $\boldsymbol{H}_{0,0} = \boldsymbol{I}$, θ is the maximum value among the absolute values of the diagonal elements of $\boldsymbol{T}$ and

$$\gamma_n = \int_0^{\infty} \exp\{-\theta t\} \frac{(\theta t)^n}{n!} dG(t). \tag{32}$$

It can be proved that the sequence $\boldsymbol{G}_k$ converges to $\boldsymbol{G}$ monotonically.

Let $\boldsymbol{n}_g$ denote the expected number of tasks being processed during the fundamental period. It is easily found that

$$\boldsymbol{n}_g = \boldsymbol{e} + \sum_{k=1}^{\infty} \boldsymbol{A}_k \sum_{l=0}^{k-1} \boldsymbol{G}^l \boldsymbol{n}_g. \tag{33}$$

We also define the probability vector $\boldsymbol{g}$ satisfying

$$\boldsymbol{g}\boldsymbol{G} = \boldsymbol{g}, \qquad \boldsymbol{g}\boldsymbol{e} = 1. \tag{34}$$

Using the probability vector $\boldsymbol{g}$ and the relationship;

$$\left\{\boldsymbol{I} - \sum_{k=1}^{\infty} \boldsymbol{A}_k \sum_{l=0}^{k-1} \boldsymbol{G}^l \right\} \{\boldsymbol{I} - \boldsymbol{G} + \boldsymbol{e}\boldsymbol{g}\} = \boldsymbol{I} - \boldsymbol{A}^*(1) + (\boldsymbol{e} - \boldsymbol{\nu})\boldsymbol{g}, \tag{35}$$

the expected number of tasks being processed during the fundamental period is given by

$$\boldsymbol{n}_g = \{\boldsymbol{I} - \boldsymbol{G} + \boldsymbol{e}\boldsymbol{g}\} \{\boldsymbol{I} - \boldsymbol{A}^*(1) + (\boldsymbol{e} - \boldsymbol{\nu})\boldsymbol{g}\}^{-1} \boldsymbol{e}, \tag{36}$$

where $\boldsymbol{\nu}$ is the number of transitions of the phase, *i.e.*

$$\boldsymbol{\nu} = \frac{d}{dz} \boldsymbol{A}^*(z) \bigg|_{z \to 1} \boldsymbol{e}. \tag{37}$$

3.3 Formulation of Power Effectiveness

The power effectiveness can be defined as Equation (2) and is derived by Proposition 2.1. Thus, in the similar way to the single-user operating system, the power effectiveness is formulated as the ratio of the mean length of available time during one cycle to the expected amount of electrical power consumption during one cycle. Since the (i,j)-element of the matrix $\boldsymbol{K}(t_0)$ represents the probability that the phase transfers from i to j at the time point when the number of requests becomes zero, the stationary probability vector is given by

$$\boldsymbol{\pi}(t_0) = \boldsymbol{\pi}(t_0)\boldsymbol{K}(t_0), \qquad \boldsymbol{\pi}(t_0)\boldsymbol{e} = 1. \tag{38}$$

From Equation (38), we have

$$\boldsymbol{\pi}(t_0) = \boldsymbol{aG}\Big(\boldsymbol{I} + \exp(\boldsymbol{T}t_0)\boldsymbol{ea}\left[\boldsymbol{I} - \exp\{(\boldsymbol{T}+\boldsymbol{\lambda aG})\tau\}\right]\boldsymbol{G}\Big)^{-1}. \tag{39}$$

Denoting the expected number of tasks being processed during one cycle by $\boldsymbol{n}_c(t_0)$, the mean length of available time during one cycle is given by

$$\gamma(t_0) = (1/\mu)\boldsymbol{n}_c(t_0). \tag{40}$$

From the conservation law in the queueing theory [7], it follows that

$$\frac{(1/\mu)\boldsymbol{\pi}(t_0)\boldsymbol{n}_c(t_0)}{\boldsymbol{\pi}(t_0)\boldsymbol{\beta}(t_0)} = \rho, \tag{41}$$

where $\rho = \boldsymbol{a}(-\boldsymbol{T})^{-1}\boldsymbol{e}/\mu$ is the traffic intensity and $\boldsymbol{\beta}(t_0)$ is the mean length of one cycle. The mean length of one cycle can be expressed in the following form:

$$\begin{aligned}\boldsymbol{\beta}(t_0) &= \int_0^{t_0} \exp(\boldsymbol{T}t)\boldsymbol{\lambda}\left(t + (1/\mu)\boldsymbol{an}_g\right) dt \\ &\quad + \int_{t_0}^{\infty} \exp(\boldsymbol{T}t)\boldsymbol{\lambda}\left(t + \tau + (1/\mu)\boldsymbol{a}\sum_{k=0}^{\infty}\boldsymbol{P}(k,\tau)\sum_{l=0}^{k}\boldsymbol{G}^l\boldsymbol{n}_g\right) dt \\ &= \{(-\boldsymbol{T})^{-1} + \tau\boldsymbol{I}\}\boldsymbol{e} + (1/\mu)\boldsymbol{n}_c(t_0) - \tau\int_0^{t_0}\exp(\boldsymbol{T}t)\boldsymbol{\lambda}dt.\end{aligned} \tag{42}$$

Equation (41) yields

$$\boldsymbol{\pi}(t_0)\boldsymbol{\beta}(t_0) = \frac{1}{1-\rho}\left\{\boldsymbol{\pi}(t_0)(-\boldsymbol{T})^{-1}\boldsymbol{e} + \tau - \tau\boldsymbol{\pi}(t_0)\int_0^{t_0}\exp(\boldsymbol{T}t)\boldsymbol{\lambda}dt\right\}. \tag{43}$$

Thus, the mean length of available time during one cycle can be derived as follows.

$$\begin{aligned}\boldsymbol{\pi}(t_0)\boldsymbol{\gamma}(t_0) &= \rho\boldsymbol{\pi}(t_0)\boldsymbol{\beta}(t_0)\\ &= \frac{\rho}{1-\rho}\left\{\boldsymbol{\pi}(t_0)(-\boldsymbol{T})^{-1}\boldsymbol{e} + \tau - \tau\boldsymbol{\pi}(t_0)\int_0^{t_0}\exp(\boldsymbol{T}t)\boldsymbol{\lambda}dt\right\}. \end{aligned}\tag{44}$$

Similarly, the expected amount of electrical power consumption during one cycle is given by

$$\begin{aligned}\boldsymbol{\alpha}(t_0) &= \int_0^{t_0}\exp(\boldsymbol{T}t)\boldsymbol{\lambda}\left(P_1 t + P_1(1/\mu)\boldsymbol{a}\boldsymbol{n}_g\right)dt\\ &\quad + \int_{t_0}^{\infty}\exp(\boldsymbol{T}t)\boldsymbol{\lambda}\left(P_1 t_0 + P_2\tau + P_1(1/\mu)\boldsymbol{a}\sum_{k=0}^{\infty}\boldsymbol{P}(k,\tau)\sum_{l=0}^{k}\boldsymbol{G}^l\boldsymbol{n}_g\right)dt\\ &= P_2\tau\boldsymbol{e} + P_1(1/\mu)\boldsymbol{n}_c(t_0) - \left\{P_2\tau\boldsymbol{I} + P_1\boldsymbol{T}^{-1}\right\}\int_0^{t_0}\exp(\boldsymbol{T}t)\boldsymbol{\lambda}dt. \end{aligned}\tag{45}$$

Using the traffic intensity ρ, Equation (45) becomes

$$\begin{aligned}\boldsymbol{\pi}(t_0)\boldsymbol{\alpha}(t_0) = \frac{1}{1-\rho}\Bigg\{&\{\rho P_1 + (1-\rho)P_2\}\tau + \rho P_1\boldsymbol{\pi}(t_0)(-\boldsymbol{T})^{-1}\boldsymbol{e}\\ &- \boldsymbol{\pi}(t_0)\Big(\{\rho P_1 + (1-\rho)P_2\}\tau\boldsymbol{I}\\ &+ (1-\rho)P_1\boldsymbol{T}^{-1}\Big)\int_0^{t_0}\exp(\boldsymbol{T}t)\boldsymbol{\lambda}dt\Bigg\}. \end{aligned}\tag{46}$$

The power effectiveness can be obtained by Equations (44) and (46). The problem is then to find the optimal shutdown policy t_0^* maximizing the power effectiveness.

3.4 Optimal shutdown policies

Consider the homogeneous Poisson arrival process with the rate λ (> 0). Since the phase-type distribution for the inter-arrival time has the following parameters;

$$\boldsymbol{T} = -\lambda \qquad \text{and} \qquad \boldsymbol{a} = 1, \tag{47}$$

the power effectiveness is given by

$$W(t_0) = \frac{\gamma(t_0)}{\alpha(t_0)}, \tag{48}$$

where

$$\gamma(t_0) = \frac{1}{1-\rho}\Big\{1/\lambda + \tau\exp\{-\lambda t\}\Big\} \tag{49}$$

and

$$\alpha(t_0) = \frac{1}{1-\rho}\Bigg\{P_1/\lambda + \{\rho P_1 + (1-\rho)P_2\}\tau\exp\{-\lambda t_0\} \\ - (1/\lambda - 1/\mu)P_1\exp\{-\lambda t_0\}\Bigg\}. \tag{50}$$

Theorem 3.1. Suppose that $\rho < 1$.

Case (i): If $P_2/P_1 < 1 + 1/(\lambda\tau)$, the optimal shutdown policy is $t_0^* = 0$ and

$$W(0) = \frac{1/\mu + \rho\tau}{P_1/\mu + \{\rho P_1 + (1-\rho)P_2)\}\tau}. \tag{51}$$

Case (ii): If $P_2/P_1 < 1 + 1/(\lambda\tau)$, then $t_0^* \to \infty$ and

$$W(\infty) = \frac{\rho}{P_1}. \tag{52}$$

The proof of Theorem 3.1 is omitted. Figure 5 depicts the above result. Similar to the result on the single-user operating system, the optimal shutdown policy is zero or infinity. That is, the simple on-off switching policy for the multi-tasking operating system is also optimal in the Poisson arrival case. It is seen that the optimal shutdown policy for the multi-tasking operating system is more likely to be zero than that for the single-user operating system.

We also consider the optimal shutdown policy in the general arrival case. Since the power effectiveness converges to a constant as $t_0 \to \infty$, we can derive the condition on which a finite optimal shutdown policy exists.

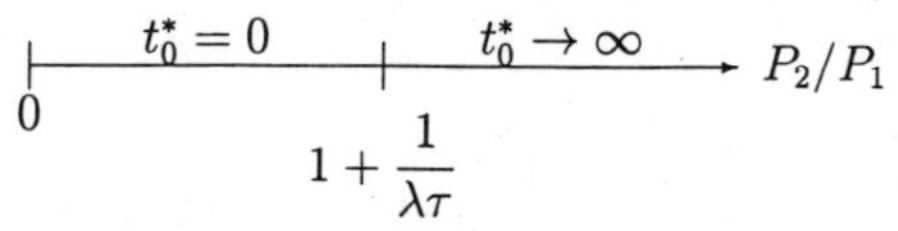

Figure 5: Optimal shutdown policy for the multi-tasking operating system in the Poisson arrival case.

Theorem 3.2. There exists a finite optimal shutdown policy which maximizes $W(t_0)$, if and only if there exits a finite $t \in [0, \infty)$ satisfying

$$\boldsymbol{\pi}(t)\{(P_2 - P_1)\tau \boldsymbol{I} + P_1 \boldsymbol{T}^{-1}\} \exp(\boldsymbol{T}t)\boldsymbol{e} < 0. \tag{53}$$

Corollary 3.1. If

$$(P_2 - P_1)\tau < P_1 \boldsymbol{\pi}(0)(-\boldsymbol{T})^{-1}\boldsymbol{e}, \tag{54}$$

there exists a finite optimal shutdown policy $t_0^* \in [0, \infty)$ maximizing $W(t_0)$.

Since the proofs of Theorem 3.2 and Corollary 3.1 are quite similar to those of Theorem 2.2 and Corollary 2.1, we omit to show them. Corollary 3.1 is the sufficient condition on which a finite optimal shutdown policy exists. Comparing Corollary 3.1 with Corollary 2.1, it can be found that both results depends on the useless electrical power consumption on the idle state, that is,

$$P_1 \times \text{(the mean length of sojourn time in the idle state)}, \tag{55}$$

and the overhead power consumption on the sleep state. However, it should be noted that sufficient conditions to exist a finite optimal shutdown policy in both cases are different. From Corollary 2.1 and Corollary 3.1, since $(P_2 - P_1)\tau$ is strictly smaller that $P_2\tau$, the optimal shutdown policy in the multi-tasking operating system is generally shorter than that in the single-user operating system.

4 A Numerical Example

In this section, we investigate the performance of the dynamic power management through a numerical example. Suppose that the inter-arrival time obeys the 2-hyperexponential distribution, where

$$\boldsymbol{\alpha} = \left[\, p \quad 1-p \,\right], \tag{56}$$

$$\boldsymbol{T} = \begin{bmatrix} -\lambda_1 & 0 \\ 0 & -\lambda_2 \end{bmatrix}. \tag{57}$$

When the hyperexponential distribution is used to the inter-arrival distribution, the arrival process consists of the Poisson processes with two kinds of arrival rates. Using the above notation, λ_1 and λ_2 are the respective arrival rates and p is a ratio of the occurrence for two kinds of arrival patterns. This stochastic process can be characterized by burst and dormant arrivals. The processing time distribution is assumed to be the exponential distribution with mean 1.0, where

$$\begin{array}{lll} \lambda_1 = 1.0, \ldots, 5.0, & \lambda_2 = 0.1, & p = 0.9, \\ P_1 = 1.0, & P_2 = 5.0, & \tau = 0.5. \end{array}$$

Table 2 presents the optimal shutdown policies in the single-user operating system when λ_1 varies from 1.0 to 5.0. Similarly, Table 3 presents the optimal shutdown policies in the multi-tasking operating system with $\lambda_1 = 1.0, \ldots, 5.0$. The columns in both tables consist of the traffic intensity (ρ), the coefficient variance of the inter-arrival time (CV), the optimal shutdown policy (optimal), the associated maximum power effectiveness (max-peff) and the efficiency of the optimal policy (efficiency), where the efficiency is defined by

$$(\text{efficiency}) = \frac{(\text{the maximum power effectiveness})}{(\text{the power effectiveness without shutdown})} \times 100 \quad (\%). \tag{58}$$

Table 2: Dependence of the optimal shutdown schedule on the arrival stream in the single-user operation system.

λ_1	ρ	CV	optimal	max-peff	efficiency
1.0	0.53	1.42	2.59	0.377	32.4%
2.0	0.69	1.97	1.53	0.429	59.2%
3.0	0.77	2.23	1.07	0.432	77.9%
4.0	0.82	2.39	0.81	0.423	91.2%
5.0	0.85	2.49	0.66	0.410	100.9%

Table 3: Dependence of the optimal shutdown schedule on the arrival stream in the multi-tasking operation system.

λ_1	ρ	CV	optimal	max-peff	efficiency
1.0	0.53	1.42	1.05	0.691	31.3%
2.0	0.69	1.97	0.13	0.878	27.3%
3.0	0.77	2.23	0.00	0.926	20.4%
4.0	0.82	2.39	0.00	0.947	16.0%
5.0	0.85	2.49	0.00	0.958	13.0%

From Tables 2 and 3, it can be seen that the optimal shutdown policy in the single-user operating system is always longer than that in the multi-tasking operating system. This result is same as the conclusions in Corollary 2.1 and Corollary 3.1. Also, the efficiency by application of the shutdown function tends to be higher in the case of the single-user operating system. In addition, as the coefficient variance of the inter-arrival time is larger, the shutdown function becomes more effective. On the other hand, in the multi-tasking operating system, the efficiency decreases gradually as the coefficient variance is larger.

5 Concluding Remarks

In this paper, we have considered the stochastic shutdown model for the dynamic power management. The underlying stochastic process has been modeled by the arrival process with the phase-type distribution. In both single-user and multi-tasking operating systems, the optimal shutdown policies have been considered under the power effectiveness criterion. In the Poisson arrival case, it has been shown that the optimal policies are the simple on-off shutdown policies. In the general arrival case, the existence of a finite optimal shutdown policy has been proved. Finally, we have calculated the shutdown policies numerically and performed the sensitivity analysis of model parameters.

A Appendix: Proof of Proposition 2.1

Define the following random variables;

$T(t_0)$: the length of one cycle,

$C_t(t_0)$: the instantaneous amount of electrical power consumption at time t.

Consider the following expected value concerning with the electrical power consumption;

$$\xi_i(r;t_0) = \mathrm{E}\left[\int_0^\infty e^{-rt} C_t(t_0)dt \,\middle|\, J_0 = i\right], \tag{59}$$

where J_t is the phase process at time t. Note that Equation (59) is the Laplace transform of $\mathrm{E}[C_t(t_0)|J_0 = i]$ with respect to t. By using $T(t_0)$, the expected value $\xi_i(r;t_0)$ is reduced to the form;

$$\begin{aligned}
\xi_i(r;t_0) &= \mathrm{E}\left[\int_0^\infty e^{-rt} C_t(t_0)dt \,\middle|\, J_0 = i\right] \\
&= \mathrm{E}\left[\int_0^{T(t_0)} e^{-rt} C_t(t_0)dt + \int_{T(t_0)}^\infty e^{-rt} C_t(t_0)dt \,\middle|\, J_0 = i\right] \\
&= \mathrm{E}\left[\int_0^{T(t_0)} e^{-rt} C_t(t_0)dt \,\middle|\, J_0 = i\right] \\
&\quad + \sum_{j=1}^m \mathrm{E}\left[e^{-rT(t_0)} \chi(J_{T(t_0)} = j)\right. \\
&\quad \times \left.\mathrm{E}\left[\int_0^\infty e^{-rt} C_t^*(t_0)dt \,\middle|\, J_{T(t_0)} = j\right] \,\middle|\, J_0 = i\right] \\
&= \mathrm{E}\left[\int_0^{T(t_0)} e^{-rt} C_t(t_0)dt \,\middle|\, J_0 = i\right] \\
&\quad + \sum_{j=1}^m \mathrm{E}\left[e^{-rT(t_0)} \chi(J_{T(t_0)} = j) \,\middle|\, J_0 = i\right] \xi_j(r;t_0),
\end{aligned} \tag{60}$$

where $C_t^*(t_0) = C_{t+T(t_0)}(t_0)$ and $\chi(A)$ is the indicator function for an event A, which is defined by

$$\chi(\omega) = \begin{cases} 0 & \omega \notin A, \\ 1 & \omega \in A. \end{cases} \tag{61}$$

From the Markov property, it is easily found that the stochastic processes C_t^* and C_t have the same distribution, so that

$$\mathrm{E}\left[\int_0^\infty e^{-rt} C_t(t_0)dt \middle| J_0 = i\right] = \mathrm{E}\left[\int_0^\infty e^{-rt} C_t^*(t_0)dt \middle| J_0 = i\right]. \tag{62}$$

Now, define the joint density function of the length of one cycle and the phase at the end of one cycle, *i.e.* for $i = 1, \ldots, m$ and $j = 1, \ldots, m$,

$$p_{ij}(t; t_0) = \Pr\{T(t_0) \in dt,\ J_{T(t_0)} = j | J_0 = i\}. \tag{63}$$

Taking Laplace transform of $p_{ij}(t; t_0)$, we have

$$p_{ij}^*(r; t_0) = \mathrm{E}[e^{-rT(t_0)}\chi(J_{T(t_0)=j})|J_0 = i]. \tag{64}$$

Let $\boldsymbol{T}^*(r; t_0)$ denote a matrix whose (i, j)-element is $p_{ij}^*(r; t_0)$. For the electrical power consumption, we define $q_{ij}(x, t; t_0)$ as the joint density function of the instantaneous electrical power consumption during one cycle and the phase at the end of one cycle, *i.e.*,

$$q_{ij}(x, t; t_0) = \Pr\{C_t(t_0) \in x,\ T(t_0) > t,\ J_{T(t_0)} = j | J_0 = i\}. \tag{65}$$

Also, $\boldsymbol{C}^*(r; t_0)$ denotes a matrix whose (i, j)-element is $q_{ij}^*(r; t_0)$ which is given by

$$q_{ij}^*(r; t_0) = \mathrm{E}\left[\int_0^{T(t_0)} e^{-rt} C(t; t_0)dt\chi(J_{T(t_0)=j}) \middle| J_0 = i\right]. \tag{66}$$

By using the column vector $\boldsymbol{\xi}(r; t_0)$ with its i-th element $\xi_i(r; t_0)$, we have

$$\boldsymbol{\xi}(r; t_0) = \boldsymbol{C}^*(r; t_0)\boldsymbol{e} + \boldsymbol{T}^*(r; t_0)\boldsymbol{\xi}(r; t_0). \tag{67}$$

Here, we attempt to derive the stationary electrical power consumption from Equation (67). Let $\boldsymbol{\pi}(r;t_0)$ be the left eigenvector of $\boldsymbol{T}^*(r;t_0)$, with the maximum eigenvalue $\mathrm{sp}(\boldsymbol{T}^*(r;t_0))$. Multiplying Equation (67) by $\boldsymbol{\pi}(r;t_0)$, it is seen that

$$\boldsymbol{\pi}(r;t_0)\boldsymbol{\xi}(r;t_0) = \boldsymbol{\pi}(r;t_0)\boldsymbol{C}^*(r;t_0)\boldsymbol{e} + \mathrm{sp}(\boldsymbol{T}^*(r;t_0))\boldsymbol{\pi}(r;t_0)\boldsymbol{\xi}(r;t_0). \tag{68}$$

We therefore obtain

$$\{1 - \mathrm{sp}(\boldsymbol{T}^*(r;t_0))\}\boldsymbol{\pi}(r;t_0)\boldsymbol{\xi}(r;t_0) = \boldsymbol{\pi}(r;t_0)\boldsymbol{C}^*(r;t_0)\boldsymbol{e}. \tag{69}$$

Taking $r \to 0$ in Equation (69) and using the well-known Tauberian theorem [5], it can be obtained that $\lim_{r\to 0} r\boldsymbol{\pi}(r;t_0)\boldsymbol{\xi}(r;t_0)$ converges to the expected electrical power consumed per unit time in the steady state. Since $\mathrm{sp}(\boldsymbol{T}^*(r;t_0))$ converges to unity as $r \to 0$, we have

$$r^{-1}\{1 - \mathrm{sp}(\boldsymbol{T}^*(r;t_0))\} \to -\lim_{r\to 0}\frac{d}{dr}\mathrm{sp}(\boldsymbol{T}^*(r;t_0)). \tag{70}$$

Taking account of $\boldsymbol{\pi}(r;t_0)\boldsymbol{T}^*(r;t_0) = \mathrm{sp}(\boldsymbol{T}^*(r;0))\boldsymbol{\pi}(r;t_0)$, it can be seen that

$$\lim_{r\to 0}\frac{d}{dr}\mathrm{sp}(\boldsymbol{T}^*(r;t_0)) = \boldsymbol{\pi}(0;t_0)\lim_{r\to 0}\frac{d}{dr}\boldsymbol{T}^*(r;0)\boldsymbol{e}. \tag{71}$$

Consequently, the expected electrical power consumed per unit time in the steady state $V(t_0)$ is given by

$$\begin{aligned} V(t_0) &= \lim_{r\to 0} r\boldsymbol{\pi}(r;t_0)\boldsymbol{\xi}(r;t_0) \\ &= \frac{\boldsymbol{\pi}(0;t_0)C^*(0;t_0)\boldsymbol{e}}{-\boldsymbol{\pi}(0;t_0)\lim_{r\to 0}\frac{d}{dr}T^*(r;t_0)\boldsymbol{e}}. \end{aligned} \tag{72}$$

It is obvious that the denominator and numerator in Equation (72) are the expected amount of electrical power consumption during one cycle and the mean length of one cycle in the steady state, respectively. Also, $\boldsymbol{\pi}(0;t_0)$ represents the stationary probability vector at the beginning of one cycle in the steady state. Thus, the following result can be derived;

$$V(t_0) = \frac{\boldsymbol{\pi}(t_0)\boldsymbol{\alpha}(t_0)}{\boldsymbol{\pi}(t_0)\boldsymbol{\beta}(t_0)}, \tag{73}$$

where $\boldsymbol{\beta}(t_0)$ is the mean length of one cycle.

Next, we consider the power effectiveness. From the definition of power effectiveness, we have

$$\begin{aligned} W(t_0) = \lim_{t\to\infty} & \frac{\mathrm{E}[\text{ the total length of available time in } [0,t)\ ]}{t} \\ & \times \frac{t}{\mathrm{E}[\text{ the total amount of electrical power consumed in } [0,t)\ ]}. \end{aligned} \tag{74}$$

Since the second term in Equation (74) is equivalent to the inverse of the stationary electrical power consumption, we obtain

$$\lim_{t\to\infty} \frac{t}{\mathrm{E}[\text{ the total amount of electrical power consumed in } [0,t)\ ]} = \frac{1}{V(t_0)}. \tag{75}$$

Therefore, we focus on the proof of the following equation;

$$\lim_{t\to\infty} \frac{\mathrm{E}[\text{ the total length of available time in } [0,t)\ ]}{t} = \frac{\boldsymbol{\pi}(t_0)\boldsymbol{\gamma}(t_0)}{\boldsymbol{\pi}(t_0)\boldsymbol{\beta}(t_0)}. \tag{76}$$

Let $A_t(t_0)$ denotes the instantaneous available rate at time t. Also, let $\boldsymbol{\psi}(r;t_0)$ be a column vector whose i-th element is

$$\psi_i(r;t_0) = \mathrm{E}\left[\int_0^\infty e^{-rt}A_t(t_0)dt \,\middle|\, J_0 = i\right] \tag{77}$$

and $\boldsymbol{A}^*(r;t_0)$ is a matrix whose (i,j)-element is

$$\gamma^*_{ij}(r;t_0) = \mathrm{E}\left[\int_0^{T(t_0)} e^{-rt}A_t(t_0)dt\chi(J_{T(t_0)} = j) \,\middle|\, J_0 = i\right]. \tag{78}$$

Similar to the previous discussion in this appendix, the vector $\boldsymbol{\psi}(r;t_0)$ can be formulated as

$$\boldsymbol{\psi}(r;t_0) = \boldsymbol{A}^*(r;t_0)\boldsymbol{e} + \boldsymbol{T}^*(r;t_0)\boldsymbol{\psi}(r;t_0). \tag{79}$$

Using the left eigenvector $\boldsymbol{\pi}(r;t_0)$ of $\boldsymbol{T}^*(r;t_0)$, we have

$$\lim_{t\to\infty}\frac{\text{E[the total length of available time in } [0,t) \text{]}}{t}$$
$$= \lim_{r\to 0} r\boldsymbol{\pi}(r;t_0)\boldsymbol{\psi}(r;t_0)$$
$$= \frac{\boldsymbol{\pi}(0;t_0)\boldsymbol{A}^*(0;t_0)\boldsymbol{e}}{-\boldsymbol{\pi}(0;t_0)\lim_{r\to 0}\frac{d}{dr}\boldsymbol{T}^*(r;t_0)\boldsymbol{e}}. \tag{80}$$

It is clear that the denominator and numerator in Equation (80) are the mean length of available time during one cycle and the mean length of one cycle, respectively. Also, $\boldsymbol{\pi}(0;t_0)$ represents the stationary probability vector at the beginning of one cycle in the steady state. Thus, the proof is completed. □

Acknowledgements

This research was partially supported by the Ministry of Education, Science, Sports and Culture, Grant-in-Aid for Scientific Research (B), 13480109, 2001 and Nanzan University Pache Research Subsidy I-A.

Exercises:

1. Let X be an inter-arrival time following the phase-type distribution with parameters $\boldsymbol{\alpha}$ and $\boldsymbol{T}$. Derive the first two moments, *i.e.* $\mathrm{E}[X]$ and $\mathrm{E}[X^2]$.

2. Prove Theorem 2.1.

3. Consider an ordinary *M/G/1* queueing system. Let ζ denote the time length of a busy period. Show

$$\mathrm{E}[\exp\{-s\zeta\}] = \int_0^\infty \exp\{-\lambda t(1-\mathrm{E}[\exp\{-s\zeta\}])\}\, dG(t), \tag{81}$$

where λ and $G(\cdot)$ are the arrival rate and the service distribution, respectively.

4. Derive the expected time length of a busy period in the ordinary *M/G/1* queueing system (Hint: use the result of Exercise 3).

5. Consider an ordinary *PH/G/1* queueing system. Given the fundamental matrix $\boldsymbol{G}$, derive the probability vector, $\boldsymbol{\pi}$, for the phase at the end of a busy period and the distribution function for the length of an idle period.

References

[1] L. Benini and G. De Micheli, *Dynamic Power Management: Design Techniques and CAD Tools*, Kluwer Academic Publishers, New York, 1997.

[2] L. Benini, A. Bogliolo, G. A. Paloologo, and G. De Micheli, Policy optimization for dynamic power management, *IEEE Trans. on Computer-Aided Design of Circuits Systems*, vol. 18, no. 6, pp. 813–833, 1999.

[3] L. Benini and G. De Micheli, System-level power optimization: techniques and tools, *ACM Trans. on Design Automation of Electronic Systems*, vol. 5, no. 2, pp. 115–192, 2000.

[4] D. R. Cox, *Renewal theory*, John Wiley & Sons Inc., London, 1962.

[5] W. Feller, *An Introduction to Probability Theory and Its Applications*, vol. 2, 2nd ed. John Wiley & Sons Inc., New York, 1957.

[6] C. Hwang and A. C. Wu, Predictive system shutdown method for energy saving of event-driven computation, *ACM Trans. on Design Automation of Electronic Systems*, vol. 5, no. 2, pp. 226–241, 2000.

[7] L. Kleinrock, A conservation law for a wide class of queueing disciplines, *Naval Research Logistics Quarterly*, vol. 12, pp. 181–192, 1965.

[8] R. Kravets and P. Krishnan, Power management technique for mobile communication, *Proc. 4th ACM/IEEE Annual Int'l Conf. on Mobile Computing and Networking*, pp. 157–168, IEEE Computer Society Press, Los Alamitos, 1998.

[9] J. R. Lorch and A. J. Smith, Reducing processor power consumption by improving processor time management in a single-user operating system, *Proc. 2nd Int'l Conf. on Mobile Computing and Networking*, pp. 143–154, ACM Press, New York, 1996.

[10] J. R. Lorch and A. J. Smith, Software strategies for portable computer energy management, *IEEE Personal Communications*, vol. 5, no. 3, pp. 60–73, 1998.

[11] D. M. Lucanoni, K. S. Meier-Hellstern, and M. F. Neuts, A single-server queue with server vacations and a class of non-renewal arrival processes, *Adv. Appl. Prob.*, vol. 22, pp. 676–705, 1990.

[12] D. M. Lucantoni and V. Ramaswami, Efficient algorithms for solving the non-linear matrix equations arising in phase type queues, *Stochastic Models*, vol. 1, pp. 29–51, 1985.

[13] M. F. Neuts, Renewal processes of phase type, *Naval Research Logistics Quarterly*, vol. 25, pp. 445–454, 1978.

[14] M. F. Neuts, *Matrix-Geometric Solutions in Stochastic Models: An Algorithmic Approach*, Johns Hopkins University Press, Baltimore, 1981.

[15] M. F. Neuts, *Structured Stochastic Matrices M/G/1 Type and Their Applications*, Marcel Dekker, New York, 1989.

[16] G. R. Newsham and D. K. Tiller, The energy consumption of desktop computers: measurement and saving potential, *IEEE Trans. on Industrial Applications*, vol. 30, no. 4, pp. 1065–1072, 1994.

[17] H. Okamura, T. Dohi and S. Osaki, The auto-sleep scheduling based on power effectiveness for a computer system (in Japanese), *Transactions of IEICE (A)*, vol. J82–A, no. 12, pp. 1808–1818, 1999.

[18] H. Okamura, T. Dohi and S. Osaki, A power-effective design of auto-sleep mode for computer hard disks (in Japanese), *Transactions of SICE*, vol. 36, no. 1, pp. 108–115, 2000.

[19] H. Okamura, T. Dohi and S. Osaki, The phase type approximation for the optimal auto-sleep scheduling, *Proc. First Western Pacific and Third Australia-Japan Workshop on Stochastic Models in Engineering, Technology and Management*, pp. 426–435, Technology Management Centre, The University of Queensland, Brisbane, 1999.

[20] H. Okamura, T. Dohi and S. Osaki, A structural approximation method to generate the optimal auto-sleep schedule for a computer system, *Proc. Int'l Conf. on Applied Stochastic System Modeling*, pp. 180–187, 2000.

[21] H. Okamura, T. Dohi and S. Osaki, Performance analysis of a transaction based software system with shutdown, *Proc. 2nd International Workshop on Software and Performance*, pp. 168–179, ACM Press, New York, 2000.

[22] M. Pedram, Power minimization in IC design: principles and applications, *ACM Trans. on Design Automation of Electronic Systems*, vol. 1, no. 1, pp. 3–56, 1996.

[23] H. Sandoh, H. Hirakoshi, and H. Kawai, An optimal time to sleep for an auto-sleep system, *Computers & Operations Research*, vol. 23, pp. 221–227, 1996.

[24] M. Srivastava, A. Chandrakasan, and B. Brodersen, Predictive system shutdown and other architectural techniques for energy efficient programmable computation, *IEEE Trans. on Very Large Scale Integrated Systems*, vol. 4, no. 1, pp. 42–55, 1996.

[25] B. W. Suessmith and G. Paap, Power PC603 microprocessor power management, *Communication of ACM*, vol. 37, no. 6, pp. 43–46, 1994.

[26] D. K. Tiller and G. R. Newsham, Switch off your office equipment and save money, *IEEE Industry Applications Magazine*, vol. 2, no. 4, pp. 17–24, 1996.

Local Search heuristics For Combinatorial Optimization Problems

Dr Ashok K Mittal
Professor
I.I.T.Kanpur
Kanpur 208016

Abstract

Over the years combinatorial optimization problems have become of considerable importance and have been studied in literature extensively. In this chapter we describe a unified structure for such problems and concentrate on local solutions with respect to a given neighborhood. Such problems can be structured as search problems on hypercubes.

Key words : **Combinatorial Optimization, Local Search , Heuristics**

1.0 Combinatorial Optimization Problems

Optimization problems can be divided into two major categories:

(1) Problems whose solution consist of a set of continuos variables.

(2) Problems whose solution consist of a set of discrete variables.

Problems in the second category are generally referred to as combinatorial optimization problems. Typically in these problems, we are selecting a combination of objects from the set of finite or possibly countable infinite objects. Examples of such set of objects are, integers, permutations, sequences, vertices of graphs etc. Linear Programming is a problem, which can be viewed both as continuos or discrete optimization problem and hence forms a natural bridge between two categories of problems. Generally by identifying a suitable set of inequalities, it is possible to structure combinatorial optimization problems as linear programming problem. However such transformation may require adding a very large number of inequalities, and may not be amenable to solution in a reasonable time.

Over the years a very large number of applications have been identified which give rise to combinatorial optimization problems. Some such examples are: Knapsack, Assignment, Travelling Salesman, Graph Coloring, Vehicle Routing, Sequencing and Scheduling problems.

Definition:

An instance of an optimization problem is a pair (**S**, C) where **S** is any set, the domain of feasible solutions; C is the cost function, a mapping C: **S** – R.

Problem is to select a f$\in$ **S** s.t , C(f) $\leq$C (y) for any y$\in$ **S**

Such a solution f is called a globally optimized solution to the given instance of the problem.

An optimization problem is a set of instances of optimization problem with common structure of (**S**,C). Each specific instance of such a problem can be identified by specifying as input , all the data needed to distinguish this instance from all others .

However for most of the real life problems , it is not possible to list all the members of the set **S** explicitly, as there are large number of members in set **S** . For example in a travelling salesman problem (TSP) on n cities there are n! such members. Another representation of such problems will be through a set of object **F** and a constraint set **G.** In this representation a subset of **F** is a feasible solution (i.e. member of **S**), if it satisfies the constraints in **G**. For example for TSP, let **F** $\equiv \{a_1, a_2,\ldots, a_n\}$, be the list of the cities. **G** consist of constraint that only those subsets of **F**, which form a tour are feasible. The feasible tours are evaluated using an evaluation function C: f $\rightarrow$R for all f which are feasible.

Thus we can also consider structure (**F, G** , C) for defining combinatorial optimization problems.

1.1 Some examples of Combinatorial Optimization Problems

Knapsack Problem:

Let there be a set of n objects. Let the weight of ith object be w_i and its value c_i. Problem is to select a subset of objects of maximum value s.t. the weight of selected objects is less than or equal to a given weight W. For this problem ,

Let **F** = $\{a_1, a_2, \dots a_n\}$ be a set of objects , f a subset of **S**

$$G = \{ \sum_{a_i \in f} w_i \le W \} \text{ Constraint set}$$

$$C = \sum_{a_i \in f} c_i$$

Where w_i and c_i are weight and value of the object a_i

Travelling Salesman problem:

Given a set of n vertices (cities) $u_1, u_2, \dots u_n$ and distance dij between city ui and u_j problem is to identify a sequence (a cycle) of minimum distance such that each vertex (city) is visited exactly once.

Here **F** = $\{u_1, u_2, \dots u_n\}$

G = { sequence is a tour}

C(f) ={$\sum a_{ij}$, s.t. u_i, uj are two consecutive cities in the sequence f}

Set Covering Problem

F is a set of objects. Let **P** be a set of subsets of **F**. Problem is to select a subset of **F** s.t. all subsets in **P** are covered and cardinality of selected subset is minimum.

Here **F**= $\{ x_1, x_2, \ldots x_n \}$

G = { subset of **F** which covers all subset in **P**}

$C(f) = |f|$

Sequencing Problem:

Let n jobs are required to be completed on a single machine, p_i is the process time and d_i the duedate of the job i. Problem is to identify a sequence which maximizes (minimizes) a given function of completion times of the jobs.

Here **F** $\equiv \{J_1, J_2, \ldots J_n\}$

G $\equiv$ {Sequence of $J_1, J_2, \ldots J_n$ s.t. each job appears only once in the sequence}

C(f) = Function of completion times of the jobs, where completion time of the job J_i is c_i.

Some such function are:

$C(f) = \sum c_i w_i$

$C(f) = \sum w_i \max(c_i - d_i, 0)$

Vehicle Routing Problem

Let **F** be a set of n objects (load), with weight w_i of the ith object . These objects are to be loaded in k vehicles, such that load in each vehicle is less than or equal to its capacity. Each load is to be delivered to a specified location. Problem is to assign loads to the vehicles, such that the total cost of all the k tours(each vehicle will start from a center point and will come back to this point after delivering loads) is minimized.

Hence **F** = {x_1, x_2, … x_n} List of loads

G = { If a collection of k subtours is such that , the weight assigned to each of the vehicles is less than or equal to its capacity}.

C (f)= cost of the k tours

There are many other problems which give rise to similar structure. These problems can be further classified in to three categories as follows:

(A) **Problems of subset selection** :

Consider a set of object **F** ≡ {a_1, a_2, …a_n} and a function C (evaluation function) which maps subset of **F** into R. Problem is to select the subset f of **F** , which satisfies a given set of constraints **G** and has the best possible value (Maximum or minimum) of C(f) among all such subsets.

Examples of such problems are , Knapsack , Set Covering, Set Packing , Graph Coloring, etc.

(B) Path and Cycle Problems

Given a set .of objects, paths are defined as a sequence of objects. If a path starts and ends with the same object, it is called a cycle. Let **G** be the set of constraints which determines the feasibility of each such sequence for a problem and C a mapping which maps each feasible such sequence f into C(f) (value of the sequence).Problem is to select ,a feasible sequence f , which maximize(minimizes) the value of C(f).

Examples are Travelling Salesmen , Sequencing , Vehicle Routing problems.

(C) Combination Problems

Problems such as Transportation , Assignment , Max-flow, Minimum Cost-flow are examples of the problems which can be viewed as problems of identifying best (minimum or maximum value) subset of sequences (path, cycles) defined over a set **F** and satisfying constraints in **G** .

This categorization is not extensive, as the problems on cycles/paths can also be structured as problem of selection of subset. In fact all such problems can be visualized as optimization problems over hypercubes of suitable dimensions. For example, n toy Knapsack Problem can be visualized as optimization problem over a n-dimensional hyper cube. Similarly n city symmetric Travelling Salesman Problem can be visualized as optimization problem over n (n-1)/2 dimensional hypercube.

1.2 A combinatorial optimization problem is three problems:

As stated earlier we can consider (**F**, **G**, C) to define a combinatorial optimization problem. With this notation ,combinatorial optimization problems can be viewed as the sequence of following three problems:

Given **F**, **G** and C, and an integer L,

(P1) Is there a feasible solution f s.t. $C(f) \leq L$? (Feasibility problem)

(P2) Find the cost of the best (optimization) solution (Evaluation Problem)

(P3) Find f such that $C(f) \geq C(y)$ for all $y \in \mathbf{F}$. (Optimization Problem)

For most of the combinatorial optimization problems, it is not difficult to get an upper bound on C(f). In that case P2 can be solved using P1 iteratively by changing value of L. Further P3 can be solved using answer to P2 and then solving P1 with this value for L.

Thus one can define a sequence of these problems as P1 → P2 → P3; where P3 is most difficult to solve.

1.3 Complexity of Problems

Combinatorial optimization problems are generally solved by iterative procedures referred to as algorithm. An algorithm takes as input specific data, required to specify an instance of the optimization problem, operates a set of instructions and come to halt after a finite number of execution of such instructions , giving an output. For example the input for the travelling salesman problem will be number of cities , and distances between all pair of cities. Similarly for knapsack problem the input will be number of toys , weight and value for each toy.

Generally algorithms are developed to solve a set of instances of combinatorial optimization problem (all instances, with similar structure of (**F**, **G**, C)).

Let p be a specific instance of the problem under consideration, and let Ip be the input string required to code this instance for the algorithm A. Let m be the length of this input string. Then m denotes size of the problem. It may be noted that the length of input will depend on the method for coding. Let $t(I_p)$ be the computational time required for solving this problem instance p using algorithm A. If H(m) is a function such that ,

$t(I_p) \leq H(m)$ for all instances of size m for this problem, then H(m) is an upper bound on the computational times of all instances of the problem with input size m using algorithm A. If H(m) can be bounded by a polynomial function of m ,p(m) for all m> k (a fixed constant), then this algorithm is considered polynomial time efficient.

A combinatorial optimization problem is considered polynomial time solvable if there exist a polynomial time algorithm to solve the problem. Only a few combinatorial optimization problems are known to be in this category. Examples are, Assignment, Matching , Chinese Postman , Shortest Path , Minimum Spanning Tree problem etc. However, for most of the combinatorial optimization problems, no such polynomial time algorithm is known to exist. In theory of complexity, problems for which a polynomial time algorithm exists are classified in class P. For almost all of the problems which are not known to belong to this class, it can be shown that the recognition version (feasibility problem P1)of the problem belongs to the class NPC. This class has the property, that if any of the problem in the class can be solved by a polynomial time algorithm, then all the problems in this class can be solved in polynomial time. Generally any algorithm for solving the optimization version P3 of the problems in NPC can be shown to have worst case computational bound which grows as exponential function of the size of the problem. As for these problems finding global optimal solution is computationally time consuming (except for small size problems) methods are developed which can provide a reasonably good solution in a reasonable computational time even for large size problems. Such methods are called Heuristics.

2.0 Heuristics to solve Combinatorial Optimization Problem:

A large number of problem specific heuristics have been proposed in the past for various problems. The literature distinguishes two broad classes of heuristic algorithms : Constructive and Local search algorithms. In this chapter we shall concentrate on Local Search heuristics.

2.1 Local Search Heuristics:

To construct a local search heuristic for an optimization problem ,one superimposes a neighborhood structure on the solutions. That is, one specifies for each solution a set of neighboring solutions. The heuristics starts from an initial solution, and from then on it keeps on moving to a better neighbor as long as there is one. If there is no such neighbor it terminates at a locally optimal solution i.e. a solution which does not have a better neighbor.

The local search in combinatorial optimization has been extensively used since late fifties and early sixties.

2.2 Definitions

The use of local search algorithms presupposes definitions of a problem and a neighborhood. In this section we shall use the definition of the optimization problem as structure (**S**, C) .

The **problem** is to find a globally optimal (minimal) solution, i.e., an i* $\in$ ***S*** such that

$$C(i^*) \leq C(y) \quad y \in \mathbf{S}$$

Furthermore

$$C^* = C(i^*)$$

denotes the **optimal cost**, and

$$\boldsymbol{S}^* = \{ \; \mathbf{i} \in \boldsymbol{S} \;\; C(i) = C^* \; \}$$

denotes the set of **optimal solutions** (**Aarts & Lenstra** [1]).

It is important to distinguish between a problem and an instance of a problem. Informally, in an instance of a problem we are given the "input data" and have enough information to obtain a solution; a problem is a collection of instances, which usually are generated in a similar a fashion.

Definition I Neighborhoods :

Let *(S,C)* be an instance of a combinatorial optimization problem. A neighborhood function is a mapping

$$\mathbf{N}: S \rightarrow 2^{S}$$

which defines for each solution $i \in S$ a set $\mathbf{N(i)} \subseteq S$ of solutions that are in some sense close to i. The set **N (i)** is the neighborhood of solution i and each $j \in \mathbf{N(i)}$ is a neighbor of i. It is assumed that $i \in \mathbf{N(i)}$, for all $i \in S$ (**Aarts & Lenstra** [1]).

Definition II Local Optimality :

Let *(S,C)* be an instance of a combinatorial optimization problem and let N be a neighborhood function. A solution $\hat{i} \in S$ is **locally optimal** (minimal) with respect to N,

if $$C(\hat{i}) \leq C(j) \quad \text{for all } j \in \mathbf{N}(\hat{i}).$$

We denote the set of locally optimal solutions by $\hat{S}$ (**Aarts & Lenstra** [1]).

Definition III Exact Neighborhoods

Let *(S ,C)* be an instance of a combinatorial optimization problem and let N be a neighborhood function. N is exact if $\hat{S} \subseteq S^{*}$ (**Aarts & Lenstra** [1]). In other words the neighborhood N is said to be exact if, whenever f is locally optimal with respect to N, it is also globally optimal.

After defining the neighborhoods the next section discusses the various types of neighborhoods.

2.2 Neighborhoods

Neighborhoods depend on the problem under consideration, and finding efficient neighborhood functions that lead to high quality local optima can be viewed as one of the challenges of local search. Discrete neighborhoods must be large enough to include some discrete variants of the current solution and small enough to be surveyed within reasonable computation times. A class of more intricate neighborhood functions are described next.

- **Unit Neighborhood** : for a given solution x^k, the unit neighborhood about x^k is the one formed by complementing components of x^k one at a time, i.e. ,

$$N_1(x^k) \quad = \quad \{\text{binary } x: \sum_j |x_j - x^k_j| = 1\}. \quad \text{(Parker [21])}.$$

- **t – Change Neighborhood**. The t –change neighborhood generalizes the unit neighborhood by allowing complementation of upto t solution components. Specifically,

$$N_t(x^k) \quad = \quad \{\text{binary } x: \sum_j |x_j - x^k_j| \le t\}. \quad \text{(Parker [21])}.$$

- **Pair-wise interchange**. Pair-wise interchange neighborhoods change two binary components at a time, but in a complementary fashion.

$$N_{2p}(x^k) \quad = \{\text{binary } x: \sum_j |x_j - x^k_j| = 2, \sum_j (x_j - x^k_j) = 0 \}. \quad \text{(Parker [21])}.$$

- **t–Interchange Neighborhood**. A t–Interchange neighborhood changes up to t values of the solutions in the same complementary manner as pair-wise interchange.

$$N_{tp}(x^k) \;=\; \{\textbf{binary x: } \sum_j |x_j - x^k_j| \le t,\ \sum_j (x_j - x^k_j) = 0\ \}.$$ **(Parker** [21]).

In the next subsection performance measures related to optimization problems are discussed.

2.3 Analysis and Complexity

Unless a local search algorithm employs an exact neighborhood function, it is generally not possible to give non-trivial bounds on the amount by which the cost of a local optimum deviates from the optimal cost. However, in practice, many example of local search algorithms are known that converge quickly and find high quality solutions.

In the performance analysis of combinatorial algorithms one usually distinguishes between the average case and the worst case. The performance of a heuristic can be quantified by its running time and its solution quality. The running time is usually given by the number of CPU seconds the algorithm requires to find a final solution on a specified computer and operating system. The solution quality is typically measured by the ratio of its cost value to that of an optimal solution or to some easily computed bound on that optimal value.

3.0 A Model of Local Improvement Algorithm for Hypercube

This section discusses a model as proposed by **Tovey**[29] for local improvement algorithms for hypercube. As stated earlier almost all combinatorial optimization problems can be visualized as optimization problems on hypercube.

Consider the problem of maximizing a real valued function C whose domain is the set of vertices of the n-cube. It is assumed here for simplicity that all the values of C are distinct. The domain of the function can be thought of as a set of Boolean decision variables. Such a function induces a unique priority ordering on the vertices of the n-cube.

The distance between two vertices of the n-cube is the number of components in which they differ. This distance is a metric and is known as the **Hamming distance**. If x and y are at a distance of zero, then x = y: is x and y are at a distance of one, they share an edge and are said to be adjacent or neighbors. A vertex whose function value is greater than any of its n neighbors is called a local maximum, if C has the property that all its local maximum are also global maximum we say that C is Local-Global (LG).

A natural implementation of local improvement algorithm is the Optimal Adjacency (OA) algorithm, which may be stated as:

Step1 Start with any vertex x.

Step2 If x is locally optimal, stop with x the solution. Otherwise proceed to 3.

Step3 Let y be the best vertex adjacent to x. Set x equal to y and go to 2.

3.1 Optimal Adjacency Trees:

If a particular local global function f is given, a directed tree to show how many iterations the optimal adjacency algorithm will require can be constructed as follows

Step1 Each vertex of the n-cube corresponds to a node of the tree.

Step 2 The father of a vertex is its optimal adjacent vertex: if a vertex is local optimum, it has no father.

The tree is called **Optimal Adjacency Tree, or OAT**. Its root is the local optimal vertex. The OAT displays the path followed by the algorithm by moving from son to father on the tree.

Figure 1 illustrates the notion of adjacency trees for the two cases. All solutions for the case when n = 3 as shown in part (b) belong to a single tree, so the local improvement will always yield a optimal solution. Starting search at $x^1 = (0,1,1)$, for example, the tree indicates local search would proceed to $x^2 = (0,1,0)$ and then to optimal $x^3 = (0,0,0)$.

Part (a) shows a case of local optima that are not global for n=2. Search for any of (1,0) , (0,1) or (0,0) leads to (0,0) solution. But (1,1) is a separate local optimum.

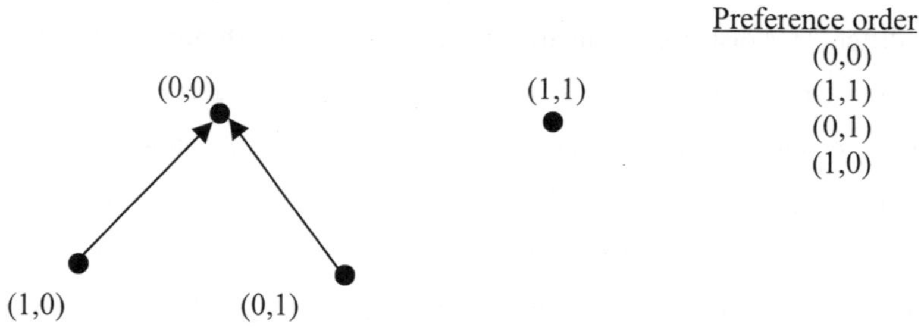

Figure 1(a) Optimal Adjacency Forest

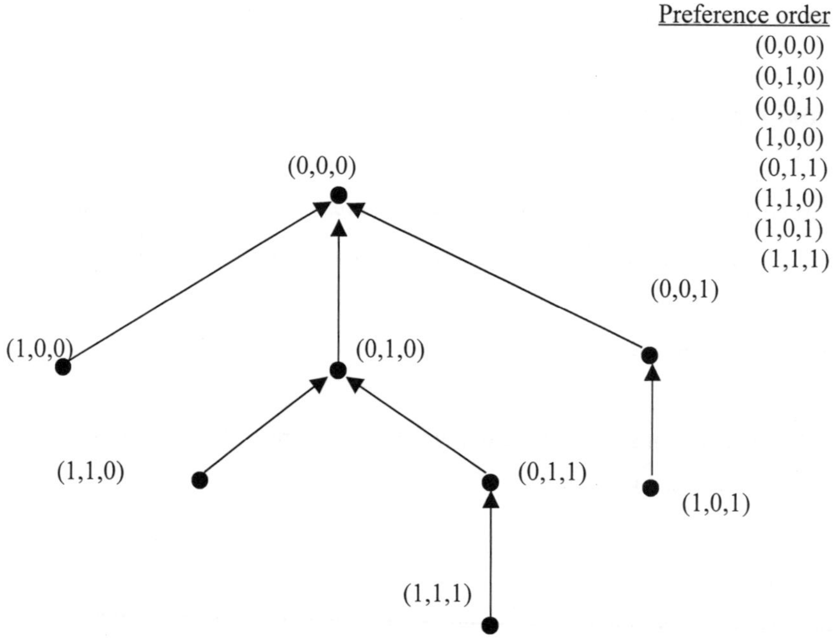

Figure 1(b) Optimal Adjacency Tree

The number of iterations required to complete the OAT of figure 1 (b) is computed next. If the starting vertex is chosen at random, there would be an equal probability of starting at each of the eight vertices. In general for each starting vertex, the path to the root in the OAT is by definition the path the OA algorithm will follow. Thus the height or path-length of each vertex in the tree is the number of iterations the algorithm would need to reach the optimum from the vertex. The mean path length of the tree is precisely equal to the expected number of iterations the algorithm would need to reach the optimum from that vertex.. Thus for the problem having structure as in figure 1(b), the OA algorithm would be expected to take

$$(1*1 + 3*2 + 3*3 + 1*4)/8 = 5/2 \quad \text{iterations.}$$

If C is not LG, the rules for producing the OAT will instead produce an OAF, or Optimal Adjacency forest, as shown in figure 1(a), with one tree per local optimum.

3.2 Expected Duration

For the Worst Case maximum number of iterations required by the Optimal Adjacency Algorithm in any optimal adjacency search of the vertices on the n-hypercube is at least

$O(2^n / n)$ as shown by (**Tovey**[30])

Instead of the worst case performance an average or expected number of search iteration is more useful. This requires some possible forms of probability distributions to be studied over the possible adjacency trees for the n-cube. This is equivalent of introducing probability distributions over the orderings of the vertices of the n-cube.

3.3 Better adjacency trees

The OA algorithm always chooses the best neighbor to go to. If this condition is relaxed, and it is only required that the algorithm proceeds from a vertex to a better adjacent vertex then Better Adjacency Algorithm results which is described below:

Step1 Start at random vertex x.

Step2 Search through x's neighbors until a better one, y, is found or all neighbors have been tried. In the former case set x = y and iterate Step2 ; in the later case stop, with x optimal.

4.0 Some Other Examples of Neighborhood Search Heuristics

Plant layout Problem:

Given a set of facilities $\{x_1, x_2, \ldots x_n\}$ and n locations, facilities are to be assigned to locations s.t. each facility is assigned exactly to one location. Assignment of the location is to be made such that the total material handling between the facilities (distance * weight) is minimized). This problem can be viewed as path optimization problem with $C(f) = \sum$ (Load between x_i and x_s) * distance between location of x_i and x_s.

A simple neighborhood search heuristics can be described as follows:

(a) Start with any sequence.

(b) Exchange a pair of adjacent facilities and calculate material handling for each such new sequences .

(c) Select the sequence which has the smallest material handling movement among all such sequences obtained by adjacent exchange. If no such sequence is found stop, else repeat step (b),with the selected sequence.

Another variation of this heuristic can be constructed as by defining as neighbors all those allocation which can be obtained by pairwise interchanges, instead of adjacent interchanges.

Travelling Salesman Problem

(a) Start with an arbitrary tour

(b) Delete two edges from this tour. This will result in two disconnected paths. Join these paths in such away as to form a new tour.

(c) Compute the weight of the new tour, and if it is smaller than the current tour, then make it the current tour and repeat the step (b), otherwise select another pair and repeat step (b). If no such tour is found stop.

A 3-opt heuristic can be constructed by selecting three edges to be removed. It will result in three disconnected paths and four possible tours.

5.0 Advance Search Strategies:

One of the major problem with the local search heuristics is that heuristics will stop after finding a local optimal solution. An extension of the local search heuristic is to repeat the heuristic with several random start point and to keep the best solution obtained. This approach has not resulted in any major successes. However this has lead to development of Meta-Heuristics, in which cost-positive i.e. inferior neighbor are also selected with some probability. Some such heuristics are Simulated Annealing , Genetic Algorithm , and Tabu Search. These heuristics have been able to solve large size problems in reasonable computational time. Aarts & Lenstra [1], Laporte & Osmen [8], Goldberg[9],Osmen & Laporte [20] and Reeves [26] are excellent references on these Heuristics.

Bibliography

1 E.H.L. Aarts, J.K. Lenstra (Eds.) (1997). Local Search in Combinatorial Optimization, Wiley, Chichester
2 A.V. Aho, J.E. Hopcroft, J.D. Ullman (1974). The Design and Analysis of Algorithms. Addition-Wesley, Reading, MA.
3 A Bronsted (1993). An Introduction to Convex Polytopes, Sppringer, New York.
4 N. Christofides (1985). Vehicle routing. E.L. Lawler, J.K. Lenstra, A.H.G. Rinnooykan, D.B. Shmoys (Eds.). The Travelling Salesman Problem: A guided Tour of Combinatorial Optimization, Wiley, Chichester, 431-448.
5 M. Dell' Amico, F. Maffioli, S. Martello (Eds.) (1997). Annotated Bibliographies in Combinatorial Optimization, Wiley, Chichester.
6 A.T. Fischer (1995). A note on the complexity of local search problems. Information Processing Letters 53, 69-75.
7 M.R. Garey, D.S. Johnson (1979). Computers and Intractability: A Guide to the Theory of NP-Completeness, Freeman San Francisco, CA.
8 G.Laporte, I.H. Osman (Eds.). Metaheuristics in Combinatorial Optimization, Annals of Operations Research 63, Baltzer, Amsterdam, 489-509.
9 D.E. Goldberg (1989). Genetic Algorithms in Search, Optimization, and Machine Learning, Addition-Wesley, Reading, M.A.
10 B.L. Golden, A.A. Assad (Eds.) (1988). Vehicle Routing: Methods and Studies, North-Holland, Amsterdam.
11 L.K. Grover (1982). Local Search and the Local Structure of NP-Complete Problems, Operations Research Letters 12, 235-243.
12 D.S. Jonson, C.H. Papadimtriou, M. Yannakakis (1988). How easy is local search? Journal of Computer and System Sciences 37, 79-100.
13 P.Joshi (2000),Performance of genetic algorithm and greedy heuristic with respect to distribution of local solutions on hypercubes , M.Tech thesis I.I.T.Kanpur
14 R.M. Karp (1972). Reducibility among combinatorial problems. R.E. Miller, J.W. Thatcher (Eds.). Complexity of Computer Computations, Plenum Press, New York.
15 M.W. Krentel (1989). Structure in local optimal solutions. 30th Annual Symposium on Foundations of Computer Science, JEEE Computer Society Press, Los Alamitos, CA, 216-222.
16 M.W. Krentel, (1990). On finding and verifying locally optimal solutions, SIAM Journal on Computing 19, 742-751.
17 S. Lin, B.W. Kernighan (1973). An effective heuristic algorithm for the traveling-salesman problem. Operations Research 21, 498-516.
18 B. C. Llewllyn, C.A. Tovey, M.A. Trick (1989, 1993). Local optimization on graphs, Discrete Applied Mathematics 23, 157-178. Erratum. Discrete Applied Mathematics 46, 93-94.

19 O. C. Martin, S.W. Otto (1996). Combining simulated annealing with local search heuristics. G. Laporte, I.H. Osman (Eds.). Mathematics in Combinatorial Optimization, Annals of Operations Research 63, Baltzer, Amsterdam, 57-75.
20 I.H. Osman, G. Laporte (1996). Metaheuristics: a bibliography, G. Laporte, I. H. Osman (Eds.). Metaheuristics in Combinatorial Optimization, Annals of Operations Research 63, Baltzer, Amsterdam, 513-623.
21 R.G.Parker,R.L.Rardin (1988). Discrete optimization. Academic Press
22 C.H. Papadimitriou, A.A. Schaffer, M. Yannakakis (1990). On the complexity of local search. Proceedings of the Twenty-Second Annual ACM Symposium on Theory of Computing. ACM, New York, 438-445.
23 C.H. Papadimitriou, K. Steiglitz (1982). Combinatorial Optimization: Algorithms and Complexity, Prentice Hall, New York.
24 M. Pirlot (1992). General local search heuristics in combinatorial optimization: a tutorial. Belgian Journal of Operations Research, Statistics and Computer Science 32, 719-737.
25 H.N. Psarafts (1983). K-Interchange procedures for local search in a precedence constrained routing problem. European Journal of Operational Research 13, 391-402.
26 C.R. Reeves (Ed.) (1993a). Modern Heuristic Techniques for Combinatorial optimization, Blackwell, Oxford.
27 V. Rodl, C.A. Tovey (1987). Multiple optima in local search. Journal of Algorithms 8, 250-259.
28 S.L. Savage (1976). Some theoretical implications of local optimality. Mathematical Programming 10, 354-366.
29 A.A. Schaffer, M. Yannakakis (1991). Simple local search problems that are hard to solve. SIAM Journal on Computing 20, 56-87.
30 C.A. Tovey (1983). On the number of iterations of local improvement algorithms. Operations Research Letters 2, 231-238.
31 C.A. Tovey (1986). Low order polynomial bounds on the expected performance of local improvement algorithms. Mathematical Programming 35, 193-224.
32 R.J.M. Vaessens, E.H.L. Aarts, J.K. Lenstra (1992). A local search template, R. Manner, B. Manderick (Eds.). Parallel Problem Solving from Nature 2, North-Holland, Amsterdam, 65-74.
33 M. Yannakakis (1990). The analysis of local search problems and their heuristics, C. Choffrut, t. Lengauer (Eds.), ST ACS 90: 7th Annual Symposium on Theoretical Aspects of Computer Science, Lecture Notes in Computer Science 415, Springer, Berlin, 298-311.

